Theory and Design of Digital Computer Systems

To Will, my Father – a craftsman
who taught me engineering.

Douglas Lewin

Professor of Digital Processes, Brunel University

Theory and Design of Digital Computer Systems

Si quid novisti rectius istis,
Candidus imperti; si non, his utere mecum.
HORACE

'Now, brother, if a better system's thine,
Impart it frankly, or make use of mine.'

A HALSTED PRESS BOOK

John Wiley & Sons
New York

Published in the USA
by Halsted Press, a Division of
John Wiley & Sons, Inc., New York

First published in Great Britain
by Thomas Nelson & Sons Ltd, 1972
Second edition 1980.

ISBN 0 470–26959–6

Printed in Great Britain by A. Wheaton & Co. Ltd., Exeter

Contents

Preface

Knowledge: A little light expels much darkness
Bahya ibn Paquda, Duties of the Heart.

Since this book was first published there have been major advances in LSI technology, in particular the developments in microprocessors and semiconductor memory. However, though the emergence of cheap and reliable microcomputers on a chip will certainly revolutionize computer systems architecture, the basic structure of the microcomputer itself still follows very closely the original principles of digital computer design. In fact in some respects it is rather like going back to the halcyon days of first generation computers!

Consequently, the fundamental principles described in the original edition of this book are still very relevant, and some aspects, such as machine-code programming, micro-programming and interfacing, have acquired even greater importance.

It is important to realize that the computer in its microcircuit form, has now become a system component which can be used to construct complex computer architectures. It is this aspect coupled with the availability of cheap semiconductor memory, that will change the face of computer engineering. The concepts of distributed and parallel processing, up to now barely being economically viable, will soon become standard practice. Moreover the structure of main frame computers as we know them today will drastically change, being replaced by dedicated microcomputer configurations performing both hardware and system software functions.

It is from this point of view that the second edition of the book has been written. The fundamental aspects of digital computers have been retained whilst considerable emphasis has been placed on parallel and distributed computing systems, including associative processing and the implementation of microcomputer based systems. Each chapter has been thoroughly revised and updated and the contents, where relevant, have been reorientated to take into consideration the effect of semiconductor technology. In particular important topics like stack processing, pipelining, distributed processing, interfacing, interrupt procedures and semiconductor storage systems have been completely revised and extended to include current thinking in these areas.

My sincere thanks must be recorded to all those readers of the first edition who spurred my enthusiasm to undertake this revision and to Molly Richardson, Betty Clark and Sue Lovett who unscrambled my manuscript into readable text.

1

The stored program principle

1.1 Introduction

There are three basic types of computer currently in use – the **analogue computer**, the **digital computer**, and a combination of both called the **hybrid computer**. In this book we shall be concerned primarily with the digital computer, but before we start it is worthwhile describing the characteristics of all three types.

The analogue computer represents the variables (and constants) in its calculations by physical quantities (usually voltage and time), hence the name 'analogue'. The slide rule is a very simple example, where 'length' is used to represent the actual values in a calculation. The accuracy of such calculations is of course limited by the accuracy with which we can measure the physical quantities involved. Usually the computing reference voltage is ± 10 V, and voltmeters, oscilloscopes and X–Y plotters are used to measure and record the values of the variables, generally to within an accuracy of 0·1% to 1%.

The solution to a mathematical or systems problem is obtained by setting up an analogue of the mathematical equations (or by simulating its transfer functions) using operational amplifier circuits functioning as adders, sign changers, integrators, and so on.[1] Thus each integration or addition, etc., in an equation is performed simultaneously by separate operational amplifiers working in parallel. Consequently the answer is in a continuous form; that is, the analogue computer produces a general solution to an equation which is normally displayed as a graph of voltage against time. The time required to produce a solution depends on the problem, but the computer can be suitably time-scaled (for example to allow for the response of the output equipment) to work either in **machine** or **real-time**. This has considerable advantages particularly in real-time problems when actual equipment can be included in a simulation. Another advantage is the rapport that exists between the designer and the machine, since the parameters of a problem may be easily changed by adjusting potentiometers and the results observed instantaneously.

The digital computer, on the other hand, represents its variables in a quantized or **digital** form. Thus numbers must be represented by using a discrete state or condition to represent each symbol (0–9 for decimal numbers). For example, in a decimal counting system (a car mileometer for example) gear wheels with ten teeth may be used to represent a decade, each cog corresponding to a symbol. A complete revolution

Table 1.1
Examples of the binary notation

Decimal radix of 10						*Binary radix of 2*										*Octal radix of 8*				
10^2	10^1	10^0	$\cdot\ 10^{-1}$	10^{-2}	10^{-3}	2^6	2^5	2^4	2^3	2^2	2^1	2^0	$\cdot\ 2^{-1}$	2^{-2}	2^{-3}	8^2	8^1	8^0	$\cdot\ 8^{-1}$	8^{-2}
100	10	1	$\cdot\ \frac{1}{10}$	$\frac{1}{100}$	$\frac{1}{1000}$	64	32	16	8	4	2	1	$\cdot\ \frac{1}{2}$	$\frac{1}{4}$	$\frac{1}{8}$	64	8	1	$\cdot\ \frac{1}{8}$	$\frac{1}{64}$
		0	·1	2	5							0	·0	0	1			0	·1	0
		0	·2	5	0							0	·0	1	0			0	·2	0
		0	·5	0	0							0	·1	0	0			0	·4	0
		1	·0	0	0							1	·0	0	0			1	·0	0
		2	·0	0	0						1	0	·0	0	0			2	·0	0
		3	·0	0	0						1	1	·0	0	0			3	·0	0
		4	·0	0	0					1	0	0	·0	0	0			4	·0	0
		5	·0	0	0					1	0	1	·0	0	0			5	·0	0
		6	·0	0	0					1	1	0	·0	0	0			6	·0	0
		7	·0	0	0					1	1	1	·0	0	0			7	·0	0
		8	·0	0	0				1	0	0	0	·0	0	0		1	0	·0	0
		9	·0	0	0				1	0	0	1	·0	0	0		1	1	·0	0
	1	0	·0	0	0				1	0	1	0	·0	0	0		1	2	·0	0
1	0	0	·0	0	0	1	1	0	0	1	0	0	·0	0	0	1	4	4	·0	0

of a gear wheel causes the next gear wheel to enmesh so producing the effect of a carry. To perform the same task electronically we would need either a ten-state device, such as a decimal counter, or a specially constructed device using, in the simplest sense, ten on/off switches each connected to a lamp to represent one decade. As naturally occurring ten-state devices are very rare, and when specially made tend to be very expensive in components, it would appear obvious to use a number system with fewer symbols. Consequently in electronic digital computers the **binary system**, employing the two symbols 0 and 1 only, is used to represent numbers. This is a convenient and economic engineering solution since there are numerous examples of simple two-state devices (switches – on/off; transistors – conducting/cut off) which may be used to represent the symbols. Number systems may be defined mathematically in terms of the polynomial:

$$N = a_n q^n + a_{n-1} q^{n-1} + \cdots + a_1 q^1 + a_0 q^0 + a_{-1} q^{-1} + \cdots + a_{-m} q^{-m}$$

where N is a positive real number, q a positive integer **radix**, and a represents the symbols. Thus, for example, the decimal number 27·5 may be expressed in terms of this polynomial as:

$$(27{\cdot}5) \text{ decimal} \equiv 2 \times 10^1 + 7 \times 10^0 + 5 \times 10^{-1}$$

and

$$(27{\cdot}5) \text{ binary} \equiv 1 \times 2^4 + 1 \times 2^3 + 0 \times 2^2 + 1 \times 2^1 + 1 \times 2^0 + 1 \times 2^{-1}$$

Table 1.1 shows further examples of the binary system of notation; note that the binary (or radix) point occurs after the last positive index of the radix, that is, q^0. As a consequence of using the binary notation the accuracy of a calculation depends on the number of binary digits (bits) used to represent the variables. Using a 10-bit **word** we can expect an accuracy of 1 part in 10^3, the accuracy may be increased by using more bits, but one must be careful not to exceed the accuracy of the original data. It is pointless expressing input data, accurate to only 2%, to 1 part in 10^4!

We have said that the analogue machine works in a parallel mode producing an instantaneous solution. In contrast the digital machine works **sequentially**. Thus integrations which are performed autonomously in separate units in an analogue machine must be carried out one after the other in a digital computer. Furthermore the result is a numerical one, giving a particular solution to an equation rather than a general one. Thus to produce a general solution to an equation would require many iterations of the digital computing routine. Consequently in some problems, particularly real-time applications, the digital computer is too slow!

When speed and accuracy are required the hybrid computer is used; this incorporates some of the advantages of analogue and digital computers in one machine. In particular, iterative operations may be performed, enabling a complete family of curves to be produced, by automatically changing the parameters of an equation using digital control techniques. Alternatively, analogue and digital computers may be used in conjunction, with the analogue machine providing overall control and calling upon the digital machine for function and delay generation, parameter changing, and so on. An alternative approach is to use a digital computer to simulate the operations normally executed by the analogue machine but to employ analogue sub-routines to perform integration or the solution of differential equations when high speed is required, as for example in real-time working. Another advantage of hybrid operation is that the preparation time required to program a problem (solutions are *patched* in an analogue computer by interconnecting individual sub-units) can in general be reduced. However, it is the author's contention that hybrid computer systems are a temporary means of overcoming the present limitations of digital computers; in the future when digital computers become faster (including fast storage systems) large analogue and hybrid computers may well become obsolete.

Digital computers are now being applied in all branches of technological and commercial endeavour. In particular, computers have been used to manufacture ice-cream, gas, and steel; to control road, rail, and air traffic; to set up newspaper type; to supervize stock control and insurance records; and to design engineering systems. Moreover the basic principle of digital computers, the **stored program concept**, has been employed in the design of many special purpose machines, such as telephone switching networks, and TV studio lighting.

More recently the development of cheap LSI microprocessors and microcomputer systems has revolutionized the design of digital systems. Computer techniques are now freely available to all engineers, who have employed them prodigiously in many different applications, particularly instrumentation and real-time control. Moreover, cheap programmable calculators, multi-function digital watches and even 'do it yourself' computer kits are now available for the public at large.

The object of this book is to explain the theory and design philosophy of digital computer systems. Not necessarily so as to enable people to design computers (very few will do this) but with the broader objectives of using computers or computer-like machines as modules in a large digital system. To achieve these objectives it is essential to know how to use the machines (the **software** design) as well as to understand their engineering and components (the **hardware** design).

1.2 Basic principles

The fundamental principles of organizing a digital computer were first set forth by Babbage[2] in the early part of the nineteenth century. As a result of building calculating machines to compute mathematical tables, Babbage conceived the idea of an Analytical Engine and, though it was never built, laid the foundation for modern automatic computing machines. These ideas were later extended, though not changed in principle, by Von Neuman[3] who applied them to the design of an actual machine.

A digital computer consists of the following functional elements:

A **store** or memory for numbers (*operands*) and instructions.
An **arithmetic unit**, where arithmetical operations are performed on the operands.
The **control unit**, a device for controlling the logical operations of the machine, and causing them to take place in the correct sequence.
The **input/output unit**, used to transfer data into and out of the computer store.

Figure 1.1 shows a block schematic for a typical digital computer.

In general a numerical problem is solved, using a digital computer, by first breaking down the calculation into a number of discrete arithmetic operations (such as add, subtract) which are performed on the binary operands. These operations, together with any necessary organizing functions, such as input, output, register transfers, and so on, are then arranged to form a **program** of instructions for the computer. This program, suitably coded in a binary form, is written into the computer store using the input and control units. Instructions are read down from the store in sequence and obeyed, again under the action of the control unit, using the arithmetic unit as and when required. The store contains both program and data, plus 'working-space' and storage for results. The final operation is to output the results of the calculation via the output unit. Note the similarity between machine and manual computation. In the latter case a desk calculating machine could be regarded as the arithmetic unit, books of tables become the computing procedure and note-pads the store, and the control unit would of course be the human operator.

The philosophy of storing the program instructions as well as the operands, that is, the stored program concept, is the basic reason for the power of the digital computer. This is because, as we shall see later, the instructions can be treated as data and arithmetic operations may be performed on them in the normal way.

1.3 Instruction and number representation

In a digital computer the program instructions and operands (variables and constants) are stored together in the same storage unit. Each location in the store contains the same number of bits (called a **computer word**) and is allocated an absolute, uniquely identifiable address. The computer words, the length of which can vary from 8 to 32 bits for practical machines, are used to represent *both* the computer instructions and the data, see Figs 1.2 and 1.3. In the simplest case the **instruction word** would be divided up into an **order** and an **address** section plus some

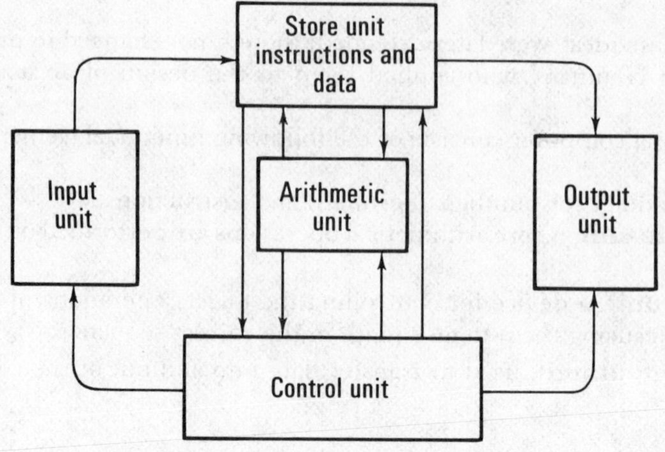

Figure 1.1 Computer block diagram.

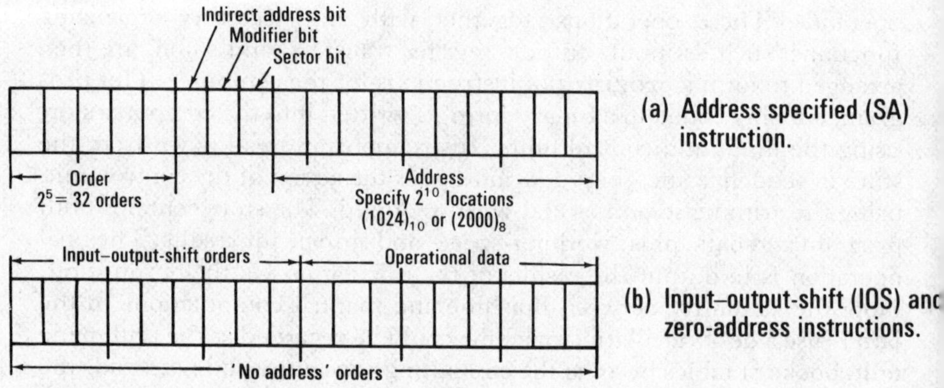

Figure 1.2 Instruction formats.

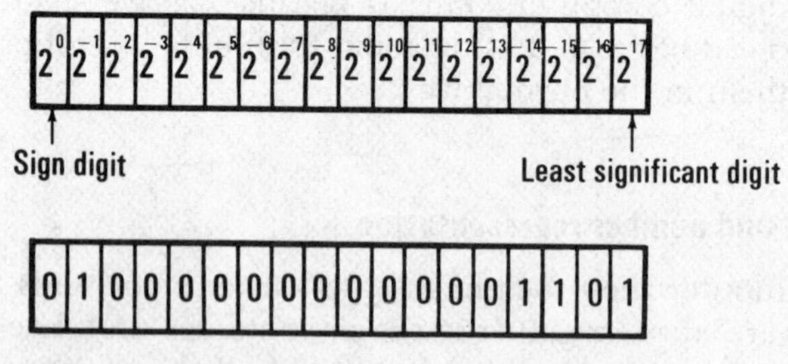

(i) Shift contents of accumulator 13 places left.
(ii) Integer 65549×2^{-17}.
(iii) Fraction 0·50009918.

Figure 1.3 Operand words.

control bits. The order part of the word has sufficient bits to allocate a unique binary or octal code to all the **machine-code orders**, likewise the range of the address section is such as to enable most of the store locations to be specified directly. Assuming an 18-bit wordlength, we can address 2^{10} (=1024) store locations, and represent up to 2^5 (=32) machine-code orders. The function of the three control bits, called the **indirect address**, **modifier**, and **sector** bits respectively, will be explained in the next chapter.

Numbers are generally represented in the computer using **fixed-point binary fractions**, such that $-1 \leqslant x < 1$, where x is the binary number; negative numbers normally take the **2's complement form** with the most significant digit indicating the sign (see Chapters 2 and 5). For representing instructions, however, the pure binary notation is too cumbersome to use. This is because of the large number of digits needed to represent a number and the necessity to convert from decimal to binary, and vice versa, when writing instructions or inspecting stored program in the computer. Consequently, the **octal** system of notation is used, which, since it is based on a radix of eight, is easily converted to binary and has the advantage of requiring fewer digits for number representation. For example:

$$(1467)_{10} \equiv (10110111011)_2 \equiv (2673)_8$$

The octal number 2673 may be represented in polynomial form as:

$$2 \times 8^3 + 6 \times 8^2 + 7 \times 8^1 + 3 \times 8^0 = (1467)_{10}$$

The conversion from octal to binary is easily accomplished by writing down the pure binary equivalent of each octal digit, for example:

$$\begin{array}{lccccc} & 2 & 4 & 6 & 6 & 7 \\ (24667)_8 = & (10 & 100 & 110 & 110 & 111)_2 \end{array}$$

The reverse process of converting from binary to octal follows directly from above.

A typical, but minimal, set of machine-code orders is shown in Table 1.2, the orders are divided into three groups consisting of:

(a) those orders which require the address of a store location to be specified, for example in the add, subtract orders, etc.;

(b) orders which do not require any form of address (ZA orders) and hence all 18 bits may be used to specify an order; examples of this type of order are the stop instruction, exchanging register contents, and incrementing registers;

(c) orders which do not require a store address but do need additional information to be specified (IOS orders), for example the shift orders which must include the number of places (n) to be shifted, and the input/output instructions.

Table 1.2

Typical machine-code order set

(**a**) Specified address instructions (SA)

Code in octal	*Mnemonic*	*Description of order*
00		Reserved for ZA orders
01		Spare
02		
03		
04		
05		
06		
07		
10		Reserved for IOS orders
11		Spare
12		
13		
14	ADDM	Add contents of address specified to modifier
15	FETM	Fetch contents of address specified to modifier
16	STRM	Store contents of modifier in address specified
17	LINK	Store contents of instruction register in address specified and jump to address specified +1
20	ADD	Add contents of address specified to accumulator
21	SUB	Subtract contents of address specified from accumulator
22	FET	Fetch contents of address specified to accumulator
23	STR	Store contents of accumulator in address specified
24	COL	AND accumulator with contents of address specified
25	OR	OR accumulator with contents of address specified
26		Spare
27		
30	JMP	Jump to address specified
31	JMPN	Jump to address specified if accumulator negative, otherwise take next instruction in sequence
32	JMPP	Jump to address specified if accumulator positive, otherwise take next instruction in sequence
33	JMPO	Jump to address specified if accumulator zero, otherwise take next instruction in sequence
34	JMPM	Jump to address specified if modifier zero, otherwise take next instruction in sequence
35	MULT	Multiply accumulator by contents of address specified, resulting 36-bit product contained in accumulator and X-register
36	DIV	Divide 36-bit dividend, contained in accumulator and X-register, by contents of address specified
37	EXCO	Perform exclusive OR function between accumulator and contents of address specified

(**b**) Zero address instructions (ZA)

Code in octal	*Mnemonic*	*Description of order*
000000	STOP	Stop the computer
000001	COML	Form 2's complement of accumulator
000002	NOOP	No operation
000003	ENI	Enable interrupts
000004	INI	Inhibit interrupts
000005	EXAM	Exchange contents of accumulator and modifier register
000006	INCM	Add +1 to modifier register
000007	EXAC	Exchange contents of accumulator and X-register
000010	EXIO	Exchange contents of accumulator and input/output register
000011	CLRA	Clear accumulator
000012 : 017777		Spare

(**c**) Input/output/shift instructions (IOS)

Code in octal	*Mnemonic*	*Description of order*
200	SLL	Logical shift left of the accumulator *n* places
201	SRL	Logical shift right of the accumulator *n* places
202	SLA	Arithmetic shift left of the accumulator *n* places
203	SRA	Arithmetic shift right of the accumulator *n* places
204	EAS	End-around shift of the accumulator *n* places
205	INA	Input word to accumulator
206	OTA	Output word from accumulator
207 : 217		Spare

The simplest method of organizing the order code structure is to allocate specific octal codes in the specified address group for the ZA and IOS orders, in the example used in Table 1.2 the codes 00 and 10 have been used respectively. Note that the restricted use of these two codes allows up to 16 IOS instructions and 8192 ZA instructions. It will be apparent that there are many possible variations of order codes that can be generated in this way and one of the chief responsibilities of the computer systems designer is to devise a viable and optimum choice of order code.

Note that in the case of specified address instructions the operations add, subtract, etc., take place between the contents of the address specified in the instruction word and the contents of the accumulator (a register in the arithmetic unit). Computers operating in this mode, and with the order-address type of instruction format, are known as **single-address** machines. Note also that the action of reading from the computer store (as required, for example, in the orders fetch, add, subtract, collate, etc.) does not affect the store contents, which remain unchanged. However, the instructions to fetch to the accumulator or modifier registers (or store in the computer store) always overwrites the previous contents. Thus in all cases the contents of the source address remains unchanged, but the contents of the destination address are overwritten. The instruction and operand words are stored in the computer as simple binary patterns and, as we shall see later, it depends on what part of the computer they are routed to as to the interpretation placed upon them. Figure 1.3 shows the computer representation of the instruction 'Shift the contents of the accumulator 13 places (decimal) to the left', note that it can also represent the binary fraction 0·50009918 or the binary integer 65549×2^{-17}.

We can now make use of these ideas to understand how a simple computer program is written. Suppose for example we wish to write a program to add together two numbers, say x_1 and x_2. A suitable program is shown in Table 1.3. The program instructions are stored in the computer in locations 1000–1003 (octal notation is used throughout) with the data in locations 1004 and 1005; the answer has to be stored away in location 0022. Note that it is necessary to specify *exactly* where the instruc-

Table 1.3
Example of a simple computer program

Location	*Instruction*	*Comment*
1000	22 1004	Fetch x_1 to accumulator
1001	20 1005	Add x_2 to accumulator
1002	23 0022	Store $x_1 + x_2$ in location 22
1003	00 0000	Stop
1004	x_1	
1005	x_2	

tions and data are to be held in the computer store, and that it is the binary equivalent of the machine-code instructions, represented in Table 1.2 in the octal-coded numeric form, that would actually be stored in the computer. At this stage it is best to ignore the obvious question of how the program got into the computer store in the first place, and simply assume that it is there. The execution of the program is apparent from Table 1.2, the first instruction brings x_1 (the contents of store location 1004) to the accumulator, and the next instruction adds x_2 (the contents of location 1005) to form the result, $x_1 + x_2$, in the accumulator. The result is then stored away in location 0022 (with the store instruction) after which the computer comes to a halt. We shall now describe in more detail how the computer performs these operations.

1.4 Digital computer operation

The block diagram of a parallel digital computer, or **central processor unit**, is shown in Fig. 1.4. As we have seen, the instructions and data

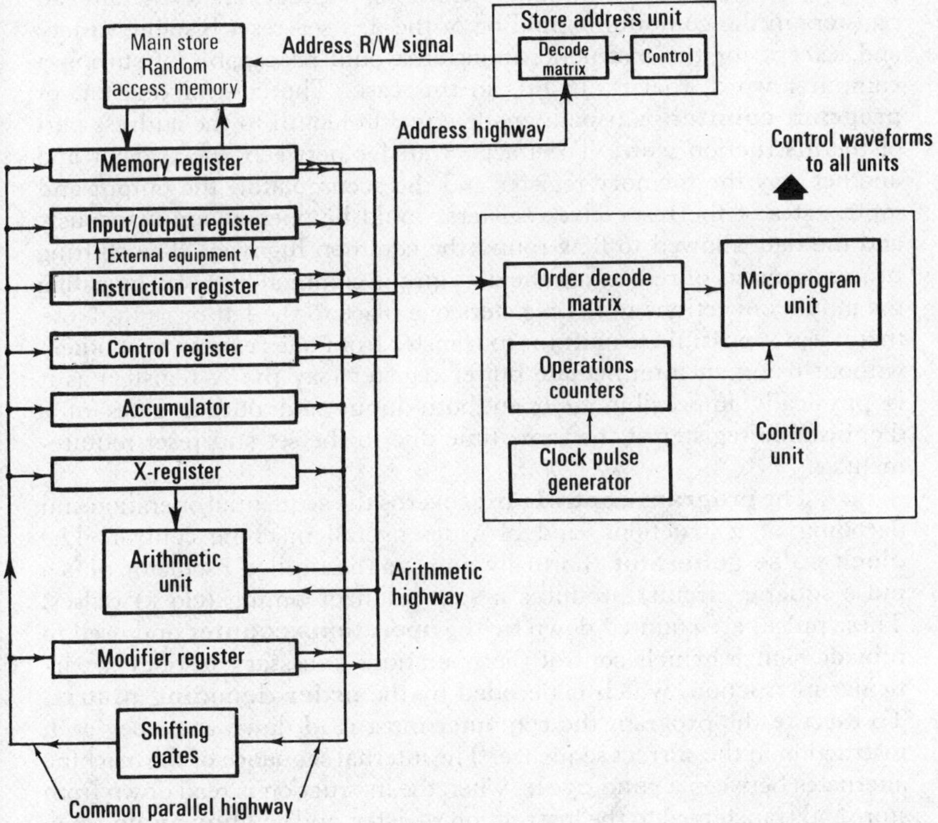

Figure 1.4 Block diagram for a parallel digital computer.

forming the computer program are initially placed in the computer store in locations specified by the programmer. The selection and reading of a store location is performed by the **store address unit**, which takes the address of the required location, either from the **instruction register** or the **control register**, and after decoding initiates a store read/write cycle (on receipt of an appropriate control waveform) which results in the contents of the selected location being placed in the **memory register**. Information may also be read into the store by transferring the data to the memory register, then, after addressing the store address unit with the relevant store location, initiating a write cycle only.

Arithmetic operations are performed by the **arithmetic unit** (using *parallel full-adder* stages) in conjunction with the **accumulator** and the auxiliary **X-register**. Shifting and register transfer operations are executed by means of shifting gates incorporated in the **common parallel highway** or **bus**. The common highway is peculiar to parallel computers (though the concept can be extended to serial machines) and consists simply of all the common interconnections between the various registers. In practice, these are the input and output terminals associated with the gating logic of the bistables forming the registers. Generally, all registers in the computer would be of the d.c. set–reset bistable variety and, except for the instruction register, would be capable of storing a computer word, that is, 18 bits in this case. The control register, or **program counter**, is usually made equal in length to the address part of the instruction word. To effect a transfer between one register and another, say the memory register and the accumulator, the output and input gates of the respective registers would be opened simultaneously and the data allowed to flow round the common highway. The shifting process consists of re-routing the data on the common highway by gating the input connections to the registers one place to the left or right. Note that it is not possible to shift (or to transfer from one register to another) without using an intermediate buffer register (say the X-register) as it is physically impossible to open both input and output gates of a d.c. bistable register at the same time due to the set and reset requirements.

The **program control unit** governs the sequential operation and decoding of instructions, and provides overall machine control. The **clock pulse generator** (normally a crystal-controlled oscillator plus a pulse squarer circuit) produces a succession of timing (clock) pulses. These pulses are counted down by the **operations counter** and used to provide signals which control the operations necessary to obey a particular instruction, which is decoded by the **order decoding matrix**. To execute the program the computer must read down and obey each instruction in the correct sequence. The internal sequence of the machine alternates between a **read cycle**, when the instruction is read down from store and transferred to the instruction register, and an **obey cycle** when the order is decoded and obeyed. A computer operating in this way is generally referred to as a **two-beat** machine.

Let us now explain in detail how the simple program described in the last section would be performed. The initial step is to place the address of the location in which the first instruction is stored (1000) in the control register; we will assume that this may be done manually using switch keys. The machine is now started and it immediately goes into the read cycle; this has the effect of transferring the contents of the control register to the address highway, where it goes to the store address unit and is decoded. This results in the contents of the location, that is, the first instruction to be obeyed, being placed in the memory register, and then transferred, via the common highway, to the instruction register. During the time it takes to read down the instruction (the read/write time of a main store is typically 0·5–2 μs) the contents of the control register are passed through the arithmetic unit and +1 added in to the least significant digit position. This ensures that the next instruction will be taken from location 1001 in the following read cycle, thus ensuring the sequential read-out of the instructions. The computer is now ready to enter the obey cycle, and the first step is to transfer the order digits of the instruction register to the order decoding matrix in the control unit. Here the order is decoded and, in conjunction with the operations counter, sets in motion the sequence of operations required to execute the instruction. In this case since it is a fetch instruction (that is, order 22) the address digits of the instruction register must be routed to the store address unit to allow the required number, x_1, to be read down and placed in the memory register. The final operation is the transfer, via the common highway, from the memory register to the accumulator. The computer has now completed the obey cycle and is ready to commence the next read cycle. In this way each instruction is read down from the computer store and obeyed, until the final stop instruction is reached when the machine comes to a halt. The internal machine operations required to read down and obey the instructions, such as opening input/output gates, setting bistables, initiating store read/write cycles, etc., are called **micro-orders**.

Note that instructions and numbers are distinguished by their final destination in the computer, that is, instructions go to the instruction register and numbers to the accumulator. Should the computer get out of step, due to a programming error or logical fault, and a number is routed to the instruction register, it will be decoded and obeyed just as if it were in fact an actual instruction. Moreover, it is possible to deliberately fetch an instruction to the accumulator where it may be *modified* by adding or subtracting a constant. This facility, combined with the conditional jump instruction (that is, the ability to choose one of two actions as determined by the result of a previous calculation) accounts for the tremendous power of the digital computer, and the reason it is colloquially referred to as an 'electronic brain'.

In the case of the microcomputer the basic structure and mode of operation is identical to that described above. The microcomputer normally consists of separate LSI chips (mounted on a printed circuit board) which perform the functions of **clock generation**, **input-output channels** (called **ports**), **read only memory** (ROM's to store the program which is normally pre-determined) **random access memory** (RAM's for temporary data storage) and a **central processing unit** (excluding main store but containing general purpose hardware registers). The wordlength is normally quite short, 8-bits being the norm with some 16-bit processors becoming available. The term **microprocessor** usually refers to the main CPU chip; note that in order to construct a microcomputer it is essential to have the other chips available, as well as the attendant interconnection hardware and power supplies. Recently microcomputer modules have been produced which contain all the basic functions, including limited ROM, RAM and input-output ports, on one LSI chip.

In this, and preceding sections of the chapter, an attempt has been made to give an overview of the use and operation of digital computers. In so doing many questions have been left unanswered or covered very briefly. However, in the following chapters the theory and design of each main section of the central processor will be reiterated in detail.

References and bibliography

1 James, M. L., Smith, G. H., and Wolford, J. C. *Analog-Computer Simulation of Engineering Systems*. International Textbook Co., Pennsylvania, 1966.

2 Morrison, P. and Morrison, E. *Charles Babbage and his Calculating Engines*. Dover Publications, New York, 1961.

3 Taub, A. H. (ed.). *John Von Neuman – Collected Works*, Vol. V. Pergamon Press, Oxford, 1963.

4 Wilkins, B. R. *Analogue and Iterative Methods*. Chapman and Hall, London, 1970.

5 Taube, M. *Computers and Common Sense*. McGraw-Hill, New York, 1963.

6 Anderson, A. R. (ed.). *Minds and Machines*. Prentice Hall, Englewood Cliffs, N.J., 1964.

7 Apter, M. J. *The Computer Simulation of Behaviour*. Hutchinson University Library, London, 1970.

8 Wiener, N. *The Human Use of Human Beings*. Sphere, London, 1969.

9 Goldstine, H. H. *The Computer from Pascal to Von Neumann*. Princeton University Press, New Jersey, 1972.

10 Randell, B. *The Origins of Digital Computers*. Springer-Verlag, New York, 1974.

11 Williman, A. O. and Jelinek, J. H. Introduction to LSI Microprocessor Developments. *IEEE Computer*, **9**, 34–36, June 1976.

Tutorial problems

1.1 Discuss the importance of the stored program concept and in particular the self-modification of instructions.

1.2 Could digital computers, given unlimited storage, ever be considered as 'thinking machines' in the same way as the human brain?

1.3 Consider the effect of computers on society in general, and then attempt to formulate a 'code of ethics' for the professional computer scientist with the objective of safeguarding the public interest.

1.4 Comment on the efficiency of the digital computer when used to solve non-numerical problems, for example the processing of symbolic data as required in language translation.

***1.5** Write the binary representation as held in the computer store, and the equivalent interpretation as integer and fractional numbers, of the following machine-code orders:

(a) 203 007
(b) 00 0006
(c) 34 0763
(d) 21 0012/MS

***1.6** Write a program, beginning in location 1500, which interchanges the contents of locations 1732 and 1607. Give both the mnemonic representation of the program, and the binary words as stored in the computer.

***1.7** Write a program to perform the operation $A + B - C$, where A, B, and C are stored in locations 1001, 1003, and 1007 respectively. Put the result in location 0073.

* Problems marked with an asterisk are those for which fully worked solutions are given at the end of the book.

2

Principles of machine-code programming

2.1 Introduction

One of the initial steps in the design of digital computer systems is to lay down the structure for the basic machine-code orders. This is done in practice by a systems engineering team comprising logic designers, software specialists, and others, who would postulate a suitable set of orders bearing in mind current technology and the user's requirements. Thus, before one can study in depth the hardware design of a computer, it is essential to acquire a full understanding of the mechanics of machine-code programming, since the machine-code orders dictate, to a large extent, the overall structure of the machine. Consequently, the objective of this chapter is to elaborate further the simple ideas of machine-code programming already introduced, and in so doing to present more sophisticated concepts of programming technique.

2.2 Number representation

Before commencing our study of programming, however, let us first consider the methods of representing numbers in the computer in more detail. We have seen that computers use the binary notation as their basic number system, with negative numbers generally represented in the 2's complement form. There are, however, two main methods of representing negative numbers in a computer, these are:

Sign and magnitude, where the number is expressed by its absolute value together with a symbol indicating the sign. This is, of course, the normal method of writing negative numbers which has been widely adopted for everyday use.

Complement notation. The 2's complement of a binary number, known generally as the **radix** complement, may be obtained by subtracting the number from the next significant power of two (2^1 in the case of binary fractions). An easy way to perform this operation is to interchange all 1's and 0's (to form the **1's complement** or **radix-1 complement**) and then add +1 to the least significant digit.

The main advantage of using 2's complement representation compared with the other methods is that the most significant digit may be used to indicate the sign of the number, which is automatically generated when performing addition and subtraction; this subject is treated fully in Chapter 5. For example, assuming 8-bit words, we have:

	$2^0 \cdot 2^{-1} \ldots 2^{-7}$
$\frac{11}{16}$	0 ·1011000
1's complement	1 ·0100111
Add +1, 2's complement	1 ·0101000 $\equiv -\frac{11}{16}$

	2^1	$2^0 \cdot 2^{-1} \ldots 2^{-7}$	
	1	0 ·0000000	
Subtract $\frac{11}{16}$	0	0 ·1011000	
	0	1 ·0101000	$\equiv -\frac{11}{16}$

$-\frac{11}{16}$	1·0101000	
Add $\frac{3}{4}$	0·1100000	
	1 ← 0·0001000	$\equiv \frac{1}{16}$

$-\frac{3}{16}$	1·1101000	
Subtract $-\frac{11}{16}$	1·0101000	
	0·1000000	$\equiv \frac{1}{2}$

$\frac{1}{4}$	0·0100000	
Subtract $\frac{15}{16}$	0·1111000	
	1 ← 1·0101000	$\equiv -\frac{11}{16}$

$\frac{3}{16}$	0·0011000	
Add $-\frac{11}{16}$	1·0101000	
	1·0000000	$\equiv -\frac{1}{2}$

Note that the carry from the most significant digit into the 2^1 column must be ignored, also that the method applies equally well to both fractional and integer systems, but in the latter case (if the same accuracy is to be obtained) an extra digit is required at the most significant end of the word to serve as a sign digit.

There are two ways of representing binary operands in the digital computer, using either a **fixed-point** or **floating-point** system. In the fixed-point scheme all numbers are treated as binary fractions within the range $-1 \leqslant x < 1$ with the radix point (binary point) to the immediate right of the most significant digit which, using the 2's complement system, is also the sign digit. Integers may be represented in this system if each number is considered (by the programmer) to be multiplied by a suitable scaling factor, for example:

2^0 $\quad\quad\quad\quad$ 2^{-17}	
010100000000000000	$= 0{\cdot}625 \times 2^0 = 81920 \times 2^{-17}$ $= 5 \times 2^{-3}$
000000000000011101	$= 0{\cdot}00022125 \times 2^0 = 29 \times 2^{-17}$ $= 7{\cdot}25 \times 2^{-15}$
101100000000000000	$= -0{\cdot}625 \times 2^0 = -81820 \times 2^{-17}$
111111111111100011	$= -0{\cdot}00022125 \times 2^0 = -29 \times 2^{-17}$
100000000000000000	$= -262144 \times 2^{-17}$
111111111111111111	$= -1 \times 2^{-17}$
011111111111111111	$= 0{\cdot}99999237060546875 \times 2^0$

Unfortunately the use of a fixed-point number system gives rise to difficulties when performing arithmetic functions, due to the results of the operations going outside the defined number range. For example, when adding or subtracting it is necessary to ensure that both numbers have the same scaling factor, and that the sum or difference is within the number range of the machine. For example:

	0011100000000000	$7 \times 2^{-4} = 0{\cdot}4375 \times 2^{0}$
Add	0111100000000000	$15 \times 2^{-4} = 0{\cdot}9375 \times 2^{0}$
	1011000000000000	$22 \times 2^{-4} = 1{\cdot}375 \times 2^{0}$

Note that the sum has gone outside the number range of the machine, which is indicated by the fact that the most significant digit has changed to one signifying a *negative* number. In the same way it is also possible to go out of range using negative numbers. For example:

	1001100000000000	$-13 \times 2^{-4} = -0{\cdot}8125 \times 2^{0}$
Add	1001100000000000	$-13 \times 2^{-4} = -0{\cdot}8125 \times 2^{0}$
$1 \leftarrow$	0011000000000000	$-26 \times 2^{-4} = -1{\cdot}6250 \times 2^{0}$

Again the number range is exceeded and this time the sum has gone positive. The same problems arise in multiplication and division; for example, if we perform the multiplication $(22 \times 2^{-17}) \times (22 \times 2^{-17})$ the resultant product is 484×2^{-34}. This is a **double-length** product requiring a 36-bit word for its representation, consequently the effect of multiplying two single-length words together, in this case, is to shift the product right out of the accumulator. To perform single-length multiplication it is first necessary to scale the multiplicand and multiplier such that the product can be represented in a single-length register; for example, $(22 \times 2^{-9}) \times (22 \times 2^{-8}) = 484 \times 2^{-17}$. Similar difficulties apply to division where it is not possible to divide by zero or any number smaller than the dividend, that is, the divisor must always be greater than the dividend so as to yield a fractional quotient. It is obvious, then, that in coding computional type problems, the programmer must take great care to ensure that the variables are correctly scaled, both before and after each arithmetic operation, otherwise erroneous results will occur. This necessity to scale each problem can be a serious handicap to the programmer who may have difficulty in keeping all the numbers involved in a computation within machine range. In many computers hardware checks in the form of logical circuits are incorporated to test and indicate when, for example, the contents of the accumulator go out of range (see Chapter 5).

To overcome the scaling problems associated with fixed-point numbers a **floating-point** system is used, where the radix point for each number varies automatically during the course of a computation thereby resulting in easier programming. Floating-point arithmetic procedures may be implemented in either software or hardware form, the former being slower in operation but cheaper to incorporate.

Floating-point numbers are expressed in the form:

$$x = a \times 2^b$$

where a is the **mantissa** (the fractional part) and b the **exponent** (or characteristic). This method of representation is analogous to expressing very large or very small decimal numbers in the form:

$$197835{\cdot}437 = 0{\cdot}197835437 \times 10^6$$
$$0{\cdot}0002569 = 0{\cdot}2569 \times 10^{-3}$$

note that each number consists of a fractional part and an exponent part. In the case of our 18-bit binary word we might choose to represent a floating-point number using a 6-bit exponent and a 12-bit mantissa, with normal 2's complement notation to represent negative quantities; note that using a 12-bit mantissa will reduce the accuracy of the number. Using 2's complement notation the range of the exponent is given by $-32 \leqslant b < 32$, that is:

$$100000\ (-32) \rightarrow 011111\ (+31)$$

by convention when the mantissa $a = 0$, the exponent b is made equal to its most negative value, that is, $b = -32$. However, this is an awkward convention to adopt, since it is more convenient to have the exponent equal to zero when the mantissa is zero, thus giving an all-zeros floating-point number. This limitation may be overcome if the range of the exponent is arbitrarily limited in such a way that the sign of the exponent is always positive. Thus a **biased exponent** $b' = b + 32$ (in this case) is used instead of b, giving an exponent range of $0 \leqslant b' < 64$, which ensures that the most negative exponent is represented by all zeros. In this case when the mantissa is zero, the biased exponent b' also equals zero; note however that it is necessary to restore the exponent to its unbiased form when performing arithmetic operations, this is done by adding or subtracting 32 to the biased exponent.

To ensure that the sum or difference of two mantissas remains within the scale of the machine and to standardize on the format of floating-point numbers, a **normalized** form is used. A further advantage is that the number of significant figures is maximized in a normalized floating-point number. To achieve this the fractional part of the number, namely a, is constrained within specified limits by adjustment of the exponent b. The range of the mantissa, in the general case and assuming a fractional number, is given by $1/R \leqslant a < 1$, where R is the radix. For binary fractions with a radix of 2 we have:

$$-1 \leqslant a < -\tfrac{1}{2} \quad \text{or} \quad a = 0 \quad \text{or} \quad \tfrac{1}{2} \leqslant a < 1$$

The mantissa range for our 18-bit machine is therefore:

positive numbers	010000000000 → 011111111111
negative numbers	100000000000 → 101111111111

Table 2.1
Floating-point numbers

Mantissa												*Exponent*	*Value*
2^0	2^{-1}	2^{-2}	2^{-3}	2^{-4}	2^{-5}	2^{-6}	2^{-7}	2^{-8}	2^{-9}	2^{-10}	2^{-11}		
0	1	0	1	0	0	0	0	0	0	0	0	100010	$0{\cdot}625 \times 2^2$ or 5×2^{-1} or $(5 \times 2^{-3} \times 2^2)$
0	1	0	0	0	0	0	0	0	0	0	0	111111	$0{\cdot}5 \times 2^{31}$ or 1×2^{30} or $(1 \times 2^{-1} \times 2^{31})$
0	1	0	0	0	0	0	0	0	0	0	0	000000	$0{\cdot}5 \times 2^{-32}$ or 1×2^{-33} or $(1 \times 2^{-1} \times 2^{-32})$
1	0	1	0	1	0	0	0	0	0	0	0	010001	$-0{\cdot}6875 \times 2^{-15}$ or -11×2^{-19} or $(-11 \times 2^{-4} \times 2^{-15})$
0	1	1	0	0	0	0	0	0	0	0	0	011110	$0{\cdot}75 \times 2^{-2}$ or 3×2^{-4} or $(3 \times 2^{-2} \times 2^{-2})$
1	0	0	1	0	1	0	0	0	0	0	0	001101	$-0{\cdot}84375 \times 2^{-19}$ or -27×2^{-24} or $(-27 \times 2^{-5} \times 2^{-19})$

Note the similarity of normalized floating-point numbers to the usual practice of expressing decimal numbers in a standard form. In this case the range is given by $1 \leqslant |a| < 10$, thus numbers such as $-0{\cdot}00602$ and $19786{\cdot}67$ would be represented as $-6{\cdot}02 \times 10^{-3}$ and $1{\cdot}978667 \times 10^{4}$ respectively.

With the above scheme it is possible to represent all fractional numbers between 2^{-32} and 2^{31} to 12-bit accuracy, thereby allowing a wide range of numbers to be handled in the computer. However, there is a loss of accuracy compared with the 18-bit fixed-point representation since digits normally used to give more significant places are now used to represent the exponent value. Thus to get the same accuracy but with an extended range it is necessary to use two computer words per floating-point number. Table 2.1 shows some typical floating-point numbers. Note that positive integers may also be represented by suitable scaling, but if this is done the usable range is reduced to $1 \times 2^{30} \rightarrow 1 \times 2^{-33}$.

2.3 Basic machine-code orders and programming techniques

Before it is possible to write, and understand, more complex computer programs, there are a number of basic programming techniques which must be discussed in detail.

2.3.1 Fetch, store, add, and subtract orders

These instructions, in some form or other, appear in the order repertoire of all machines. They are used to load the registers and store of the computer, and to perform the basic arithmetic operations. In the case of a single-address machine the instructions refer specifically to the accumulator (or to a modifier register). For example, the instruction to add (order 20 in Table 1.2) adds the contents of the address specified in the instruction to the contents of the accumulator (or the modifier register if order 14 is used) leaving the contents of the store location unchanged. In the same way the store instruction (order 23) replaces the contents of the location specified in the instruction with the contents of the accumulator, which are left unchanged.

An alternative method of loading (or storing from) the modifier register would be to include **transfer instructions** in the order set. These instructions (which do not need a specified address) transfer the contents of the accumulator to the modifier register (or vice versa) leaving the source contents unchanged. Thus to load the modifier register, the required contents would be brought down to the accumulator (using a fetch instruction) and then transferred to the modifier register. The disadvantage of this type of technique is that in many cases the contents of the accumulator may be needed (they are of course overwritten in the fetch instruction), in which case they must be stored away and retrieved after the modifier transfers; this can be a time consuming process and also uses extra storage space.

Many current machines, such as the DEC PDP/11 series and microcomputers etc., have an array of general purpose (**scratchpad**) registers in the processor itself which may be used for processing data without the need to continually refer to the main store (thereby considerably enhancing the processing rate). In this case the transfer instruction set would be expanded to include arithmetic and logic operations between registers, thus any register can effectively be used as an accumulator. Note also that a zero address instruction mode (also called register-register instructions) would be required since it is only necessary to specify the type of operation and the register(s) involved. For example, an instruction of the form 'add the contents of register R_1 to register R_2'.

2.3.2 Multiply and divide instructions

These instructions are not generally included in the basic order-code of small computers, but are often available as optional hardware extras. The reason for this is that in control and data-handling systems (the main areas of application for mini- and microcomputers) the need for fast and frequent multiplication and division is not of paramount importance; moreover, if these functions are required they can be performed by software routines. This is not the case with scientific computers, however, which include these instructions (in both fixed and floating-point mode) in the standard machine-code repertoire.

The multiplication instruction (order 35 in Table 1.2) multiplies the accumulator by the contents of the location specified (in fixed-point mode) producing a double-length 36-bit product. The most significant 18 bits of the product are retained in the accumulator itself (which in some cases is rounded-off to give a single-length product) with the least significant half being held in the auxillary X-register. The use of the auxillary arithmetic register in this order means that a transfer instruction is necessary, which exchanges the contents of the accumulator with the X-register, and vice versa. Thus, after performing a multiplication instruction it is necessary to store the contents of the accumulator, exchange with the X-register, and then store again, in order to retain the double-length product.

Division instructions (order 36 in Table 1.2) function in a similar way; the contents of the accumulator and X-register (containing a 36-bit double-length dividend) are divided by the contents of the location specified. The resultant single-length quotient is held in the accumulator, with the remainder being left in the X-register.

2.3.3 Transfer of control instructions

These are the so-called **jump** instructions (see Table 1.2) and their purpose is to allow the normal sequential execution of instructions in the computer to be bypassed, for example, to transfer control from one program (or part of the store) to another. When a jump instruction is carried out the contents of the control register are replaced by the address portion of the jump instruction. Consequently, the next instruction is not taken in sequence (that is, from the following location) but from the new address as specified in the jump instruction. This is in fact how a computer program may be initiated, by manually obeying a jump instruction to the first location (the start) of the program.

The instruction may also be used to effectively stop the computer, for if the jump instruction is directed to jump to its own location, for example:

1000 JMP 1000

the computer will go into a dynamic loop (or stop) continually replacing the contents of the control register with the jump address. The difference between a dynamic stop and the stop instruction (order 000000) is that with a dynamic stop the machine can only be restarted by executing another jump instruction manually, whereas with the stop instruction the machine will proceed in sequence when the start button is pressed.

There are many different forms of jump instructions in the order set of a computer, the one described above is an **unconditional jump**, order 30 in Table 1.2. The other types of jump instruction make up a general class of **conditional jumps**, where the transfer of machine control is determined by the state of various registers and bistables in the computer. For example, order 31 (JMPN) in Table 1.2 is a jump instruction which is conditional on whether or not the contents of the accumulator are negative. If the contents are negative a transfer of control is effected, in the usual way, to the address specified in the instruction; if the contents are non-negative, program control proceeds in sequence. Many variations on this type of instruction are possible depending, for instance, on the state of the accumulator or modifier and scratchpad registers, and whether zero, positive, or negative, etc.

An alternative method of incorporating conditional jumps in a computer order code is to use a **skip** instruction. This type of instruction, which normally has the advantage of not requiring an address to be specified in the instruction, is of the form: 'Skip the next instruction if condition fulfilled, otherwise take the next instruction in sequence'. The conditions may be of the usual type – for example, tests on the accumulator for zero, positive, or negative contents. Alternatively the state of various control bistables or switches (used, for example, in the interrupt procedures, see Chapter 7) may be sensed. This type of instruction must generally be followed by an unconditional jump instruction (to another program area) if the program to be executed for the unsatisfied condition is longer than one instruction. The skip instruction may also include a store address in which case, as well as testing for zero, positive, or negative contents, more complex conditions may be used, such as inequality relationships between the contents of the address specified in the instruction and the contents of the accumulator. This type of order which allows the computer to discriminate between one of two courses of action is one of its chief distinguishing features, and one which makes it so much more powerful than any other form of machine.

2.3.4 Instruction modification

Perhaps the most important aspect of the stored program concept is the fact that a computer can automatically *modify* its own instructions. This may be done very simply by fetching an instruction from store to the accumulator, where all the relevant machine-code instructions may be performed on it, that is, it is treated exactly as if it were an operand.

Note however that it is bad programming practice to have self-modifying software, since it gives rise to non-re-entrant code (see section 3.9) and the program listing does not correspond to the instructions actually being executed.

The principle reason for utilizing order modification in programming is to economize on storage space in the computer. Consider, for example, the problem of adding together 10 numbers. This may be accomplished simply by extending the program shown in Table 1.3 in a linear fashion, by adding nine extra add instructions, giving a total of twelve instructions altogether for the program, plus eight more locations to store the variables.

An alternative way of adding together ten numbers is to use the program shown in Table 2.2. In this case after each number has been added to the sum store (location 0020), the actual add instruction in location 1001 (that is, 20 0021) is brought to the accumulator. A constant which is equivalent to the last addition order to be obeyed (that is, 20 0031) is then subtracted from the accumulator and if the result is zero, as determined by the jump instruction, we know that the program has been completed. If it is not zero we must add back what we subtracted, plus an additional 1 in the least significant digit position of the address, to ensure that the next number is taken in sequence and added to the sum store. The modified instruction (with its address incremented by 1) is then returned to its correct position in the program, and the process repeated. The reader is advised to work carefully through all the program examples as this is the only way of becoming familiar with the ideas involved. The reader is also warned that all the instructions, including addresses and constants, are in the **octal** notation.

In this program (see Table 2.2) we have established the important concept of a **program loop**, that is, a number of instructions (in this case nine, held in locations 1000 → 1010) which are performed many times over during the execution of the program. The **loop constant** determines the number of times the loop is traversed, and it is necessary to perform a **loop test** each time the set of instructions is obeyed in order to determine when the program has finished. Note the use of the unconditional jump instruction to transfer control back to the start of the program. Note also the conditional jump which is used to test the accumulator, after the subtraction of the loop constant, to determine the end of the program loop.

The number of the stored instructions required for the program still totals twelve as before, but using this technique the program may be

Table 2.2

Addition of ten numbers using direct order modification

Location	Instruction		Comment
→1000	22 0020		Fetch x_1 to the accumulator
1001	20 0021		Add $x_2, x_3, \ldots, x_{10}$ to accumulator
1002	23 0020		Store $(x_1 + x_2), (x_1 + x_2 + x_3)$, etc.
1003	22 1001	Loop	Fetch add instruction to accumulator
1004	21 1012	Test	Subtract loop constant (20 0031)
1005	33 1011		Jump if accumulator zero
1006	20 1013	Modify	Add modification constant (20 0032)
1007	23 1001		Replace instruction back into program
1010	30 1000		Jump to start of program
→1011	00 0000		Stop
1012	20 0031		Loop constant
1013	20 0032		Modification constant
0020	x_1		Numbers to be added
0021	x_2		
⋮	⋮		
0031	x_{10}		

easily altered (keeping the same number of instructions) to add together *any number of variables* simply by changing the program constants. However, the running time of the program has been considerably lengthened. In the simple linear program there are only twelve instructions to be executed; in the modification program, which requires that the loop instructions be repeated nine times, a total of 82 instructions must be executed. Thus it is essential to keep the number of instructions in the loop as small as possible, and in some cases where execution time is more important than storage space, it may be necessary to resort to a simple linear program.

The efficiency of the modification process can be considerably improved if **modification registers** (originally called B Line registers when first used in the Manchester University computer) are available. The instruction word of our running example includes a **modifier bit**, and a modification register is included in the block diagram of the computer (see Fig. 1.4). The effect of the modifier bit being set to 1 in the instruction word (for specified address instructions only) is to indicate to the control unit of the computer that the contents of the modifier register must be *added* to the instruction *before* it is obeyed. It is also possible to design a machine-code order structure such that *all* instructions may be modified. To use the modifier it is necessary to write the instruction in the form, for example, 20 0231/M (note that when writing instructions the presence of any of the control bits I, M, or S signifies that the bit is set to 1 in the instruction, otherwise it will be interpreted as being zero). When the instruction is executed the presence of the modifier bit initiates

Table 2.3
Addition of ten numbers using modifier register

Location	*Instruction*	*Comment*
1001	15 1010	Load modifier register with loop constant
1002	22 0020	Fetch x_1 to accumulator
→1003	20 0031/M	Add $x_2, x_3, x_4, \ldots, x_{10}$
—1004	34 1007	Jump if modifier zero (loop test)
1005	00 0006	Add +1 to modifier register
—1006	30 1003	Jump back to start of loop
→1007	00 0000	Stop
1010	37 1770/IMS	−10 octal as pseudo-order (loop constant)
0020	x_1	Numbers to be added
0021	x_2	
⋮	⋮	
0031	x_{10}	

the action of adding the contents of the modifier register (say octal 24) to the contents of the instruction register to produce the new instruction, 20 0255/M, which is then decoded and obeyed as usual. The original instruction, 20 0231/M, is left unchanged in the computer store; note that the modifier bit is only examined once, at the start of the execute sequence, and that the modification may extend right through to the order digits. To facilitate the use of the modifier register specific instructions, such as fetch to the modifier, etc. (these are listed in Table 1.2), must be included in the machine-code structure.

If we rewrite our program example of adding together ten numbers using the modifier register, see Table 2.3, we get an immediate improvement in both storage and running time. The preliminary operations in the program are concerned with setting the loop constant (−10 octal) in the modifier register (this register must be refurbished each time the program is entered). Note the use of a pseudo-instruction format to represent the program constants, for example −10 octal (−8 decimal) is binary 1111111111111000; this is written as an instruction 37 1770/IMS, the last three characters after the oblique stroke indicating that the indirect address, modifier, and sector bits are all set to 1. The add instruction, 20 0031/M, is modified each time round the loop – observe that the loop count is negative (that is, −10, −7, −6, . . . , 0) – so that the first add instruction will be obeyed as 20 0021.

The next step in the program is the loop test performed on the contents of the modifier register. If the contents are non-zero +1 is added to the modifier and the program jumps back to the beginning of the loop and the process repeated. The modified program now has four instructions in the loop, which are repeated nine times, giving a total of 39 orders to be obeyed and the total storage required has been reduced to 19 locations. This is still less efficient in *time* however than the simple linear program, and in general this execution time/storage trade-off problem will apply to all programming systems.

The number of modification registers incorporated in the structure of a computer is generally a good indication of the computing (or programming) power of the machine. In practice, however, the number of distinct registers that can be employed is limited by technical (word-length) and/or economic considerations. An alternative approach is to design the computer such that *any storage location* may be used as a modification register. This may be done simply, but at the expense of storage space and execution time, by using a specified address instruction of the form: 'Modify the *next* instruction with the contents of the address specified'.

To obey this order the contents of the address specified in the modify instruction are read down from store and held in a temporary register. The next instruction in sequence is then read down, added to the contents of the temporary register, and the resultant instruction obeyed in the usual way. Thus to use this method of modification two instructions need to be stored and obeyed for each modified order; for example, the instructions held in locations 1002 and 1003 in Table 2.4 are both necessary to perform the modified add operations. It is also apparent from Table 2.4 that the number of instructions contained within the loop is excessive (in fact in this case the same number as required by the basic program of Table 2.2). This comes about because it is necessary to fetch the modifier from the store to the accumulator (each time round the loop) in order to increment the count and perform the loop test. Thus, to use

Table 2.4
Addition of ten numbers using 'modify next instruction' order

Location	*Instruction*		*Comment*
→1001	22	0020	Fetch x_1, $x_1 + x_2$, etc. to accumulator
1002	MNI	1013	Modify next instruction order
1003	20	0031	Add $x_2, x_3, \ldots, x_{10}$
1004	23	0020	Store $x_1 + x_2$, $x_1 + x_2 + x_3$, etc.
1005	22	1013	Fetch loop constant to accumulator
1006	33	1012	Jump if accumulator zero
1007	20	1014	Add +1 to loop constant
1010	23	1013	Store away loop constant
1011	30	1001	Jump back to start of program
→1012	00	0000	Stop
1013	37	1770/IMS	Loop constant −10 octal
1014	00	0001	+1 constant
0020		x_1 (Sum store)	Numbers to be added
0021		x_2	
⋮		⋮	
0031		x_{10}	

Table 2.5
Addition of ten numbers using 'modify next instruction' order in conjunction with skip and increment orders

Location	*Instruction*		*Comment*
1001	22	0020	Fetch x_1 to accumulator
→1002	MNI	1010	Modify next instruction with contents of location 1010
1003	20	0031	Add $x_2, x_3, \ldots, x_{10}$
1004	SKP	1010	Skip next instruction if location 1010 negative
1005	00	0000	Stop
→1006	INC	1010	Increment location 1010 by +1
1007	30	1002	Jump back to start of loop
1010	37	1770/IMS	Loop constant −10 octal
0020	x_1		Numbers to be added
0021	x_2		
⋮	⋮		
0031	x_{10}		

this method of modification efficiently it is essential to include specified address orders which allow the contents of any store location to be incremented (or decremented) by +1. Moreover, skip instructions of the form: 'Skip next instruction if contents of the address specified are negative' must also be included in the order repertoire. Table 2.5 shows the same program as before, but this time utilizing the increment and skip instructions.

An alternative approach to using a special modifier instruction is to use the control bit, normally reserved to indicate a *particular* modifier register, to specify the modifier action. In this case if the modifier bit in an instruction is set to 1, then the contents of the address specified in the *next* instruction (sometimes the address itself is used) are added to the original instruction before execution. In practice commercial computers use many variations of the modification methods described above but as well as the actual modification process the basic precepts of loop counting and testing must always be obeyed.

Moreover, since in practice it is only the address portion of the instruction which needs to be changed other means such as indirect addressing and indexed addressing, which are explained in the next section, are normally employed.

2.3.5 Indirect addressing

Another technique, which in most cases can replace the use of a modifier register in loop counting, etc., is to specify operands by an *indirect* rather than a direct address. In this case (usually signified by an indirect address bit in the instruction word being set to 1) the address part of an instruction specifies the address of a location where the address of the actual operand is to be found (called a *pointer*). Consider, for example, the indirectly addressed add instruction 20 1023/I, where location 1023 contains the octal address 0025. The instruction would be obeyed with an effective address of 0025, resulting in the contents of location 0025 being added to the accumulator. Note that two store accesses are required to execute this instruction.

However, to make effective use of this technique in loop counting it is necessary to incorporate in the order structure the instructions to increment (or decrement), and test, the contents of any store location as described in Section 2.3.4. Table 2.6 shows how indirect addressing may be used to perform our running example.

Alternatively, for example in microprocessors, a CPU scratchpad register is used as a pointer to the RAM or ROM store; in this case the register to register instructions can be used to set up the address modification and loop count procedures

Since in indirect addressing it is the address part only of the store location which is relevant, there is no reason why, if the indirect bit of the first level address is also set, this should not in turn refer to another address and so on, resulting in **multiple-level** indirect addressing. For instance in our earlier example, if location 1023 contained 0025/I, and location 0025 contained the octal address 0333, the instruction would be obeyed as 20 0333. The use of this facility is somewhat restricted in practice, though it is incorporated in many commercial computers, owing to the difficulties encountered in its application (for example, endless loops must be avoided, which could happen if location 0025 above contained 1023/I!) and the fact that each indirect level of address requires two store accesses.

Table 2.6
Addition of ten numbers using indirect addressing and increment order

Location	*Instruction*		*Comment*
→1001	22	0020	Fetch x_1, $x_1 + x_2$ etc. to accumulator
1002	20	1023/I	Add x_2, x_3 . . . . x_{10} indirectly addressed
1003	23	0020	Store $x_1 + x_2$, $x_1 + x_2 + x_3$ etc.
1004	22	1023	Fetch location 1023
1005	21	1024	Subtract octal 31
1006	33	1011	Jump if accumulator zero
1007	INC	1023	Add +1 to location 1023
1010	33	1001	Jump to start of loop
→1011	00	000	Stop
1023	00	0021	
1024	00	0031	
0020		x_1 (and Sum Store)	Numbers to be added
0021		x_2	
⋮			
0031		x_{10}	

Indexed addressing is a cross between instruction modification and indirect addressing. In this method the address field of the instruction, normally an offset value (such as the displacement from a predetermined memory location which, for instance, could be the start of a data set) is added (or subtracted) to the contents of an index register to obtain the actual store address. The index register must of course be initially loaded with the start address of the relevant store area (sometimes called the **block address**). The index registers can also be automatically decremented or incremented (in some cases they may be general purpose CPU registers) after each operation. Note also that though the index registers are normally a full computer word length long (and hence capable of directly addressing any store location) the instruction offset field may be considerably shorter.

2.3.6 Sub-routine links

During the course of a large computer program it is often necessary to execute the same set of program instructions many times over, for example, the evaluation of trigonometrical functions in mathematical problems. It is obviously possible to write this set of instructions into the main program each time the computation is required, but this repetition is very wasteful of storage space. A better method is to write, for example, a sine–cosine routine as a completely separate program, called a **sub-routine**, and let the main program jump into this sub-routine each time the calculation is required. To operate this scheme, however, there must be some means of allowing the sub-routine to initiate a jump back to the correct place in the main program after the sub-routine has finished, this is known as the **sub-routine link**.

The simplest method of marking a place in the main program is to provide an order which stores the contents of the control register (which contains the address of the next instruction) into a specified location, usually the first location of the sub-routine; alternatively, a special modifier register may be used for this purpose. This function is usually combined with a jump facility to give the link instruction (order 17 in Table 1.2). The process is illustrated in Table 2.7; the action of the link instruction is to place the current contents of the control register (in our example the address 1002, since at this stage the control register would already have been incremented by +1 (see Section 1.4) into the location specified in the instruction address (in this case location 0300, the first location of the sub-routine) and then jump to the next location (0301). After the sub-routine is completed the link address is brought to the modifier register, and a modified jump instruction back to the main program is performed. The disadvantage of this technique is that the modifier register may be required for some other purpose and in performing the link back to the main program the contents would, of course, be destroyed. This may easily be overcome if indirect addressing is available when the single instruction 30 0300/I would suffice.

Table 2.7
Sub-routine linking

	Location	Instruction		Comment
Main program	1000	22	1043	Fetch variable to accumulator
	1001	17	0300	Link to sub-routine
	→1002	23	1245	Continuation of main program
	1003	31	0753	
	etc.			
Sub-routine	0300	(	)	Blank location reserved for link address
	→0301			Start of sub-routine
	0302			Continuation of sub-routine
	⋮			
	0374	22	0340	Fetch result to accumulator
	0375	15	0300	Fetch link address to modifier
	0376	30	0000/M	Jump back to main program

Table 2.8
Data handling instructions

(a) Shift instructions

Instruction		*Contents of accumulator*	*Comment*
		101010100000000101	Initial contents
200	003	010100000000101000	Logical shift left 3 places
201	003	000101010100000000	Logical shift right 3 places
202	003	010100000000101000	Arithmetic shift left 3 places, error indicated after 1st shift
203	003	111101010100000000	Arithmetic shift right 3 places
204	003	101101010100000000	End-around shift 3 places

(b) Logical instructions

Location	*Instruction*		*Comment*
1000	22	1004	Fetch variable A to accumulator
1001	24	1003	Collate with variable B
1002	37	1003	Exclusive OR of AB and B
1003	07	1400/IMS	Variable B
1004	27	1657/M	Variable A

Contents of accumulator, A	101110101110101111
Variable B	001111111100000000
Result left in accumulator after collation	001110101100000000
Result left in accumulator after exclusive OR	000001010000000000

When using sub-routines it is essential to define the parameters of the program, by this we mean the registers or locations which contain the variables to be used in the computation and the locations where the results are to be stored. In the example in Table 2.7 the sub-routine is entered with the variable in the accumulator and after the sub-routine is finished the result is also brought to the accumulator; this occurs just before the link back to the main program is initiated.

By far the best method of handling sub-routine linkages, particularly when routines are required to be nested (one sub-routine calling another, or itself) is to use **push-down stack registers** – these are described in section 3.9.

2.3.7 Data-handling instructions

The majority of computers make some provision in their machine-code order set for instructions which are primarily concerned with the manipulation of non-numerical data within the machine. All computers would include, for instance, some form of **shift instruction**; for example, in Table 1.2 we have five different shift instructions. These instructions normally refer to the contents of the accumulator and although they do not require a store address it is necessary to specify the number of digit places to be shifted through the binary word. There are three basic forms of shift instruction:

(a) The **logical** shift, which shifts the binary word either to the left or right, filling up the vacated digit positions, from the least significant or most significant ends respectively, with zeros.

(b) The end-around shift, also known as a **right rotation**, shifts digits out from the least significant end of the binary word and inserts them back in at the most significant end; it is also possible to have a left rotation which performs the reverse operation.

(c) **Arithmetic** shifts where, in the case of a right shift, the sign digit is repeated in the vacated positions to retain the correct number sense. In an arithmetic left shift an error is indicated (by setting an appropriate bistable) when the sign of the number changes as a result of the shifting operations.

These shifting functions are illustrated in Table 2.8(a). In practice, the shifting operations would be used for doubling (left shift) or halving (right shift) binary numbers, normalization of floating-point numbers, assembling binary data into words, and so on.

Logical operations are also included in the machine-code order set, for example the AND (or **collate**) and OR functions. The AND order can be used for the extraction (or masking) of certain groups of digits within a computer word. For example, if the ten least significant digits of a word have to be separated, the word would be brought to the accumulator and collated using the AND instruction (order 24) with the constant 000000001111111111. Table 2.8(b) shows in detail how the logical function $R = \overline{A}B$ might be generated using the AND and exclusive OR instructions. The initial contents of the accumulator are A, which are collated with B to give AB, this result is in turn exclusively OR'd with B, thus:

$$R = AB \oplus B = \overline{A}\overline{B}B + AB\overline{B} = (\overline{A} + \overline{B})B$$

Thus $R = \overline{A}B$

The same result could have been obtained by forming the inverse or true complement of A (using the 2's complement instruction, order 000001, and subtracting +1) and then collating with B; for example:

Contents of accumulator, A	101110101110101111
2's complement of A	010001010001010001
Subtract +1	010001010001010000
Result after collation with B	000001010000000000

It will be obvious by comparing the computation above with Table 2.8(b) that the results are identical. The technique above, however, requires two additional storage locations.

In some computers, particularly those designed for real-time process control applications, a full range of logical operations, including logical inversion and the compare operation, are provided.

2.4 Alternative instruction formats

The choice of an instruction format is a crucial decision in the system design of a digital computer, and predetermines to a large extent the resultant structure of the machine. So far we have only considered the simple but well proven single-address system. There are, however, many forms of instruction layout and these may be broadly classified into **single-** and **multi-address** formats. It is also possible to have **zero-address** instructions, which refer to operations between specific registers contained in the hardware of the central processor.

In the **two-address** systems the instruction word contains two addresses (A_1 and A_2) as well as the operation order code. These two addresses are usually employed to specify the locations of two separate operands; for example, in an addition instruction the sum of the two specified operands would be placed in the accumulator. Alternatively, the addresses could be used such that the result of a functional operation between the contents of the accumulator and A_1 would be placed in location A_2. In earlier machines, however, where the main storage unit was of the circulating type (for instance, a delay line store, see Chapter 6) the second address was used to specify the location of the next instruction, thus each instruction was associated with an automatic jump to location A_2. Using this scheme (also called 1 + 1 address format) maximum programming efficiency could be obtained by synchronizing the instruction execution time with the access time of the storage unit; with the advent of random access storage 'optimum programming' of this form is no longer relevant.

Three-address instruction formats are also possible, where three addresses, A_1, A_2, and A_3 are included with the operation code. The usual arrangement with this scheme is that two of the addresses, A_1 and A_2, specify the operands to be used in a given operation, with the third address, A_3, specifying the location in which the result is to be stored.

It is difficult to appraise the relative advantages and disadvantages of each type of instruction format since in most cases the final efficiency is dependent solely on the problems to be solved. The single-address system is by far the most commonly used, and has the advantages of simplicity in use and economic hardware implementation. Multi-address systems, on the other hand, require fewer instructions to solve a particular problem and hence can be expected to operate faster. However, they can require a longer wordlength, and consequently more storage space, than single-address machines; furthermore, the control logic is more complex.

The main speed advantage with multi-address systems comes about because a number of operations can be specified in a single instruction. For example, if it were required to add the operand in address A_1 to that contained in address A_2, putting the resultant sum in location A_3, this could be completely specified by a single three-address instruction. The execution time for this instruction would be approximately four store access times – one to read the initial instruction and three more to obey the orders. The single-address instruction performing one operation at a time requires six store access times (one read access and one obey access per operation) for the same computation. Thus the overall speed advantage is not overlarge, particularly if we take into account the extra storage that is required. If, for example, the wordlength of the single-address machine was extended to that of the three-address machine, we could include *two* single-address instructions per word. In this case the computation could be performed in five store accesses, with an additional instruction already available for the subsequent part of the problem. The multi-address systems are also very inefficient when jump and zero-address instructions are performed, since in these cases full advantage cannot be taken of the extra addresses.

However, with the availability of cheap LSI registers and logic modules multi-address systems have become cost effective and are coming back into favour. For instance, the PDP 11 computers have both single and double operand addressing modes. The single address mode is used for orders such as clear, increment, test shift etc., while the double address mode is used for operations which imply two operands, such as arithmetic, compare and transfer. In the PDP 11 there are eight general purpose hardware registers which can be used for addressing purposes and as accumulators; arithmetic and logic operations can be performed between registers, memory and registers, and memory locations.

The instruction word of the PDP 11 is 16-bits long, and for the single operand address mode 10-bits are allocated to the operation code and 6-bits for the destination address field. Three bits are used in the destination address to specify one out of the eight general purpose registers which are used as pointers to the main store (or operands in their own right), and 3-bits are used for indicating the address mode (for instance, direct or indirect addressing). The double address mode employs 4-bits for the operation code and 6-bits each for source and destination addresses. The source address field is used to select the source location, the first operand, while the destination address locates the second operand and the end result. For example, the instruction ADD A,B adds the contents of location A (source) to the contents of location B (destination); after execution location B will contain A + B and the contents of A will be unchanged.

Note that using the general registers as pointers does not entail an increase in word length and the more complex control logic can be efficiently implemented in hardware. There is still the disadvantage of increased store accesses with indirect addressing, but in many cases this is compensated for by using register to register operations whenever possible.

The instruction (and operand) words discussed above have all been of fixed wordlength and although this is ideal for most purposes many modern computers (such as the PDP 11 and in particular microprocessors) now employ a variable wordlength structure. In these machines the basic wordlength is an 8-bit binary word called a **byte**; longer wordlengths can be obtained by taking multiples of bytes, for example, a **half-word** would be two bytes, that is, 16 bits, with four bytes forming a **full-word**. As a consequence of using a byte-organized structure, it is common practice to use a **hexadecimal** notation (using base 16)

Table 2.9
Hexadecimal notation

Decimal	*Binary*	*Hexadecimal*
0	0000	0
1	0001	1
2	0010	2
3	0011	3
4	0100	4
5	0101	5
6	0110	6
7	0111	7
8	1000	8
9	1001	9
10	1010	A
11	1011	B
12	1100	C
13	1101	D
14	1110	E
15	1111	F

16-bit word

0000	0001
0002	0003
0004	0005
0006	0007
0008	0009
000A	000B
000C	000D
000E	000F
0010	0011
0012	0013
0014	0015
0016	0017
0018	0019
001A	001B
001C	001D
001E	001F

etc.

Figure 2.1 **Layout of byte organised store.**

for the representation of all binary information, including the order-code and store addresses. In this system (see Table 2.9) four binary bits of information can be expressed by a single hexadecimal digit (note the need to use letters of the alphabet to represent the decimal numbers 10, 11, 12, 13, 14, and 15). The conversion from a hexadecimal number to its pure binary equivalent is simply a question of writing down the binary number corresponding to each hexadecimal digit, starting from the least significant end, in an analogous manner to octal conversion.

Bytes, half-words, and full-words can all be effectively addressed though in practice the wordlength contained in the main store would be standardized to 16 bits for example, with selection or extra store accesses being performed as required. Each 8-bit byte in the main store would be numbered consecutively (as shown in Fig. 2.1). Thus if two bytes were used to represent the address field of the instruction, 65,536 bytes could be directly addressed, that is, $(0000)_{16}$ through to $(FFFF)_{16}$.

In a variable wordlength machine bytes may be handled independently or in groups of 2, 4, 8, etc., depending on the function being performed. The computer order-code would include separate instructions for operations on bytes, half-words, and full-words and these may be intermingled in a program. For instance, fixed-point arithmetic functions would use either full-words or half-words, with floating-point operations taking place on full-words and double full-words. Furthermore, each byte can be used to store an alphanumeric character (or two binary-coded decimal numbers) which may then be handled as strings of data. The method of addressing the byte-orientated store is determined by the instruction (whether it refers to bytes, half-words, etc.) and the specified hexadecimal address. If, for example, a four-digit hexadecimal address was employed, the address $(0014)_{16}$ could be intepreted (see Fig. 2.1) as:

(a) byte 0014,
(b) a half-word comprising bytes 0014 and 0015, or
(c) the full-word consisting of bytes 0014, 0015, 0016, and 0017.

Thus, though any address may be used to select a byte, half-word addresses must have 0,4,6,8,A,C, or E as the least significant hexadecimal digit, and similarly full-word addresses must end in 0,4,8, or C. Each store access would read down the contents of two bytes (that is, a 16-bit word) and then, depending on the address and instruction, route either the left- or right-hand 8-bit byte, or the entire 16-bit word, to the accumulator or control unit; in the case of a full-word address another store cycle must be initiated.

In the same way that multiples of bytes may be used as operands, so instructions can consist of up to 2, 4, or even 6 bytes, depending on their complexity. Instructions such as stop, shifts, skip, etc. (and some input/output orders) which do not require an address or additional information to be specified can be very short, for example, two bytes. Add, store, and fetch instructions, however, must specify a store address and hence need an additional two bytes for the address field. The computer then will have instructions of different lengths according to the operations to be performed; this means that the control unit requires, in some cases, two store accesses to fetch the complete instruction to the control register. In some machines fast registers are available for immediate working space (**scratch pad** stores) and instructions specifying operations between registers are provided; in this case also the instructions can be very short since no store address is required.

Many microprocessors, such as the Intel 8080 and Motorola 6800, use a basic 8-bit wordlength (due to ROM limitations) but since this is restrictive for both addressing and instruction purposes a variable byte system is adopted. Thus instructions can be either 1-byte or 2-bytes in length; note that since the basic wordlength is 1-byte the control register (program counter) will indicate a byte address. Consequently the control logic must determine (by the coding of the first byte) whether a 1- or 2-byte instruction is intended.

The instruction sequence commences with the byte address held in the control register being used to access the ROM and fetch down the first byte; the control register is also incremented at this stage. If the byte read down corresponds to a 1-byte instruction the obey cycle is entered and the instruction executed. If, however, it is a 2-byte instruction another fetch cycle must be initiated and the obey cycle commences using the two bytes. Because the control register is automatically incremented after each fetch cycle, the instructions can be stored contiguously in the ROM and obeyed sequentially. Note that for large memory sizes, if direct addressing is required, additional bytes will be required in the instruction.

2.5 Expansion of storage capacity

The problem here, which is particularly pertinent with small machines, is how to increase the storage capacity of a fixed wordlength computer which may, as in our example, have only 10 address bits allowing 1024 locations to be directly referenced. With a variable wordlength machine this problem is easily overcome, if it exists in the first place, simply by using additional bytes in the instruction word. The solution which has generally been adopted, however, for small machines is to use a **sector** (or **page**) control bit in the instruction word; for example, in Fig. 1.2(a) the instruction word has 10 address bits and a sector bit. The obvious way of using this additional bit, that is, by simply including it as part of the general address scheme thus allowing 2048 directly addressed locations, does not give such a flexible solution as using a sector bit in conjunction with the indirect addressing operation described in Section 2.3.5.

The sectoring technique entails dividing the main store (for the purpose of addressing only – not physically) into sectors (or pages) the sizes of which are determined by the maximum number of directly addressed locations. In our example the sectors would comprise 1024 locations; small machines are available, however, which contain 128 and 512 words per page. The sectors are numbered 0,1,2,3, etc., in sequence through the store, thus sector 0 would contain locations 0000 → 1777; sector 1, locations 2000 → 3777; sector 2, locations 4000 → 5777; etc. The sector bit is used in the following way:

(a) If the sector bit is set to 1 the address referenced by the instruction is in the same sector of the store (also called the **current sector** or **page**) as the instruction itself.

(b) If the sector bit is set to zero the address referenced by the instruction is in sector 0, that is, locations 0000 → 1777.

Thus it is possible to address any location in sector 0 or the current sector directly; the method of addressing locations in other sectors of the store is by indirect addressing. For example (Table 2.6) if one wanted to refer to the contents of a location in sector 2, from a program residing in sector 1 of the store, it would be necessary to plant the address of this location in sector 0 (or sector 1 if preferred) and indirectly address it via sector 0. Note from the example in Table 2.10 that, except when referring to sector 0, the sector bit must always be set to 1 (signified by writing /S after the instruction). Moreover, the address referenced indirectly may exceed the normal address part of the instruction and for this reason they are specified as a 6-digit octal number. The maximum size store that can be addressed in this way depends on the length of the control register (and the store unit address register); if this is made equal in size to the standard computer word, that is, 18 bits in this case, it would be possible indirectly to address up to 2^{18} locations.

In general the address field in the instruction set of a small computer does not allow direct addressing of the main memory, for this reason it is sometimes called the **displacement field** since it is often

Table 2.10
Addressing using sectoring method

Location	*Sector*	*Instruction*		*Comment*
1000	0	00	4000	Address of variable in Sector 2
1001		00	4001	Address of sum in Sector 2
2000	1	22	0024/S	Fetch variable held in Sector 1 to accumulator
2001		20	1000/I	Add variable from Sector 2
2002		23	1001/I	Store sum in Sector 2
2003		31	0705/S	Jump, if accumulator negative, to location 2705 in Sector 1
4000	2		x_n	Variable
4001		(	)	Sum location

used to specify the displacement or offset from a specific memory location. The address mode (i.e. indirect and indexed addressing etc.) is used in conjunction with the displacement field to determine the **effective address** of the location in main memory. The address range of a particular address mode is defined as the number of words in main memory that can be addressed; this is usually expressed as a multiple of $1024(2^{10})$ where K is used to represent 1024, thus in our example the address range for the 18-bit wordlength (using indirect addressing) is $256K$.

In processors which have hardware scratchpad registers (of an adequate length) such as the PDP 11, these may be used either as pointers or index registers to provide the required memory addressing modes, otherwise special registers must be provided. For example, in the Intel 8080 microprocessor there are six 8-bit scratchpad registers, one 8-bit accumulator, a 16-bit program counter and a 16-bit stack pointer. In this case two of the 8-bit registers or the stack pointer can be used to address the main memory using a single byte instruction; note that the 16-bit program counter gives an address range of $64K$. (Note also that since the 8080 is a byte machine direct addressing can always be obtained by using a 3-byte instruction, however this is a slow process requiring three fetch cycles).

For larger machines more sophisticated methods of memory management are required, these are described in Chapter 6.

References and bibliography

1 Gear, C. W. *Computer Organisation and Programming*. McGraw Hill, New York, 1974.
2 Stone, H. S. *Introduction to Computer Organisation and Data Structures*. McGraw Hill, New York, 1972.
3 Soucek, B. *Microprocessors and Microcomputers*. John Wiley, New York, 1976.
4 McCracken, D. D. *Digital Computer Programming*. John Wiley, New York, 1967.
5 Peatman J. B. *Microcomputer Based Design*. McGraw Hill, New York, 1977.
6 Kilburn, T., Tootill, G., Edwards, D., and Pollard, B. Digital computers at Manchester University. *Proc. I.E.E.*, 1953, **100**, 487–500.
7 Church, C. Computer instruction repertoire – time for a change. *AFIPS Conf. Proc. S.J.C.C.*, 1970, **36**, 343–349.
8 Bell, C. G. and Cody, R. A New architecture for minicomputers – The DEC PDP 11. *AFIPS FJCC* **36**, 657–675, 1970.
9 Cluley, J. C. *Programming Mini-computers*. Crane, Russak & Co. Inc., New York, 1978.

Tutorial problems

Write computer programs to perform the following operations, using the basic machine-code language given in Table 1.2; place your programs in locations 0000 → 0500 of the computer store.

2.1 Add the operands stored in locations 1011, 1017, and 1032; if a negative answer results put the sum in location 0045, otherwise store it in location 0067.

***2.2** (a) Clear locations 1000 → 1777 of the main store to zero.
(b) Fill location 0500 with 1, 0501 with 2, 0502 with 3, etc. up to location 1000.

Use the modifier register in both cases and then repeat with indirect addressing.

***2.3** Determine the number of locations with zero contents that are held in a table occupying addresses 1000 → 1700 of the computer store, note also the addresses of the empty locations in a new table starting in location 0500.

***2.4** Interchange the contents of locations 1000 → 1200 of the store with locations 0500 → 0700.

2.5 (a) Multiply two single-length operands together, putting the double-length product in locations 0703 and 0704 of the store.
(b) Divide a double-length dividend by a single-length divisor, putting the quotient in location 0707 and the remainder in location 0710.

2.6 Write a sub-routine to multiply any number n by a given power of 2, the answer being required as a double-length product.

***2.7** Convert 4-bit binary-coded decimal numbers in the 8421 code, held as 4 decimal digits in the 16 least significant places of an 18-bit word, into a pure binary representation.

***2.8** Write a program to perform the logical function

$$S = \bar{a}\bar{b}c + abc + \bar{a}b\bar{c} + a\bar{b}\bar{c}$$

2.9 Repeat question 2.3 but this time assume the table is stored in locations 2000 → 2700.

***2.10** Repeat question 2.4 interchanging the contents of location 2000 → 2200 with location 4500 → 4700.

3

Input compilers and procedure orientated languages

3.1 Introduction

The preceding chapters have assumed that the machine-code instructions and data have, in some way, already been read into the main store of the central processor. In this chapter we consider the means by which instructions and operands (data-words) are inserted into (and extracted from) the computer store. As we have seen, the computer functions internally using binary-coded patterns of 1's and 0's to represent information; but the operator thinks in terms of decimal and alphabetical notations, that is, a symbolic system. Moreover, since the basic operations of the computer (the machine-code) are very simple, the symbolic representation used by the operator must be compiled into machine-code form. It is thus necessary to provide an interface (either in hardware or software form) between man and machine to enable both to operate efficiently. The function of this input/output interface is primarily one of encoding and decoding (translating) the basic information. We will also consider the technique of controlling input/output operations, and the evaluation and hierarchy of programming languages, including their influence on computer structures.

3.2 Input/output codes

The usual operator methods of data input into the computer are via the media of punched paper tape, punched cards, keyboards (normally associated with tele-typewriters), and push-buttons situated on the control console (see Fig. 3.1). These devices will be described in more detail, together with more sophisticated systems such as graphic input, in Chapter 7. In all cases information must be represented in binary form, using either pure binary or binary-coded alpha-numerics.

The minimum requirements for an input symbol set are to represent the numbers 0–9, the alphabet A–Z, control symbols such as carriage return, line-feed (for the teleprinter), space, and so on, and normal punctuation and mathematical symbols. Punched paper tape normally has eight data channels (though 5-, 6-, and 7-hole tape can be obtained) and a control channel. With 5-bit punched tape there is a maximum of 32 binary-coded symbols, and it is obviously not possible to have a distinct code for each character (see Table 3.1). So to obtain a useful symbol set recourse must be made to figure-shift and letter-shift characters to

Table 3.1
5-bit teleprinter code

Decimal	*Binary*	*Character*	
		Figure shift	*Letter shift*
0	00000	Blank	
1	00001	A	1
2	00010	B	2
3	00011	C	*
4	00100	D	4
5	00101	E	$
6	00110	F	£
7	00111	G	7
8	01000	H	8
9	01001	I	'
10	01010	J	,
11	01011	K	+
12	01100	L	:
13	01101	M	—
14	01110	N	.
15	01111	O	%
16	10000	P	0
17	10001	Q	(
18	10010	R	)
19	10011	S	3
20	10100	T	?
21	10101	U	5
22	10110	V	6
23	10111	W	/
24	11000	X	@
25	11001	Y	9
26	11010	Z	=
27	11011	Figure shift	
28	11100	Space	
29	11101	Carriage return	
30	11110	Line feed	
31	11111	Letter shift	

distinguish between identical 5-bit codes. For example, A and 1 are both represented by 00001, but A would be preceded by a letter-shift and 1 by a figure-shift. Once a letter- or figure-shift has occurred all succeeding characters would be interpreted as belonging to the same group until another control character appears. Thus a major disadvantage of this system is that if an error occurs in these control symbols all succeeding characters are corrupted. This is a common fault in standard Post Office Telex equipment, which also ,uses punched paper tape to transmit messages. Note that the control symbols carriage return, line-feed, etc., must be common to both shifts, and furthermore that the numbers 0–9 include an odd-parity check digit.

Table 3.2
Examples of ASCII input/output codes

Character	*Binary*	*Octal*	*Hex*	*Character*	*Binary*	*Octal*	*Hex*
A	10100001	241	A1	8	01011000	130	58
B	10100010	242	A2	9	01011001	131	59
C	10100011	243	A3	:	01011010	132	5A
D	10100100	244	A4	;	01011011	133	5B
E	10100101	245	A5	<	01011100	134	5C
F	10100110	246	A6	=	01011101	135	5D
G	10100111	247	A7	>	01011110	136	5E
H	10101000	250	A8	?	01011111	137	5F
I	10101001	251	A9	blank	01000000	100	40
J	10101010	252	AA	!	01000001	101	41
K	10101011	253	AB	″	01000010	102	42
L	10101100	254	AC	#	01000011	103	43
M	10101101	255	AD	$	01000100	104	44
N	10101110	256	AE	%	01000101	105	45
O	10101111	257	AF	&	01000110	106	46
P	10110000	260	BO	′	01000111	107	47
Q	10110001	261	B1	(	01001000	110	48
R	10110010	262	B2	)	01001001	111	49
S	10110011	263	B3	∗	01001010	112	4A
T	10110100	264	B4	+	01001011	113	4B
U	10110101	265	B5	,	01001100	114	4C
V	10110110	266	B6	—	01001101	115	4D
W	10110111	267	B7	.	01001110	116	4E
X	10111000	270	B8	/	01001111	117	4F
Y	10111001	271	B9	@	10100000	240	A1
Z	10111010	272	BA	[	10111011	273	BB
0	01010000	120	50	\	10111100	274	BC
1	01010001	121	51	]	10111101	275	BD
2	01010010	122	52	↑	10111110	276	BE
3	01010011	123	53	←	10111111	277	BF
4	01010100	124	54				
5	01010101	125	55				
6	01010110	126	56				
7	01010111	127	57				

In many cases (for example when high-level languages are used) 5-hole punched tape with a total symbol set of some 52 characters is not large enough, and it is now current practice to use an 8-bit punched tape with a maximum symbol set of 256 uniquely coded characters. In some codes, however, a parity check digit is included on all characters, thereby reducing the total set symbol to 128. A typical 8-bit input code, a version of the ASC II teletype code, is shown in Table 3.2. Punched-cards generally have 80 columns and 12 rows, giving a maximum of 4096 binary characters; however, the 256 symbol set is normally retained to give compatibility with punched paper tape and teletype equipment. A typical example of a punched-card code is shown in Table 3.3.

Table 3.3
Example of punched-card code

Character	*Card code-rows*											
	12	11	0	1	2	3	4	5	6	7	8	9
A	1	0	0	1	0	0	0	0	0	0	0	0
B	1	0	0	0	1	0	0	0	0	0	0	0
C	1	0	0	0	0	1	0	0	0	0	0	0
D	1	0	0	0	0	0	1	0	0	0	0	0
E	1	0	0	0	0	0	0	1	0	0	0	0
F	1	0	0	0	0	0	0	0	1	0	0	0
G	1	0	0	0	0	0	0	0	0	1	0	0
H	1	0	0	0	0	0	0	0	0	0	1	0
I	1	0	0	0	0	0	0	0	0	0	0	1
J	0	1	0	1	0	0	0	0	0	0	0	0
K	0	1	0	0	1	0	0	0	0	0	0	0
L	0	1	0	0	0	1	0	0	0	0	0	0
M	0	1	0	0	0	0	1	0	0	0	0	0
N	0	1	0	0	0	0	0	1	0	0	0	0
O	0	1	0	0	0	0	0	0	1	0	0	0
P	0	1	0	0	0	0	0	0	0	1	0	0
Q	0	1	0	0	0	0	0	0	0	0	1	0
R	0	1	0	0	0	0	0	0	0	0	0	1
S	0	0	1	0	1	0	0	0	0	0	0	0
T	0	0	1	0	0	1	0	0	0	0	0	0
U	0	0	1	0	0	0	1	0	0	0	0	0
V	0	0	1	0	0	0	0	1	0	0	0	0
W	0	0	1	0	0	0	0	0	1	0	0	0
X	0	0	1	0	0	0	0	0	0	1	0	0
Y	0	0	1	0	0	0	0	0	0	0	1	0
Z	0	0	1	0	0	0	0	0	0	0	0	1

Character	*Card code-rows*											
	12	11	0	1	2	3	4	5	6	7	8	9
0	0	0	1	0	0	0	0	0	0	0	0	0
1	0	0	0	1	0	0	0	0	0	0	0	0
2	0	0	0	0	1	0	0	0	0	0	0	0
3	0	0	0	0	0	1	0	0	0	0	0	0
4	0	0	0	0	0	0	1	0	0	0	0	0
5	0	0	0	0	0	0	0	1	0	0	0	0
6	0	0	0	0	0	0	0	0	1	0	0	0
7	0	0	0	0	0	0	0	0	0	1	0	0
8	0	0	0	0	0	0	0	0	0	0	1	0
9	0	0	0	0	0	0	0	0	0	0	0	1
Space	0	0	0	0	0	0	0	0	0	0	0	0
Delete	1	0	0	0	0	0	0	0	0	1	0	1
New line	0	1	0	0	0	0	0	1	0	0	0	1
@	0	0	0	0	0	0	1	0	0	0	1	0
(	1	0	0	0	0	0	0	1	0	0	1	0
)	0	1	0	0	0	0	0	1	0	0	1	0
,	0	0	1	0	0	1	0	0	0	0	1	0
etc.												

The standard teletype keyboard has some 40 keys, plus a spacebar and shift keys, this gives a symbol set of some 80 characters. Consequently there is a restriction on the number of symbols that can be used in manual input (and printed output) – for example, upper and lower case characters are normally not provided. This restriction, however, can be overcome by other forms of input/output equipment, for example, line printers and electric typewriters. Keyboard equipment can be used off-line to prepare punch tape (or cards) for input to the computer, or alternatively teletype terminals (or typewriters) can be connected on-line allowing direct input to the computer.

The operator also has at his disposal the console keyboard. This is the most primitive form of input/output device and comprises push-buttons (or switches) and indicator lights. Using this facility data and instructions may be set up on the switches in binary form and read directly into the computer, and the contents of registers (and specified store locations) may be monitored directly on the indicator lamps. Note that in this case the actual binary format must be used, whereas in the other types

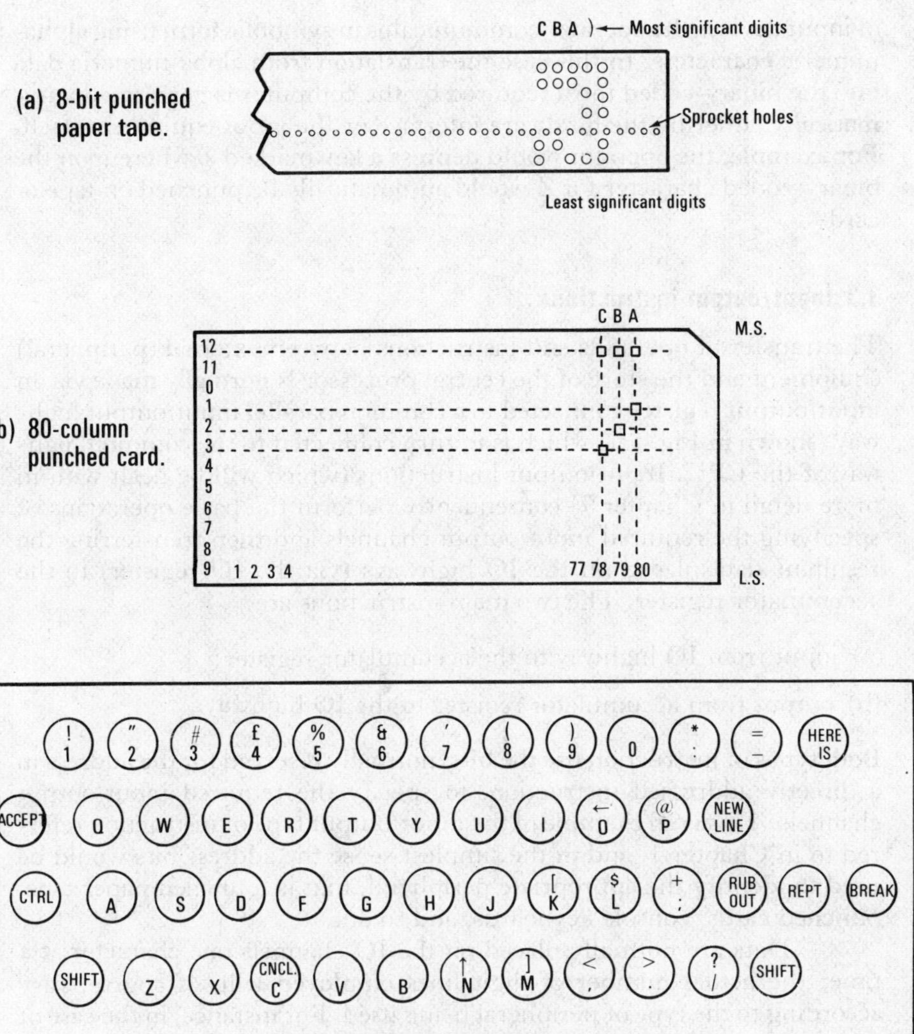

Figure 3.1 Input media.

of input devices the operator communicates in symbolic form using alpha-numeric characters. In this case the translation from alpha-numeric data into the binary-coded form required by the computer is performed automatically either by the hardware interface or the input equipment itself. For example, the operator would depress a key marked Z whereupon the binary-coded character for Z would automatically be punched on tape or card.

3.3 Input/output instructions

The transfer of operands and instructions between external (peripheral) equipment and the store of the central processor is normally made via an input/output register connected to a common parallel input/output highway (shown in Fig. 1.4) which is in turn connected to the common highway of the CPU. Input/output instructions (which will be dealt with in more detail in Chapter 7) consequently perform the basic operations of specifying the required input/output channels and then transferring the resultant data placed on the IO highways (via the IO register) to the accumulator register. The two main instructions are:

(a) input from IO highway to the accumulator register;

(b) output from accumulator register to the IO highway.

Both types of instruction use the bits, normally reserved for the address in a directly addressed instruction, to specify the required input/output channels. This is an example of the input/output type of instruction referred to in Chapter 1, and in the simplest sense the address bits would be used to identify the appropriate peripheral, that is, punched paper tape, punched cards, console keyboards, and so on.

Data are normally placed on the IO channels one character at a time; the actual number of digit lines employed will, of course, vary according to the type of peripheral being used. For instance, in the case of punched paper tape only 5–8 bits are used to represent a character, and only the least significant 5–8 bits of the common IO highway would be used in data transfers. Since input (and output) takes place one character at a time, with 8-bit punched paper tape a character would be fed in to the least significant 8 bits of the accumulator (leaving the original n–8 most significant digits unchanged, where n is the wordlength) on each input operation. An alternative to merging old and new data in the accumulator during the input operation is to clear the accumulator automatically on each instruction. Punched cards are usually read (and punched) one column at a time, corresponding to a character read-out, but in some cases they are processed row by row which necessitates reading and storing a complete card and then reassembling into the correct format.

One of the major problems encountered in the transfer of data between the central processor and the peripheral equipment is the large difference in operating speeds. The average processor, for example, executes an instruction in 2 μs, whereas a fast tape reader operates at speeds of up to 1000 characters per second (1 ms per character) giving a 500:1 disparity in speed. Consequently, if after initiating an input instruction the processor had to wait for the tape reader to input a character before commencing the next instruction, a considerable amount of processor time would be wasted. This is inevitable if only input/output instructions are being performed but in many cases other instructions, not requiring peripheral equipment, could be executed at the same time as the input (or output) instruction. Ideally then, if we can execute program instructions such as add, store, fetch, etc., interleaved with input/output instructions the overall speed of the processor could be considerably increased. This is achieved by using an **interrupt** mode of operation.

The interrupt facility enables an external signal from a peripheral device to interrupt a currently running program and to force the computer into a different program, the **interrupt routine**. Thus an input (or output) instruction may be initiated in the main program and then immediately followed by other instructions; when the input device has read a character and is ready to transfer the data to the computer it indicates this by passing an interrupt signal back to the computer. This signal causes the computer automatically to link to a supervisory sub-routine which examines and identifies the interrupting source and directs control to the appropriate input routine. At the finish of the input routine another input instruction is initiated (if required) and program control restored back to the original main program. Figure 3.2 shows a block diagram of the interrupt procedure. Note that the supervisory routine, as well as identifying the interrupting devices, stores away the contents of registers used in the main program in order to be able to restore them at the end of the input/output operation. It is also necessary with this system to establish some system of priority for the peripheral devices to sort out simultaneous interrupts. The use of interrupt and priority systems will be covered in much greater detail in Chapter 7.

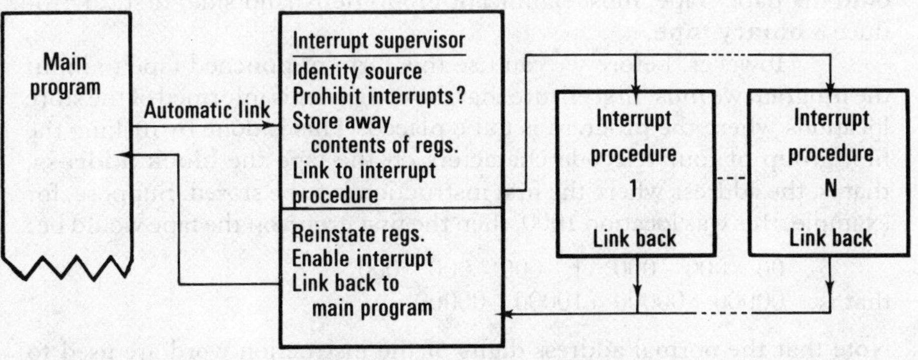

Figure 3.2 Interrupt procedure.

3.4 The basic program loader

Let us now consider the problems involved in reading programs into the computer store, a process known as **loading**, and in particular when faced with a virgin computer with a completely empty store. One method of loading would be to set up the program's instructions and operands on the console keyboard and then manually transfer them, one by one, to their assigned store locations. This is a tedious and time consuming process and as such is seldom used, since all computers have some form of rudimentary input program prewired into a protected area of the main store which inputs paper tape punched in pure binary code. In some computers, however, the simple loader itself must be manually keyed into reserved store locations which are protected during run-time only.

The function of the simple loader is to read pure binary-coded punched tape, one character at a time, into the accumulator where it is assembled into the required instruction format and stored away. Thus in order to input data with the simple loader the alpha-numeric instructions comprising the program must first be converted into appropriate telecode characters, such that when they are punched on a tape the pure binary equivalent of the instruction will result. For example, consider the instruction 22 1004 – the first instruction in the simple program shown in Table 1.3. To convert this alpha-numeric form into the pure binary representation as stored in the computer we simply write down the binary equivalent of the octal numbers, remembering at the same time to insert the modifier, indirect address, and sector bits, that is,

```
 2    2         1    0    0    4
10   010  000   1   000  000  100
```

This 18-bit binary word is then divided into four 5-bit groups (assuming that 5-bit punched paper-tape is to be used) starting from the least significant end:

```
00100  10000  10000  00100
```

The telecode character equivalent of each 5-bit group is then punched onto the paper tape, most significant group (left-hand side) first, to produce a **binary tape**.

However, before we can use this form of punched tape to input the program we must first ensure that the computer is informed of the store locations where the program is to be placed. This is done by making the first group of four telecode characters on the tape the **block address**, that is, the address where the first instruction is to be stored. Suppose, for example, this was location 1000, then the first group on the tape would be:

```
          00  000  000  1  000  000  000
that is,  00000  00000  10000  00000
```

Note that the normal address digits of the instruction word are used to specify the start location of the program. Figure 3.3 shows a schematic of

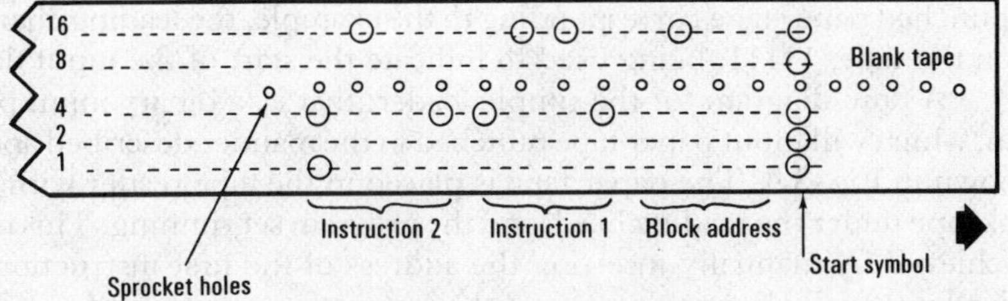

Figure 3.3 Binary input tape.

Start

Set character count to −4
Clear modifier (address count)
Clear accumulator

Search for start of tape.
Input character to acc.

Is char. zero?
Yes
No

Shift acc left 5 places
Input character to acc.
Add +1 to character count

Is char. count negative?
Yes
No

Is modifier zero?
No
Yes

Store acc in address held in modifier.
Add +1 to modifier

Store acc. in modifier (block address)

Set char. count to −4

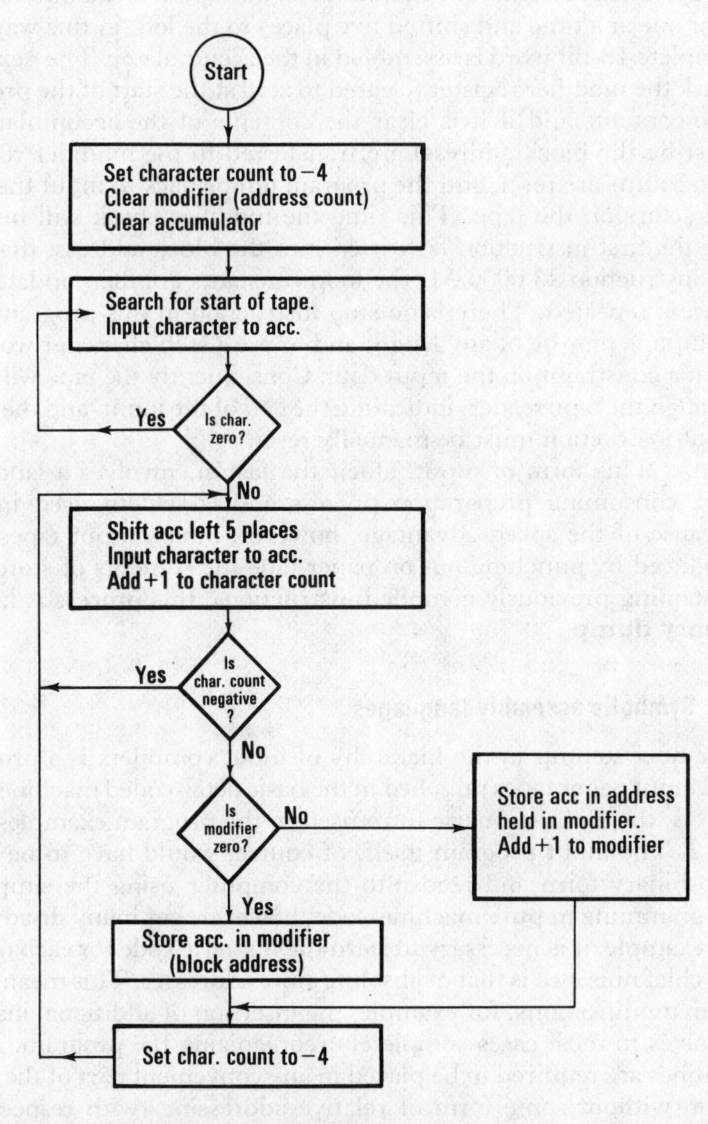

Figure 3.4 Flow-diagram for binary loader.

the punched paper tape corresponding to the example, the leading character on the tape, 11111, being used to indicate the start of the input data.

A flow-diagram for the simple loader, that is, a binary input program, which will input paper tape punched in the manner described above is shown in Fig. 3.4. The paper tape is placed in the tape reader with the blank tape under the reading head, and the program set running. This may be achieved by manually inserting the address of the first instruction of the loader (usually location zero) into the instruction register and pressing the run button to start the first read cycle. The tape is read, character by character, until the first non-zero character is found, that is, the all 1's group. Then the next four characters on the tape are read into the accumulator one at a time and shifted five places to the left, in this way the com-complete 18-bit word is assembled in the accumulator. The next step is to check the modifier register (cleared to zero at the start of the program) for zero contents and if it is clear the contents of the accumulator (which must be the block address) are transferred to the modifier register, the loop counts are reset, and the program jumps back to input the next four characters on the tape. This time the modifier check will be non-zero and the first instruction is written into the block address, that is, using the instruction 23 0000/M; the loop constants are then updated and the process repeated. There is no stop instruction in this program since the input tape may be of any length and to use a stop character would be too great a constraint on the input data. Consequently the tape will run right through the tape reader, indicating the end of the input, and the unobeyed input instruction must be manually reset.

This form of input, albeit the fastest, involves a laborious and time consuming preparation process and is seldom used in practice. Because of the speed advantage, however, binary input tapes are often produced by punching out on paper tape the contents of store locations containing previously compiled instructions, this process is known as a **binary dump**.

3.5 Symbolic assembly languages

The next step up in the hierarchy of input compilers is a program that will input paper tapes punched in the basic octal-coded machine language, that is, the alpha-numeric form used in the program examples of Chapter 2. The input program itself, of course, would have to be converted into binary form and read into the computer using the simple loader. Programming in pure machine-code, however, has many disadvantages – for example, it is necessary to learn the numeric code for each order – but the chief nuisance is that of absolute store addresses. This means that program modifications, for example, the insertion of additional instructions, involves in most cases completely reorganizing the program. Also, subroutines are required to be placed in any convenient part of the store, and again without some form of relative addressing (with respect say to a block address) the program needs to be rewritten each time.

As a result of the inherent handicaps of octal-coded machine language, most current computers have some form of **symbolic assembly language** as the basic and lowest form of computer input code. In this type of input code a **mnemonic** form is substituted for the alphanumeric codes with a direct one-to-one correspondence between them. For example, from Table 1.2 the octal code for add is 20; this would be replaced when programming by the mnemonic ADD which would be punched directly on the input tape. Programs written in this symbolic code, called the **source** program, are translated into the binary coded machine representation (the **object** program) by the input assembly or translator program. Moreover, in this form of coding there is no need to specify addresses as absolute locations, since they can be represented symbolically. For example, in Table 3.4 the numbers to be added are stored in NUM, NUM+1, NUM+2, etc., the actual store locations being assigned by the assembly program. Similarly, storage for the instructions (including **labels**, that is, specific locations used in jump instructions) is allocated automatically by the program.

Source program translation and the allocation of storage locations (that is, the replacement of symbolic addresses by corresponding absolute addresses) is usually performed in two phases. In the first operation the source tape is read and the symbolic addresses [labels and variables (**identifiers**)] are extracted from the source program, assigned absolute addresses, and a **symbol address table** (also called a **dictionary**) is constructed. In the second stage the source instructions are again examined one by one and the mnemonic codes (such as ADD) are used as arguments to search through a stored **operations-code table** to obtain the actual binary-coded order which is inserted in the first part of the object instruction. The symbolic address portion of the source instruction is examined next; this is used to search the symbol address table to find the appropriate absolute address which is then inserted into the second part of the object instruction. This procedure often involves a second reading of the source tape (depending on the availability of storage space), hence the term **two-pass compiler**.

Table 3.4
Symbolic assembly program to add together ten numbers

```
        SUM OF TEN NUMBERS
  NUM   DAT 10
        FETM = −8
START   FET     NUM
        ADD     NUM +9/M
        JMPM    FINI
        INCM
        JMP     START
 FINI   STOP
        END
```

Table 3.5
Typical assembly code pseudo operations

DAT	Used to reserve store locations
=	Interprets the number following the equals sign as a decimal literal and automatically generates and assigns a location for the value
DEC	Used to define a store location containing a constant
OCT	As above but octal literals may be defined
END	End of assembly program statements

As well as allowing symbolic notation to be used for the machine-code orders most assembly languages also include **pseudo-operations**. These pseudo-operations have no direct equivalent in the machine-code order set and in general represent several lines of basic machine-code instructions, for this reason they are sometimes called **macro-generators**. The pseudo-operations are usually concerned with data allocation (including relative addressing, that is, **relocatable** programs), data formats, and the specification of literal values for operands. Table 3.5 shows a typical selection of such pseudo-operations. The instruction DAT is used in Table 3.4 to reserve symbolic locations NUM through to NUM+9. The equals operation (=) as well as generating a binary literal from the decimal statement also allocates a location and inserts its address into the address portion of the machine-code instruction. For example, the source code instruction FETM = −8 would generate and store the binary equivalent of −8 and then insert its absolute location into the address part of the object code fetch instruction. This operation is often extended to allow other forms of literal statement – if, say, an apostrophe sign follows the equals sign (=') an octal literal may be used. The DEC instruction is used to define a store location containing a constant: for example, the statement CONS DEC 25 would place the binary equivalent of decimal 25 into a store location called CONS; this operation may also be extended to octal constants using the instruction OCT.

A symbolic assembly language is a very sophisticated input routine and as such constitutes a sizeable program. To write this program without the benefits of such facilities as symbolic addresses and so on (the very facilities included in the symbolic assembly program!) would be a very difficult task. Consequently, in practice a step by step approach (called bootstrapping) is adopted in which the assembly program is written in progressively complex stages. For example, routines for handling basic mnemonic-coded instructions and a simple relative addressing scheme might be programmed first, encoded using the pure binary technique, and inputted with the simple loader. This program would in turn be used to input a program for handling symbolic addresses, in this way the complete range of facilities would be gradually built up. Once the full program is in the store it may be punched out using a binary dump routine and any future inputs performed using the simple loader.

3.6 Procedure orientated languages

Though the use of symbolic assembly programs make machine-code programming very much easier, it is still necessary to specify in considerable detail the majority of program steps. In some cases, for example in non-numerical problems, this is essential if maximum efficiency (in terms of program running time and storage space) is desired. However, for a large number of mathematical and business problems (mainly 'one-off' calculations) the program running time is not so important as the program preparation time. Moreover, programming mathematical problems in machine-code, particularly if fixed-point arithmetic is used, can give rise to some nasty scaling problems. Many of these difficulties can be overcome, however, by using floating-point arithmetic, either in hardware or software form. Another area which often causes trouble is the organization of input and output routines and the general control of peripheral equipment. Other problems which arise, particularly in large software systems, are concerned with keeping track of store allocation, up-dating of loop counts, sub-routine organization, and so on – that is, problems involving the general 'housekeeping' procedures.

In order to overcome many of these difficulties, and in an endeavour to make programming easier and available to a larger section of the technical and scientific community, high level computer languages have been developed. These computer languages, called generally **procedure orientated languages** (such as ALGOL[1,2] and FORTRAN[3]) allow the program operations to be written in a more easily understood form. The computer itself, or rather the software compiler, assembles together the necessary machine-code instructions required to perform a particular operation. In this way mathematical problems, for example, may be expressed using standard mathematical formulae including both real and integer numbers. Complete input/output routines may be initiated (called up) by simply writing 'read' or 'print' in the program statement, and all the necessary 'housekeeping' requirements are organized automatically by the compiler. The detailed description and use of procedure orientated languages[4,5,6] is outside the scope of this book, but typical

Table 3.6
Example of an ALGOL program

```
Comment   Sum of 10 numbers;
   begin  Integer x, Sum;
          Integer array n [1:10];
          Sum : = 0;
          for x : = 1 step 1 until 10 do
          begin  Read n[x];
                 Sum : = Sum + n[x];
          end;
          Print Sum
   end
```

Table 3.7
Example of FORTRAN program

```
SUM OF 10 NUMBERS
DIMENSION X(10)
1    FORMAT (I6)
     J = 1
     SUM = 0
     DO 2  J = 1, 10
     READ 1, X(J)
2    SUM = SUM + X(J)
3    PRINT 1, SUM
     END
```

examples of ALGOL and FORTRAN programs are shown in Tables 3.6 and 3.7.

Before we leave this subject it is worthwhile mentioning one other important aspect of high level languages – the advantages they offer for interchanging programs between different machines. Unlike assembly languages which are governed by the machine structure, high level languages are basically machine independent (except for details of input/ouput procedures). This means that programs written in FORTRAN, for example, may be run on any machine with an appropriate FORTRAN compiler. Moreover, the languages themselves form an admirable means of communicating computing algorithms from one person to another, particularly in scientific journals.

3.7 List processing languages[7,8]

Non-numerical data processing problems are primarily concerned with bit and symbol manipulation, and in particular the sorting, examination, and modification of large lists or tables of symbols. The main characteristics of a programming system for this 'information manipulation', rather than computing, function are:

(a) The manipulation of symbols that have no numerical meaning.
(b) Storage requirements cannot be specified in advance – for example, in writing an assembly language compiler it is impossible to state in advance the exact number of locations required for the symbol-address table.
(c) The storage organization and data relationships are continually changing as the program proceeds.
(d) Frequent modification of the data and program will be required.

Many of these problems can be overcome by using a **list-processing language**, particularly the major difficulties of storage organization. In a normal computer program storage space is contiguously allocated; thus we would store the symbolic addresses from a source program, say, in a block of consecutive stores (see Fig. 3.5(a)). This means that we must know beforehand exactly how much storage will be required, and if this is not known an estimate must be based on the worst possible conditions that could arise in a program. Furthermore, additional blocks of storage must be reserved for any data generated in the program. It is obvious that in a complex problem the estimation of the storage requirement is very difficult, and often results in uneconomical store utilization owing to the use of too generous safety factors. Thus, we require some means of allowing extra storage to be automatically allocated when required (while the program is running) and according to problem size. This facility, sometimes known as **dynamic storage allocation**, is in fact available to a limited extent in some versions of the ALGOL compiler.

The concept of a **list memory**, the basis of all list-processing languages, was developed to alleviate this store allocation problem. In a list each symbol is stored in a separate computer location, not necessarily consecutive, together with a link or 'tag' address. Each store location, then, contains two items of information:

(a) the actual list symbol, and
(b) the address of the next symbol on the list, that is, the link,

with the final word on the list containing an 'end' symbol in place of the link address. Thus, with this form of storage structure it is only necessary to know the address of the first symbol, or 'head' of the list, in order to gain access to the stored information. Furthermore the list may be easily extended, using any available storage space, by modifying the link part of the word. This technique is illustrated in Fig. 3.5(b), the five members of the list, *a*, *b*, *c*, *d*, and *e*, are stored in locations 50, 55, 100, 51, and 102. Note that symbol *a*, for example, indicates that the next symbol on the list is held in store location 55, and so on. The last symbol is stored in location 102, and contains an 'end' symbol in place of the link. To modify the list so that a new symbol, *f*, may be inserted between symbols *c* and *d*, the link of symbol *c* must be altered to the address of the new symbol *f*, and the word containing *f* must have a link address to symbol *d*; this is shown in Fig. 3.5(c). To operate this list structure it is also necessary to have an 'available space' list, recording the free storage in the computer, and some means of gathering the used, but no longer required, storage – this process is known as 'garbage collection'.

It is possible for list structures to have many levels of sub-lists – that is, the main list link will name the head of another sub-list, and so on. Furthermore, in many cases the routine being performed on the main list (for example symbol comparison and modification) may also be required to be executed on a sub-list should one be encountered during the routine.

(a)

Store locations	
50	Symbol a
51	Symbol b
52	Symbol c
53	Symbol d
54	Symbol e
55	End

(b)

Store locations		
50	Symbol a	55
51	Symbol d	102
52		
53		
54		
55	Symbol b	100
100	Symbol c	51
101		
102	Symbol e	End

(c)

Store locations		
50	Symbol a	55
51	Symbol d	102
52	Symbol f	51
53		
54		
55	Symbol b	100
100	Symbol c	~~51~~ 52
101		
102	Symbol e	End

Figure 3.5 List processing structures.

In this case the routine itself is used as a sub-routine, the process being known as **recursion**.[9] This facility is not usually available in conventional computer order codes, though again some versions of the ALGOL compiler have recursion procedures. The reason for this is that, as we have seen, sub-routines are conventionally stored in a set of locations which include the actual program, input variables, intermediate results, and a link address back to the main program. Consequently, each time the sub-routine is executed the previous results are either destroyed or lost, thereby preventing any recursive action. Thus, if recursive operation is required, special precautions must be taken to prevent the corruption of variables and results; this is achieved by using the technique of **push-down storage**. In this method every store location which holds intermediate results, or link addresses, is replaced by a 'push-down' store, in which data are always added to, or taken from, the top of a list. Thus data may be 'stacked' and used as and when required.

A typical list processing language would include such operations as: delete items from a list; add items to a list; locate specific entries; combine lists; extract items from a list. Though a conventional computer can be programmed to perform all these functions, and the use of high-level list processing languages such as LISP[10] create the impression that an effective programming system is readily available, nevertheless the numerically orientated computer is totally unsuited for this type of processing. Consequently, the computer will often perform long sequences of operations merely to execute a single list-processing statement and as a result the efficiency of these programs leaves much to be desired. The problems of information processing, and a possible hardware solution, will be discussed in detail in Chapter 9.

3.8 Influence of high level languages on computer design

It is a fact that there has been very little change in computer system design since the original ideas of Babbage and Von Neuman – the machines have simply got faster and more powerful as hardware techniques have advanced. In the past, machines were developed by engineers generally without regard to the manner in which they would be employed and with little or no cooperation with programmers. (An exception perhaps is the IBM Stretch machine,[11] which seemed to go to the other extreme with programmers defining a system structure without considering the engineering difficulties involved.) As a result we have the development of high level computer languages such as FORTRAN, ALGOL, and LISP, which alleviate the problems involved in program preparation and machine usage. These languages are really a software simulation of some advanced structured special purpose computer, involving many thousands of machine-code instructions. In other words, the programmers have taken the engineers' embryonic machine and produced a sophisticated adult designed to solve a particular problem, in most cases a numerical one.

Consequently a situation exists today where, as the complexity of computer systems increases, the ability of the manufacturer to produce them in a reliable and error free form (and to deliver them on schedule!) has diminished. This 'complexity barrier'[12] may well prove to be a more realistic obstruction to systems growth than the often quoted minimal time for the propagation of signals at the speed of light. A new generation of computing machines, which must emerge if advances in data manipulation and control are to be forthcoming, will require a major breakthrough in the form of new concepts of systems design, and these in turn will almost certainly depend on an amalgam of hardware/software technology. To design the complex computer systems of the future it is essential that the design commence with the user and/or problem specification, working downwards to processor hardware plus software requirements. Barron and Hartley[13] have also stated that the correct approach to computer system design is to start with the problem orientated language and to attempt, where feasible from an engineering point of view, to implement these requirements in terms of hardware.

Let us now consider how and to what extent a procedure orientated language can, and should, influence computer design. Computer systems have already been designed which implement in hardware form some of the basic characteristics of languages such as ALGOL and FORTRAN.[14,15] There is, however, a certain danger in the direct implementation of a specific computer language owing to the lack of flexibility. A machine which is designed to execute FORTRAN statements only would be of little use when presented with a non-numerical problem, say, requiring some list-processing facilities. Moreover, high-level languages themselves appear to be in a continual state of development and consequently a too exact hardware replica of the high level software procedures would be extremely difficult, if not impossible, to modify.

One way out of this difficulty is to ascertain the common structural properties of procedure orientated languages and to implement in hardware form only those general manipulative features necessary to compile and execute high-level statements. To illustrate this approach let us consider a compiling technique, known as precedence analysis,[16] which is employed in many high level language compilers. A technique for compiling arithmetic expressions based on this method has been described by Dijkstra[17] and utilizes the **reverse Polish notation**.

The problem we are faced with is how to transform an arithmetic expression written in the high-level language form:

$$(A * B) + (C/((D-E) * F))$$

into a suitable storage structure that can easily generate the corresponding machine-code instructions to evaluate the formula. Dijkstra's technique involves translating the high-level expressions into a parenthesis free notation – the reverse Polish form. The transformation process is carried out with the aid of a stack-store; this is a storage structure (or device, see

Table 3.8
Translating formulae into reverse Polish

(The top of the lists are at the right-hand side)

(A * B) + (C/((D − E) * F))

Polish list	*Operator stack*	*Symbol scanned*
	(	(
A	(	A
A	(*	*
AB	(*	B
AB*		)
AB*	+	+
AB*	+(	(
AB*C	+(	C
AB*C	+(/	/
AB*C	+(/(	(
AB*C	+(/((	(
AB*CD	+(/((	D
AB*CD	+(/((−	−
AB*CDE	+(/((−	E
AB*CDE−	+(/(	)
AB*CDE−	+(/(*	*
AB*CDE−F	+(/(*	F
AB*CDE−F*	+(/	)
AB*CDE−F*/	+	)
AB*CDE−F*/+		

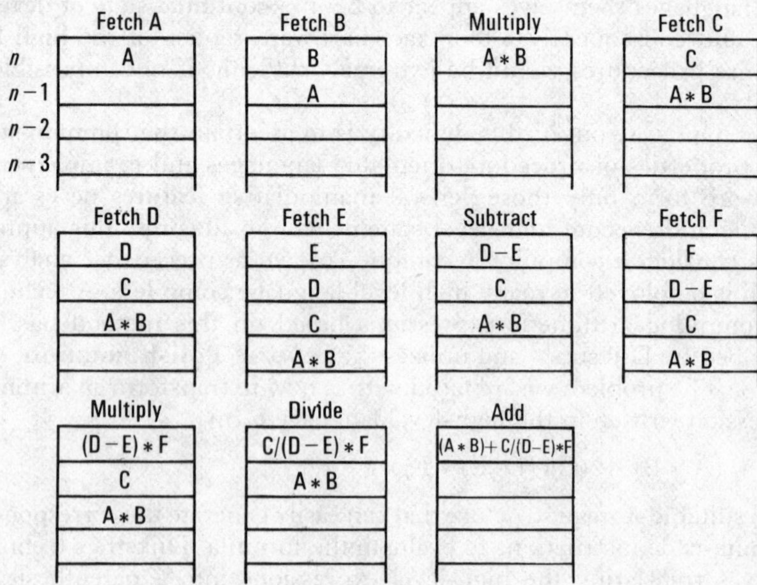

Figure 3.6 Use of a 'stack-store' in compiling.

Chapter 6) which stores information such that the item stored last is the first one retrieved – this is of course the push-down store mentioned earlier. To perform the translation the elements of the expression are examined one at a time commencing from the left-hand side, and then placed on to stacks according to the following rules:

1 Identifiers are added immediately to the Polish list.

2 Operators, including brackets, are compared with the top item on the operator stack and placed according to the following order of precedence:

0	(	
1	+ −)	
2	* /	(multiply, divide)
3	↑	(exponentiation)

(a) If the precedence of the stack symbol is equal to or greater than the precedence of the scanned symbol, the stack symbol is transferred to the Polish list; the comparison is repeated until the stack symbol precedence is less than the scanned symbol precedence.

(b) Open brackets are placed directly on the stack; closed brackets are not stacked but simply compared, which has the effect of unstacking down to the corresponding open brackets when both brackets are discarded.

(c) When all symbols are scanned the remaining stack symbols are transferred to the Polish list.

This process is shown in detail in Table 3.8 for the expression (A * B) + (C/((D − E) * F)); note that the equivalent reverse Polish form is AB*CDE − F*/+. Figure 3.6 shows how the Polish form may be interpreted (and evaluated) using stack storage to give the correct order of precedence. In interpreting Polish expressions the operator symbol is understood to operate on the two operands immediately to its left, the result of each operation is then considered to be an operand for subsequent operations. It is important to note that for non-commutative operators, for example ' − ', the top item on the stack must be subtracted from the second.

Once the Polish list has been compiled it is a relatively simple operation to generate the equivalent machine-code instructions to evaluate the arithmetic expressions. However, the programs generated in this way would be somewhat inefficient since conventional computers have neither push-down storage, nor zero-address instructions of the type indicated in Fig. 3.6. Consequently, several basic instructions would be required for each Polish operation. Moreover, the stack-storage used in compiling must be set up as a data-structure using conventional contiguous storage which would also require a considerable amount of complicated programming.

After examining this basic problem in compiling high-level languages we are now in a position to suggest some ways in which the hardware structure of the computer may be modified in order to improve the overall efficiency of the process. It is also worth bearing in mind that the principles described above are not just restricted to arithmetic expressions but can be extended to include all forms of statements, and in some cases the entire execution of the program.

The first and most obvious requirement is to provide a hardware version of stack-storage consisting of special push-down registers. If only arithmetic expressions have to be evaluated the stack does not need to be very deep – some 16 registers would be sufficient – but if the compiling procedures are extended some 10^2–10^3 locations may be required. Secondly, if the Polish form is to be executed economically, it is also necessary to include zero-address instructions in the machine-code order set of the computer, which operate on the top two registers of the stack only. For example, the order 'subtract' would subtract the contents of stack register n from the contents of n–1 (see Fig. 3.6) putting the result in n and moving all the stack contents one place up. Similarly, the orders add, multiply, and divide would operate in the same way. In addition, if only a small amount of stack-storage is available, normal single address instructions such as 'Fetch contents of address specified to register n' and 'Store contents of register n in address specified' will also be required.

Another aspect of this design philosophy is the use of micro-programming techniques (see Chapter 4) to perform the direct interpretation of high-level codes.[18] This method suffers to some extent from inflexibility (depending on the method of micro-programming) caused by the necessity to choose a particular high-level language.

A more practical development in this area is the Burroughs B1700 machine which is dynamically microprogrammable and has a stack mechanism controlled by microcode.

The use of stacks in computers was first described by Randell and Russell[19] who presented a detailed stack structure for ALGOL 60 implementation and Brooker[20] who considered the advantages of stacks for dynamic storage allocation and also the use of data descriptors (employed in accessing arrays). Wirth and Weber[21] also proposed a stack architecture for implementing the EULER language.

One of the first computers to utilise a stack data structure was the English Electric KDF 9 machine,[22,23] which had a hardware nesting store consisting of 16 one word registers. Other machines which incorporated stack principles were the ATLAS computer and the Burroughs B6500, B5700 and B6700 computers[24,25]. More recently the MU5[26], Hewlett Packard's HP 3000 and the PDP 11 series have appeared with stack structures, and the principle now seems to be well established as a basic computer architectural feature (microprocessors also incorporate some form of stack processing facility).

3.9 Stack architecture [27,28]

The most usual method of incorporating stack storage in a computer structure is to set aside an area of contiguous storage in the main memory, which is used in conjunction with hardware **stack pointer** (SP) registers, to effect a last-in, first-out data structure. This arrangement is shown in Fig. 3.7(a) where the main store locations 004 → 010 are used to store the items, and the SP register contains the address of the last item on the stack.

Stack storage can be accessed using the basic operations of PUSH and POP, which can be defined as assignment functions as follows:

PUSH – transfer contents of register X to next location on stack S (S: = X).

POP – transfer contents of last location on stack S to register X (X: = S).

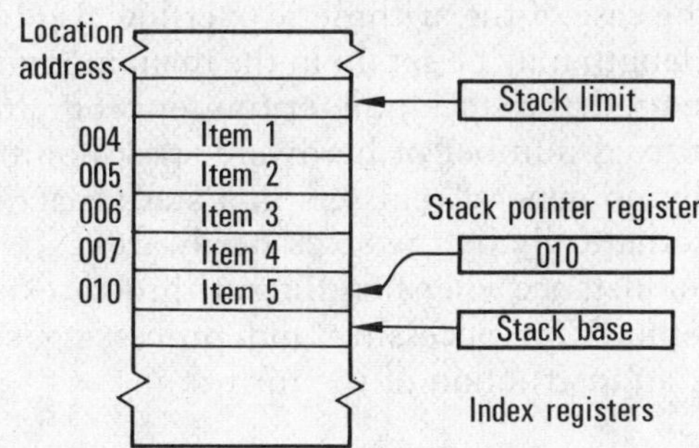

(a) Stack store

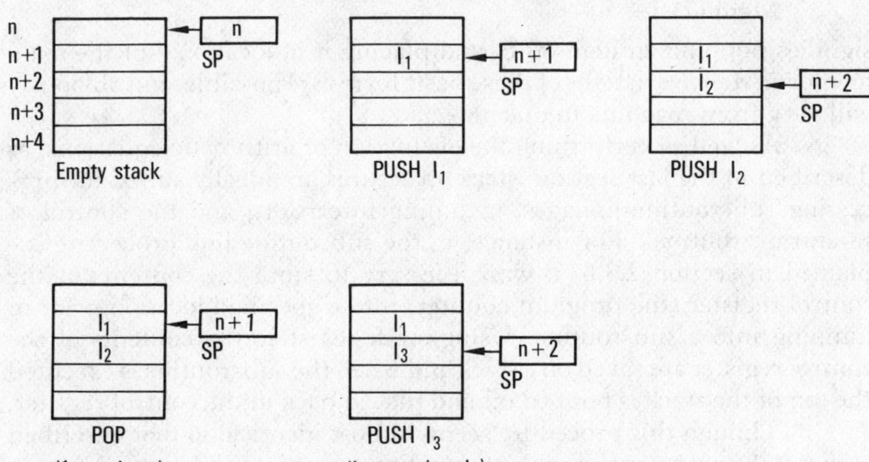

(b) PUSH and POP operations.

Figure 3.7 Stack structures.

Note that in each case the address of the stack location is obtained from the SP register, which must then be incremented or decremented depending on whether a PUSH or POP operation is being performed. These operations are executed automatically in the hardware, thus the PUSH and POP operation (or their equivalent) are basic machine-code instructions. The PUSH and POP operations are shown diagrammatically in Fig. 3.7(b).

It will be obvious that when the stack is empty the operation POP has no significance, similarly the stack must have a finite length and a PUSH operation on a full stack is meaningless. These two conditions, called **stack underflow** and **stack overflow** respectively, must be detected in the hardware and suitable corrective measures initiated. This can be performed using stack limit and base registers (shown in Fig. 3.7(a)) which are initially set under program control. During the PUSH (POP) operations the contents of the SP register is compared with the base (limit) register and the result used (if necessary) to generate an interrupt signal (as in the case of the arithmetic overflow flag being set). Note that stacks of any length may be set up in the main memory, moreover any register can be used as a stack pointer (under program control) but there is normally a fixed number of hardware stack pointers. Note that there are essentially no general registers in a stack machine, their function is fulfilled automatically by the stack hardware.

In some cases, for instance when handling arithmetic expressions, more sophisticated methods for accessing and processing stacks are required. For example, an instruction of the form:

$$S_i := S_i + S_i$$

would imply that the top two items of Stack S_i are popped off, added together and the result pushed back onto S_i. Similarly the instruction:

$$\text{Mem (7)} := S_i$$

signifies popping an item off S_i and placing it in location 7 of the main memory. Many variations of these basic forms are possible, and obviously will vary from machine to machine.

As well as performing the evaluation of arithmetic equations, as described in the last section, stack structures are ideally suited for processing sub-routine linkages, including interrupts, and the control of re-entrant routines. For instance, in the subroutine link procedure explained in section 2.3.6, it was necessary to store the contents of the control register (the program counter) into a specified location prior to jumping into a sub-routine. Using stack registers the contents of the control register are saved on a stack and when the sub-routine is executed the top of the stack is popped off and placed back in the control register.

Though this procedure seems almost identical to that described earlier it does have major advantages. For example, subroutine links can be nested in the stack storage thereby enabling sub-routines to call each other to any depth and in any order. Moreover, since the links are stored and processed in time sequence recursive procedures are easily handled.

Note that the stack employed for this purpose, sometimes called the **control stack**, must be separate from, and independent of other stacks.

In some systems it is necessary to move a subroutine in memory during program execution time, for example in multi-access working called **dynamic relocation** (see section 6.10.2). Normally machine-code for a subroutine is held in absolute locations fixed in memory and in order to relocate it the addresses must be modified with the contents of an index register. The main program then only has to keep a record of the starting location of the sub-routine (usually written as if stored in location zero onwards) which is indexed (either by hardware register or under program control) as required. Difficulties arise when nesting subroutines and the possibility that routines could be relocated during operation; stack storage can still be used effectively in these conditions.

Suppose the absolute address of the entry points for the subroutines are held in a system table (this would be compiled and updated by the operating system during the relocation procedures). The main program would now hold an index value for each subroutine which points to the appropriate starting address held in the system table. Should a subroutine be entered, and then another subroutine is called from it, all that is necessary is to place on the stack the index value of the calling routine and the position in the program relative to the entry point of that routine. Thus, the subroutine call would take the form:

$$S_i := (P, PC - TAB(P)) \qquad PC := TAB(R)$$

where P is the index value of the calling subroutine, R the index value of the subroutine being called, TAB(P) the contents of location P in the system table and PC is the program counter (control register).

Another major advantage of stack organization is the ease with which several tasks can be handled concurrently. The ability to share a single copy of a program among different users or processes is called **re-entrancy** the technique being known as **re-entrant programming**. The advantage of using this method is that the amount of main memory required is considerably reduced – the alternative of course is to store separate copies of the subroutine for each user.

Re-entrant programs differ from ordinary subroutines in that a routine can undertake a new task before it has finished processing its current one. Thus, a re-entrant program can be used concurrently by several processes and can be at different stages of execution in each process. In order to effect a re-entrant routine it is necessary to write the program in 'pure code', that is code which contains only instructions and constants and which cannot modify itself during execution. Thus instruction modifying programming techniques cannot be used for writing re-entrant programs.* (A re-entrant program is identical in concept

*Though it was stated earlier that one of the most powerful attributes of a computer is its ability to modify its own code, there are strong practical reasons for not using this facility. Not the least being that the executed program is not the one given on the listing and in general the technique is less efficient than other methods (such as the use of external index registers).

to a control program written for a read only memory.) It will also be apparent that a re-entrant program must not store data local to itself, thus each process must have its associated data storage area.

In order to allow a subroutine to be re-entered by a new process before its completion of the previous task it is essential that any intermediate results and linkage and control information (called the **processor state**) are saved and restored. The use of stack storage to perform this function (a stack segment consisting of a number of contiguous locations would be allocated for each process) has considerable advantages and ensures that the routine isolates its instructions and data and thus maintain its re-entrancy.

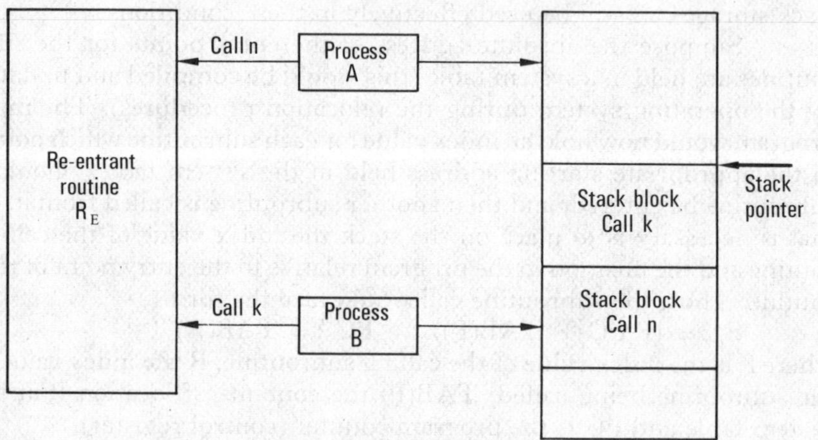

Figure 3.8 Re-entrant routines.

The re-entrant process is illustrated in Fig. 3.8 and proceeds as follows:

(a) Process A calls the re-entrant routine R_E and the necessary return information for call *n* is pushed onto the control stack.

(b) Process B interrupts R_E before it has finished processing Process A's request. The current processor state and return linkages for call *k* are pushed onto the stack and Process B takes over control of R_E.

(c) Process B finishes processing (or is interrupted in turn) and pops off return information from the top of the stack (as indicated by the stack pointer). Processor state is restored and Process A carries on processing, using R_E, from where it left off.

The sequencing of these operations is normally performed on a priority interrupt basis with processing being terminated (and returns initiated) at the end of the current instruction.

One of the major advantages of stack architecture is that it is ideally suited for implementing high-level languages. In particular this comes about because of the ability of the stack machine to produce extremely compact code, some 50 % less than its conventional counterparts. In a conventional register machine data addresses fall into two categories, **explicit addresses** which refer to variables defined by the program and **implicit addresses** which are required by the compiler for intermediate results. Stack computers however do not require implicit addresses, as such, and explicit addresses can be expressed using fewer bits due to the inherent hardware structure. The register machine usually holds implicit variables in temporary storage which is used once and then forgotten, the stack computer however automatically allocates and deallocates space using the POP/PUSH facilities. As well as saving storage space the overall speed is also enhanced since fewer memory access cycles are required.

Many microprocessors, such as the Motorola 6800, Intel 8080 etc., incorporate some form of stack capability. In general this is limited to a stack pointer, which points to an area of memory used as a stack, or a small hardware stack in the microprocessor unit itself.

3.9.1 Tagged Architecture[29]

The concept of tagged data is an important adjunct to stack organization. One of the difficulties that can arise in using stacks is due to dissimilar data types, each having different widths and formats, occurring in the same process (for example, in double and single precision working). One solution is to allow the larger data width to determine the width of the stack and to have data type conversion routines and special instructions incorporated into the stack evaluation algorithms. A second approach is to provide a separate evaluation stack for each data type.

Alternatively a small tag of some 3–6 bits can be associated with each word in computer storage which enables each data type to be uniquely identified. In this way, rather than having a separate machine instruction for each different data type (fixed, floating-point, decimal etc) and for each different width operand, the associated tag-bits can be used to recognize the data type and a single instruction can proceed accordingly with the required operation.

As well as eliminating the difficulties of performing stack arithmetic the tagged data approach also significantly reduces the number of bits needed to represent the instructions, thus giving an even greater degree of code compactation.

3.10 Hardware/software duality

The process of programming a digital computer, that is, preparing the list of instructions in machine-code or some higher level language which the computer then obeys in sequence, is a completely analogous process to that of logical design. In fact, since the computer is itself a logic complex, the instructions comprising the computer program are merely a means of selecting particular logical paths through the machine. Thus there exists a direct relationship between logic hardware and computer software or program. Glaser[30] has also made the point that software and hardware are almost functionally indistinguishable and that a close relationship exists between computing languages, machine design, and logic design. In fact, it is highly probable that the theory of computer programming (at the moment there is very little theory, only expert practitioners!) lies somewhere in switching and automata theory.

An approach which could solve many of the complex computer systems problems is to replace conventional software functions with hardware. The cost of software production, using essentially human effort, is certain to increase in the future whereas hardware costs, particularly of LSI circuits, will decrease rapidly as technology advances. Thus if custom designed logical hardware could be produced quickly and cheaply using computer-aided design techniques[31] considerable advantages in cost and reliability would result.

The major areas where hardware could replace software are the input/output, compilation, and supervisory functions of the computer system. An elementary example of this is the simple loader: the operations associated with the assembly of characters into instruction words could be performed externally with hardware and only complete words transferred to the CPU via the input/output highway. These ideas may be extended to the compilation function by including hardware compilers, and to the supervisory and executive control in multi-programming systems (see Chapter 7). Hardware table-look-up for function generation and the use of firmware techniques (see Chapter 4) are other examples of software replacement techniques.

Though progress to date has been rather slow in this area, computer architects are now beginning to take advantage of the availability of cheap microcomputers to perform various functions within the central computer. Many software processes now lodged in the central processor can be performed by microprocessors acting autonomously within the memory system. For example, garbage collection in list processing, associative searching in an otherwise non-associative memory, control of virtual memory etc. Microprocessors can also be used in collaboration with the microprogram control store (see Chapter 4) to allow complex logic functions to be obtained at a very low cost.

In addition the tendency now is to distribute many of the required system software functions, for example input/output processing, around the total system as in the provision of intelligent terminals employing microprocessors[32].

3.11 The Iverson programming language[33,34]

One of the most interesting and potentially useful problem orientated languages that has arisen on the computing scene in recent years is the

Iverson language (APL). This language has been used extensively by IBM to define computer systems at all levels (both hardware and software), its principle characteristic being the facility to provide a succinct description of algorithms. The power of the language lies in its ability to express manipulations on *entire arrays* of operands simply and consistently. Thus complex operations may be described in Iverson using far fewer steps (and branching operations) than are needed to specify the same task in other languages. The basic characteristics of the language may be summarized as follows:

(a) It allows a clear and simple representation of the sequence in which the steps of an algorithm are performed.
(b) It provides a concise and consistent notation for the operations occurring in a wide range of processes.
(c) It permits the description of a process to be independent of the choice of a particular data representation.
(d) It allows economy in operation symbols.
(e) It provides convenient subordination of data without loss of detail.

The complete language includes facilities to operate on scalars, vectors, and matrices, but in this brief description we shall confine ourselves to a subset of the language operating in the main on scalars and vectors only; the notation used in this description is defined in Table 3.9. At the present time a full compiler for the Iverson language is not available, the notation has, however, been implemented to some extent in the APL interpreter[35, 36, 37] written for the IBM 360 and 1130 series of machines.

Let us now consider the structure of the language in more detail. Scalar and vector operands are represented with both lower and upper case characters, with vector quantities being printed in bold text face (**x** or **X**). Upper case characters only are used for matrix operands and they are printed in bold sans serif (**X**)*. The term vector is used to denote an entire list or array of operands by a single name. Each element of the vector may be identified by the use of a subscript, that is, the **index**; for example:

$$\mathbf{x} \equiv x_0, x_1, x_2, x_3, \ldots, x_7$$

The number of elements in a vector is called the **dimension** of a vector and is represented by $\nu(\mathbf{x})$. Also, in this text, we shall use **0-origin indexing** with the first element having the index 0, thus:

$$\mathbf{x} \equiv x_0, x_1, x_2, \ldots, x_{\nu(\mathbf{x})-1}$$

The assignment symbol (←) denotes specification in the same sense as

* In the handwritten version of the language vector quantities are usually represented by underlined lower case characters only, with upper case being reserved for matrix operands.

Table 3.9
Definition of Iverson notation

Operation	*Notation*	*Definition*
1. *Operands*		
Scaler	x	
Vector	$\mathbf{x}$ or $\mathbf{X}$	ith component $\mathbf{x}_i$, dimension $\nu(\mathbf{x})$
Matrix	$\mathbf{X}$	ith row $\mathbf{X}^i$, jth column $\mathbf{X}_j$, ijth element $\mathbf{X}^i_j$ dimension row vectors $u(\mathbf{X})$ dimension column vectors $v(\mathbf{X})$
2. *Logic and arithmetic*		
Arithmetic ($+\ -\ \times\ \div$)	$y \leftarrow s\mathrm{A}t$	usual definition
*		exponentiation
Residue modulo m	$k \leftarrow mn$	remainder after dividing n by m
AND	$y \leftarrow s \wedge t$	$y = 1$ if and only if $s \not\equiv 0$ and $t \not\equiv 0$
OR	$y \leftarrow s \vee t$	$y = 1$ if and only if $s \not\equiv 0$ or $t \not\equiv 0$
Negation	$y \leftarrow \overline{s}$	$y = 1$ if and only if $s = 0$
Relation ($< \leqslant > \geqslant = \neq$)	$y \leftrightarrow s\mathscr{R}t$	$y = 1$ if and only if $s\mathscr{R}t$ is true
3. *Shifting*		
Left circular	$\mathbf{y} \leftarrow k \uparrow \mathbf{s}$	cyclic rotation left (right) by k places
Right circular	$\mathbf{y} \leftarrow k \downarrow \mathbf{s}$	
Left shift	$\mathbf{y} \leftarrow k \underset{\circ}{\uparrow} \mathbf{s}$	shift k places left or right bringing
Right shift	$\mathbf{y} \leftarrow k \overset{\circ}{\downarrow} \mathbf{s}$	zeros into vacated positions
4. *Base value*		
Base 2 value	$y \leftarrow \perp\mathbf{s}$	y is the base 2 value of vector $\mathbf{s}$
Base 10 value	$y \leftarrow 10 \perp \mathbf{s}$	y is the base 10 value of vector $\mathbf{s}$
Represent base 2	$\mathbf{y} \leftarrow (n) \top j$	$\nu(\mathbf{y}) \equiv n$ and $\perp\mathbf{y} \equiv 2^n \mid j$
Represent base 10	$\mathbf{y} \leftarrow 10(n) \top j$	$\nu(\mathbf{y}) \equiv n$ and $10 \perp \mathbf{y} \equiv 10^n \mid j$
5. *Selection and reduction*		
Catenation	$\mathbf{y} \leftarrow \mathbf{s}, \mathbf{t}$	$\mathbf{y} \equiv \mathbf{s}_0, \ldots, \mathbf{s}_{\nu(\mathbf{s})-1}, \mathbf{t}_0, \ldots, \mathbf{t}_{\nu(\mathbf{t})-1}$
Vector compression	$\mathbf{y} \leftarrow \mathbf{u}/\mathbf{s}$	$\mathbf{y}$ obtained by suppressing from $\mathbf{s}$ each $\mathbf{s}_i$ for which $\mathbf{u}_i = 0$
Mask	$\mathbf{y} \leftarrow /\mathbf{s};\mathbf{u};\mathbf{t}/$	$\overline{\mathbf{u}}/\mathbf{y} \equiv \overline{\mathbf{u}}/\mathbf{s}$; $\mathbf{u}/\mathbf{y} \equiv \mathbf{u}/\mathbf{t}$
Indexing	$\mathbf{y} \leftarrow \mathbf{s}_m$	$\mathbf{y}_i \equiv \mathbf{y}_{m_i}$
Reduction	$y \leftarrow \odot/\mathbf{s}$	$y \equiv \mathbf{s}_0 \odot \mathbf{s}_1 \ldots \odot \mathbf{s}_{\nu(\mathbf{s})-1}$
6. *Special vectors*		
Full	$\mathbf{w} \leftarrow \boldsymbol{\varepsilon}(n)$	vector of n 1's
Characteristic	$\mathbf{w} \leftarrow \boldsymbol{\varepsilon}^j(n)$	vector of n elements all 0 except for w_j
Prefix	$\mathbf{w} \leftarrow \boldsymbol{\alpha}^j(n)$	vector of n elements: left most j elements are 1, the rest 0
Suffix	$\mathbf{w} \leftarrow \boldsymbol{\omega}^j(n)$	vector of n elements: right most j elements are 1, the rest 0
Infix	$\mathbf{w} \leftarrow i \downarrow \boldsymbol{\alpha}^j$	vector having i leading 0's followed by j 1's
Interval	$\mathbf{w} \leftarrow \mathbf{i}^j(n)$	vector of n elements consisting of the numbers $j, j+1, j+2$, etc.

the symbol '=' in FORTRAN and ':=' in ALGOL (in some languages, as we shall see later, the symbol $\rightarrow$ is also used).

The basic operations employed in the language include arithmetic and logical functions and the residue and inequality relationships occurring in elementary number theory; these are summarized in Table 3.9. Any operation defined on a scalar operand can be extended, component by component, to dimensionally compatible vectors. If one of the operands is a scalar it is treated as a vector of the appropriate dimension. Table 3.10 shows some typical examples of the use of these basic operations. The logical operations AND, OR, and NOT are defined in the usual way upon the logical variables 0 and 1, and are augmented by the relational statements, for example:

$$y \leftarrow s < t \;\equiv\; y \leftarrow \bar{s} \wedge t$$
$$y \leftarrow s \leqslant t \;\equiv\; y \leftarrow \bar{s} \vee t$$
$$y \leftarrow s = t \;\equiv\; y \leftarrow (s \wedge t) \vee (\bar{s} \wedge \bar{t})$$
$$y \leftarrow s \neq t \equiv\; y \leftarrow (\bar{s} \wedge t) \vee (s \wedge \bar{t})$$

Where the relation $s \mathscr{R} t$ is satisfied the result is specified by a 1, otherwise it is specified by a 0. Note that the last two relationships are, in fact, the exclusive OR functions.

The **shifting** operations, or rotations, defined on a vector **s** are quite straightforward and correspond to normal logical shifts of k places left or right, and an end-around shift; the shift left rotation is defined, for example, by:

$$\mathbf{y} \leftarrow k \uparrow \mathbf{s}$$

The **base value function** derives a number y from a specification of the radix r and a digit string **s** representing the number. In general we have the expression:

$$y \leftarrow r \perp \mathbf{s}$$

which means that y is specified as the value obtained by treating the elements of the vector **s** as digits in the number system of radix r. For example (see also Table 3.10):

$$563 \equiv 10 \perp 5, 6, 3$$
and $$15 \equiv \perp 0, 1, 1, 1, 1$$

It will be apparent that to obtain the number represented by a digit string a polynomial evaluation must be performed, for example:

$$15 = 1 \times 2^0 + 1 \times 2^1 + 1 \times 2^2 + 1 \times 2^3 + 0 \times 2^4$$

The inverse operation is also available and is specified in the general form as:

$$\mathbf{s} \leftarrow r(n) \top y$$

where $\mathbf{s}$ is the digit string, r the radix, n the number of digits in the string, and y the number to be represented.

When an operation such as addition is applied to all the components of a vector to produce a simpler structure (usually a scalar) it is called a **reduction**. The $\odot$-reduction of a vector $\mathbf{s}$ is denoted by $\odot/\mathbf{s}$ and is defined as:

$$y \leftarrow \odot/\mathbf{s} \equiv \mathbf{s}_0 \odot \mathbf{s}_1 \odot \cdots \odot \mathbf{s}_{\nu(\mathbf{s})-1}$$

where $\odot$ represents any binary operation (that is, applying to two operands).

Another facility available in the Iverson language is the generalized **vector product**, this is represented by the expression:

$$y \leftarrow \mathbf{s} \overset{\odot_1}{\underset{\odot_2}{}} \mathbf{t} \quad \text{which is equivalent to} \quad y \leftarrow \odot_1/\mathbf{s} \odot_2 \mathbf{t}$$

Thus the operator $\odot_2$ is applied to $\mathbf{s}$ and $\mathbf{t}$ in the usual way, element by element, and then the operator $\odot_1$ is used as a reduction on the result.

Table 3.10
Examples of Iverson notation

Arguments used in the examples:

$\mathbf{a} = 2, 4, 6, -5$	$\mathbf{b} = 3, 6, 9, 12$	$\mathbf{c} = 4, 5, 3$
$\mathbf{d} = 1, 0, 1, 1, 0, 1$	$\mathbf{e} = 1, 1, 0, 1, 1, 0$	$\mathbf{f} = 1, 0, 1, 1$

1. *Logic and arithmetic*	
$y \leftarrow \mathbf{a}_1 + \mathbf{b}_1 \equiv 10$	$\mathbf{y} \leftarrow \mathbf{a} + \mathbf{b} \equiv 5, 10, 15, 7$
$y \leftarrow \mathbf{a}_1 \times \mathbf{b}_1 \equiv 24$	$\mathbf{y} \leftarrow \mathbf{a} \times \mathbf{b} \equiv 6, 24, 54, -60$
$\mathbf{y} \leftarrow \mathbf{a}_0 \times \mathbf{b} \equiv 6, 12, 18, 24$	$y \leftarrow \mathbf{a}_2 \leqslant \mathbf{b}_2 \equiv 1$
$\mathbf{y} \leftarrow \mathbf{a} < \mathbf{b} \equiv 1, 1, 1, 1$	$y \leftarrow \mathbf{d}_1 \vee \mathbf{e}_1 \equiv 1$
$y \leftarrow \mathbf{d}_1 \wedge \mathbf{e}_1 \equiv 0$	$\mathbf{y} \leftarrow \mathbf{d} \vee \mathbf{e} \equiv 1, 1, 1, 1, 1, 1$
$\mathbf{y} \leftarrow \mathbf{d} \wedge \mathbf{e} \equiv 1, 0, 0, 1, 0, 0$	$\mathbf{y} \leftarrow \mathbf{d} \neq \mathbf{e} \equiv 0, 1, 1, 0, 1, 1$
2. *Shifting and base values*	
$\mathbf{y} \leftarrow 2 \uparrow \mathbf{a} \equiv 6, -5, 2, 4$	$\mathbf{y} \leftarrow 2 \underset{\circ}{\uparrow} \mathbf{a} \equiv 6, -5, 0, 0$
$\mathbf{y} \leftarrow 4 \downarrow \mathbf{d} \equiv 1, 0, 1, 1, 1, 0$	$\mathbf{y} \leftarrow 4 \overset{\circ}{\downarrow} \mathbf{d} \equiv 0, 0, 0, 0, 1, 0$
$y \leftarrow \perp\mathbf{d} \equiv 45$	$y \leftarrow \perp\mathbf{e} \equiv 54$
$\mathbf{y} \leftarrow (6) \top 29 \equiv 0, 1, 1, 1, 0, 1$	$\mathbf{y} \leftarrow (3) \top 29 \equiv 1, 0, 1$
$\mathbf{y} \leftarrow 10(4) \top 53 \equiv 0, 0, 5, 3$	$y \leftarrow 10 \perp \mathbf{c} \equiv 453$
3. *Selection and reduction*	
$\mathbf{y} \leftarrow \mathbf{c}, \mathbf{b} \equiv 4, 5, 3, 3, 6, 9, 12$	$\mathbf{y} \leftarrow \mathbf{f}/\mathbf{b} \equiv 3, 9, 12$
$\mathbf{y} \leftarrow /\mathbf{a}; \mathbf{f}; \mathbf{b}/ \equiv 3, 4, 9, 12$	$\mathbf{y} \leftarrow \mathbf{f}/\mathbf{a} \equiv 2, 6, -5$
$y \leftarrow +/\mathbf{a} \equiv 7$	$y \leftarrow \times/\mathbf{c} \equiv 60$
4. *Special vectors*	
$\mathbf{w} \leftarrow \boldsymbol{\alpha}^3(6) \equiv 1, 1, 1, 0, 0, 0$	$\mathbf{w} \leftarrow \boldsymbol{\varepsilon}(6) \equiv 1, 1, 1, 1, 1, 1$
$\mathbf{w} \leftarrow \boldsymbol{\omega}^2(6) \equiv 0, 0, 0, 0, 1, 1$	$\mathbf{w} \leftarrow \bar{\boldsymbol{\varepsilon}}(6) \equiv 0, 0, 0, 0, 0, 0$
$\mathbf{w} \leftarrow \boldsymbol{\varepsilon}^{1,3,5}(6) \equiv 0, 1, 0, 1, 0, 1$	$\mathbf{w} \leftarrow \boldsymbol{\varepsilon}^3(6) \equiv 0, 0, 0, 1, 0, 0$
$\mathbf{w} \leftarrow \boldsymbol{\alpha}^2(4)/\mathbf{a} \equiv 2, 4, 0, 0$	$\mathbf{w} \leftarrow \bar{\boldsymbol{\varepsilon}}^2(6) \equiv 1, 1, 0, 1, 1, 1$
$\mathbf{w} \leftarrow 5 \downarrow \boldsymbol{\alpha}^3(8) \equiv 0, 0, 0, 0, 0, 1, 1, 1$	$\mathbf{w} \leftarrow \mathbf{i}^0(4) \equiv 0, 1, 2, 3$

To make effective use of structured operands the facility to specify and select certain elements or groups of elements must be provided. Single elements may of course be selected by the use of indices, as for example: $\mathbf{s}_i$, $\mathbf{M}^i$, $\mathbf{M}_j$, and $\mathbf{M}^i_j$. Consequently, **selection operators** are provided in the language which are used to form new arrays by moving and selecting elements from existing arrays. The **catenation** operator forms a new array from all the elements of two other arrays. For vectors this is achieved by simply joining the two strings together which is formally expressed as:

$$\mathbf{y} \leftarrow \mathbf{s}, \mathbf{t}$$

Another operation is that of **compression** which forms a vector **y** from a vector **s** by deleting certain elements of **s**, this operation is denoted by:

$$\mathbf{y} \leftarrow \mathbf{u}/\mathbf{s}$$

where **u** is a logical vector (also called a **selection map**) of dimension $\nu(\mathbf{s})$ and $\mathbf{s}_i$ is deleted if, and only if, $\mathbf{u}_i = 0$. For example

$$\mathbf{u} = 1, 0, 1, 1, 0, 1, 0, 1$$
$$\mathbf{s} = 0, 1, 1, 1, 1, 0, 1, 0$$

Hence $\mathbf{y} = 0, 1, 1, 0, 0$

A more complex selection operation is that of **masking**. In this case the selection vector **u** is used to select elements from a vector **t**, while its inverse $\overline{\mathbf{u}}$ is used to select elements from another vector **s**. The resulting elements are then merged to give the final vector **y**. This operation is expressed formally as

$$\mathbf{y} \leftarrow /\mathbf{s}; \mathbf{u}; \mathbf{t}/$$

For example:

$$\mathbf{s} = 1, 0, 1, 1, 1, 0, 0, 0 \qquad \mathbf{t} = 1, 0, 0, 0, 1, 1, 1, 1$$
$$\overline{\mathbf{u}} = 1, 1, 0, 1, 1, 1, 1, 0 \qquad \mathbf{u} = 0, 0, 1, 0, 0, 0, 0, 1$$
$$\overline{\mathbf{u}}/\mathbf{s} = 1, 0, 1, 1, 0, 0 \qquad \mathbf{u}/\mathbf{t} = 0, 1$$

Hence $\mathbf{y} \leftarrow /\mathbf{s}; \mathbf{u}; \mathbf{t}/ \equiv 1, 0, 0, 1, 1, 0, 0, 1$

Certain types of vectors occur frequently enough in programming to warrant direct data generation rather than program specification. Thus special vectors are included in the language which allows particular data structures to be specified directly in the program; for example, the **prefix** vector $\boldsymbol{\alpha}^j(n)$ is a logical vector of dimension n whose first j components are unity. The **suffix** vector $\boldsymbol{\omega}^j(n)$ is defined similarly, with the last j components being unity. The **full** vector $\boldsymbol{\varepsilon}(n)$ denotes a logical vector of dimension n whose components are all unity, and $\overline{\boldsymbol{\varepsilon}}(n)$ denotes the zero vector. An **infix** vector, having i leading 0's followed by j 1's, is also defined as $i \downarrow \boldsymbol{\alpha}^j$. Examples of all these special vectors are shown in Table 3.10.

These vectors may of course be used to considerable advantage in the selection and masking operations described earlier:

$$\mathbf{s} = 0, 1, 1, 1, 0, 1, 0$$
$$\boldsymbol{\alpha}^3(7) = 1, 1, 1, 0, 0, 0, 0$$
$$\mathbf{y} \leftarrow \boldsymbol{\alpha}^3(7)/\mathbf{s} = 0, 1, 1$$

$$\mathbf{s} = 0, 1, 1, 1, 0, 1, 0 \qquad \mathbf{t} = 1, 1, 1, 0, 1, 0, 1$$
$$\overline{\boldsymbol{\omega}}^4(7) = 1, 1, 1, 0, 0, 0, 0 \qquad \boldsymbol{\omega}^4(7) = 0, 0, 0, 1, 1, 1, 1$$
$$\overline{\boldsymbol{\omega}}^4(7)/\mathbf{s} = 0, 1, 1 \qquad \boldsymbol{\omega}^4(7)/\mathbf{t} = 0, 1, 0, 1$$

$$\mathbf{y} \leftarrow /\mathbf{s};\, \boldsymbol{\omega}^4(7);\, \mathbf{t}/ \equiv 0, 1, 1, 0, 1, 0, 1$$

Before we conclude this very brief description of the Iverson language there is one very important aspect that still needs to be discussed and that is the **branching** facilities. The various forms that may be used are shown in Table 3.11; in all cases it is assumed that the statements involved in a branching operation have been uniquely labelled in the program. The simplest form of branching operation is, of course, the unconditional jump to a statement, which may be used by itself or appended to the right of any statement by using a semi-colon as a separator. This form of branching instruction may also be extended to a specified jump to any one of a number of statements selected by the value of a parameter. Note that these jump instructions correspond to the GO TO statement in FORTRAN; for instance:

GO TO 6 $\equiv \rightarrow (6)$
GO TO (6, 15, 21), I $\equiv \rightarrow (6, 15, 21)_i$

The general conditional branching statement takes the form:

$$s{:}y;\, R \rightarrow (n)$$

which denotes a jump to statement n_i of the program if the relationship $s\, \mathscr{R}_i\, y$ holds good; the parameters $\mathscr{R}$ and n may themselves be defined in other parts of the program. This form of jump instruction is equivalent to the FORTRAN IF statement; for example:

IF $(X - Y)$ 5, 9, 31 $\equiv (x - y){:}0;\, (<, =, >) \rightarrow 5, 9, 31$

We are now in a position to write complete programs in the Iverson notation and the next section will be devoted to some typical examples of its use and application.

3.12 Use and advantages of Iverson notation

Let us now consider some programming examples. Table 3.12 shows the Iverson program for our running example of adding together ten numbers, while Table 3.13 shows a typical program for sorting numbers into descending order (highest value first). Both these programs are straightforward examples and their operation should be apparent from the preceding text. Note that the sorting program relies on the principle of

Table 3.11
Branching operations in Iverson

Notation	*Definition*
$\rightarrow(n)$	Unconditional jump to line n of the program
$y \leftarrow s + t; \rightarrow(n)$	y is specified by $s + t$, then program jumps to n
$\rightarrow(n_1, n_2, n_3, n_4)i$	Jump to line n_1 or n_2 or n_3 or n_4 as the value of i is 0, 1, 2, or 3 respectively
$s : y; (=, <) \rightarrow (n_1, n_2)$	s is compared to y, if $s = y$ jump to n_1 or if $s < y$ jump to n_2, otherwise carry on

Table 3.12
Iverson program to add ten numbers

Step no.	*Statement*	*Comment*
0	$y \leftarrow 0$	y contains the final sum
1	$i \leftarrow 10$	i is the count location
2	$i : 0; (=) \rightarrow (5)$	
3	$i \leftarrow i - 1$	
4	$y \leftarrow y + \mathbf{s}_i; \rightarrow (2)$	
5	$\rightarrow (5)$	Stop

Table 3.13
Program to sort numbers into descending order

Step no.	*Statement*	*Comment*
0	$j \leftarrow -1$	allow for 0—origin indexing
1	$j \leftarrow j + 1$	main loop count
2	$j : \nu(\mathbf{x}) : (\geqslant) \rightarrow (2)$	Stop, end of program
3	$k \leftarrow 0$	
4	$k \leftarrow k + 1$	subsidiary loop count
5	$k : \nu(\mathbf{x}); (=) \rightarrow (1)$	
6	$\mathbf{x}_j : \mathbf{x}_{j+k}; (>) \rightarrow (4)$	compare numbers
7	$\mathbf{x}_j \leftarrow \mathbf{x}_{j+k}; \rightarrow(4)$	exchange locations

Table 3.14
Program for Hamming code check

Step no.	*Statement*	*Comment*
0	$\mathbf{a}_2 \leftarrow 2 \mid (^+/(\boldsymbol{\varepsilon}^{0,2,4,6}(7)/\mathbf{h}))$	least significant error digit
1	$\mathbf{a}_1 \leftarrow 2 \mid (^+/(\boldsymbol{\varepsilon}^{1,2,5,6}(7)/\mathbf{h}))$	
2	$\mathbf{a}_0 \leftarrow 2 \mid (^+/(\boldsymbol{\omega}^4(7)/\mathbf{h}))$	
3	$y \leftarrow \perp\mathbf{a}$	base 2 value of $\mathbf{a}$
4	$y{:}0; (=) \rightarrow (n)$	Correct, jump to (n)
5	$\rightarrow(5)$	Error Stop

exhaustively comparing each element of the vector **x** with all other elements, interchanging values such that the highest valued element always takes precedence in the vector. The process requires $n - 1$ passes through the data to sort n items into order.

The real power of the language, however, can be more readily seen from a non-numerical example. For instance, the operation of checking a vector **a** for even parity can be accomplished in the one line statement:

$$2 \mid ({}^{+}/\mathbf{a}):0;\ (=) \rightarrow (n)$$

An even better example is a program to perform the checking functions associated with Hamming codes.[38] Hamming codes are single error detection and correction codes which employ check digits distributed throughout the message word. These check digits provide even parity checks on particular message digit positions in such a way that if the parity checks are made in order with successful checks being designated by 0 and failure by 1 the resulting binary number gives the position of the incorrect digit. For example, a 7-bit Hamming coded word would be represented as:

1	2	3	4	5	6	7
C	C	M	C	M	M	M

where C is a check digit, and M a message bit. To encode, the sum of the 1's in the digit groups (1, 3, 5, 7), (2 3, 6 7), and (4 5 6 7) are arranged to be even by inserting the appropriate digit in the check digit position. To decode, the word parity checks are performed in order on the same group of digits, the results being used to form a 3-bit word the decimal equivalent of which is the digit position in error. For example:

1	2	3	4	5	6	7	
0	1	1	0	0	1	1	Correct word
0	1	1	0	0	0	1	Incorrect word

Performing the parity checks we have:

First check (1, 3, 5, 7) is correct	0
Second check (2 3, 6 7) is incorrect	1
Third check (4 5 6 7) is incorrect	1

The decimal equivalent of the binary word 110 is 6, therefore the digit in position 6 is incorrect!

To program this procedure in machine-code or FORTRAN is a lengthy and involved process but in the Iverson notation it reduces to five statements (see Table 3.14). In this program the vector **h** is the Hamming word, and **a** the error word giving the position of the incorrect digit; note also the use of special vectors to select the check digits and the way in which the operations may be nested together. It is also interesting to observe that there is no reason why all the statements should not be merged together, to form the single statement:

$$\perp[2 \mid (^{+}/(\boldsymbol{\varepsilon}^{0,2,4,6}(7)/\mathbf{h})), 2 \mid (^{+}/(\boldsymbol{\varepsilon}^{1,2,5,6}(7)/\mathbf{h})), 2 \mid (^{+}/(\boldsymbol{\omega}^{4}(7)/\mathbf{h}))]:0; (=) \rightarrow (n)$$

thereby demonstrating the highly parallel nature of the language. This property of the language has been exploited by the Control Data Corporation of America (CDC) in the development and programming of highly parallel computing systems.[39]

One other major advantage of the Iverson language is that it may be used to represent a logical algorithm which may then be implemented in either hardware or software form. This capability not only illustrates once again the duality of these two forms of logical implementation, but it also provides a general technique for specifying logical systems which might well be an answer to the 'complexity barrier' problem mentioned earlier. For example, the program for parallel to serial conversion shown in Table 3.15 may also be implemented in hardware form. In this case

Table 3.15
Program for parallel to serial conversion

Step no.	*Statement*
0	$i \leftarrow 0$
1	$i : 4; (=) \rightarrow (n)$
2	$\mathbf{c} \leftarrow \boldsymbol{\varepsilon}^{i}(4)$
3	$s \leftarrow \mathbf{r} \overset{\vee}{\wedge} \mathbf{c}$
4	$i \leftarrow i + 1; \rightarrow (1)$

r would represent a 4-bit register containing the parallel inputs, the vector **c** would simulate the clock timing pulses (used to gate the parallel inputs), and s the serial output.

Another important advantage of the language is that it is comparatively easy to represent any other high level language, such as FORTRAN or ALGOL, in the notation. Thus it is an excellent vehicle for compiler writing and language specification. Unfortunately, though, with conventional computers the Iverson language must itself be compiled into the basic object language, thereby lowering the efficiency of operation considerably. If, however, the basic assembly language (machine-code) of the computer was, in fact, the Iverson notation,[40] program compiling times and running efficiencies would be considerably enhanced. This is, of course, another example of the influence of procedure orientated languages on computer architecture, where effectively the Iverson operations have been implemented in hardware.

The Iverson notation has also found considerable acclaim in the teaching of digital systems and at least three textbooks have been based on APL[41,42,43]. In particular AHPL (A Hardware Programming Language) developed by Hill and Peterson has been accepted for teaching purposes (as well as for specification and design) and a number of compilers are in existence[44].

References and bibliography

1 Naur, P. (ed.). Revised report on the algorithmic language ALGOL 60. *Computer Journal*, 1963, **5**, 349–367.

2 Wirth, N. A generalisation of ALGOL 60. *Comm. ACM*, 1963, **6**, 547–554.

3 Rabinowitz, I. N. Report on the algorithmic language FORTRAN II. *Comm. ACM*, 1962, **5**, 327–337.

4 Dijkstra, E. W. *A Primer of ALGOL 60 Programming*. Academic Press, New York, 1960.

5 McCracken, D. D. *A Guide to FORTRAN IV Programming*. John Wiley, New York, 1965.

6 Higman, B. *A Comparative Study of Programming Languages*. Macdonald, London, 1967.

7 Green, B. F. Computer language for symbol manipulation. *IRE Trans. Electron. Comput.*, 1961, **EC10**, 729–735.

8 Foster, J. M. *List Processing*. Macdonald, London, 1967.

9 Barron, D. W. *Recursive Techniques in Programming*. Macdonald, London, 1967.

10 McCarthy, J. Recursive functions of symbolic expressions and their calculation by machine, Part I. *Comm. ACM*, 1960, **3**, 184–195.

11 Buchholz, W. *Planning a Computer System*. McGraw Hill, New York, 1962.

12 Glaser, E. L. Breaking the complexity barrier. *Proc. IEEE*, Winter Meeting, 1969.

13 Barron, D. W. and Hartley, D. F. Influence of automatic programming on machine design. B.C.S. Conference: *The Importance of Users' Needs on the Design of D.P. Systems*, Edinburgh, 1964.

14 Gram, C. *et al*. GIER – A Danish computer of medium size. *IEEE Trans. Electron. Comput.*, 1963, **EC12**, 629–650.

15 Melbourne, A. J. and Pugmire, J. M. A small computer for the direct processing of FORTRAN statements. *The Computer Journal*, 1965, **8**, 24–27.

16 Hopgood, F. R. A. *Compiling Techniques*. Macdonald, London, 1969.

17 Dijkstra, E. W. Making a translator for ALGOL 60. *A.P.I.C. Bulletin*, 1961, **7**, 3–11.

18 Bashkow, T. R., Sasson, A., and Kronfield, A. System design of a FORTRAN machine. *IEEE Trans. Electron. Comput.*, 1967, **EC16**, 485–499.

19 Randell, B. and Russell, L. *ALGOL 60 Implementation*. Academic Press, New York, 1964.

20 Brooker, R. A. Influence of high level languages on computer design, *Proc. IFIPS Congress*, Edinburgh, 1968.

21 Wirth, N. and Weber, H. EULER: a generalisation of ALGOL and its formal definition Part 1, *Comm. ACM*, **9** 13–25, 1966.

22 Haley, A. C. The KDF9 computer system, *Proc. AFIPS Conf.*, **22**, 108–120, 1962.

23 Allmark, R. H. and Lucking, J. R. Design of an arithmetic unit incorporating a Nesting Store, *Proc. AFIPS Conf.*, **22**, 694–698, 1962.

24 Hauck, E. H. and Dent, B. A. Burroughs' B6500/B7500 stack mechanism, *Proc. AFIPS Conf.*, **32**, 245–251, 1968.

25 Organick, E. I. *Computer System Organisation – the B5700/B6700 Series*, Academic Press, New York, 1973.

26 Kilburn, T., Morris D., Rohl J. S. and Sumner F. H. A System Design Proposal, *Proc. IFIPS Congress*, 76–80, Edinburgh, 1968.

27 Special issue on Stack Machines, *IEEE Computer*, **10**, 14–52, May 1977.

28 Doran, R. W. Architecture of Stack Machines, *High-Level Language Computer Architecture*, ed. Y. Chu. Academic Press, New York, 1975.

29 J. K. Iliffe. *Basic Machine Principles*. Macdonald, London, 1968.

30 Glaser, E. L. Hardware-software interaction. First Congress of the Information Sciences, Homestead, Va., 1962.

31 Lewin, D. *Computer-Aided Design of Digital Systems*. Crane Russak, New York, 1977.

32 Raymond, J. and Banerji, D. K. Using a microprocessor in an intelligent graphics terminal, *IEEE Computer*, **9**, 18–25, April 1977.

33 Iverson, K. E. *A Programming Language*. John Wiley, New York, 1962.

34 Iverson, K. E. A programming language. *Proceedings of the AFIPS Conference*, 1962, **21**, 345–351.

35 Falkoff, A. D. and Iverson, K. E. *APL/360 : User's Manual*. IBM Watson Research Centre, Yorktown Heights, N.Y., 1968.

36 Carberry, R. S. *et al. APL/1130*. IBM Contributed Library, 1130 03.3.001, 1968.

37 Foster, G. H. APL: a perspicuous language. *Computers and Automation*, 1969, Nov., 24–28.

38 Hamming, R. W. Error detecting and correcting codes. *Bell System Tech. J.*, 1950, **29**, 147–160.

39 Lincoln, N. R., Jones, P. D., and Thornton, J. E. A parallel approach to information processing. IEE Conference: *Computer Science and Technology*, July, 1969. IEE Publication, **55**, 316–324.

40 Thurber, K. J. and Myrna, J. W. System design of a cellular APL computer. *IEEE Trans. Electron. Comput.*, 1970, **C19**, 291–303.

41 Hellerman, H. *Digital Computer System Principles*. McGraw Hill, New York, 1967.

42 Hill, F. J. and Peterson, G. R. *Digital Systems : Hardware Organisation and Design*. John Wiley, New York, 1973.

43 Blaauw, G. A. *Digital System Implementation*. Prentice Hall, Englewood Cliffs, 1976.

44 Gentry, M. A Compiler for AHPL Control Sequences, PhD Dissertation, University of Arizona, June 1971.

45 Ledley, R. S. *Programming and Utilizing Digital Computers*. McGraw Hill, New York, 1962.

46 Barron, D. W. *Assemblers and Loaders*. Macdonald, London, 1969.

47 Levison, M. and Sentance, W. A. *Introduction to Computer Science*. Oldbourne, London, 1968.

48 Carlson, C. B. The mechanism of a push-down stack. *Proc. AFIPS FJCC*, 1963, **24**, 243–250.

Tutorial problems

***3.1** Write a machine-code program to output the contents of store locations n to $n+k$ on 5-bit punched paper tape using the binary format required by the simple loader described in the text.

3.2 Write an interrupt routine, using machine-code, to input 4-digit decimal numbers encoded in the teleprinter code shown in Table 3.3 from punched paper tape. Describe, with the aid of flow charts, how the interrupt procedure operates and in particular the supervisory routine.

***3.3** Consider the problems involved in writing an input compiler for the numeric coded form of computer instruction, that is, of the form 20 1460. Illustrate your answer, where appropriate, with flow diagrams.

3.4 Write a simple engineering test program to check out the operation of the input/output instructions associated with punched paper tape equipment.

***3.5** Describe a machine code program, giving flow diagrams, to evaluate the Polish expression:

AB*CD − E/ +

Discuss the problems involved in modifying this program for the general case, that is, a Polish expression of any length or complexity.

3.6 Assuming that a hardware stack-store is available, use zero-address instructions to show how the expression in question 3.5 may be evaluated. Contrast this approach with the normal software technique of evaluation.

***3.7** Using as a basis the flow diagram shown in Fig. 3.4, write a program for a simple binary loader using Iverson notation.

3.8 Rewrite the 'binary dump' program of question 3.1 using the Iverson notation.

***3.9** Write an Iverson program to convert from a serial to parallel mode of operation, and show how the algorithm may be implemented in hardware.

3.10 Devise a program, using the Iverson notation, to generate the check digits for a 7-bit Hamming code (four information bits and three check digits) including the subsequent assembly into the complete encoded word.

4

Control unit organization

4.1 Introduction

As we have already seen in Chapter 1, the main function of the control unit is to decode the order digits of an instruction word, thereby generating the necessary sequence of control waveforms (micro-orders) to allow the instruction to be executed during the obey (execute) cycle. The other basic requirement is to provide the signals which allow the instructions to be read down in sequence from the main store, that is, the read (load) cycle operations. Thus the control unit must generate as many different sequences of gating or switching waveforms (micro-programs) as the number of different arithmetic, logical, or transfer operations the computer is required to perform. These properties are not peculiar to the control unit of a computer, however, since they apply equally well to any controller (or programmer) of special purpose digital systems.

To perform these functions the control unit utilizes logic circuits such as shift registers, counters, decoders, and so on, which will be described in the early sections of this chapter. We will go on to consider the overall structure of the control unit, using the concepts of **static** and **dynamic** micro-programming, and the effect of orders such as modification, interrupt, etc., on the design of the control unit. Finally, we consider the use of high-level languages (in particular Iverson) to represent the micro-program.

4.2 Serial-parallel operation

In common with all digital systems, computers may function in either the **parallel** or **serial** mode of operation. In parallel working the binary digits, represented by voltage levels, are each allocated a separate bistable store or connecting wire. Thus all the digits of an n-bit operand or instruction word appear simultaneously on n different wires, or bistable outputs. In serial operation the bits are represented by voltage levels on a single wire, but displaced in time. Thus an n-bit operand would appear in time sequence on a single output wire, with the least significant digit occurring first. In practice this serial pulse train would originate from the output of a shift register (see later) which must be clocked sequentially n times to produce the complete word. Because in the parallel method logical operations are performed separately on each bit of the operand, approximately n times as much hardware is required as with the serial technique, but the operations are performed nearly n times faster. Serial systems, operating on a bit by bit process, are considerably slower but require less hardware.

The mode of operation also affects the control circuits – for example, in a parallel machine the timing is simpler since an operand can be transferred from one part of the machine to another by the application of a single control waveform to a set of gates. In a serial machine a set of timing (clock) pulses synchronized to individual digit positions must be generated and distributed to various control gates.

First generation computers were invariably serial in operation, this meant that with an average 40-bit wordlength and a digit time of 2 μs (equivalent to a clock rate of 500 kHz) the wordtime (WT) was 80 μs. Since in a single-addressed machine a typical directly addressed instruction, such as store or add, requires two accesses to the store, the average order time for these machines is three wordtimes, that is, for the example above, 240 μs. This allows for a typical machine: 1WT to fetch the instruction from store, 1WT to fetch the contents of the location specified, and a further wordtime to execute the order.

To increase the speed of operation second generation machines were designed with a parallel structure. They incorporated a random, parallel access, core store (see Chapter 6) with a read/write cycle time of about 10 μs and 1 MHz clock rate. With this configuration a typical order will have an execution time slightly greater than two store access times, (read/write cycle times) since the basic logical operations, such as addition, are performed in 1 bit time. Third and subsequent generation machines still use a parallel organization when high speed is required but, with the advent of fast integrated circuits with clock rates up to 100 MHz, serial and serial/parallel computers are coming back into use. This is particularly so with calculator chips and process controllers requiring only a short wordlength in the order of 8–12 bits. To bring the cost down serial arithmetic and control circuits are used in conjunction with a fast (1 μs R/W cycle) RAM store. This technique can also be an attractive engineering solution when designing small stored-program controllers for use in instrumentation, etc. For example, a machine with a 12-bit word and using a 10 MHz clock rate (100 ns digit rate) would have addition times in the order of 5μs.

Microprocessors normally employ a parallel mode of operation since in general they tend to be slow in operation (especially MOS types); typically they operate with a 1 MHz clock and a 2 μsec instruction time.

4.3 Synchronous/asynchronous operation

Synchronous computers are clocked by a basic square wave or **digit pulse** (produced by a square wave oscillator) that is distributed throughout the system to control the timing of operations. Other pulses defining wordtimes, etc., are generated from the basic clock by counters or by reading a timing track prerecorded on the surface of a drum or disc store (see Chapter 7). This means that the inputs, outputs, and internal logical states of a circuit are sampled only at regular (or specified) intervals of time. In this way inherent delays in the logic devices and propagation delays in data transmission paths can be easily overcome – that is, the circuits have time to settle out to their final value before the arrival of the next clock pulse. Thus the duration (and timing) of a logical function is determined by the longest operation to be performed, this in turn depends on the length of the signal paths and the speed of the slowest element.

Synchronous organization considerably limits the overall speed of operation of the system, which could reach its maximum if it were free-running or **asynchronous** in operation. With this type of system, logical operations proceed at their own speed, the next operation being initiated when the preceding one has finished. Thus the speed of operation is entirely determined by the average speed of all the relevant logic circuitry. Asynchronous machines are in general only compatible with parallel storage systems and asynchronous arithmetic; synchronous organization, however, may be employed with any mode of working.

In practice, computers or digital systems in general are a mixture of both types of circuit – synchronous and asynchronous. Theoretically since the overall computer system is generally clocked it falls into the category of a synchronous machine, however many sections of the computer (for example the adder) are completely asynchronous. By far the majority of computers are fully synchronous in operation; this is mainly owing to the difficulties inherent in asynchronous machines in providing logic signals to indicate when the current operation has finished and the problems caused by propagation delays and circuit hazards.

4.4 Logic circuits used in the control unit

The control unit is comprised of the following basic logic sub-systems:

Storage and shift registers – used to store and manipulate data, usually of one wordlength capacity.

Counters and dividers – used for counting or dividing clock pulses to generate timing waveforms, or to count the operation cycles within a micro-program.

Decoders and encoders – used to transform data from one coded form to another, for example in instruction decoding.

Comparators – used to compare the contents of registers and to generate a signal according to whether the result is 'equal to', 'greater than', or 'less than' some predetermined quantity.

Timing control circuits – an assembly of square wave oscillator, delay units, counters, gates, etc., used to generate the basic clock pulses for a synchronous system.

The majority of these circuits are now available as standard modules in integrated circuit logic systems and logic circuits for all the sub-systems, using basic logic elements, are given in manufacturers' application notes.[1, 2, 3]

Storage and shift registers may be either static, consisting of bistable elements, or dynamic, using circulating, CCD devices or MOS type storage (see Chapter 6). Dynamic registers operate in the serial mode only but static registers may have either parallel or serial access of data.

In a parallel computer system the control registers would normally consist of d.c. set–reset (SR) bistables with input and output gating to and from a common parallel highway. However, both phases of output (Q and $\overline{Q}$) would normally be distributed in order to economize on interconnections (an important consideration with high speed logic); some systems distribute only one signal phase and generate the complements locally. Typical logic circuits for a parallel register, using d.c. SR and clocked JK bistables, are shown in Fig. 4.1; note that the SR bistables consist of cross-coupled NAND units. The operation of the circuit shown in Fig. 4.1(a) is such that when a data transfer is required between registers M and C, the control waveforms M_o and C_i are generated, which enable the gates and allow the outputs of bistable M to set bistable C in the usual manner. The circuit shown in Fig. 4.1(b) operates similarly, except that in this case the input gating is performed by clocking the JK bistables.

With serial organization, clocked shift registers of either static

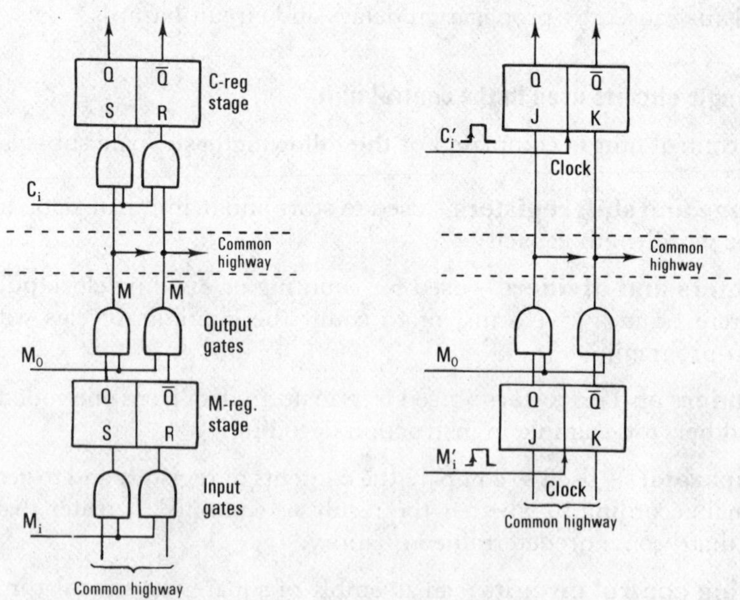

Figure 4.1 Single parallel register stages.

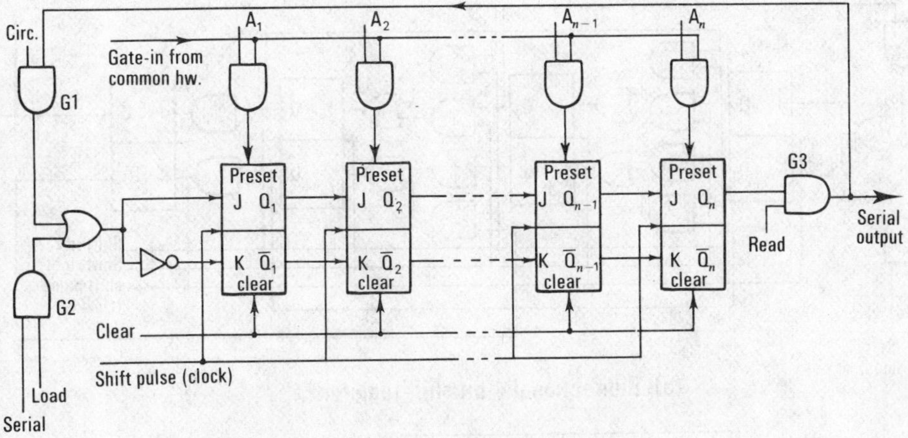

Figure 4.2 Static shift register circuit.

or dynamic type would be employed; a typical circuit using JK bistables is shown in Fig. 4.2. In this circuit the contents of stage Q_1 would be transferred to stage Q_2, and stage Q_2 to stage Q_3, and so on, on the arrival of the shift pulse. Note that the final stage of the register is fed back to the input stage via a gating network; this is known as an **end-around** shift. When reading data from the register, gates G1 and G3 would be enabled and the circuit clocked with n shift pulses, when n is the number of bits (stages) in the computer word. To load data into the register gate, G1 would be inhibited and the new data entered via gate G2.

It is possible with this circuit to load the register serially, or to enter parallel data from a common highway using gates and the d.c. preset inputs of the bistables. Serial access registers of this type provide a very convenient method of converting from serial to parallel data mode, and vice versa. They could be used, for example, in a serial/parallel machine using core storage, the parallel output of the store would be entered into the register (the memory buffer) and then clocked serially into the processor system.

Shift registers may also be used to manipulate data as well as performing the storage function. For example they can be made to shift left or right (a **bidirectional** shift register) by n places thereby multiplying or dividing the contents of the register by 2^n. The circuit for a bidirectional serial shift register using static JK bistables is shown in Fig. 4.3(a). Note that the direction of shift is controlled by the waveforms 'shift right' (SR) and 'shift left' (SL), which must be set appropriately before the actual shift pulse occurs. The number of places the contents of the registers are shifted left or right is controlled by the number of shift pulses applied to the circuit. With this type of shift register on a right shift operation the least significant bit is lost and the most significant bit set to zero; the reverse happens on a left shift operation.

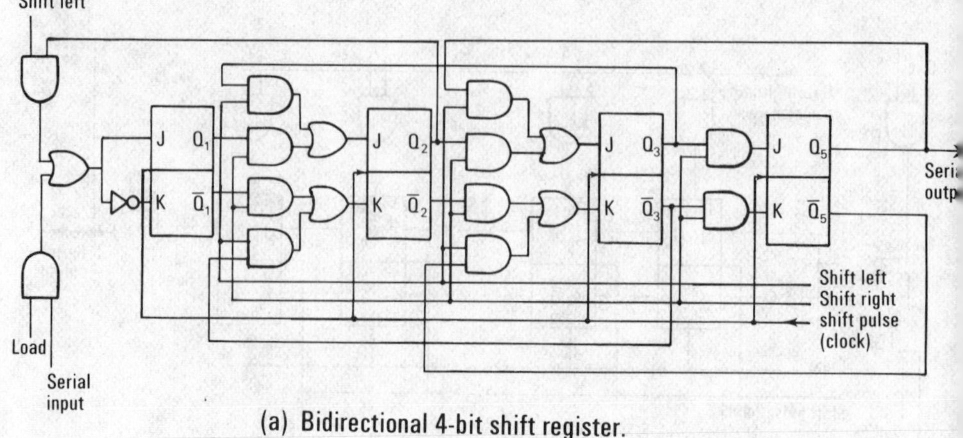

(a) Bidirectional 4-bit shift register.

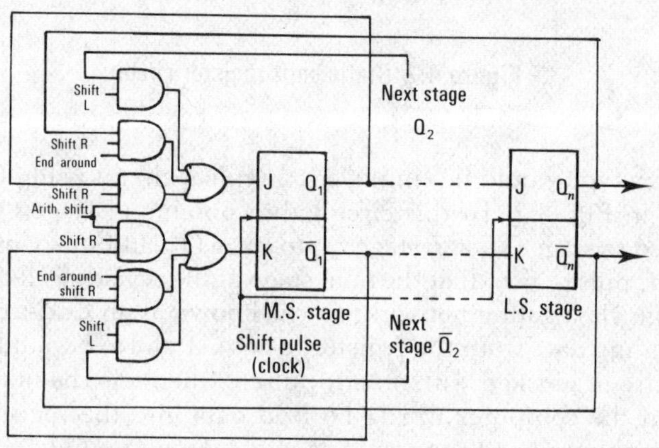

(b) Combined end-around and arithmetic shift register.

Figure 4.3 Special purpose shift registers.

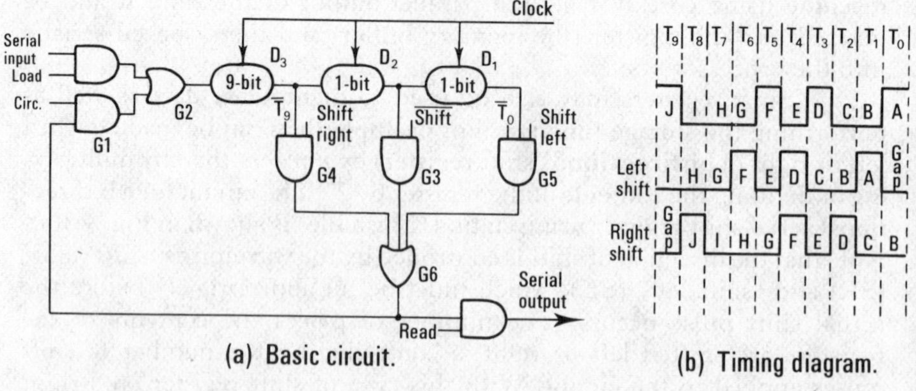

(a) Basic circuit.

(b) Timing diagram.

Figure 4.4 Dynamic shift register.

As we have seen in Chapter 2, there are basically three modes of shifting required in a digital computer, these are:

(a) a logical shift, when the most significant or least significant digit stages are filled with zeros;
(b) an end-around or circular shift, where the MSD is inserted in place of the LSD on a left shift operation, and vice versa;
(c) the arithmetic shift when the sign digit of the number held in the register is set into the MSD position on a right shift operation.

All these operations may be implemented using the basic shift register circuit, modified as shown in Fig. 4.3(b). Note the inclusion of additional gating on the MSD stage to enable both circular and arithmetic shift operations to take place.

The same shifting functions may be performed using dynamic shift registers, for example MOS circuits, but the logic circuit is slightly more complicated because the information flow through dynamic registers is usually unidirectional. In these circuits shifting is achieved by increasing or decreasing the normal one wordlength (n-bit) delaying path through the circuit by 1 bit time, to obtain a left or right shift respectively. The logic schematic and timing diagram for a 10-bit dynamic shift register is shown in Fig. 4.4; a tabulated description of its operation is given in Table 4.1. The normal circulation path when there is no shift order is via gates G1, G2, through the 9- and 1-bit delay elements and

Table 4.1
Dynamic shift register

(a) Shift left 1 place

Shift register contents								D_3	D_2	D_1	*Timing period*
J	I	H	G	F	E	D	C	B	A	J	T_0
gap	J	I	H	G	F	E	D	C	B	A	T_1
H	G	F	E	D	C	B	A	gap	J	I	T_9
I	H	G	F	E	D	C	B	A	gap	J	T_0

(b) Shift right 1 place

Shift register contents								D_3	D_2	D_1	*Timing period*
J	I	H	G	F	E	D	C	B	A	J	T_0
B	J	I	H	G	F	E	D	C	B	A	T_1
J	I	H	G	F	E	D	C	B	J	I	T_9
gap	J	I	H	G	F	E	D	C	B	J	T_0

gates G3 and G6, back to gate G1. During this time (that is, one wordtime T_0–T_9) the control waveforms SL, SR, shift, input, and output are all at zero, with the circulate waveform at one. For a normal circulation (that is, a storage cycle) the least significant digit of the word appears in time T_0 (see Fig. 4.4(b)) at the output of delay D_2 and gate G3. Note also from Table 4.1 that at this time delay D_1 contains the most significant digit, that is, J.

When a shift left operation is performed the waveforms SL and shift are made equal to one in time T_0 and stay up for one wordtime (T_0–T_9). This causes gate G3 to be inhibited and gate G5 enabled, thus giving an 11-digit circulating path including the additional delay stage D_1. Note that the left shift gate G5 is inhibited in time T_0, thus inserting a zero in the least significant position. Similarly, for the right shift operation, waveforms SR and shift go to one for one wordtime giving a 9-digit circulation path through gate G4. In this case gate G3 is inhibited in time T_9 if a logical shift is required; alternatively, for an arithmetic shift the sign digit (J) may be reproduced by opening the normal circulation in time T_9.

Operands stored in the parallel mode may be shifted left or right using a **two-rank** parallel shift register circuit, employing d.c. set–reset bistables (Fig. 4.5). The circuit consists of two parallel registers with interconnecting gates, one of which is used as a buffer. The buffer register is necessary since, unless external delays are used, it is not possible to use the output of a d.c. bistable (to set another stage) at the same

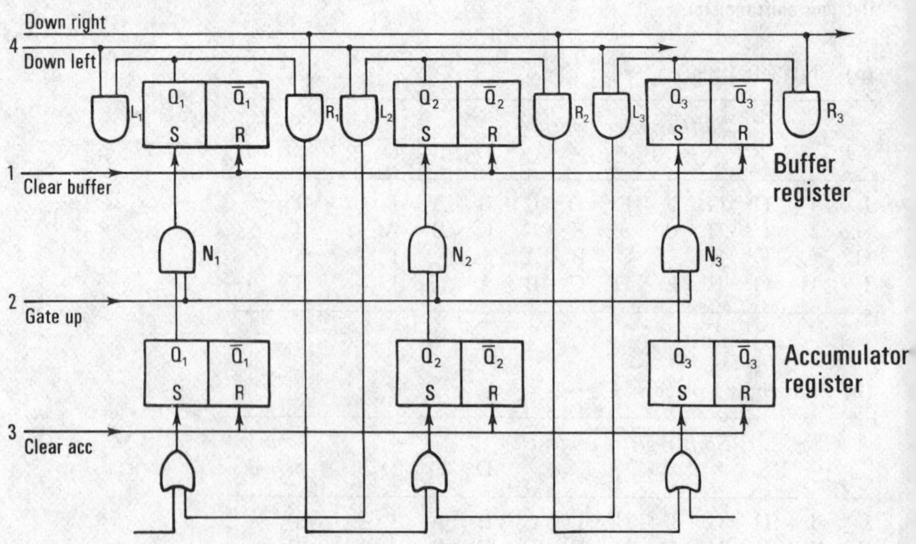

Figure 4.5 Two-rank parallel shift register.

time as its input is being changed. The shifting function requires four sequential operations to take place:

1 The stages of the buffer register are reset to zero; that is, they are cleared.
2 Gates N are enabled, allowing the contents of the accumulator to be set (unshifted) into the buffer register.
3 The accumulator register is cleared.
4 Gates L or R are enabled according to whether a left or right shift is to be performed. This allows, for example, on a right shift, the Q_1 output of the buffer register to be set into the Q_2 stage of the accumulator.

In order to economize on interconnections and gates only the Q outputs are transferred; this requires the registers to be reset to zero before setting in the actual data. Note also that only one place is shifted each time, and that a control unit must be provided to generate the required number of control waveform sequences.

The circuit described above would be used as a wired-in shifting circuit in a special-purpose parallel system. In a parallel digital computer the same function would be performed but using registers already available as part of the central processor structure. The shifting would be effected by taking the output of the accumulator register via a common parallel highway and shifting gates (shown in Fig. 4.6) to a temporary buffer register, and then back again through the gates to the accumulator. Each pass through the shifting gates shifts the operand one place left or right. Again provision must be made for overall control of the operations.

Other essential parts of the control unit are the **counter circuits** which are used in conjunction with a clock generator and decoding circuit to produce the system timing waveforms. A simple asynchronous (ripple through) counter circuit is shown in Fig. 4.7 together with its waveform diagram. This form of counter has the disadvantage of introducing delays between successive outputs owing to the signal rippling through the stages. For example, when counting from 011 → 100 the change in output from stage A is used to trigger stage B, the output of which triggers stage C to produce the final output. It is obvious that the outputs A, B, C cannot change together and that they are accumulatively delayed by the switching time of the bistable devices. As a consequence, when the counter outputs are decoded to produce timing waveforms, for example, because the edges of the waveforms are not coincident troublesome voltage spikes are produced which could easily switch a following device; Fig. 4.7(c) shows diagrammatically how this can occur. More serious, though, is the fact that the decoded pulses are delayed by varying amounts with respect to the input. If the counter is being used to generate control waveforms, with the system clock as input, the inherent delays make them useless for gating waveforms synchronized to the clock.

One method of overcoming this problem is to use a synchronous counter circuit (see Fig. 4.7(d)) in which the input is taken in parallel to

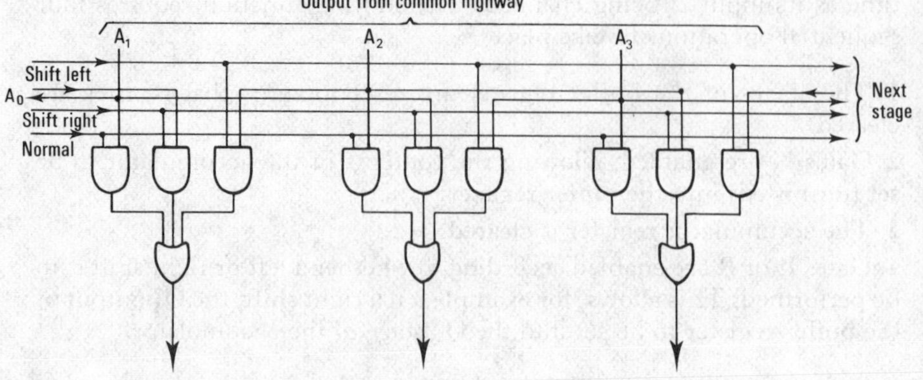

Figure 4.6 Parallel shifting gates.

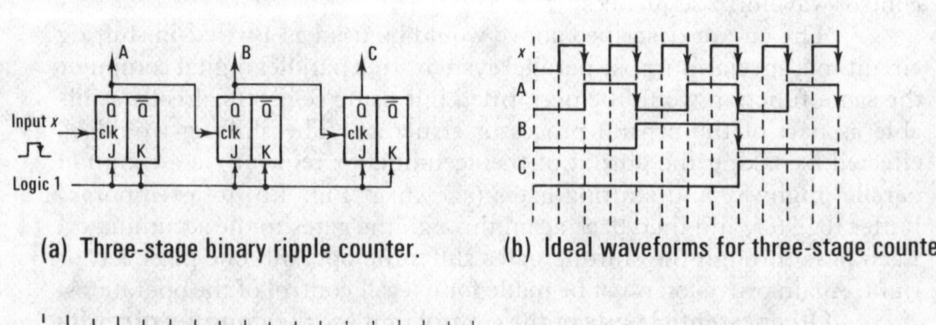

(a) Three-stage binary ripple counter.

(b) Ideal waveforms for three-stage counter.

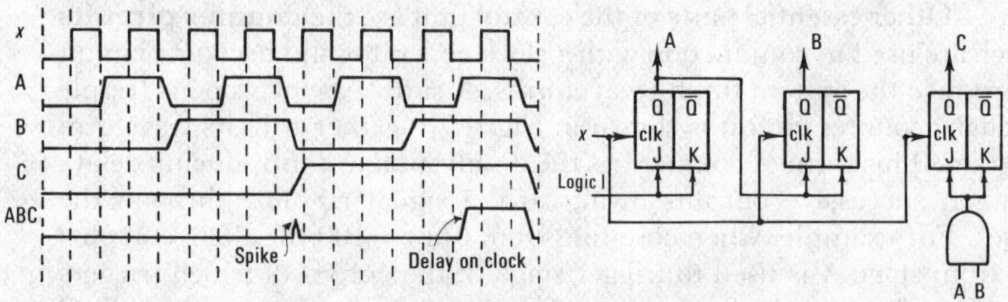

(c) Real waveforms for three-stage counter.

(d) Three-stage binary synchronous counter

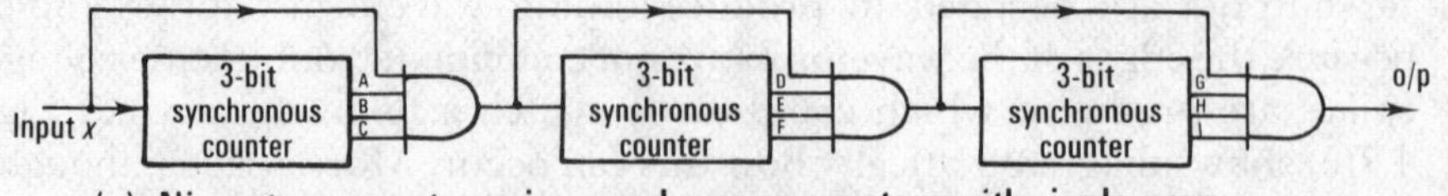

(e) Nine-stage counter using synchronous counters with ripple-carry.

Figure 4.7 Binary counter stages.

each stage. With this type of circuit coherent outputs are produced (since all bistables switch at the same time) which may be decoded without generating delayed waveforms. They also have the advantage of being easily designed, using switching theory[4], for any radix. Unfortunately they require more logical inputs (and hence logic gates) which rapidly increase with the number of stages. For instance, in a 9-bit counter the required JK inputs are:

1 A AB ABC ABCD ABCDE ABCDEF
ABCDEFG ABCDEFGH

This rather excessive amount of hardware may be reduced to some extent (without forfeiting the coherent outputs) by using, for example, 3-bit counter stages with 'carry–bypass' gates and ripple carry between stages; the circuit is shown in Fig. 4.7(e). The action of the counter is such that if a stage is registering all 1's the input is immediately applied, via the carry–bypass gate, to the next stage. An alternative method of generating delay-free counter outputs, which may also be gated together without producing spikes, is to use a **master–slave** counter (see Fig. 4.8) or a **Gray-code counter** (Fig. 4.9) both of which produce out-of-phase outputs. In the case of the Gray-code counter it is possible to decode all the outputs without using any waveforms with coincident edges; this is shown in Fig. 4.9(c).

Another type of counter circuit which is used extensively in coding and timing applications is the **chain-code counter**.[5] This is basically a shift-register circuit with selected outputs added modulo 2 (the exclusive OR operation) and fed back to the input. It produces a pseudo-random bit sequence with a period of $2^n - 1$, and the same number of unique parallel output combinations, where n is the number

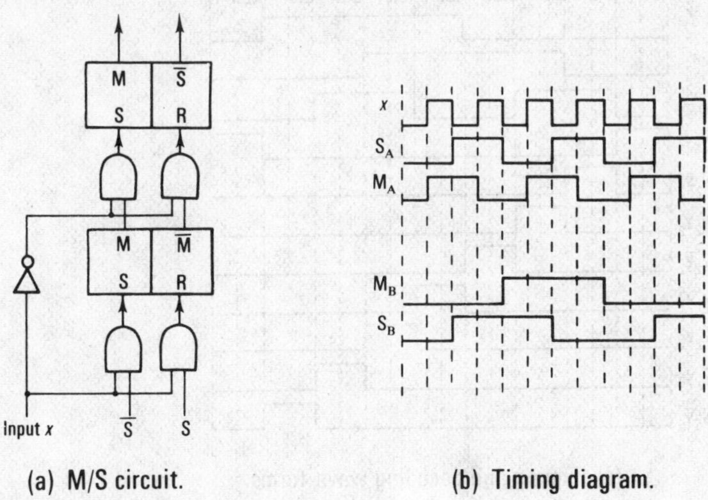

(a) M/S circuit.

(b) Timing diagram.

Figure 4.8 Master-slave binary counters.

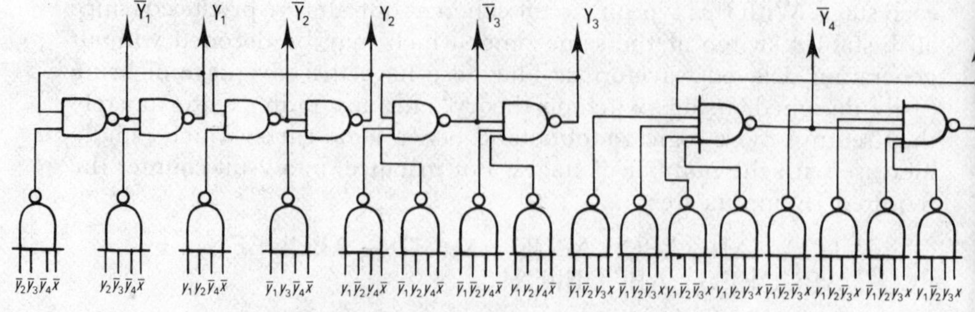

(a) Asynchronous three-bit Gray-code counter

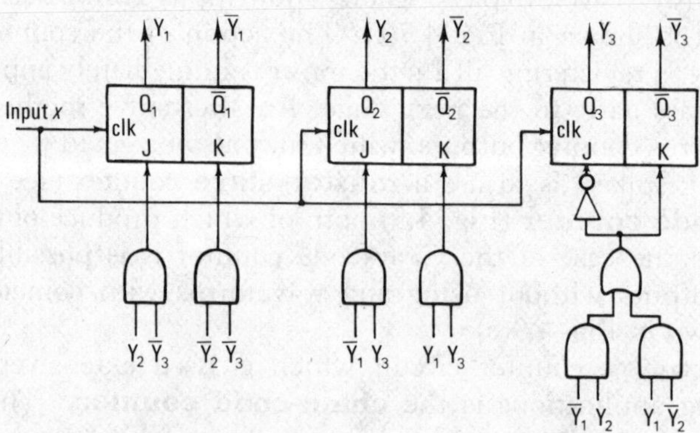

(b) Synchronous three-bit Gray-code counter

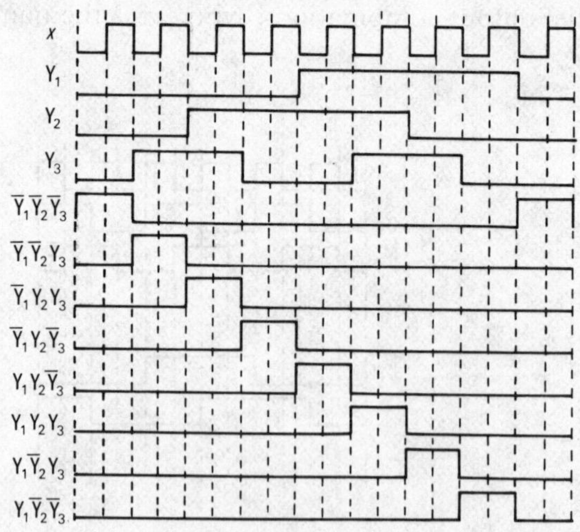

(c) Timing and decoding wave-forms

Figure 4.9 Gray-code counters.

Table 4.2
Chain code sequences

	D_1	D_2	D_3	D_4
Starting state	0	0	0	1
	1	0	0	0
	1	1	0	0
	1	1	1	0
	1	1	1	1
	0	1	1	1
	1	0	1	1
	0	1	0	1
	1	0	1	0
	1	1	0	1
	0	1	1	0
	0	0	1	1
	1	0	0	1
	0	1	0	0
	0	0	1	0
Repeats	0	0	0	1

of register stages. A typical circuit is shown in Fig. 4.10; note that only D-type bistables (or JK shift register stages) and exclusive OR circuits are required in the implementation. This type of circuit is also autonomous (that is, no external inputs are required except for clock) and will start automatically (with clock) provided an initial non-zero pattern is set into the register stages. The four-stage circuit shown in Fig. 4.10 generates the repetitive serial output sequence 100011110101100, and the states of the shift register stages (that is, the parallel outputs) on each clock pulse are shown in Table 4.2.

Dividers are counter type circuits which are used to divide a serial input by a fixed number, for example ÷3, ÷5, ÷10 etc. The basic dividers may also be cascaded to give division by any number – for

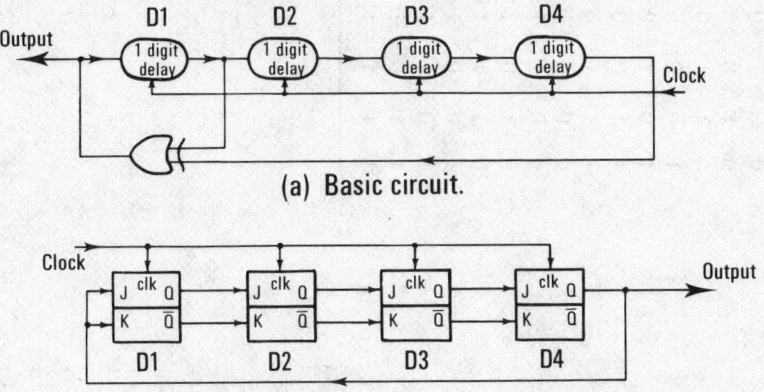

(a) Basic circuit.

(b) JK version–exclusive OR performed between input and output of JK.

Figure 4.10 Chain code counters.

Table 4.3
3-bit up/down counter

(a) Transition table

C	B	A	S	C_+	B_+	A_+
0	0	0	0	0	0	1
0	0	1	0	0	1	0
0	1	0	0	0	1	1
0	1	1	0	1	0	0
1	0	0	0	1	0	1
1	0	1	0	1	1	0
1	1	0	0	1	1	1
1	1	1	0	0	0	0
0	0	0	1	1	1	1
0	0	1	1	0	0	0
0	1	0	1	0	0	1
0	1	1	1	0	1	0
1	0	0	1	0	1	1
1	0	1	1	1	0	0
1	1	0	1	1	0	1
1	1	1	1	1	1	0

(b) K-Maps

AS \ CB	00	01	11	10
00			×	×
01	1		×	×
11			×	×
10		1	×	×

$J_C = AB\overline{S} + \overline{A}\overline{B}S$

AS \ CB	00	01	11	10
00	×	×		
01	×	×		1
11	×	×		
10	×	×	1	

$K_C = AB\overline{S} + \overline{A}\overline{B}S$

AS \ CB	00	01	11	10
00		×	×	
01	1	×	×	1
11		×	×	
10	1	×	×	1

$J_B = \overline{A}S + A\overline{S}$

AS \ CB	00	01	11	10
00	×			×
01	×	1	1	×
11	×			×
10	×	1	1	×

$K_B = \overline{A}S + A\overline{S}$

AS \ CB	00	01	11	10
00	1	1	1	1
01	1	1	1	1
11	×	×	×	×
10	×	×	×	×

$J_A = 1$

AS \ CB	00	01	11	10
00	×	×	×	×
01	×	×	×	×
11	1	1	1	1
10	1	1	1	1

$K_A = 1$

(a) Divide by three.

(b) Divide by six.

(c) Divide by five.

(d) Divide by ten.

(e) Divide by seven.

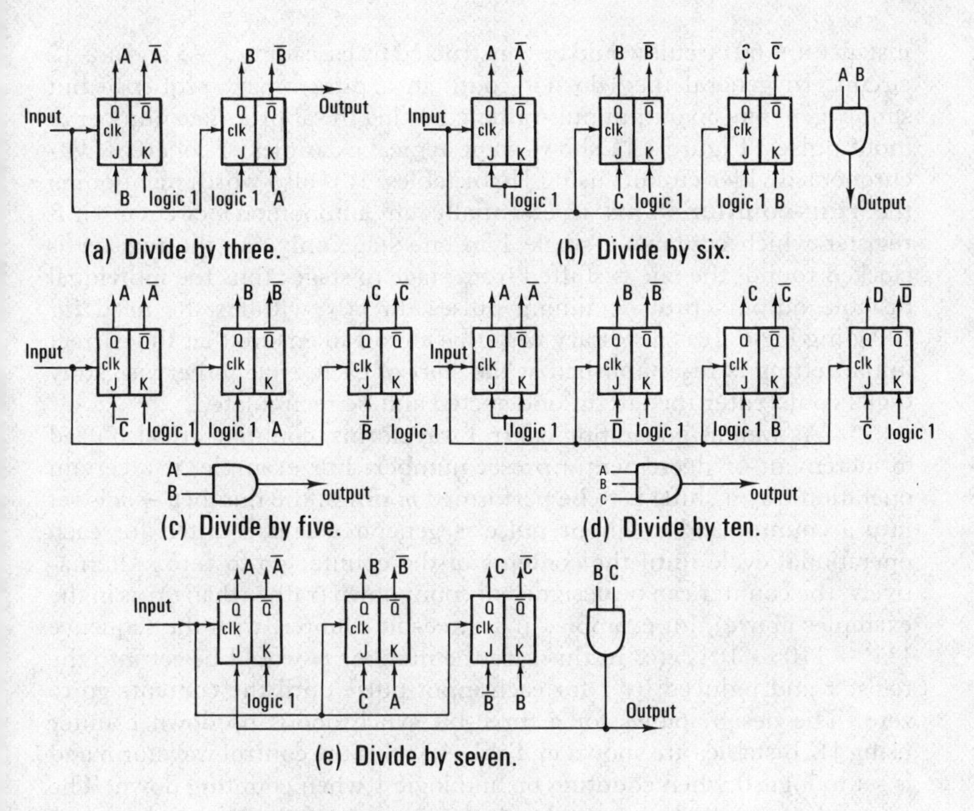

Figure 4.11 Coherent divider counters.

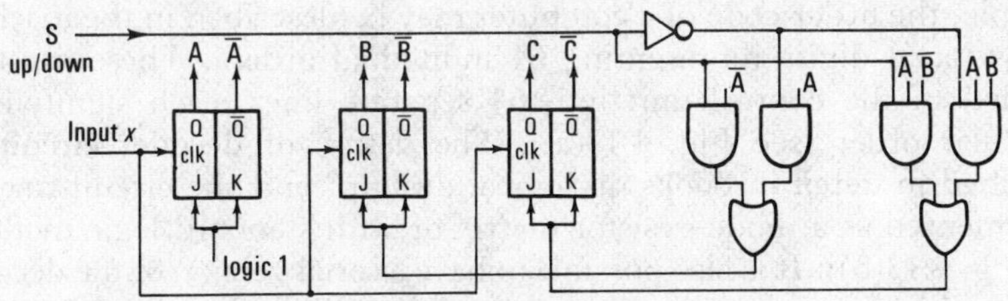

Figure 4.12 Three-bit up/down counter.

instance a ÷60 circuit would be constructed by cascading a ÷5 and a ÷12 circuit. In general they do not count in a pure binary sequence but simply generate an output pulse after counting the appropriate number of input pulses. Figure 4.11 shows some typical examples of coherent synchronous divider circuits using JK bistables. It is also worth mentioning the **ring-counter**. This is essentially an autonomous circular shift register which contains a single 1 in one stage only. As the register is clocked round, the one is shifted from stage to stage; thus the individual bistable outputs provide timing pulses directly without the need for decoding logic. It is necessary with this circuit to ensure that the correct initial setting is re-established at the start of each cycle, otherwise stray digits could enter the circuit undetected and be recirculated.

As well as generating control waveforms, counters are also used to increment or decrement a preset number. For example, if a certain operation (say a shift) is to be performed m times, the number $-m$ is set into a counter and an input pulse is generated and counted for each operational cycle until the contents of the counter go to zero. Alternatively, the counter can be designed to count down (rather than up as in the examples above), for example, in a three-bit counter, with the sequence 111 → 110 → 101, etc. In this case the number m would be set into the register and reduced by 1 for each input pulse until the contents go to zero. The design tables for a three-bit synchronous up/down counter using JK bistables are shown in Table 4.3; S is the control waveform and is set to logic 0 when counting up and logic 1 when counting down. The counter input which goes to the clock terminals of the JK bistables (see Fig. 4.12) may be ignored in the design tables since a transition only occurs when the clock input goes to one.

A **binary decoder circuit** (called a '1-out-of-M' circuit) is one with m input lines and 2^m output lines and is so designed that for each of the possible 2^m input conditions only one output line is energized. For example, the order code of a computer may be described in the machine by two octal digits representing 64 individual orders. These must be decoded in the control unit into 64 separate lines, each signifying a particular order (see Fig. 4.13(a)). The design of decoder circuits is described in detail in books on logical design,[6] and the circuit may be implemented as a diode–resistor matrix or with standard logic modules (see Fig. 4.13(b)). It is also possible to have a serial version of the decoder circuit which performs the operation of decoding a serial m-bit binary number into 2^m separate lines with the individual outputs occurring in the most significant digit time and lasting for one clock pulse. A three-bit serial decoder circuit is shown in Fig. 4.14; the clocked JK shift-register is used to delay and separate the digits by 1 digit time before they are gated (with clock pulse T_2) in the usual way to give the individual output pulses.

A binary encoder circuit translates or converts from one code to another, for example from binary-coded decimal to pure binary. The design of such circuits is a straightforward exercise in combinational logic design.

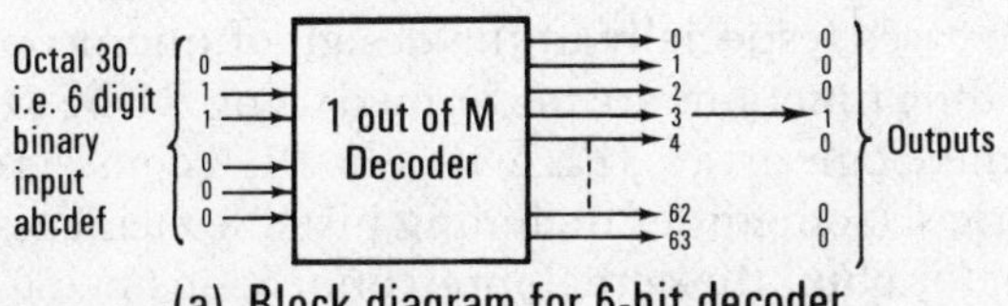

(a) Block diagram for 6-bit decoder.

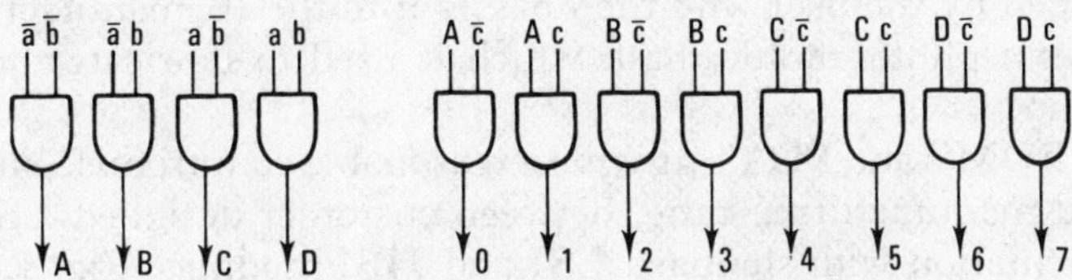

(b) Three-bit decoder using cascaded AND gates.

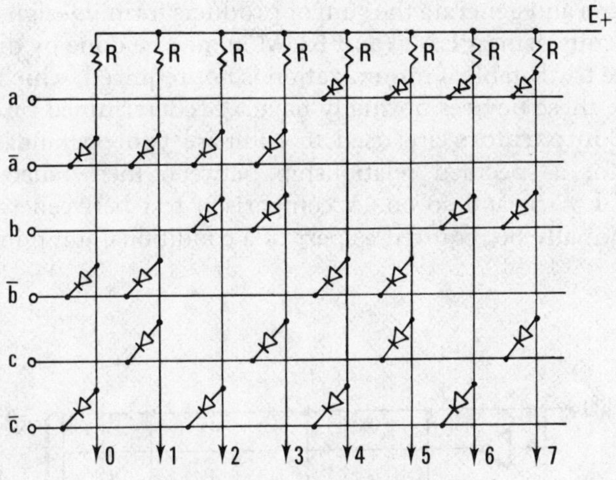

(c) Three-bit diode matrix decoder.

Figure 4.13 Decoding circuits.

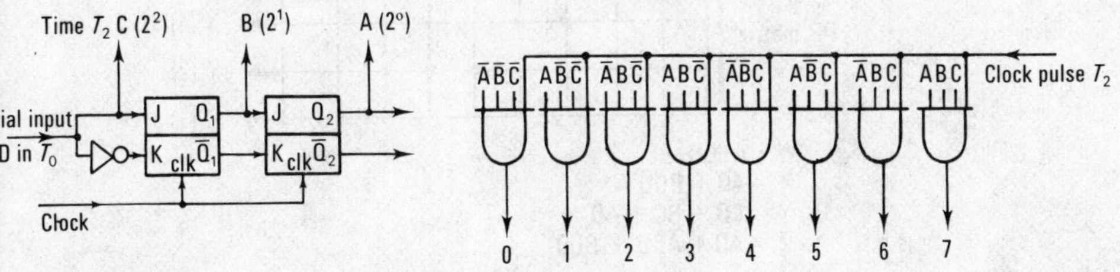

Figure 4.14 Three-bit serial decoder.

In many cases (especially in the design of microcomputers) decoding and encoding functions are performed using ROM's (see Chapter 6) and programmed logic arrays (PLA's). The PLA consists of combinational logic matrices (sometimes including bistable elements) fabricated on a semiconductor chip, the actual interconnections being determined during the final processing steps. The required interconnection pattern is specified by the user who then passes it to the IC manufacturer who produces a photographic mask which is used to fabricate the specific PLA chip.

ROM's and PLA's provide a reasonable compromise, in terms of cost and manufacturing time, between customer designed LSI and an implementation with standard LSI and MSI modules. Fig. 4.15 shows a PLA realization for a binary to 5421 code encoder; note that the matrices are effectively ROM circuits (normally using the same MOS technology) and generate the sum of products form. Design of combinational circuits using PLA's (and ROM's) may be done by direct inspection of the truth-table if minimization is not required, which is often the case since these devices normally have a predetermined bit capacity.

Comparators are used to compare two operands (A and B) looking for a specified relationship between them, such as $A \geqslant$ B, $A = B$, $A \neq B$, and so on. A comparison test between two operands would normally be required as part of a conditional jump instruction or

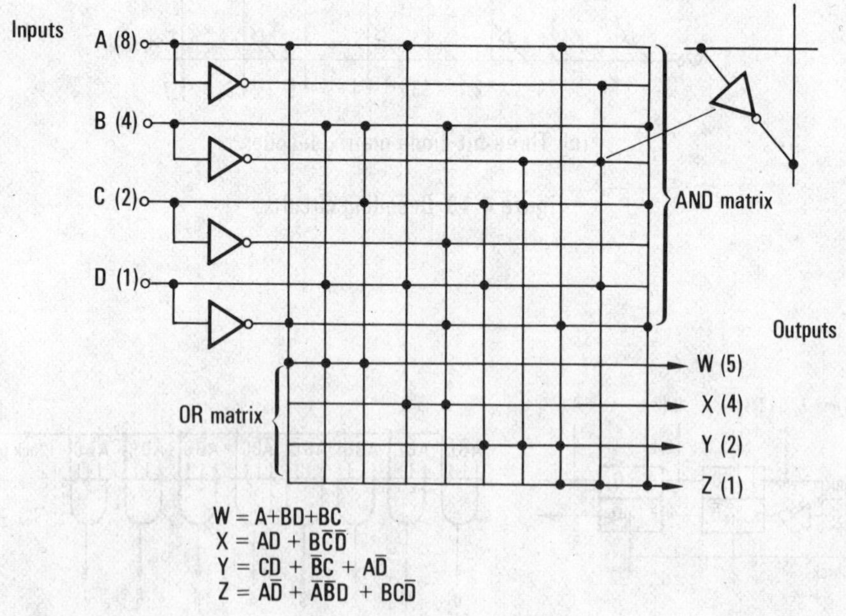

Figure 4.15 PLA decoder.

to determine loop conditions in a micro-program or arithmetic operation. Logical comparisons are also included in some computers, for example $A \supset B$ would be true if B contained 1's only in those places where there was also a 1 in A (there could be 1's in A where there are 0's in B). The detection of these conditions can be performed using micro-programmed mask or collate operations followed by arithmetic checks, or by special logic circuitry connected between the operand registers.

Two parallel operands may be compared for equality by using the relationship:

$$E = (A_0B_0 + \bar{A}_0\bar{B}_0)(A_1B_1 + \bar{A}_1\bar{B}_1)\ldots(A_nB_n + \bar{A}_n\bar{B}_n)$$

The function $AB + \bar{A}\bar{B}$ is in fact the inverse of the exclusive OR function, that is, $\overline{\bar{A}B + A\bar{B}} = (A + \bar{B})(\bar{A} + B) = AB + \bar{A}\bar{B}$, consequently the circuit may be implemented using AND/NOR logic as shown in Fig. 4.16(a). This circuit is also known as a **coincidence** unit. Serial operands may be compared using the circuit shown in Fig. 4.16(b). The operands A and B (which may be of any wordlength) are entered to the circuit least significant digit first, the state of the JK bistables indicating when $A = B$, $A > B$, or $A < B$. Table 4.4 shows the state-table and corresponding assignment for the circuit.

Timing circuits consist of a standard frequency oscillator (usually crystal controlled), a pulse shaper or squarer, a power amplifier,

Table 4.4
Serial comparator circuit
(a) State-table

	Inputs AB							
	Next state				*Output Z_1Z_2*			
Present state	00	01	11	10	00	01	11	10
1	1	3	1	2	00	01	00	10
2	2	3	2	2	10	01	10	10
3	3	3	3	2	01	01	01	10

(b) Assigned table

		Inputs AB							
Present state		*Next state*				*Output Z_1Z_2*			
	CD	00	01	11	10	00	01	11	10
1	00	00	01	00	10	00	01	00	10
2	10	10	01	10	10	10	01	10	10
3	01	01	01	01	10	01	01	01	10
Unused	11	×	×	×	×	×	×	×	×

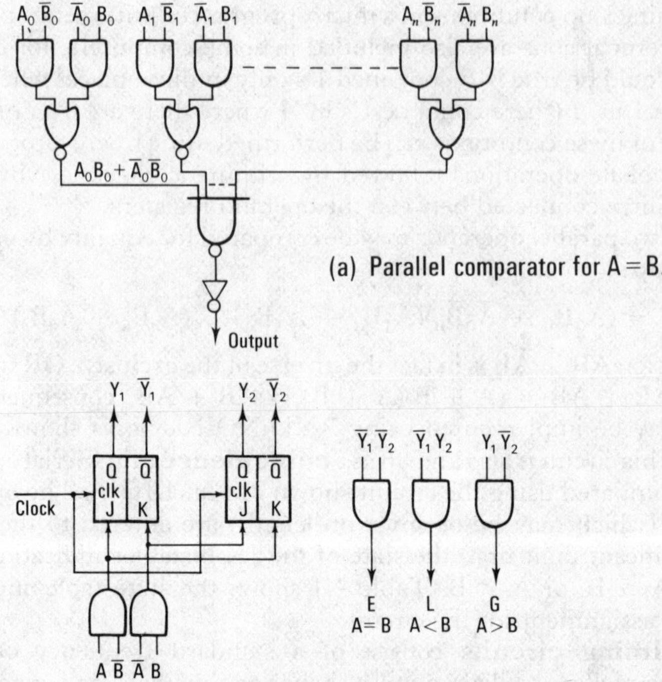

(a) Parallel comparator for A = B.

(b) Serial comparator for A = B, A < B and A > B.

Figure 4.16 Comparator circuits.

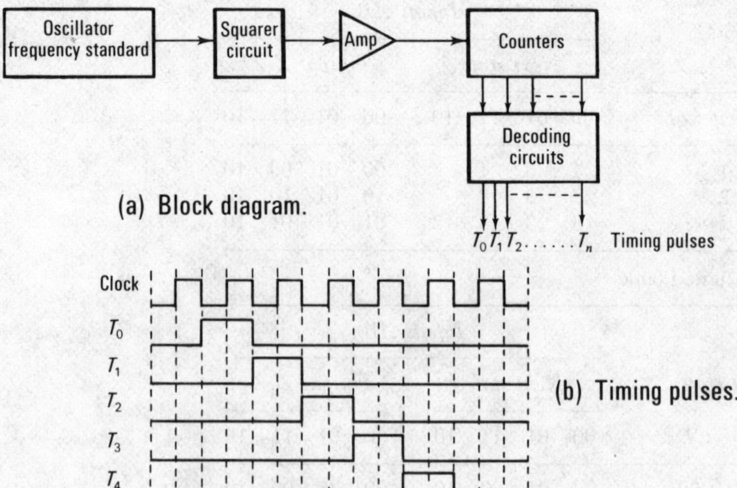

(a) Block diagram.

(b) Timing pulses.

Figure 4.17 Timing circuits.

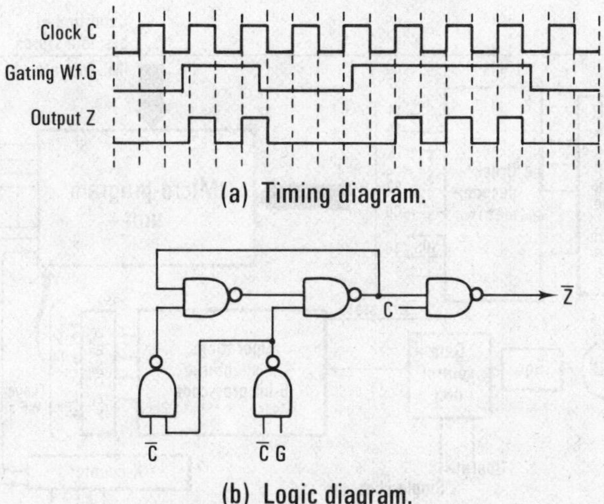

Figure 4.18 Gated-clock circuit.

and suitable counting and decoding logic (see Fig. 4.17(a)). The function of the circuit is to generate the individual timing pulses $T_0, T_1, \ldots, T_n$; these are shown in Fig. 4.17(b).

In some applications, to overcome circuit delays or to speed up the operations by spacing the control waveforms at half clock pulse periods, a 'two-phase' clock (that is 180° out of phase) is used. Another technique, used in compensating for circuit delays, is to generate progressively delayed versions of the clock waveform by passing the basic clock through a lumped constant delay line with appropriately tapped half-sections.

The clock waveform is often required to be gated before and after distribution to the system. Figure 4.18 shows a suitable NAND circuit to perform this function; note that though the gating waveform can be of variable duration and may occur at any time, only full-width clock pulses appear at the output.

4.5 The control unit

A block diagram of a typical control unit for a parallel computer is shown in Fig. 4.19. Note that this comprises an **order decoding matrix**, **micro-program unit**, **loop counter K**, and **timing control**, with the latter unit consisting of a **clock generator** and **operations counter**. The order decoding matrix decodes the instruction word and produces a separate output for each order.*

* For example, the most significant five bits would generate an output for each of the 17 specified address orders, while the octal combinations, 00 and 10, would be used in conjunction with the remaining bits to produce outputs for the zero-address and input/output-shift orders respectively.

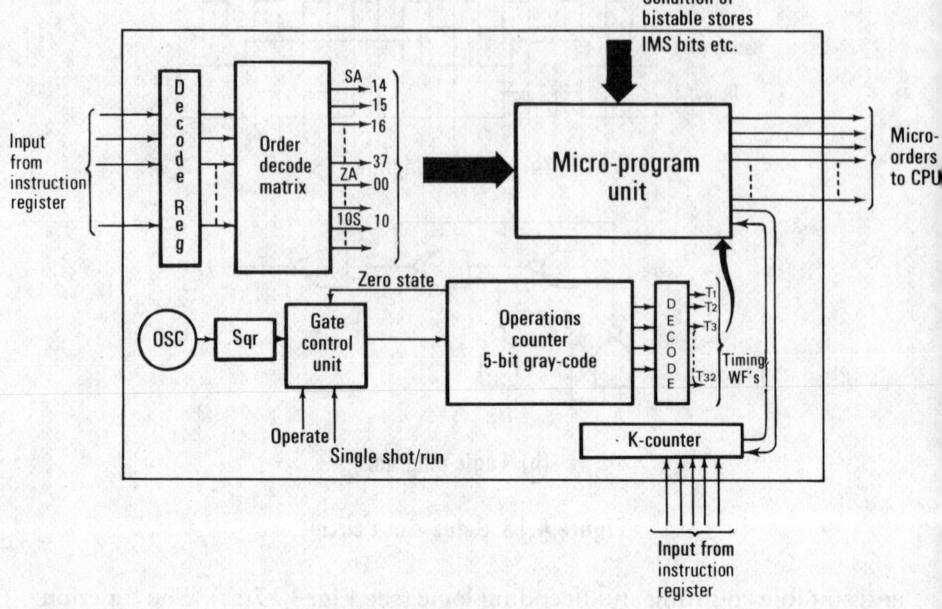

Figure 4.19 Block diagram of control unit.

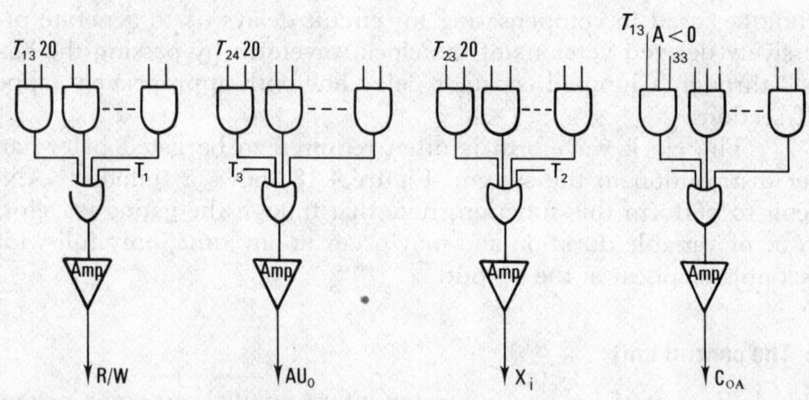

Figure 4.20 Micro-program circuits.

With the basic clock rate (typically 1–10 MHz) as input, the 5-bit Gray-code operations counter with decoding matrix produces 32 pulses or **timing-slots**. To execute an instruction the decoded order bits are used to select and generate gating waveforms in a specified time sequence (determined by the timing unit) and dictated by the relevant algorithm. Thus the gating waveforms are used to control the necessary logical operations (such as opening and closing gates, setting bistables, and so on) required to carry out an instruction. The loop counter is used for controlling the repetitive steps of an algorithm, for example in multiplication and shift instructions.

These internal machine operations are called **micro-orders**, and they can be represented symbolically and used to describe the operations necessary in the execution of a machine-code instruction; a collection of such orders is called a **micro-program**. Thus a computer would have a basic set of micro-orders and each machine-code instruction would be described by a suitable sub-set of this basic micro-order set. A typical but simplified set of micro-orders with their symbolic representation is shown in Table 4.5. The micro-orders and their function are in general self-explanatory, each order corresponding to a distinct waveform in the central processor; the structure of the computer is identical with that described in Chapter 1 for an 18-bit wordlength parallel machine shown in Fig. 1.4.

Table 4.6 shows a micro-program utilizing the basic micro-orders for the machine code instruction: 'Add the contents of the address specified to the accumulator register'. The two-beat (read/obey) nature of the computer is obvious from the micro-program, the read portion being common to all instructions. Note that the operation is synchronous, the micro-orders occurring in successive time-slots each, for example, of 100 ns duration (it is also possible to have asynchronous operation with the next micro-orders commencing when the present ones have finished). In timing-slot 1 the next instruction to be obeyed is read down from store by gating the contents of the control register (program counter) to the store address highway and initiating a read/write cycle. Meanwhile the contents of the control register are incremented by +1 (in time-slot 2) by transferring the contents of the control register to the X-register via the common highway, and setting the least significant digit stage (carry input) of the adder to +1; note that the X-register is permanently connected to the adder. The operation is completed in time-slot 3 when the output of the arithmetic unit is transferred to the control register via the common highway. After this addition process nothing happens until time-slot 11 when the read/write store cycle is completed (assumed to be 1 μs) and the new instruction word is placed in the M-register. Thus in time-slot 11 the contents of the M-register are transferred to the instruction register and the read cycle is now complete.

The obey cycle commences in time-slot 12 with the order digits of the instruction word being decoded by the order decoding matrix. The output from the decoder generates the necessary micro-orders to address

Table 4.5
Typical micro-orders

M_i	Input to M register from common highway
M_o	Output from M register to common highway
M_{AU}	Output from M register to arithmetic highway
$\overline{M}_{AU}$	Output complement of M-register to arithmetic highway
$\overline{M}_O$	Output complement of M-register to common highway
MD_o	Output from modifier register to common highway
MD_i	Input to modifier register from common highway
MD_{AU}	Output from modifier register to common highway
E_i	Input to control register from common highway
E_o	Output from control register to common highway
C_i	Input to instruction register from common highway
C_o	Output from instruction register to common highway
C_{OA}	Output address digits from instruction register to common highway
A_i	Input to accumulator A from common highway
A_o	Output from accumulator A to common highway
$\overline{A}_O$	Output complement of accumulator to common highway
X_i	Input to X-register from common highway
X_o	Output from X-register to common highway
IO_i	Input to input/output register from common highway
IO_o	Output from input/output register to common highway
IT_o	Output from interrupt register to common highway
C_{AD}	Output address digits from instruction register to address highway
C_{ADK}	Output the 5 least significant address digits from instruction register to K counter
E_{AD}	Output from control register to address highway
L	Open left shift gates
R	Open right shift gates
2^{-17}_{AU}	Gate +1 to the least significant stage of the adder logic
AU_o	Sum output of arithmetic unit to common highway

Table 4.6
Micro-program for the instruction
'add contents of the address specified to ACC'

Timing (100ns)	*Micro-order*	
1	E_{AD}, R/W	Read cycle
2	E_oX_i, 2^{-17}AU	
3	AU_oE_i	
11	M_oC_i	
12	C_{OD}	Obey cycle
13	C_{AD}, R/W	
23	A_oX_i, M_{AU}	
24	AU_oA_i, END	

C_{OD}	Output from instruction register to order decoding matrix
M_C	
E_C	
C_C	
A_C	Clear (or reset) appropriate registers to zero
X_C	
K_C	
IO_C	
Write	Initiate main store read/write cycle but inhibit read strobe
R/W	Initiate main store read/write cycle
END	End micro-program and reset operations counter
Rep n	Reset operations counter to timing-slot n to repeat micro-order n
K_0	Decrement contents of K counter by +1
K_1	
K_2	
K_4	Set K counter to number specified
K_8	
K_{16}	
I_{AD}	Gate wired-in interrupt link location to address highway
E_I	Set control register to address of interrupt routine
OF_TC	Clear overflow trigger
E_TC	Clear error trigger
M_TC	Clear modifier trigger
I_TC	Clear interrupt trigger
E_TS	Set error trigger
M_TS	Set modifier trigger
I_TS	Set interrupt trigger
I_M	Inhibit Q outputs of M from common highway (AND)
$\bar{I}_M$	Inhibit Q outputs of M from common highway (OR)

Table 4.7

Typical micro-program conditions

Accumulator register $= 0$

Accumulator register $\geqslant 0$ sign digit $= 0$

Accumulator register < 0 sign digit $= 1$

$M_T = 1$	modifier trigger
$I_T = 1$	interrupt trigger
$E_T = 1$	error trigger
$OF_T = 1$	overflow trigger
$K = 0$	state of K counter
$K \neq 0$	
R/W = 1	read/write cycle in progress
R/W = 0	read/write cycle not in progress
ID = 1	Indirect address bit
S = 1	Sector bit

the store with the instruction register address digits and initiate a read/write cycle in time-slot 13. Once again the micro-program must pause waiting for the completion of the store cycle until time-slot 23 when the contents of the accumulator are transferred to the X-register and the output of the M-register is gated directly to the arithmetic unit. In time-slot 24 the output of the arithmetic unit is transferred to the accumulator and the operations counter reset to zero ready for the start of another read/obey cycle. One method of generating the micro-orders is shown in Fig. 4.20; the gating circuits may be implemented using a diode resistor matrix, NAND/NOR logic, or read-only storage (see next section).

The timing of order decoding and execution in a computer is intimately related to the cycle time or 'access time' of the main store. Suppose for example we have a ferrite core store which has a destructive read-out (see Chapter 6) necessitating a read/write cycle time of say 1 μs, then the basic timing of the micro-program unit would be chosen to be 100 or 50 ns pulses to synchronize with the store. As each instruction in a single address machine requires at least one visit to the store, and instructions such as add, subtract, store, and fetch need two visits, it is apparent that the limiting factor as far as speed is concerned (assuming the basic logic elements can switch in nanoseconds) is the store access time. If in the micro-program example above a store with 100 ns read/write cycle time was used, the execution time for the add instruction could be reduced from 2·4 μs to 0·8 μs, since it would no longer be necessary for the micro-program to wait for the completion of a read/write cycle. This increase of processor speed by simply substituting a faster store unit (without modifying the basic logic) has led to the concept of fast store options in many computer systems.

In addition to considering the timing and the specific algorithm for a machine-code order in establishing a micro-program, it is also necessary to take into account the *condition* of various bistables and switches within the machine logic. These would include, for example, overflow and error indication, modifier and interrupt flags, and the sign of the accumulator register. A typical set of micro-program conditions is shown in Table 4.7. As an example of the use of conditions consider the micro-program for a conditional jump instruction shown in Table 4.8. In this case it is necessary to take into account the state of the accumulator, whether zero, positive, or negative, before performing the appropriate steps of the micro-program. Thus, from Table 4.8, in timing-slot 13 if $A < 0$ (that is, negative) the address digits of the instruction register are transferred to the control register; if $A \geqslant 0$ there is no further action. In practice the inclusion of the condition terms simply means gating the condition with the relevant timing and order waveforms in the micro-order logic, see for example E_i in Fig. 4.20.

One further facility, and an essential one, which must be incorporated into the control unit is the ability to repeat a certain set of micro-orders a specified number of times. This necessitates providing a counter (in this case the K-counter) the condition of which ($K = 0$,

K $\neq$ 0) determines whether or not a 'jump back' to earlier steps in the micro-program will take place. The jump operation is obtained by resetting the operations counter to *n* (in Gray-code) where *n* is the time-slot required to be repeated. Thus when the next clock pulse occurs the operation counter will cause timing-slot *n* to reappear, allowing the micro-order to be repeated; the jump operation is represented in Table 4.5 by the micro-order Rep *n*.

As an example of its use let us consider the micro-program for instruction 200, that is, 'shift the contents of the accumulator left *n* places', shown in Table 4.9. The parallel shift operation requires the use of a buffer register and in this example the X-register performs this function. As a consequence of this, however, an odd number of shifts will leave the result in the X-register and it must then be transferred to the accumulator; even shifts do not of course require this final transfer. The read part of the micro-program is standard as before. The first step

Table 4.8
Micro-program for the instruction 'jump to address specified if ACC negative'

Timing (100ns)	*Condition*	*Micro-order*	
1		E_{AD}, R/W	Read cycle
2		E_oX_i, $2^{-17}AU$	
3		AU_oE_i	
11		M_oC_i	
12		C_{OD}	Obey cycle
13	A < 0	$C_{OA}E_i$, END	
	A $\geqslant$ 0	END	

Table 4.9
Micro-program for the instruction 'Shift accumulator left *n* places'

Timing (100ns)	*Condition*	*Micro-order*	
1		E_{AD}, R/W	Read cycle
2		E_oX_i, $2^{-17}AU$	
3		AU_oE_i	
11		M_oC_i	
12		C_{OD}	Obey cycle
13		C_{ADK}	
→14	K $\neq$ 0	A_oX_i, L, K_o	
	K = 0	END	
15	K = 0	X_oA_i, END	
16		X_oA_i, L, K_o	
17	K = 0	END	
	K $\neq$ 0	Rep 14	

of the obey cycle, in time-slot 13, consists of transferring the five least significant digits of the instruction (which contains n, the number of places to be shifted) to the K-counter. In time-slot 14 if the contents of the K-counter are non-zero the highway gates are set to shift one place left and a transfer takes place between the accumulator and the X-register, at the same time the contents of the K-counter are decreased by +1. If $K = 0$ the micro-program is concluded, this accounts for the instruction 'shift left zero places'. In time-slot 15 if $K = 0$ the contents of the X-register are transferred to the accumulator and the micro-program concluded; this allows for the case of an odd number of shifts. In time-slot 16 the highway gates are once again set to shift left and a transfer between the X-register and the accumulator takes place, decrementing the K-counter by 1 as before. If $K = 0$ in time-slot 17 (after an even number of shifts) the micro-program is brought to an end, if however $K \neq 0$ the micro-order Rep 14 is generated which sets the operations counter to 13, consequently the next timing-slot is 14 which causes the micro-program to loop back and repeat.

In most computers there would be an additional common section at the beginning of each micro-program which would check if the interrupt, error, overflow triggers, and so on, were set and if so initiate the necessary action. These conditions could either be dealt with individually or treated as a common interrupt; in the latter case the interrupt sub-routine must establish the identity of the interrupting source. In the case of an interrupt, a wired-in automatic sub-routine link to a specific part of the program store (containing an appropriate routine for dealing with the interrupt) would be initiated. A suitable micro-program for performing this automatic link operation is shown in Table 4.10.

In time-slot 1 if an interrupt is present (all interrupting devices or conditions would set a bistable, the outputs of which are OR'd to give I_T) the contents of the control register would be transferred to the M-register. A read/write cycle is initiated in time-slot 2, the address of the interrupt link location being prewired to the address highway and gated by I_{AD}. Thus the address of the next instruction in the *main* program is written into the interrupt routine, thereby enabling the sub-routine to link back. The final step, in time-slot 3, is to clear the interrupt and set the start address (prewired) of the interrupt routine into the control register, the operation counter is then reset and the interrupt micro-program proceeds from time-slot 1. Note that in time-slot 1, if $I_T = 0$, it is also necessary to check that there is no R/W cycle in operation; this last condition is possible if an interrupt has just been dealt with. If both these conditions are satisfied then the interrupt link micro-program is bypassed by jumping to the beginning of the read cycle using the Rep 4 micro-order.

It will be obvious that the normal programmed link instruction is obeyed in a similar manner. In this case, however, the address of the current instruction is written either into a special modifier register or into the first location of the sub-routine (the address of which is given in the actual link instruction and held in the instruction register). The address (or the address +1) of the first location of the sub-routine is transferred to the control register and the micro-program concluded.

Finally, it is worthwhile at this point to consider the detailed mechanism of modification and modifier registers. As we have already seen in Chapter 2 there are two basic types of modifier instruction – those which involve the use of separate registers communicating directly with the common highway and those in which the instruction is modified by the contents of the location whose address was specified in the *previous* instruction. (The modifier or index word can also be stored in the word following the instruction, as in the PDP 11/40).

When using separate modifier or index registers there are distinct modifier or index bits included in the instruction word, these would be used to set up the modifier circuit condition (M_T) and used appropriately. Table 4.11 shows the way in which modification would be included in a

Table 4.10
Interrupt micro-program

Timing (100ns)	*Condition*	*Micro-order*	
1	$I_T = 1$	E_oM_i	
	$I_T = 0$, R/W = 0	Rep 4	
	$I_T = 0$, R/W = 1	Rep 1	
2		I_{AD}, R/W	
3		I_TC, E_I, END	
→4		E_{AD}, R/W	Read cycle
5		E_oX_i, 2^{-17}AU	Read cycle
6		AU_oE_i	Read cycle
14		M_oC_i	Read cycle
15		C_{OD}	
		etc.	

Table 4.11
Micro-program for modifier order

Timing (100ns)	*Condition*	*Micro-order*	
1 ⋮ 15			Interrupt and read cycle micro-program
16	$M_T = 1$	C_oX_i, MD_{AU}	Modification
	$M_T = 0$	Rep 18	Modification
17		AU_oC_i	Modification
→18 ⋮ etc.			Obey cycle

micro-program. Indexing is performed in a similar way but in this case the registers to be used in the address modification must be specified in the instruction. In the case of the modification order the instruction would be decoded in the normal way (and sometimes obeyed), the contents of the address specified being stored in a register (used to modify the next instruction), and the modifier trigger set to one.

The control unit may be operated in either automatic (run) or manual mode. For automatic working the read/obey cycles occur in sequence as determined by the resetting of the operations counter from the micro-program. The normal operating requirement for manual working is to transfer operands, set up on a console keyboard, to the control and instruction registers and then step through the read/obey cycles one by one by depressing a switch. Thus in this mode the computer stops after each complete order has been executed; the procedure is known as '*single-shot*' operation. Control of this operating sequence is effected in the clock gating unit (see Fig. 4.19) by detecting the 'all-zero' state of the operations counter. In some machines it is also possible to step the operations counter so that individual micro-steps may be observed; this technique is an invaluable aid in fault-finding and for educational purposes.

4.6 The micro-program concept[7,8]

The micro-program control unit described above is implemented in terms of wired-in logic gates (**hardwired** micro-program) which severely restricts the value of the micro-program approach. At the present time many computers are still constructed in this way; by careful partitioning of the logic on to circuit boards it is sometimes possible to modify and add extra micro-programs, but in many cases this is extremely difficult. The micro-program philosophy, first proposed by Wilkes,[9,10] is an extremely powerful one, particularly if implemented using read-only memory to store the micro-programs which are read down from store and decoded in the usual fashion; this technique is known as **static micro-programming**. There are a number of advantages to be gained using this approach:

It is possible to delay design decisions until the last moment.

Design faults may be easily rectified, and modifications incorporated, with the minimum of trouble.

Routines, usually programmed in assembly or high-level languages, may be implemented at basic machine level (with attendant increase in speed of execution) using micro-programs; the implementation of software routines in this form is called **firmware**.

The maintenance of the computer is simplified since only sequences of simple operations (the micro-orders) are implemented in the hardware. Moreover it is also possible to store special diagnostic programs in the micro-program store to give fault detection at the basic machine level.

Overall processor design is simplified, particularly for large machines, since programming skills as well as engineering skills may be utilized in the design and evaluation of the control unit.

A further economic advantage of micro-programming is that compatibility with pre-existing systems may be achieved by micro-programmed **emulation**.[11] This is a technique whereby static micro-programmed systems can have another set of micro-programs added to 'emulate' the machine-code of an earlier computer, thus alleviating the reprogramming problems. In general as much ROM is required to perform the emulation as would be needed for the basic instruction set.

The method of micro-programming employed in most second generation computers was a variation of the original Wilke's scheme, shown in block diagram form in Fig. 4.21. In this method the instruction word is used to address the matrix which generates a micro-order and also generates the address where the next micro-order is to be found on the selected bit-lines. The micro-orders are read down in the correct micro-program sequence and taken directly to the logic to be controlled. Program branching is executed by allowing the state of the conditional bistables to select the appropriate row of the address matrix, the output of which is then used to re-address the matrix. The matrix itself consisted of diode/resistor matrices, ferrite core plug-boards, or transformer coupled read-only arrays.

An alternative technique of addressing the matrix is to increment the control address register (wired as a counter) by + 1 each time a micro-order is read down from the matrix. Thus the micro-orders are obtained from contiguous locations unless a new address (obtained from the matrix) is inserted into the control address register as the outcome of a conditional branching micro-order.

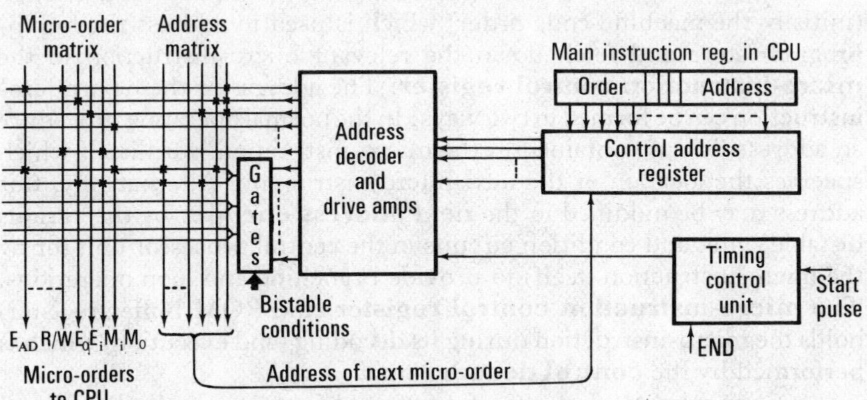

Figure 4.21 Block diagram of Wilke's scheme.

The system must of course be timed and this is done by the clock control circuit which controls the operation of the address decoder unit. In this method the necessary synchronization between the micro-program timing and the read/write cycle of the central processor's main store is obtained by gating the clock control unit with the R/W store condition which inhibits the timing output. Similarly it is also possible to obtain autonomous operation by generating a suitable micro-order which is used temporarily to cancel the R/W store condition. Note that all the micro-orders for a micro-programmed step appear simultaneously in a single time-slot, this is called **vertical** micro-programming. The alternative technique, **horizontal** micro-programming, is to generate all the micro-operations for several time-slots (or a complete micro-program) at the same time. This method, though in some cases faster (since the vertical method requires more store accesses), requires a larger matrix array and external buffer storage. Moreover the timing control tends to be more complicated since the micro-orders must be re-generated in a pre-determined time sequence before distribution to the system.

Though the design approach described above is more flexible than the hardwired system, depending on the technology in which the matrix is implemented, there are still a number of disadvantages, the major ones being the limitations on the size of the matrix and the difficulties in modifying the micro-programs. Many of these disadvantages may be overcome by using read-only memory to store the micro-programs. The ROM technique, however, is not just considered as a direct replacement for the matrix, the entire design philosophy of the micro-program unit is re-appraised to produce a control unit on a **micro-machine** scale.

The generalized block diagram for a static micro-programmed control unit is shown in Fig. 4.22. The **micro-program store** (or **control store**) is a read-only memory, for example a semiconductor ROM store utilizing 128 × 8-bit word modules, or a capacitive punched-card scheme as used in the IBM System/360 series.[12] The **micro-instruction address register** stores the address of the micro-instruction (initially the machine-code order) which is used to address the micro-program store and fetch down the relevant micro-instruction to the **micro-instruction control register**. The address of the next micro-instruction can be formed in two ways: In the normal operating procedure an address field is contained in the micro-instruction word itself which specifies the location of the next micro-instruction. Alternatively, this address may be modified in the **next address encoder** by the various bistable states and condition circuits in the central processor unit (or by the micro-instruction itself) to provide branching and loop operations. The **micro-instruction control register** (the ROM buffer register) holds the micro-instruction during its decoding (and execution) which is performed by the **control decoder** unit.

The normal sequence of operations is to access the ROM and fetch down the first micro-instruction (the address of which is determined by the machine-code order) to the micro-instruction control register. The

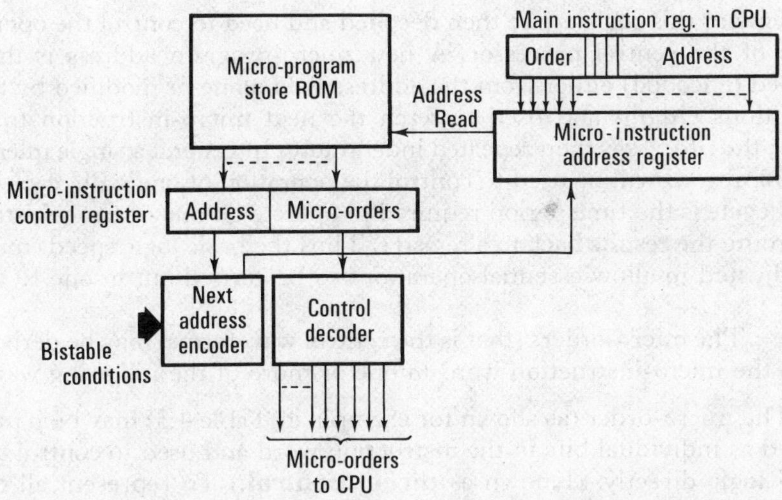

Figure 4.22 Block diagram of static micro-program unit.

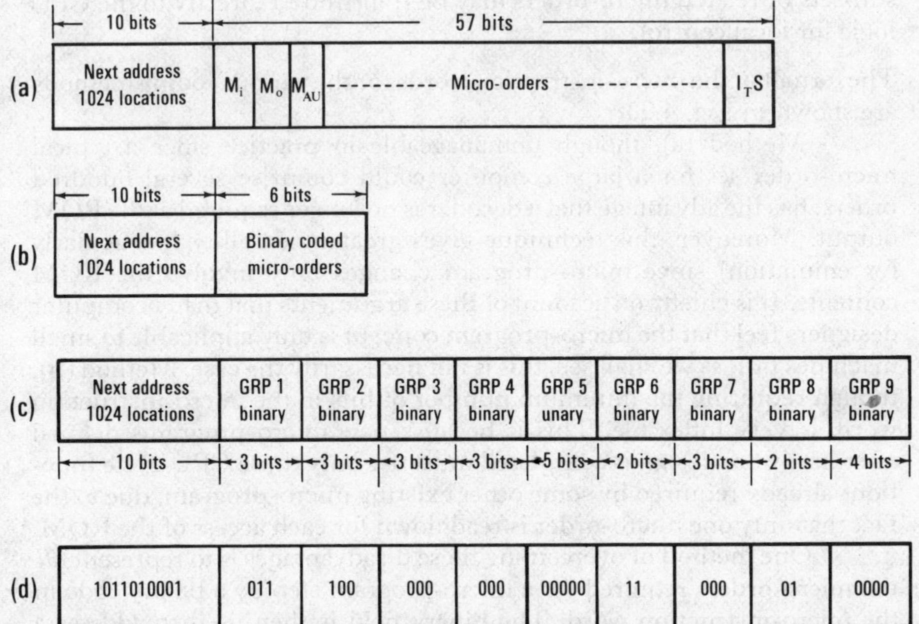

Figure 4.23 Micro-instruction word formats.

contents of this register are then decoded and used to control the operations of the central processor. A new micro-program address is then formed (encoded) either from the address field alone or modified by the conditions circuit, and used to fetch the next micro-instruction from store; the process is then repeated indefinitely. In general a single micro-program word will be used to control the operation of one CPU cycle; a CPU cycle is the time period required to operate on one or two registers and route the results back to a register. Thus the basic logic speeds must be adjusted to allow essential operations to be carried out in one ROM cycle.

The micro-orders (that is the control waveforms) may be derived from the micro-instruction word in one or more of the following ways:

(a) The micro-order (as shown for example in Table 4.5) may be represented as individual bits in the instruction word and used to control the CPU logic directly. (Known as **direct control**.) To represent all the micro-orders (and conditions) in this way would, however, be uneconomic on account of the very large instruction word (some 67 bits for the example above) that would be required.

(b) An m-bit micro-order control field can be decoded to yield 2^m distinct control signals to drive the CPU. That is, the set of 57 micro-orders shown in Table 4.5 would be represented by a 6-bit field in the instruction word and decoded to produce the individual signal.

(c) Binary-coded groups of digits in the instruction word representing sub-sets of related micro-orders may be transmitted directly to the CPU logic for local control.

The format of the micro-instruction words for the various coding methods are shown in Fig. 4.23.

Method (a), though unmanageable in practice since a typical micro-order set for a large computer could comprise several hundred orders, has the advantage that a decoder is no longer required at the ROM output. Moreover, this technique gives greater flexibility (particularly for emulation) since micro-program changes only involve the ROM contents. It is chiefly on account of these arguments that many computer designers feel that the micro-program concept is only applicable to small machines but, as we shall see, this is not necessarily the case. Method (b), though requiring the minimum number of bits in the micro-instruction word, is very inflexible. This is because new micro-programs defined after hardware design is completed must use only those CPU cycle functions already required by some other existing micro-program, due to the fact that only one micro-order is read down for each access of the ROM.

One method of overcoming these disadvantages is to represent *all* the micro-orders required for a micro-program step by a binary code in the micro-instruction word. The binary field is then used to address a separate micro-order store (on a table-look-up principle) which produces all the micro-orders in one access time. In this way the basic micro-instruction word can be kept small (hence reducing the cost of the control store) while still retaining the advantages of direct control.

An alternative to coding all the micro-orders, as in (b) above, is to partition the micro-orders into groups of related orders and to code these separately. This method, though used extensively in practice, poses the problems of finding an encoding which minimizes the dimension of the micro-order field, while at the same time ensuring maximum flexibility of micro-programming. This problem has been studied in detail by Schwartz[13] and Grasselli[14] who have both proposed minimizing algorithms. In practice, the number of bits per micro-instruction word ranges typically from 10 to 12 depending upon the design objectives.*

Thus, the fundamental design questions which must be resolved in the engineering of a static micro-program unit are as follows:

How many different micro-orders are to be specified in the micro-instruction word?

Are the micro-orders to be performed sequentially, concurrently, or a combination of both?
How are the micro-instruction words to be encoded?
How is the address of the next micro-instruction word to be derived?

All these points must be carefully weighed together since they all directly affect both the cost and performance of the micro-program unit.

A possible group encoding of the micro-order set of Table 4.5 is shown in Table 4.12; note that both binary and unary (single bit representation) has been used. To ensure continuity the same set of micro-orders as used in previous examples has been employed; in practice, however, the control signals would be derived anew bearing in mind the structure of the central processor and the different methods of control. This would affect in particular the branching type instructions (obtained in the example set by using the K-counter) since it is possible with the new technique to perform these operations by modifying the micro-instruction address register.

The grouping of micro-orders must be made in such a way that all the members of a group are **mutually exclusive**. This is essential since more than one micro-order has to be specified in the micro-instruction word. For instance in Table 4.9 timing-slot 14 we have the micro-instruction:

$$K \neq 0 \quad A_oX_i,\ L,\ K_o$$

which performs the simultaneous operations of shifting the contents of

* Some confusion appears to exist in the literature on the use of the terms vertical and horizontal microprogramming. In many cases the term horizontally structured is used to refer to direct control, as in method (a) above, where the micro-instruction word is minimally encoded and necessitates long control store words. When a highly encoded micro-instruction is employed it is called a vertically structured system. Note that the common factor is the length of the control store words – the horizontal method (whichever definition is used) always necessitates long control words.

Table 4.12
Encoding of micro-order set

Group 1
Register transfer out (3 bits)

000	No output
001	E_o
010	C_o
011	A_o
100	X_o
101	MD_o
110	M_o
111	IO_o

Group 2
Register transfer in (3 bits)

000	No input
001	E_i
010	C_i
011	A_i
100	X_i
101	MD_i
110	M_i
111	IO_i

Group 3
Arithmetic and logic orders (3 bits)

000	do nothing
001	AU_o
010	2_{AU}^{-17}
011	MD_{AU}
100	M_{AU}
101	$\overline{M}_o$
110	$\overline{M}_{AU}$
111	$\overline{A}_o$

Group 4
Control orders (3 bits)

000	do nothing
001	END
010	Write
011	Rep n
100	R/W
101	IT_o
110	Unassigned
111	Unassigned

Group 5
K-register setting (5 bits)

K_1 K_2 K_4 K_8 K_{16}
(unary code, all zeros, do nothing)

Group 6
K-register orders (2 bits)

00	do nothing
01	K_c
10	C_{ADK}
11	K_o

Group 7
Address and order highways (3 bits)

000	do nothing
001	C_{AD}
010	C_{OD}
011	I_{AD}
100	E_{AD}
101	C_{OA}
110	Unassigned
111	Unassigned

Group 8
Gating orders (2 bits)

00	normal
01	L
10	R
11	E_I

Group 9
Clear and set orders (4 bits)

0000	do nothing	1000	M_TC
0001	M_c	1001	I_TC
0010	E_c	1010	OF_TS
0011	C_c	1011	E_TS
0100	A_c	1100	M_TS
0101	X_c	1101	I_TS
0110	IO_c	1110	I_M
0111	E_TC	1111	$\overline{I}_M$

the accumulator one place left, putting the result in the X-register, and decrementing the K-counter by +1. The layout and binary coding of the micro-instruction word for this micro-order statement is shown in Fig. 4.23(d).

To facilitate the grouping of micro-orders into mutually exclusive sets it is convenient to split the orders into two basic types:

(a) Those orders which are *inherently* mutually exclusive due to hardware constraints, for example, simultaneous right and left shift operations, and

(b) *Implicitly* mutually exclusive orders which are logically constrained in the sense that it would be illogical to use them together, for example, transferring into a register from two different sources.

In general the grouping of micro-orders into inherently mutually exclusive classes does not place any restrictions on the micro-programming. However, care must be exercised when encoding implicit mutually exclusive micro-orders to ensure that future micro-programming activities are not limited in any way.

In the straightforward system described above the time required to perform the micro-order operations in the central processor represents a fraction of the total time required to execute each micro-instruction, since it is necessary to fetch the micro-instruction from store and decode it before initiating the actual CPU micro-orders. However, depending upon the access time and detailed design of the ROM, it is possible to reduce the overall operation time by fetching the next micro-instruction shortly after the current micro-orders are initiated in the CPU. Thus the access and decoding time of the next micro-instruction can be concurrent with the execution of the present micro-instruction. This method of *overlapping* the read and obey cycles of the micro-program unit introduces two major design problems:

New data may be placed in the micro-instruction control register before the current CPU operations are completed. Thus, in order to allow autonomous operation of the control decoder unit, it is necessary to add an extra control buffer register.

The next micro-instruction address may be a function of circuit conditions being generated by the current CPU operations. In this case it is necessary to initiate an extra CPU cycle (a waiting period) to allow the conditions to settle out before re-addressing the ROM.

It will be obvious that to minimize instruction execution times, extra logic and complicated micro-programs must be designed into the computer. Since the definition of new micro-programs (after a computer is built) must use the existing hardware, there is considerable motivation for providing well thought out general purpose micro-orders at the initial systems design stage.

Another design variation of the micro-program unit is to use hardwired micro-program control for the read part of the Read/Obey cycle, which we have seen is common to all micro-programs, and static control for the rest of the instruction set. This philosophy was used for the IBM360/85 system which has two independent control units, a hardwired I-unit to fetch the instruction, and a micro-programmed E-unit (containing both ROM and RAM storage) to execute the instructions.

The next logical step in the development of the micro-program concept is to use a read/write store for the micro-instructions. This more general approach is called **dynamic** micro-programming,[15,16] and allows the micro-program to be set up and modified under normal program control. This technique has an important advantage over static systems in that the computer can be restructured to represent *any* machine code instruction set that exists, or can be conceived of, by simply writing and loading its micro-programs.

In addition, it is also possible to store comprehensive testing and diagnostic programs in the read/write store, enabling a complete logical check-out at the micro-order level to be obtained. Moreover, long term reliability (as required, for example, in an airborne satellite computer) may be enhanced by re-programming the micro-instructions such that inoperative parts of the hardware are bypassed using standby units. The modified system will often be slower in operation but nevertheless will remain operational, a condition of 'graceful degradation'[17] being achieved. It is interesting to note that the first example of a dynamically micro-programmed computer was the STC ZEBRA machine[18] which used a combination of drum and core storage.

Though both hardwired and static micro-programming techniques were used in the early 1960s (notably in the IBM360 series) it was not until the middle and late sixties when integrated circuit technology had advanced sufficiently to produce economic ROM and RAM stores that the method became a practical reality. Even now the number of machines which are truly *user* micro-programmable are very small, and dynamic micro-programmed machines have yet to obtain their full potential.

Notable among the machines produced to date is the Burroughs B1700[19], which was built with the object of directly interpreting high level languages (that is emulation – considered by some workers to be the prime purpose of micro-programming). The B1700 can execute micro-programs which reside either in main memory or a faster writable control store that can be loaded from main memory. Though it is possible for users to write and load their own micro-programs (a micro-program assembler is provided as a user software aid) it is Burroughs' policy to undertake this task themselves working from user specifications. The B1700 has also been used for the direct interpretation of several conventional machines, such as the IBM 1401 and the Burroughs B2500.

Another user programmable machine is the Hewlett-Packard 2100, in which part of the control store is implemented as a ROM and part as a writable memory. Resident micro-programs in the ROM realize the standard instruction set while special micro-programs may be loaded into the writable control store using a system editor. Assemblers and debugging software are provided for the user to facilitate the preparation of micro-programs. Other machines which operate on a similar principle are the Interdata 85 and the Varian 73.

A recent development which will make a major impact on the evolution of micro-programming is the LSI **bit-slice processor element** such as the Intel 3000, AMD 2901 and the Texas SBP 400 chips. Processor element chips typically have a 2 or 4-bit wide parallel bus and contain a binary arithmetic and logic unit, major registers and data paths, including a program counter, and an operations decoder (usually implemented with a PLA). The bit-slice elements operate solely at the micro-instruction (i.e. micro-order) level and require an external control store and controller in order to function as a computer. It is important to be aware of the distinction between the bit-slice processor element and the micro-processor chip. In the latter case the processor has a fixed wordlength and instruction set and is essentially an integrated circuit version of a self-contained CPU complete with system level ALU and predetermined control algorithms. The micro-processor is programmed by the user at the machine code/assembly language level, whereas in the bit-slice processor basic machine-code instructions such as add, load, store, jump etc., must be micro-programmed.

For example the SBP 400 processor element contains 8 general purpose registers (including a program counter) 2 working registers and a 16 operation ALU with carry look-ahead. A nine-bit operation select word (an encoded micro-instruction) is processed by a PLA decoder to give some 512 micro-operations such as register to register, register to data-bus and data-bus to register data transfers, arithmetic, logic and shifting operations etc. Note that these operations correspond very closely to the micro-orders defined in Table 4.5; logic conditions and more complex micro-orders (including control store address modification) are generated externally by separate micro-controller logic. The main advantages of bit-slice processor elements are:

(a) they can be interconnected to give variable wordlength computer architectures for example, a 16-bit machine can be constructed by connecting together four SBP 400 4-bit elements
(b) they can be user micro-programmed to emulate virtually any instruction set.

Since in the near future it will be more cost effective to implement computer system designs (particularly distributed architectures) using these LSI components it seems inevitable that micro-programming will assume an even more central role in computer design. For instance, it is perfectly feasible to construct a mini-computer, such as the PDP 11 or HP 2100, using processor elements.[20]

Table 4.13
Example of a micro-program language

Store address register	R_A
Instruction register	R_S
Auxiliary arithmetic register	R_X
Store output register (M)	R_M
Control register	R_C
Instruction register order bits	R_{So}
Instruction register address bits	R_{Sa}
Control unit decode register	R_D
Accumulator register	R_R
Address store and initiate R/W cycle	READ
Transfer micro-program control	JMP (n)
Conditional jumps, test on condition C	JMP, C (n)

Table 4.14
Micro-program for the add instruction

Step	*Micro-instruction*	
1	$R_A := R_C$, READ	Read cycle
2	$R_C := R_C + 1$	
3	JMP, R/W = 1 (3)	
4	$R_S := R_M$	
5	$R_D := R_{So}$	
6	$R_A := R_{Sa}$, READ	Obey cycle
7	JMP, R/W = 1 (7)	
8	$R_R := R_M + R_R$	
9	END	

Table 4.15
Iverson description of the add micro-program

Step	*Statement*	
1	$\mathbf{R}_A \leftarrow \mathbf{R}_C$, READ $\leftarrow 1$	Read
2	$\perp \mathbf{R}_C \leftarrow 2^{11} \vert (\perp \mathbf{R}_C + 1) \vert$	
3	RW: 1; $(=, \neq) \rightarrow (3, 4)$	
4	$\mathbf{R}_D \leftarrow \mathbf{R}_{So}$	
5	$\mathbf{R}_A \leftarrow \mathbf{R}_{Sa}$, READ $\leftarrow 1$	Obey
6	RW: 1; $(=, \neq) \rightarrow (6, 7)$	
7	$\perp \mathbf{R}_R \leftarrow 2^{18} \vert \perp \mathbf{R}_M + \perp \mathbf{R}_R \vert$	
8	END	

4.7 Programming the micro-program control unit[21,22]

Because designing the control unit for a computer is rather like designing another small computer (recursive design?), programming using the micro-order set is similar to normal machine code programming. Moreover, to facilitate the communication and manipulation of micro-programs and orders they are best described in the same terms as those used for machine-code orders. For instance, a simple listing of the micro-order sequence required in a micro-program is difficult to follow, as can be seen from the examples in Tables 4.6 and 4.8. An alternative method is shown in Tables 4.13 and 4.14; here the micro-programs are described using normal software symbology by identifying operands (registers) and stating the transfers and logical operations between them. Note that instead of using micro-program timing to allow for the duration of the read/write store cycle, it is determined in the program by testing the R/W store condition. Notice also that the address bits (and the order bits) of the instruction word (held in the instruction register) are considered to be available on separate lines.

This 'functional' micro-program is very necessary in order to provide a complete and precise specification of the micro-programs (at the systems design stage) before the actual logic design is undertaken. Furthermore, the description must be concise so that the operational functions can be understood by the many people who must collaborate and concur in the specification and design of the machine. Unfortunately, functional micro-programs of this type have the disadvantage that they are at too high a level to allow circuit implementation to take place directly from the algorithmic description. For example, in Table 4.14 each register transfer statement effectively represents a sub-routine of appropriate micro-orders (or Boolean equations) necessary to effect the actual logical transfers.

Consequently, in order to produce effective micro-programs, and in particular emulators, manufacturers have found it necessary to develop **micro-assembly languages** which generate the necessary micro-orders, in the form of bit patterns, from a basic mnemonic description of the micro-program. It is also essential that micro-programs should be checked out before actual implementation; this is performed using a special software simulation of the computer system.

A natural extension of this approach has been the development of logic design languages (see also Chapter 8). The majority of these are based on a **register transfer language**[23,24] which enables a digital system to be described in terms of registers and the required logical transfer operations between them. Thus it is possible to use the logic design language to declare a linguistic description of a block diagram of a machine, detailing all the necessary registers and their interconnections. The micro-programs for a particular machine-code set may then be specified as operational statements using standard software nomenclature; for example, $R + S \rightarrow T$, $R \oplus S \rightarrow T$, and so on, where R, S, and T are previously declared registers. This high-level language specification for a computer is then compiled in the usual fashion, except that the output in this case is a set of Boolean equations which may be used directly to implement the hardware version of the machine.[25,26]

One language which has been used extensively (particularly by IBM) for the description of micro-programs is the Iverson language[27,28,29] which was introduced in the last chapter. Because the Iverson (APL) notation is an algorithmic language, complex sequential logic (represented as programs of micro-events) and inter-register transfers can be concisely and accurately defined. Table 4.15 shows the micro-program for the add instruction described in terms of the Iverson notation. The program follows similar lines to that shown in Table 4.14 – the store cycle is initiated by READ ← 1 and the end is determined by testing and jumping on the R/W condition. The modulus form of addition, used in steps 2 and 7, allows for overflow occurring in the addition operations. Instead of using separate output lines for the address and order bits of the instruction word, the instruction register (R_S) may be declared full length (18 bits) and the transfers in steps 4 and 5 stated as:

$$\mathbf{R}_a \leftarrow \boldsymbol{\omega}^{10}/\mathbf{R}_S \quad \text{and} \quad \mathbf{R}_d \leftarrow \boldsymbol{\alpha}^{5}/\mathbf{R}_S$$

Unfortunately the language is at too high a level to permit direct implementation of the micro-programs or, for that matter, in its full form for use as a logic design language. This is because a large part of the language is not pertinent to machine design; for example, some of the symbols are not easily associated with hardware, and timing and sequencing is difficult to describe. Many of these problems have been successfully overcome, however, in the ALERT system[30] which uses a sub-set of the Iverson notation to depict a system structure, which is then 'compiled' into Boolean equations.

The automatic design of micro-program control units must ultimately proceed in the directions outlined above. However, at this time the methods seem more applicable to hardwired logic systems, and the major problems of timing, micro-instruction word coding and minimization (in static and dynamic systems) are still left to the logic designer.

4.8 Software influence on micro-programming

Though micro-programming first arose as a hardware implementation technique for the economical realisation of a complex instruction set, it can also be regarded as a software method to provide programmers with more flexibility when developing particular task orientated instruction sets. With the cost of software development becoming the dominant factor in computer systems design the use of micro-programming to effect a more easily programmed machine will become of increasing importance. As we have seen one of the major applications of micro-programming is in emulation, the function of which is effectively identical with a software **interpreter** that executes statements in a source program as it is presented to the machine. The interpreter usually consists of two parts, a set of routines corresponding to the set of operations in the source program language (micro-programs) and a control routine to sequence the source statements and select operations for decoding (micro-program control unit). Thus the concept of emulation and interpretation are completely equivalent, by convention however a distinction is drawn between an interpreter, which is considered as a program resident in main memory, and an emulator comprised of micro-programs in a control store.

In emulation the physical machine as defined by its micro-instructions and their functional operations is called the **host machine.** Machine-code languages for specific computers, emulated by sets of micro-programmed routines, are called **image** or **target machines**. Recently the trend in micro-programmable machines has been to design them to perform generalized emulation, that is to emulate a variety of target machines so that one host machine with several emulators can economically replace several different machines.

In this context it is worthwhile considering **directly executable languages** (DEL) which can be considered as source languages for interpreters. In essence a DEL is a form of language, which allows the efficient execution of interpretive (or emulation) programs. The main properties required of a DEL are that the arguments and operations should be completely and autonomously defined (that is when first encountered in the program) and the source statements explicitly ordered. Note that, for instance, symbolic assembly code and assembly language with macros are not directly executable.

The importance of this work, which to date has been mainly in the software area (e.g. compiler design) is that one can consider micro-instructions as the object language of a directly executable source language which can be used to describe an emulation.[31,32] Thus, using the concept of DEL's it is possible to devise a machine structure (called a **soft** computer architecture) which can emulate a wide range of image machines and has the ability to support, as yet undefined, intermediate interpretive language levels. Note however that these machines will not in general give as good a performance as machines designed specifically for a particular instruction set. Note also that using this approach it is possible to design high level language machines via the medium of micro-programming.[33] For example, the machine-code language of a computer could be made to resemble (using micro-code) the directly executable language or interpretive form of a high-level language.

It is interesting to note that one of the current difficulties in designing distributed (dedicated) microcomputer systems is the need to have resident high-level language compilers. It is highly conceivable that the use of directly executable language techniques coupled with the use of bit-slice processor elements could solve this problem.

Another software area where micro-programming techniques have been successfully applied is that of operating systems. Architectural features essential for the development of OS software, such as segmentation, procedure call and return mechanisms, process communication with semaphores etc., have been implemented using micro-programming methods. It will be obvious from above that there is a very intimate relationship between software methods and micro-programming which is still currently being explored. It would seem almost inevitable that future advances in micro-programming will follow closely the pattern already established in assembly language and systems programming. For example, programming language techniques, such as algorithm proving, directly executable languages, interpreters and optimizing compilers will be further developed with particular reference to implementation using dynamic micro-programming techniques. Moreover, with the availability of fast, high capacity control store memory it would seem likely that memory management and cache store systems will also be required.

References and bibliography

1 *The TTL Data Book for Design Engineers*. Texas Instruments Inc., 1973.

2 Blakeslee, T. *Digital Design with Standard LSI and MSI*. Wiley, New York, 1975.

3 Greenfield, J. D. *Practical Digital Design Using IC's*. Wiley, New York, 1977.

4 Lewin, D. W. *Logical Design of Switching Circuits*. Nelson, London, 1974, Chapter 5.

5 Heath, F. and Gribble, D. Chain codes and their electronic application. *Proc. IEE*, 1961, part C, **108**, Monograph M392, 50–57.

6 Kostopoulos, G. K. *Digital Engineering*. Wiley, New York, 1975.

7 Boulaye, G. G. *Microprogramming*. Macmillan Press, London, 1975.

8 Chu, Y. *Computer Organisation and Microprogramming*. Prentice Hall, Englewood Cliffs, N.J., 1972.

9 Wilkes, M. and Stringer, C. Micro-programming and the design of control circuits in an electronic digital computer. *Proc. Camb. Phil. Soc.*, 1953, **49**, 230–238.

10 Wilkes, M. Microprogramming. *Proceedings of the Eastern Joint Computer Conference*, December 1958, 18–20.

11 Tucker, S. G. Emulation of large systems. *Comm. Assoc. Comp. Mach.*, 1965, **8**, 753–761.

12 Tucker, S. G. Microprogram control for system/360. *IBM Systems Journal*, 1967, **6**, 222–241.

13 Schwartz, S. J. An algorithm for minimizing read-only memories for machine control. *IEEE 10th Annual Symposium : Switching and Automata Theory*, 1968, 28–33.

14 Grasselli, A. and Montanari, U. On the minimization of read-only memories in micro-programmed digital computers. Instituto Di Elaborazione dell' Informazione, Pisa. Nota Interna B68/17, October 1968.

15 Cook, R. W. and Flynn, M. J. Systems design of a dynamic microprocessor. *IEEE Trans. Comput.*, 1970, **C19**, 213–222.

16 Grasselli, A. The design of a program-modifiable micro-program control unit. *IRE Trans. Electron. Comput.*, 1962, **EC11**, 336–339.

17 Breuer, M. A. Adaptive computers. *Information and Control*, 1967, **11**, 402–422.

18 ZEBRA – a simple binary computer. *Proceedings of the IFIP 1959 Congress, Paris, France*. UNESCO, 1960, 361–365.

19 Wilner, W. T. Design of the Burroughs B1700, *Proc. AFIPS FJCC*, **41**, 489–497, 1972.

20 McWilliams, T., Fuller, S. H. and Sherwood, W. Designing a PDP11 with Intel 3000 bit-slices. Dept. of Computer Science, Carnegie-Mellon University. Report, June 1976.

21 Lewin, D. *Computer Aided Design of Digital Systems.* Crane Russak, New York, 1977.

22 Hardware Description Languages, *IEEE Computer*, **7**, Dec. 1974, (special issue).

23 Chu, Y. An Algol-like computer design language, *Comm. ACM.*, **8**, 607–615, 1965.

24 Schorr, H. Computer-aided digital system design and analysis using a register transfer language. *IEEE Trans. Electron. Comput.*, 1964, **EC13**, 730–737.

25 Proctor, R. M. A logic design translator experiment demonstrating relationship of language to systems and logic design. *IEEE Trans. Electron. Comput.*, 1964, **EC13**, 422–430.

26 Barbacci, M. R. A. A comparison of register transfer languages for describing Computers and digital systems, *IEEE Trans. Computers,* **C24**, 137–150, 1975.

27 Hellerman, H. *Digital Computer System Principles.* McGraw-Hill, New York, 1967.

28 Iverson, K. E. *A Programming Language.* John Wiley, New York, 1962, Chapter 2.

29 Blaauw, G. A. *Digital System Implementation.* Prentice Hall, Englewood Cliffs, N.J., 1976.

30 Friedman, T. D. and Yang, S. C. Methods used in an automatic logic design generator (ALERT). *IEEE Trans. Comput.*, 1969, **C18**, 593–614.

31 Weber, H. A Microprogrammed implementation of EULER on IBM System 360 model 30, *Comms. ACM,* **10**, 549–58, 1967.

32 Reigel, E. W., Faber, U. and Fisher, D. A. The Interpreter – a microprogrammable building block system. *AFIPS FJCC,* **40**, 705–723, 1972.

33 Wortman, D. A. A Study of Language directed computer design. PhD Dissertation. Stanford U. Calif. 1973.

34 Vandling, G. C. and Waldecker, D. E. The microprogram control technique for digital logic design. *Computer Design,* 1969, **8**, 44–51.

35 Mercer, R. J. Micro-programming. *J. ACM,* 1957, **4**, 157–171.

36 Kampe, T. W. The design of a GP microprogram controlled computer with elementary structure. *Trans. IRE Electron. Comput.*, 1960, **EC9**, 208–213.

37 Duley, J. R. and Dietmeyer, D. L. A digital system design language (DDL). *IEEE Trans. Comput.*, 1968, **C17**, 850–861.

38 Duley, J. R. and Dietmeyer, D. L. Translation of a DDL digital system specification to Boolean equations. *IEEE Trans. Comput.*, 1969, **C18**, 305–313.

39 Falkoff, A. D., Iverson, K. E., and Sussenguth, E. H. Formal description of System/360. *IBM Syst. J.*, 1964, **3**, 198–262.

40 Rosin, R. Contemporary concepts of microprogramming and emulation. *Computing Surveys*, 1969, **1**, 197–212.

41 Husson, S. *Microprogramming: Principles and Practices.* Prentice-Hall, Englewood Cliffs, N.J., 1970.

42 Langley, F. J. Small computer design using microprogramming and multifunction *LSI* arrays. *Computer Design*, 1970, **9**, 151–157.

43 Hill, F. J. and Peterson, G. R. *Digital Systems: Hardware Organisation and Design.* John Wiley, New York, 1973.

44 Agrawala, A. K. and Rauscher, T. G. *Microprogramming: Concepts and Implementations.* Academic Press, New York, 1974.

45 Marwin, R. E. *Design Automation aids to Microprogramming. Digital System Design Automation: Languages, simulation and data base.* Editor M. A. Breuer. Computer Science Press Inc., 1975, (Chapter 4).

Tutorial problems

4.1* Devise suitable input gating circuitry for the operations counter to allow the contents to be reset by the Rep *n* micro-order. Show also how a wired-in address may be set into the control register during the execution of the interrupt routine.

4.2 Design the logic circuitry for the 5-bit K-counter as described in the text for controlling loop operations. Assume that the counter has the loop constant set into it (in parallel) and that it is decreased by +1 for each input pulse. Also include in the design the necessary circuits for testing the conditions $K = 0$ and $K \neq 0$.

4.3* Devise a suitable micro-program, using the micro-order set shown in Table 4.5, for the instruction:

LINK – 'Store contents of control register in the location whose address is specified in the instruction, and then jump to address specified +1'.

Suggest an alternative method for performing the sub-routine link instruction.

4.4 Suggest how the computer structure shown in Fig. 1.4 might be modified to allow the instruction 'Modify the *next* instruction by the contents of the location whose address is specified in the instruction' to be executed; write a suitable micro-program to perform this function. What advantages has this method of modification over the more usual technique of using separate modifier registers?

4.5* Design the timing and control logic required to implement single-shot operation in a computer; also include in the circuitry the logic for automatic and manual mode working.

4.6 Discuss in detail, giving logic and block diagrams, the problems involved in addressing the ROM in a static micro-program control unit in order to fetch down the next micro-instruction.

4.7* Using the Iverson notation devise micro-programs to perform the following functions: (a) conversion of BCD to binary; (b) conversion of binary to BCD. Comment on the relative merits of software and hardware implementation of these algorithms.

5

The arithmetic unit

5.1 Introduction

One of the major sub-systems of the central processor unit is the arithmetic unit in which the binary operations of addition, subtraction, multiplication, and division are performed. All arithmetic functions can be related to addition – subtraction is performed by adding complemented numbers, multiplication is repeated addition, and division is repeated subtraction. Thus, in all these functions the basic arithmetic element is the binary adder. Consequently, the arithmetic unit consists of a number of registers which contain the operands, intermediate and final results, a basic adder (working in pure binary or binary-coded decimal) and a control unit. All these functions, including memory, can be obtained on a single LSI chip (as used in calculators) containing the equivalent of some 6000 transistor circuits.

These operations may be performed in either serial or parallel mode; serial organization requires only one binary adder and is considerably cheaper than parallel methods. However, if extremely fast arithmetic is required, parallel methods must be used, necessitating the addition of extra logic and, of course, increasing the cost. The speed of a serial system depends on the clock-rate and the computer wordlength. Thus if medium speeds are required these can feasibly be obtained at low cost by using fast clock-rates and a small wordlength. For example, serial addition takes one wordtime, so if we have a 10 MHz clock-rate (100 ns digit pulses) and a 16-bit word, the addition time for two 16-bit numbers would be 1·6 μs. Many small medium-speed real-time computers operate in the serial mode, thus enabling a cheap and effective computer to be produced.

The function of the control unit is to provide the necessary sequence of operations required to perform the arithmetic function, that is, to implement the arithmetic algorithm. The algorithm may be wired-in, utilizing a sequencing counter and logic, to generate the waveforms to provide the necessary number of shifts, additions, etc. Alternatively, the waveforms could be derived from the micro-program unit as described in Chapter 4.

In this chapter the various ways of performing binary arithmetic operations, both in fixed and floating-point mode, will be discussed in detail. The arithmetic procedures will, in general, be described using a flow-table or micro-program concept, so that they may readily be adapted for general (computer) or special-purpose applications. The chapter concludes with a discussion of error-detecting arithmetic logic. In this type of

Table 5.1
Half-adder and subtractor circuits

(a) Binary addition table

B \ A	0	1
0	0	1
1	1	(1)0

(b) Binary half-adder
S = A + B

A	B	Sum (S)	Carry (C′)
0	0	0	0
0	1	1	0
1	0	1	0
1	1	0	1

(c) Binary half-subtractor
S = A − B

A	B	Difference (D)	Borrow (B_0')
0	0	0	0
0	1	1	1
1	0	1	0
1	1	0	0

Table 5.2
Full-adder and subtractor circuits

(a) Binary full-adder

A	B	C	S	C′
0	0	0	0	0
0	0	1	1	0
0	1	0	1	0
0	1	1	0	1
1	0	0	1	0
1	0	1	0	1
1	1	0	0	1
1	1	1	1	1

(b) Binary full-subtractor

A	B	B_o	D	B_o'
0	0	0	0	0
0	0	1	1	1
0	1	0	1	1
0	1	1	0	1
1	0	0	1	0
1	0	1	0	0
1	1	0	0	0
1	1	1	1	1

(c) K-Maps

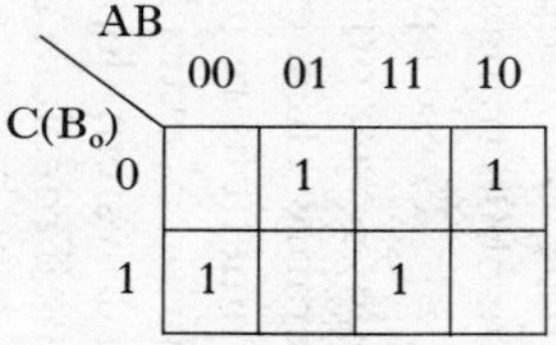

Sum and difference

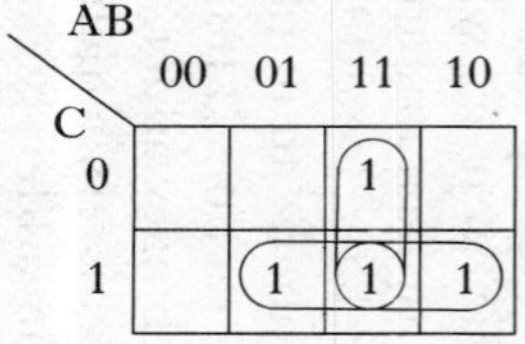

$C(B_o)$ \ AB	00	01	11	10
0		1		1
1	1		1	

Carry

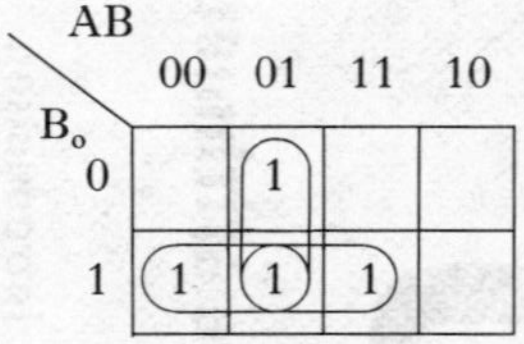

C \ AB	00	01	11	10
0			1	
1		1	1	1

Borrow

B_o \ AB	00	01	11	10
0		1		
1	1	1	1	

Table 5.3
Combined adder/subtractor circuit

(a) Truth-table

M	A	B	B_C	S_D	B_C'
0	0	0	0	0	0
0	0	0	1	1	1
0	0	1	0	1	1
0	0	1	1	0	1
0	1	0	0	1	0
0	1	0	1	0	0
0	1	1	0	0	0
0	1	1	1	1	1
1	0	0	0	0	0
1	0	0	1	1	0
1	0	1	0	1	0
1	0	1	1	0	1
1	1	0	0	1	0
1	1	0	1	0	1
1	1	1	0	0	1
1	1	1	1	1	1

(b) K-Map for B_C' output

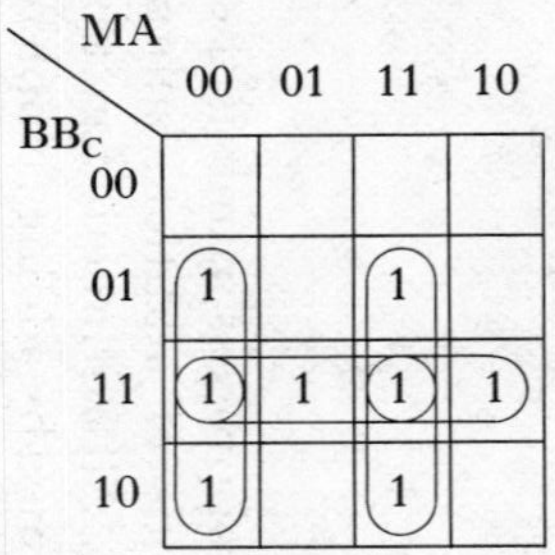

BB_C \ MA	00	01	11	10
00				
01	1		1	
11	1	1	1	1
10	1		1	

logic circuit, for example, the operands include a parity-check digit which is retained throughout the arithmetic operations and used as a final check on the correctness of the result.

5.2 Binary addition and subtraction circuits

The rules of binary addition are shown in Table 5.1(a). Note that if the carry is ignored the operation described is modulo 2 addition and subtraction (the remainder left after dividing the result of the arithmetic operation by 2) which are identical. The modulo 2 function is more commonly known as the exclusive OR relationship – that is, $S = A\bar{B} + \bar{A}B$ (or $S = A \oplus B$). The truth-table for the **half-adder** circuit, which considers two inputs A and B, and a sum (S) and carry (C′) output (that is, the direct implementation of the rules of binary arithmetic) is shown in Table 5.1(b). The Boolean equations may be read directly from the truth-table and are:

$$S = A\bar{B} + \bar{A}B \qquad C' = AB$$

A similar set of equations may be derived for the **half-subtractor** circuit (see Table 5.1(c)). These are:

$$D = A\bar{B} + \bar{A}B \qquad B'_o = \bar{A}B$$

where D is the difference and B'_o the borrow. Integrated circuit versions of the modulo 2 adder (exclusive OR) and the half-adder circuit are available as standard modules in the majority of logic systems; typical half-adder circuits are shown in Fig. 5.1.

The **full-adder** (or full-subtractor) must take into account the carry (or borrow) input from the previous digit stage, or time period if a serial device. The design tables for both full-adder and full-subtractor circuits are shown in Table 5.2. The switching equations for the full-adder are:

$$S = \bar{A}\bar{B}C + \bar{A}B\bar{C} + ABC + A\bar{B}\bar{C}$$
$$C' = \bar{A}BC + AB\bar{C} + ABC + A\bar{B}C = BC + AB + AC$$

and for the Full-subtractor:

$$D = \bar{A}\bar{B}B_o + \bar{A}B\bar{B}_o + ABB_o + A\bar{B}\bar{B}_o$$
$$B'_o = \bar{A}\bar{B}B_o + \bar{A}B\bar{B}_o + \bar{A}BB_o + ABB_o = \bar{A}B_o + \bar{A}B + BB_o$$

where C and B_o are the carry and borrow, respectively, from the previous stages.

These full-adder equations may be implemented directly in terms of NAND/NOR elements (see Fig. 5.2(a)) or alternatively the equations may be factorized into a more convenient form. For example, consider the sum equations:

$$S = (\bar{A}\bar{B} + AB)C + (\bar{A}B + A\bar{B})\bar{C}$$

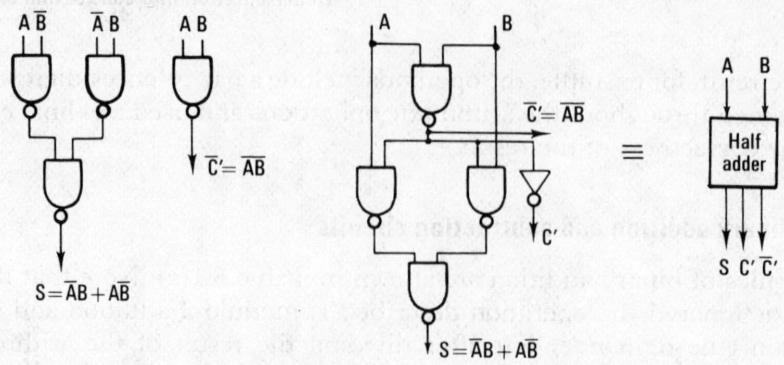

Figure 5.1 Half-adder circuits.

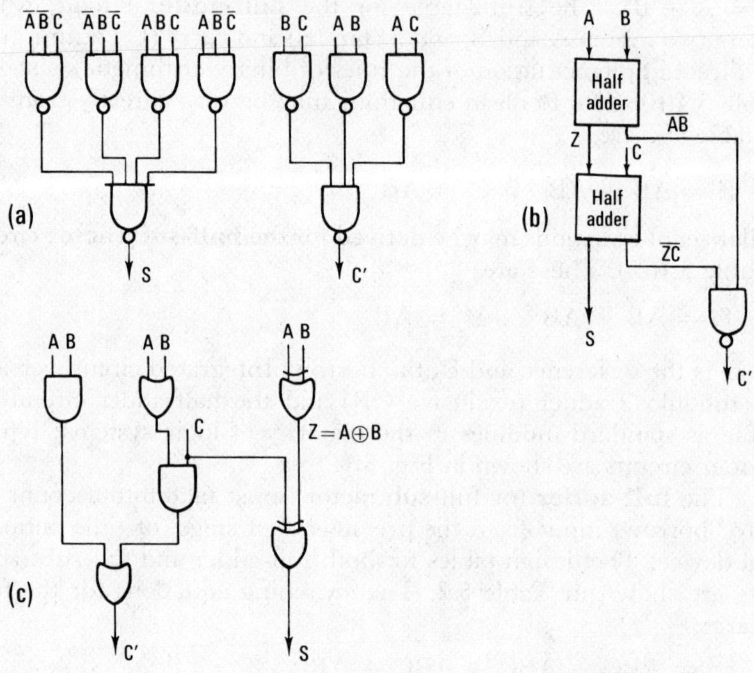

Figure 5.2 Full-adder circuits.

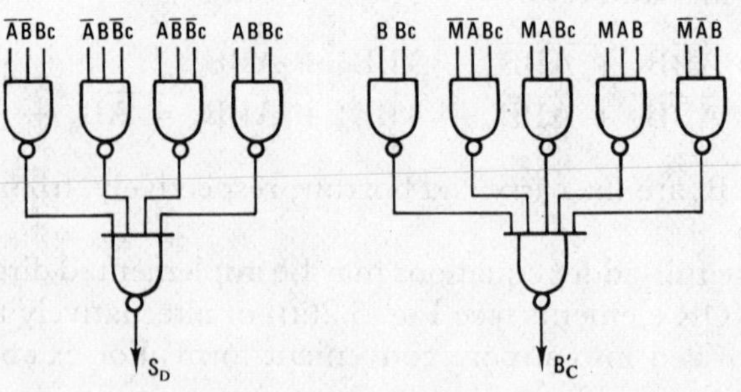

Figure 5.3 Adder–subtractor circuit.

now since

$$\overline{(\bar{A}B + A\bar{B})} = (A + \bar{B})(\bar{A} + B) = AB + \bar{A}\bar{B}$$

and letting $\quad Z = \bar{A}B + A\bar{B}$

then $\quad S = C\bar{Z} + \bar{C}Z = C \oplus Z$

which may be implemented using two cascaded exclusive OR circuits. Similarly, the carry expression may be manipulated to give:

$$C' = (\bar{A}B + A\bar{B})C + AB(C + \bar{C})$$

thus $\quad C' = ZC + AB$

Thus a complete adder may be constructed using half-adder circuits as shown in Fig. 5.2(b). Note that the original two-level circuit would be faster owing to the increased propagation delays through the exclusive OR circuits. A full-subtractor circuit may be designed using the same procedure; the corresponding set of equations are:

$$D = \bar{Z}B_0 + Z\bar{B}_0 \qquad B_0' = \bar{Z}B_0 + \bar{A}B$$

An alternative configuration for the full-adder circuit, which in some cases (depending on the logic system) is slightly more economical on hardware, is shown in Fig. 5.2(c). In this case the basic adder equations have been rearranged to the form:

$$S = C\bar{Z} + \bar{C}Z = C \oplus Z$$

and $\quad C' = G + TC$

where G is the carry generate term (AB) and T the carry transmit (A + B). Note also that in the worst case the carry has to pass through three levels of logic only, thus reducing the carry propagation time (see later in Section 5.5).

A combined adder–subtractor circuit may be designed if a control waveform *M* is included with the inputs A and B; M = 1 for addition and M = 0 for subtraction. The design tables are shown in Table 5.3 where S_D is the sum or difference output and B_C is the borrow or carry output. The tables yield the following equations:

$$S_D = \bar{A}\bar{B}B_C + \bar{A}B\bar{B}_C + A\bar{B}\bar{B}_C + AB\bar{B}_C$$
$$B_C' = \bar{M}\bar{A}B_C + MAB_C + \bar{M}\bar{A}B + MAB + BB_C$$

As one would expect, the sum and difference outputs are the same, and the waveform M controls the carry–borrow logic only; the circuit diagram is shown in Fig. 5.3.

The subtractor circuit as such is seldom if ever used in computer circuits except perhaps in special control logic, for example in controlling looping or shifting operations when a count-down to zero is required. In the majority of cases subtraction can easily be performed by adding the 2's complement of the number to be subtracted.

Table 5.4
Serial full-adder circuit

(a)

Present state	*Next state* (Inputs AB) 00	01	11	10	*Output* S (Inputs AB) 00	01	11	10
1	1	1	2	1	0	1	0	1
2	1	2	2	2	1	0	1	0

(b)

Present state	*Next state* (Inputs AB) 00	01	11	10	*Output* S (Inputs AB) 00	01	11	10
0	0	0	1	0	0	1	0	1
1	0	1	1	1	1	0	1	0

(c)

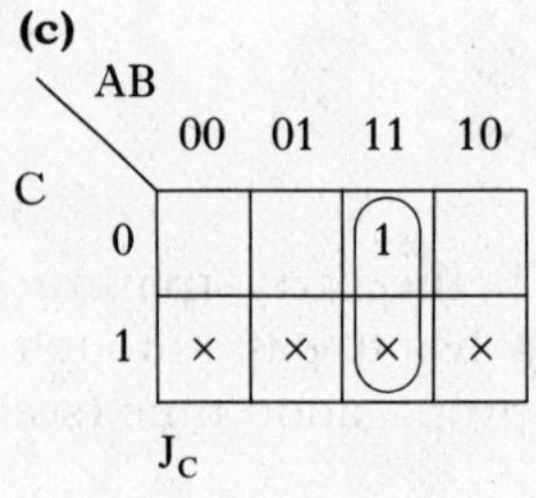

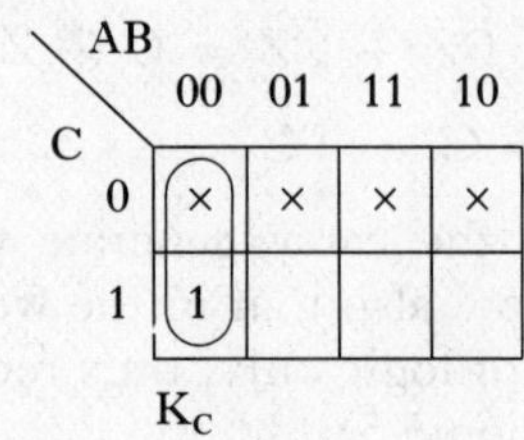

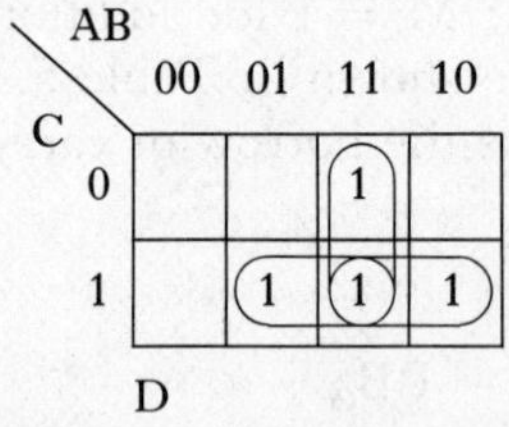

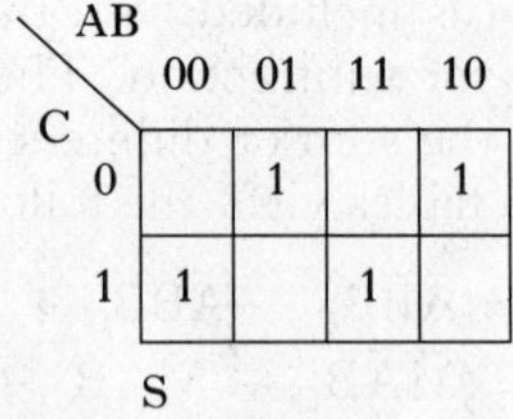

5.3 Serial full-adder circuits

The basic full-adder circuit, as described above, can be used in either the parallel or serial mode of operation. If it is used in the serial mode C′ must be stored and applied to the C input of the full-adder one digit-time later. This delay of one digit-time, controlled by the basic clock frequency of the system, may be achieved by using a clocked D-type or JK bistable. The circuit may also be designed as a synchronous sequential machine (see Table 5.4) and implemented in terms of either JK or D-type bistables. The input equations using a JK bistable are:

$$J_C = AB \qquad K_C = \bar{A}\bar{B}$$

and the output is given by:

$$S = \bar{A}\bar{B}C + \bar{A}B\bar{C} + ABC + A\overline{BC}$$

Using a D-type bistable the input equation becomes:

$$D = AB + BC + AC$$

which requires slightly extra logical gating; the sum output of course remains the same.

It is easy to see that the correct carry is generated when using a JK bistable by substituting the input equations into the characteristic equation[1] for the bistable, thus:

$$C_+ = J\bar{C} + \bar{K}C$$

where C_+ is the output of the bistable, that is, the carry, in the next clock-pulse. Now:

$$C_+ = AB\bar{C} + \overline{\bar{A}\bar{B}}C$$

$$= AB\bar{C} + AC + BC$$

Thus $C_+ = AB + AC + BC$

The logic diagram for a serial adder using shift-register storage for the operands is shown in Fig. 5.4. If there is no need to preserve both the addend or augend the sum register may be dispensed with, and the sum output of the adder taken to the input of one of the operand registers (modified to shift right only); in this form the circuit is called an **accumulator**. The circuit may easily be modified to a serial adder/subtractor circuit as shown in Fig. 5.5, but it is generally more convenient to use a full-adder and 2's complemented numbers to perform subtraction. This may be done by inverting the output of the appropriate operand register (or by taking the complementary outputs), to form the 1's complement, and then adding +1 in the least significant digit position by setting the carry bistable to one (using the set input) before the subtraction operation. Note that in all cases it takes *n* digit times or one complete wordtime (the operands must be clocked sequentially out of the registers) to add together

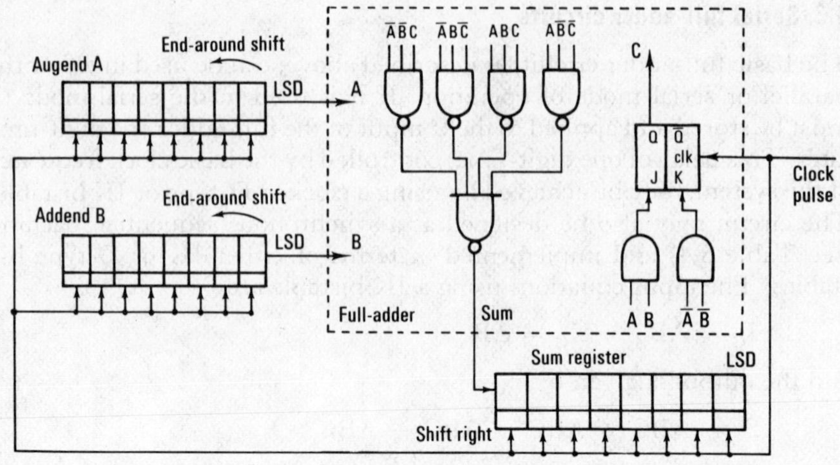

Figure 5.4 Serial full-adder circuit.

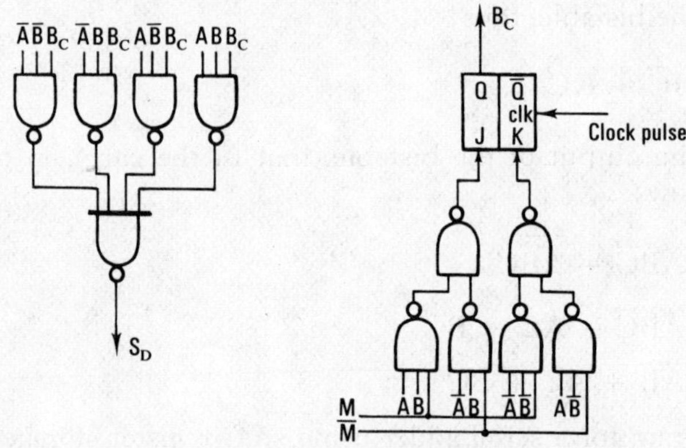

Figure 5.5 Serial adder–subtractor circuit.

two n-bit numbers, but only one full-adder circuit and a one-bit delay element is required.

5.4 Cascaded serial adder circuits

In some applications it may be required to add together, simultaneously, a number of serial binary numbers, that is, a multi-input binary adder. Conventional techniques using repeated addition tend to be either too slow if performed serially, or too costly if parallel methods (see later) are used. One solution to this problem is to use iterative circuits based on the simple two-input serial adder described in the last section. This approach has been discussed by Hennie[2] who suggested the simple cascaded system shown in Fig. 5.6(a) for six inputs ABCDEF. In this circuit the inputs AB are taken to the first adder, and then the sum output of this stage, together with input C, taken to the second adder stage, and so on. Note that each 2-input adder stage is fed with basic timing pulses (clock) and switches simultaneously. The operation of the circuit is such that, before the first clock pulse, the combinational logic in the adder stages produces the sum of the least significant digit, that is:

$$S = A \oplus B \oplus C \oplus D \oplus E \oplus F$$

Similarly, the carry producing logic presents the appropriate input conditions to the terminals of the JK carry bistable of each stage. When the clock pulse arrives the sum digit is shifted into an output register, the carry bistables are set, and the next input digits are entered to the circuit; the operation then proceeds as before.

Using this circuit q binary numbers can be added in one wordtime, that is, n clock pulse times (digit-times), where n is the number of bits in the binary word. For this particular circuit the maximum delaying path is through $2(q - 1)$ logic levels, thus the clock pulse period T must not be less than $2\alpha(q - 1)$ seconds, where α is the average propagation delay per level and q the number of binary inputs.

The simple cascaded circuit may be restructured so as to reduce the number of levels in the combinational logic, and hence increase its speed of operation; Fig. 5.6(b) shows a typical circuit. This circuit may be clocked faster since the delaying path is reduced to six logic levels allowing $T \nless 6\alpha$ seconds, furthermore no extra logic is required. Considerably faster addition times can be obtained if, instead of using 2-input adders only, a special 3-input adder is incorporated into the design[3] (see Fig. 5.6(c)). The speed of operation of this circuit is increased to $T \nless 4\alpha$ seconds (delaying path through 4 logic levels), but at the expense of slightly more logic.

The state-tables for a synchronous 3-input adder are shown in Table 5.5 and these follow directly from conventional logic design theory. The major design problem for the adder is the treatment of carries over more than one stage, that is, protracted carries. Consider the addition of

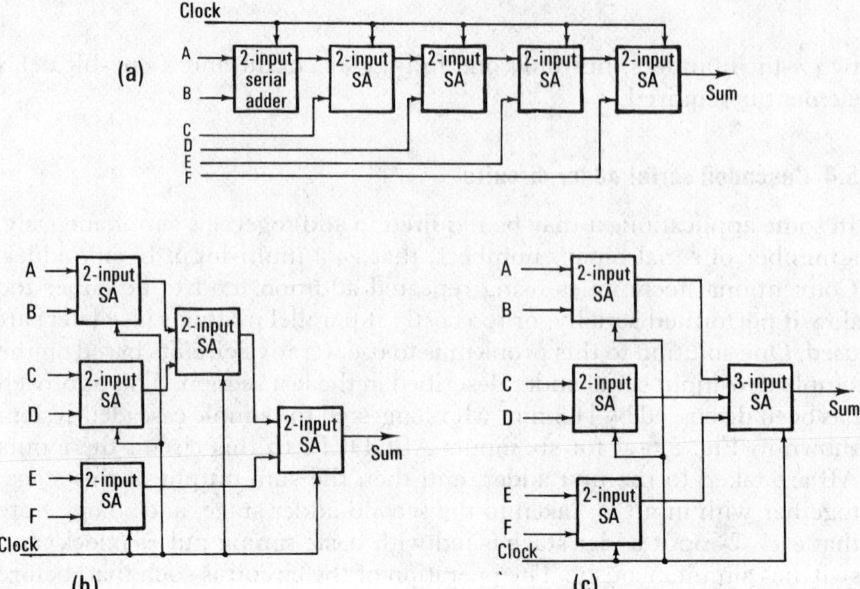

Figure 5.6 Iterative serial adders.

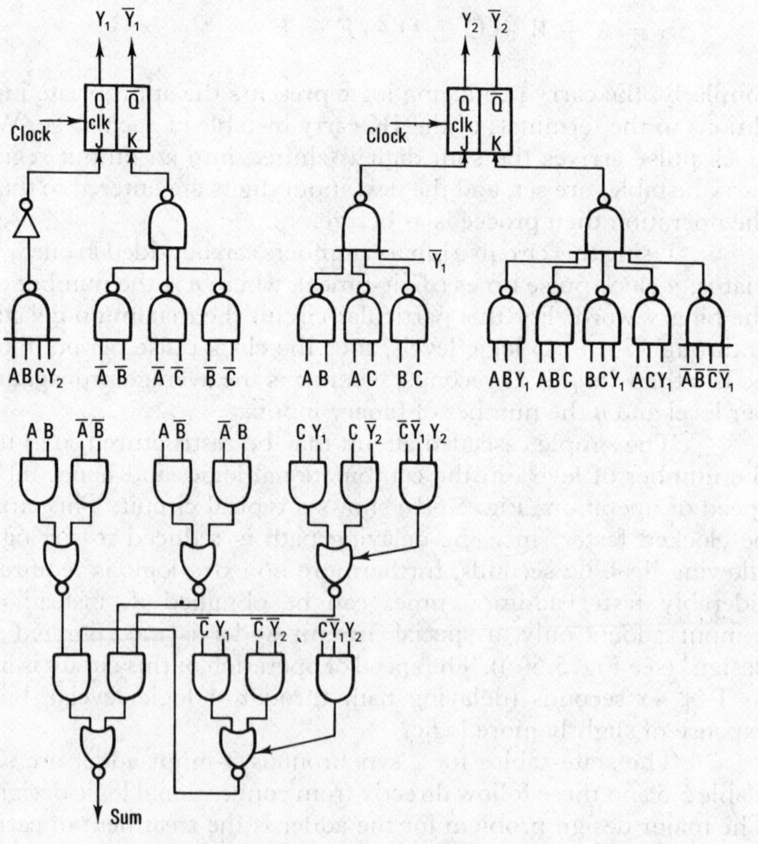

Figure 5:7 Three-input synchronous adder.

the three serial numbers A = 0110, B = 1110 and C = 1110, shown below:

Clock	5	4	3	2	1	0
A	0	0	0	1	1	0
B	0	0	1	1	1	0
C	0	0	1	1	1	0
	1	0	0	0	1	0

Note that there is a protracted carry produced in the third column from the right which must be added to the next but one digit position. As a consequence of this the circuit effectively requires two carry bistables, that is, a four-state sequential machine. From Table 5.5(a), if the machine is in present state 1 with input $A_0B_0C_0$ the circuit state is unchanged; this is the starting condition. The input condition $A_1B_1C_1$ in the next clock pulse causes a transition to state 2 (to indicate a single carry) and outputs a 1. A further input of $A_2B_2C_2$ sends the circuit to state 3 (indicating a protracted carry) and outputs a 0. In state 3, an input of $\bar{A}_3B_3C_3$ causes a transition to state 4 (carry plus protracted carry) and outputs 0. The next input will be all zeros (clock pulse 4) sending the circuit to state 2 (single carry) and outputs 0. The final input in clock pulse 5, again all zeros, returns the circuit to state 1 and outputs a 1. Note

Table 5.5
State-tables for 3-input adder
(a)

	Inputs ABC *Next states and output* S							
Present state	000	001	010	011	100	101	110	111
1	1/0	1/1	1/1	2/0	1/1	2/0	2/0	2/1
2	1/1	2/0	2/0	2/1	2/0	2/1	2/1	3/0
3	2/0	2/1	2/1	4/0	2/1	4/0	4/0	4/1
4	2/0	2/1	2/1	3/0	2/1	3/0	3/0	3/1

(b)

	Inputs ABC *Next states and output* S							
Y_1Y_2	000	001	010	011	100	101	110	111
00	00/0	00/1	00/1	01/0	00/1	01/0	01/0	01/1
01	00/1	01/0	01/0	01/1	01/0	01/1	01/1	10/0
10	01/0	01/1	01/1	11/0	01/1	11/0	11/0	11/1
11	01/0	01/1	01/1	10/0	01/1	10/0	10/0	10/1

that the circuit overflows, since the number range is exceeded, and two additional clock pulses are required to obtain the final number. A similar analysis may be performed for input sequences of any length.

The assigned state-table is shown in Table 5.5(b) and yields the following input equations for clocked JK bistables:

$$J_{Y_1} = ABCY_2 \qquad K_{Y_1} = \bar{A}\bar{B} + \bar{A}\bar{C} + \bar{B}\bar{C}$$
$$J_{Y_2} = Y_1 + AB + AC + BC$$
$$K_{Y_2} = ABY_1 + ABC + BCY_1 + ACY_1 + \bar{A}\bar{B}\bar{C}\bar{Y}_1$$

The sum equation is given by:

$$S = (\bar{A}B + A\bar{B})(\bar{C}Y_1 + \bar{C}\bar{Y}_2 + C\bar{Y}_1Y_2) + (AB + \bar{A}\bar{B})(CY_1 + C\bar{Y}_2 + \bar{C}\bar{Y}_1Y_2)$$

These equations may be implemented in the usual way to give the circuit shown in Fig. 5.7. Note the use of AND/NOR gates in the implementation, the delay through this configuration (which is available as a standard module) is equivalent to one logic level, that is, a NOR gate.

5.5 Parallel full-adder circuits

Binary numbers represented in the parallel mode may be added together using the circuit shown in Fig. 5.8. A separate full-adder circuit is required for each bit of the operand, except the least significant stage which can be a simple half-adder circuit as there is no preceding carry. It is usual practice, however, to use a full-adder for this stage as well as it is then possible to add +1 to the inputs (for example, to increment the instruction register or in the process of forming the 2's complement of a number) by initially setting the carry input of the first stage to 1. This is equivalent to setting the carry bistable to 1 in a serial adder.

The parallel adder, in the ideal case, adds two *n*-bit numbers in 1 bit time (or, in reality, the time taken to propagate through the logic), but it uses approximately *n* times as much hardware as the serial adder to realize this speed. This ideal speed is never obtained in practice owing to the carry propagation, from stage to stage, through the circuit (each

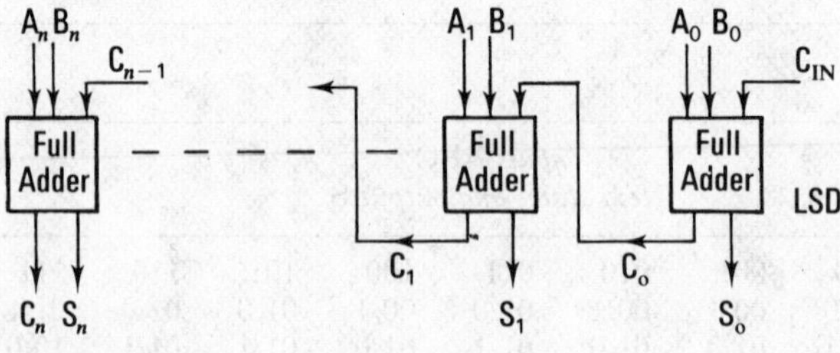

Figure 5.8 Parallel full-adder circuit.

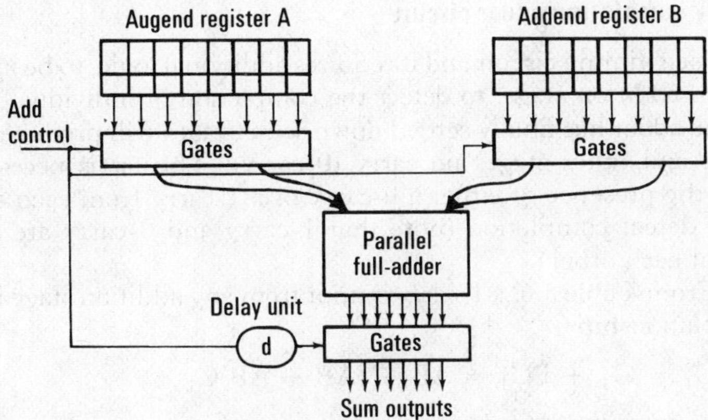

Figure 5.9 Adder timing.

stage must wait until its predecessor has determined its carry output). In the worst case the carry digit can propagate through all stages of the adder to the most significant stage; for example, in a 10-bit adder the addition of 0111111111 and 0000000001 would cause this to occur.

The simplest way to overcome this timing problem is to allow sufficient time for a full length carry propagation through the adder to be completed. This means in practice that the adder system must be clocked (this type of adder is called **synchronous** for this reason). The sum outputs must be gated by a control pulse into the sum register after the maximum time has occurred, thus the additions always take a fixed length of time irrespective of the numbers to be added. Thus if β is the carry propagation time per stage and n the number of bits in the computer word, the addition time is always $n\beta$ seconds.

One practical method of timing that can be used in an otherwise fully asynchronous system is to delay the control signal to add (that is, the gating signals to the operand registers, see Fig. 5.9) by an appropriate amount equal to the worst-case carry propagation delay, using a monostable or fixed delay element. Alternatively, if the addition operation is micro-programmed the timing is inherently obtained by the choice of the micro-order timing slots. For example, in the micro-program for addition (Table 4.6) the operands are gated to the arithmetic unit in time slot 23 and the sum output gated back to the accumulator in time slot 24.

Since a full length carry is an infrequent occurrence an excessive amount of time is wasted using this method. Von Neuman and others[4, 5] have shown, using 40-bit randomly generated numbers, that the average carry of 1's occurs over 4·6 stages. Thus a logic system that detects the completion of all carries, and then generates a control signal that gates the sum outputs into the sum register (**self-timing** or **autoasynchronous** adder), would give approximately an eight times increase in speed for 40-bit operands. In general the addition time would depend upon the size and nature of the numbers being added.

5.6 Carry completion adder circuit

This is a self-timing circuit and it requires additional logic, to be included with each addition stage, to detect the completion of individual carries. When the adder has finally settled down some stages will produce a carry (1-carry) and other stages no carry (0-carry). Thus it is necessary to indicate the presence of either a 0-carry or a 1-carry from each stage in order to detect completion (note that 1-carry and 0-carry are not the inverse of each other).

From Table 5.6, a 1-carry output from any addition stage is given by the relationship:

$$C_1' = G_1 + PC_1 = AB + (\bar{A}B + A\bar{B})C_1$$

and a 0-carry output by:

$$C_0' = G_0 + PC_0 = \bar{A}\bar{B} + (\bar{A}B + A\bar{B})C_0$$

where G_1 and G_0 are the 1-carry and 0-carry generated in the stage, and P indicates that the carry is propagated through the stage. The equations above are shown implemented in Fig. 5.10; note that the carry completed gate must have a fan-in factor equal to the number of stages in the adder. A block diagram of the carry complete adder is shown in Fig. 5.11.

Table 5.6
Carry complete circuit

A	B	C	C′	
0	0	0	0	No carry generated $G_0 = \bar{A}\bar{B}$
0	0	1	0	
0	1	0	0	Carry propagated $P = \bar{A}B + A\bar{B}$
0	1	1	1	
1	0	0	0	
1	0	1	1	
1	1	0	1	Carry generated $G_1 = AB$
1	1	1	1	

In order to prevent extraneous signals on the C_0' and C_1' lines generating premature carry complete signals it is necessary to hold off the propagated carry outputs, by means of a control signal H, until the start of the final addition operation. Note that a C_1' signal is injected at each stage where $A = B = 1$ and a C_0' signal when $A = B = 0$.

The operation of the adder commences with the augend and addend being gated to the input of the adder (in some cases these input lines may be permanently connected). At this time the H control line is low, consequently there will be no propagated carry signals and only those stages which generate internal carries (i.e. those with the input values $AB = \bar{A}\bar{B} = 1$) will have C_1' outputs, otherwise *both* C_1' and C_0' will be zero. This means, except for the case when $P = 0$ for all stages, that the carry completed signal will always be low at this time. Once the adder stages have settled out (a maximum of two passes through the

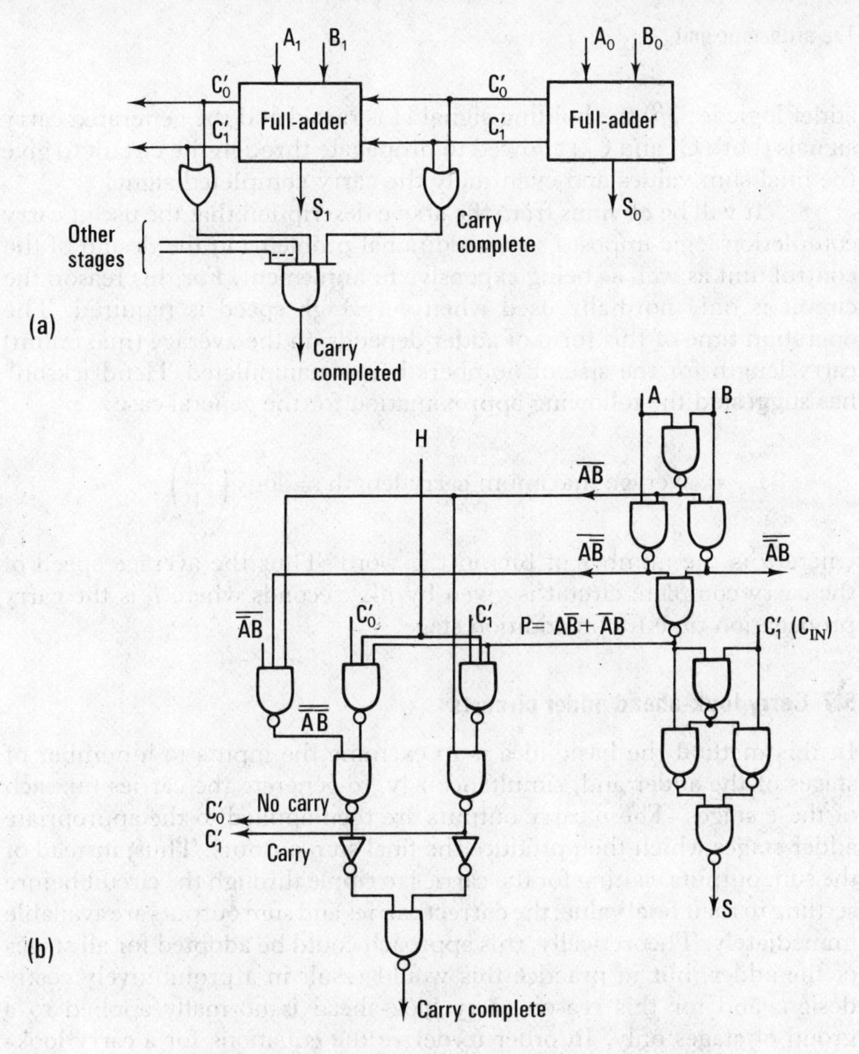

Figure 5.10 Carry complete logic, circuitry per stage.

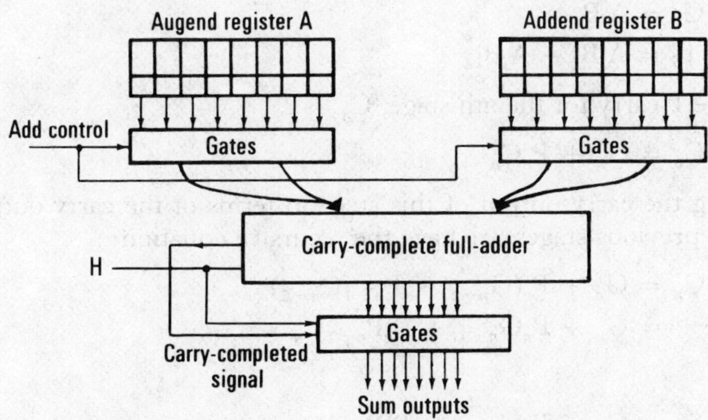

Figure 5.11 Block diagram carry-complete adder.

adder logic ie: 2β) the holding signal H is raised and the generated carry signals (both C_1' and C_0') allowed to propagate through the circuit to give the final sum values and eventually the carry completed signal.

It will be obvious from the above description that the use of carry completion logic imposes some additional problems in the design of the control unit as well as being expensive to implement. For this reason the circuit is only normally used when very high speed is required. The operation time of this form of adder depends on the average (maximum) carry length for the size of numbers being manipulated. Hendrickson[6] has suggested the following approximation for the general case:

$$L_c = \text{average maximum carry length} \doteqdot \log_2 \left(\frac{5n}{4}\right)$$

where n is the number of bits in the word. Thus the average speed of the carry complete circuit is given by βL_c seconds where β is the carry propagation time for an addition stage.

5.7 Carry look-ahead adder circuits

In this method the basic idea is to examine the inputs to a number of stages of the adder and, simultaneously, to generate the carries for each of these stages. These carry outputs are then applied to the appropriate adder stages which then produce the final sum outputs. Thus, instead of the sum outputs waiting for the carries to ripple through the circuit before settling to their final value, the correct carries and sum outputs are available immediately. Theoretically, this approach could be adopted for all stages of the adder, but in practice this would result in a prohibitively costly design, and for this reason carry look-ahead is normally applied to a group of stages only. In order to derive the equations for a carry look-ahead adder we will use the concepts of 1-carry and propagate as defined above. The general expressions are given by:

$$G_i = A_i B_i$$

and $$P_i = A_i \overline{B}_i + \overline{A}_i B_i$$

Now the 1-carry for the nth stage, C_n, is:

$$C_n = G_n + P_n C_{n-1}$$

Defining the carry output of this stage in terms of the carry outputs of the two previous stages, we have the recursive equation:

$$\begin{aligned} C_n &= G_n + P_n(G_{n-1} + P_{n-1}C_{n-2}) \\ &= G_n + P_n G_{n-1} + P_n P_{n-1} C_{n-2} \end{aligned}$$

Thus

$$C_n = G_n + P_n G_{n-1} + P_n P_{n-1} G_{n-2} + \cdots + P_n P_{n-1} P_{n-2} \ldots P_1 C_0$$

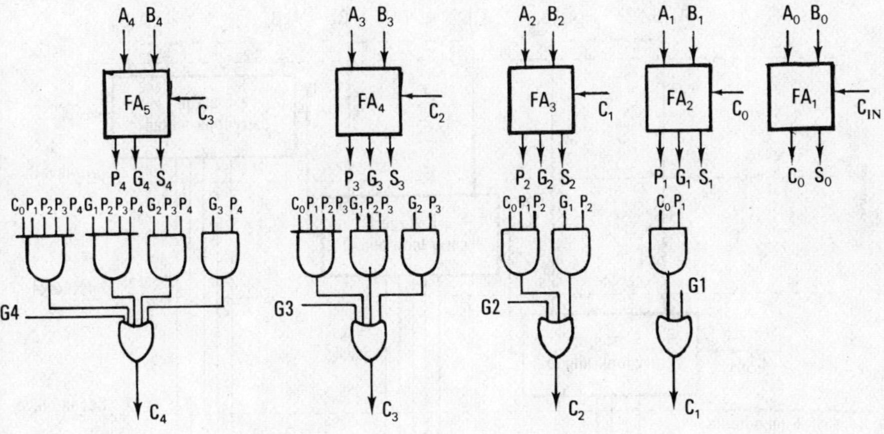

Figure 5.12 Five stage zero-level carry look-ahead.

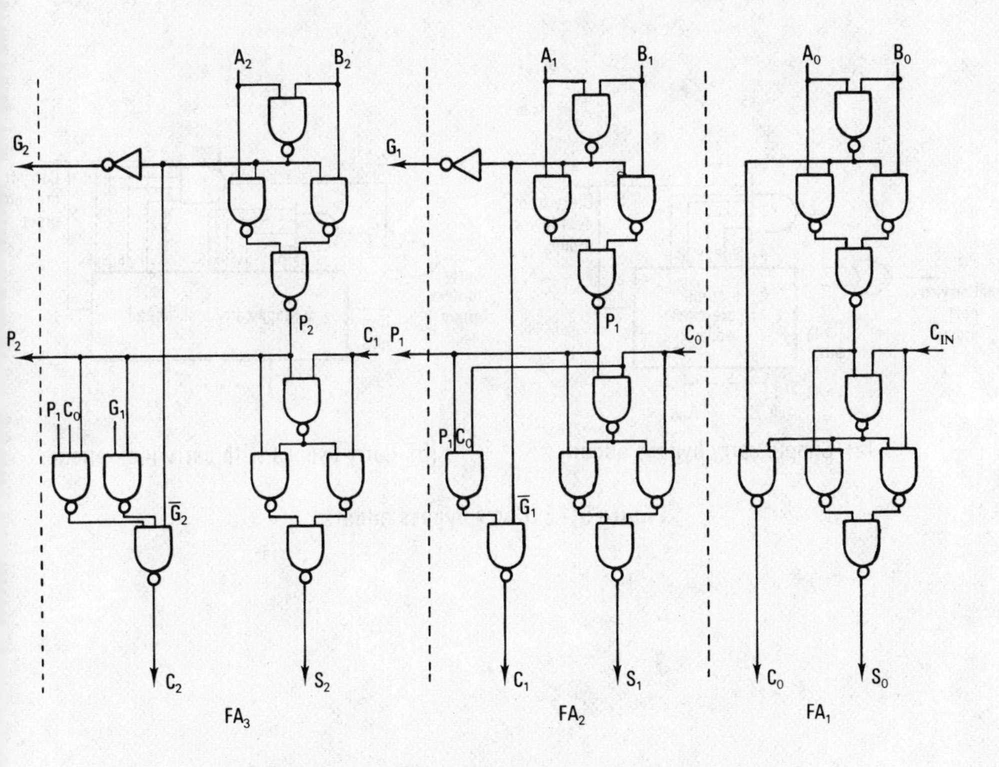

Figure 5.13 NAND implementation of carry look-ahead adder.

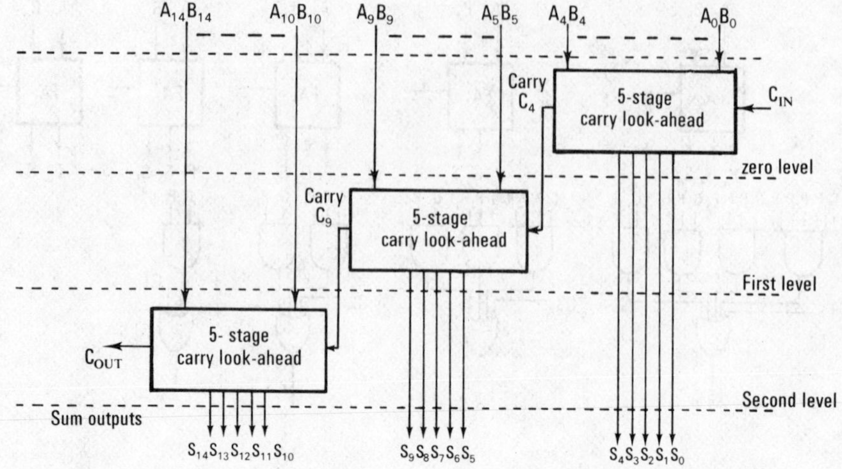

Figure 5.14 Carry look-ahead with ripple carry between groups.

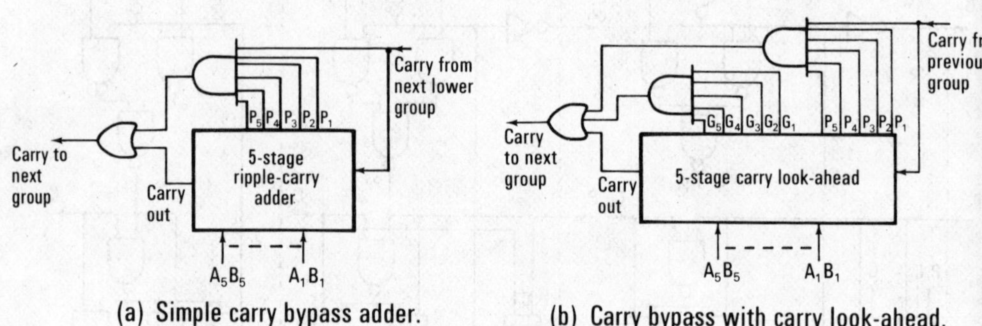

(a) Simple carry bypass adder.

(b) Carry bypass with carry look-ahead.

Figure 5.15 Carry bypass adders.

If we define $C_0 = G_0$ then we can write:

$$C_n = G_n + \left(\prod_{i=n}^{n} P_i\right) G_{n-1} + \left(\prod_{i=n-1}^{n} P_i\right) G_{n-2} + \left(\prod_{i=n-2}^{n} P_i\right) G_{n-3} + \cdots + \left(\prod_{i=1}^{n} P_i\right) G_0$$

Thus each term of the expression consists of a generate term and the product of a number of propagate terms. Thus we may say that a carry is produced in stage *n* if:

(a) stage *n* generates it itself;

(b) the previous stage $(n - 1)$ generates a carry and stage *n* propagates it, *or* if stage $(n - 2)$ generates it and it is propagated by stages *n* and $(n - 1)$, etc.;

(c) there is an initial carry and all stages propagate it.

The carry equations may also be written in the shortened form:

$$C_n = \sum_{j=0}^{n} \prod_{i=j+1}^{n} P_i G_j$$

The block diagram of a typical system is shown in Fig. 5.12, and an implemented version using NAND logic in Fig. 5.13. In the latter case full-adders composed of exclusive OR half-adders (see Fig. 5.2(b)) have been used. The propagate term P may be obtained from the Z output of the first half-adder, and the generate term G from the half-carry output of the same unit; note that it is necessary to provide both phases of the generate term.

Carry look-ahead is normally applied between bits in a group; the number of bits (or stages) in a group depends on the circuit modules available and is usually limited by the fan-in factor of the unit. Five stages are commonly accepted as a good engineering compromise. Thus the simplest form of fast adder consists of carry look-ahead within groups and ripple-carry between groups (see Fig. 5.14). In such an adder the first group (starting from the least significant end) is called the zero level and generates a carry output which must be applied to the next group (first level) as an input. Thus the first level must wait until the zero level has produced a carry output before it can produce an output itself for the second level, and so on. Consequently, all the carry and sum outputs are produced in a time proportional to the number of levels multiplied by the propagation delay of one carry look-ahead adder stage.

A similar and alternative circuit uses the **carry-bypass** scheme. In this technique the full-adder is split into groups of standard ripple-adder stages and each group has a single carry-bypass circuit consisting of all the individual stage propagate terms (see Fig. 5.15(a)). Thus, a carry generated by a lower order group which has to be passed to a higher

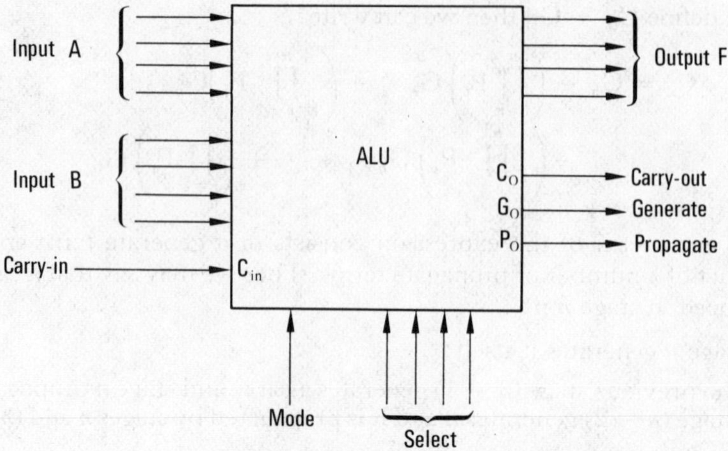

Figure 5.16 Arithmetic and logic unit.

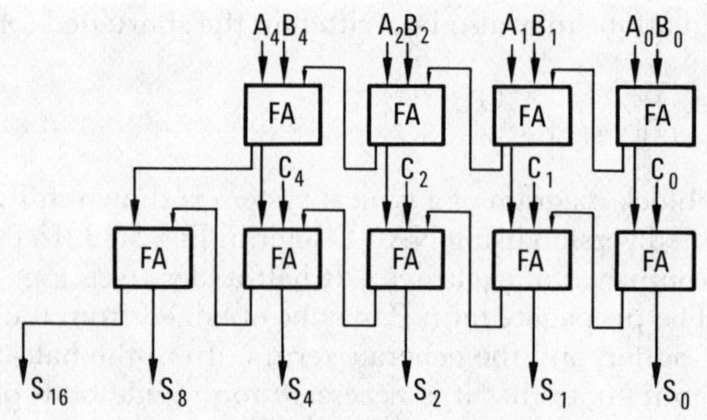

(a) Conventional addition.

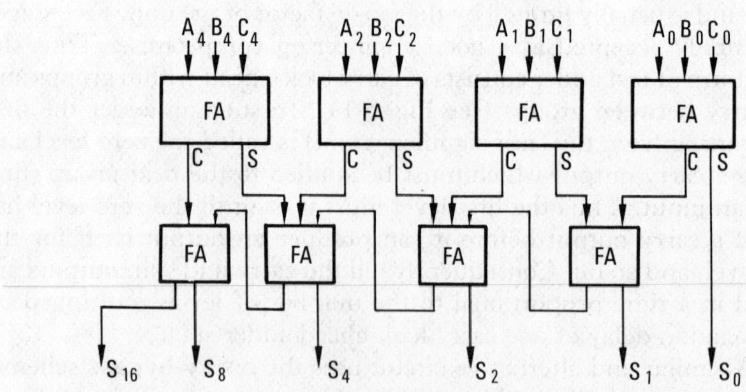

(b) Carry-save addition.

Figure 5.17 Multiple addition.

Table 5.7
ALU Functions

Selection	*Mode*		
$S_3S_2S_1S_0$	M = 1 (Logic)	M = 0 (Arithmetic)	
		$C_{in} = 1$	$C_{in} = 0$
0 0 0 0	$F = \overline{A}$	$F = A$	$F = A + 1$
0 0 0 1	$F = \overline{A \vee B}$	$F = A \vee B$	$F = (A \vee B) + 1$
0 0 1 0	$F = \overline{A} \wedge B$	$F = A \vee \overline{B}$	$F = (A \vee \overline{B}) + 1$
0 0 1 1	$F = 0$	$F = -1$ (2's comp.)	$F = 0$
0 1 0 0	$F = \overline{A \wedge B}$	$F = A + (A \wedge \overline{B})$	$F = A + (A \wedge \overline{B}) + 1$
0 1 0 1	$F = \overline{B}$	$F = (A \vee B) + (A \wedge \overline{B})$	$F = (A \vee B) + (A \wedge \overline{B}) + 1$
0 1 1 0	$F = A \oplus B$	$F = A - B - 1$	$F = A - B$
0 1 1 1	$F = A \wedge \overline{B}$	$F = (A \wedge \overline{B}) - 1$	$F = A \wedge \overline{B}$
1 0 0 0	$F = \overline{A} \vee B$	$F = A + (A \wedge B)$	$F = A + (A \wedge B) + 1$
1 0 0 1	$F = \overline{A \oplus B}$	$F + A + B$	$F = A + B + 1$
1 0 1 0	$F = B$	$F = (A \vee \overline{B}) + (A \wedge B)$	$F = (A \vee \overline{B}) + (A \wedge B) + 1$
1 0 1 1	$F = A \wedge B$	$F = (A \wedge B) - 1$	$F = (A \wedge B)$
1 1 0 0	$F = 1$	$F = A + A$*	$F = A + A + 1$
1 1 0 1	$F = A \vee \overline{B}$	$F = (A \vee B) + A$	$F = (A \vee B) + A + 1$
1 1 1 0	$F = A \vee B$	$F = (A \vee \overline{B}) + A$	$F = (A \vee \overline{B}) + A + 1$
1 1 1 1	$F = A$	$F = A - 1$	$F = A$

Where AND = $\wedge$, OR = $\vee$, ADD = + and SUB = −
*each bit is shifted to next MS place

order group has only to pass through one level of logic circuits. The carry-bypass scheme may also be used in conjunction with carry look-ahead within the groups themselves; this is shown in Fig. 5.15(b). In this case both the carry propagate and the carry generate terms are included, that is, $P_g = P_1P_2P_3P_4P_5$ and $C_g = G_1G_2G_3G_4G_5$. As well as bypassing the carry from the previous stages, the circuit also detects when all stages are generating a carry and if so a carry output is produced immediately and passed to the next stage, using one stage of logic only.

Other methods exist for designing high-speed adders utilizing the concepts described above, among them are the use of auxiliary carry look-ahead levels and the conditional adder.[7,8]

Binary arithmetic logic circuits are normally realised using integrated circuit modules. For example, 4-bit adder chips (i.e. two 2-input adders) 4-bit adder/subtraction units working in 2's complement arithmetic (longer word lengths can be obtained by cascading the chips) and carry look-ahead circuits (normally these are only required for word-lengths greater than 8-bits). Furthermore, it is possible to obtain a complete arithmetic/logic unit (ALU) on a single LSI chip. The ALU accepts two 4-bit words and, according to how the control inputs are selected (see Fig. 5.16 and Table 5.7) performs various arithmetic and logic functions. Note that it is possible to cascade these units since there are separate lines for the carry-in, carry-out, generate and propagate signals.

5.8 Carry save adders

If a sequence of binary numbers are required to be added together (accumulated) this may be done with conventional full-adders as shown in Fig. 5.17(a). An alternative and faster method, often used in multiplier circuits to accumulate partial products, is to perform a partial addition (modulo 2) and add the carries in separately – this procedure is known as **carry-save addition** and is shown in Fig. 5.17(b). For instance, consider the addition of three 4-bit numbers A, B and C, then:

```
             A    1011
             B    0111
             C    1010
                  ----
Partial sum       0110
Carries          1 11    Add with carry propagation
                 -----
Sum              11100
                 -----
```

As no more than three digits will need to be added in any one order on any one step the conventional full-adder can be used to perform the additions. Note also that only the final steps in the additions require carry propagation.

5.9 Overflow and out-of-range circuits

We have seen in Chapter 2 how numbers in a computer are represented in a fractional form with the range $1 > x \geqslant -1$, and using 2's complement notation for negative numbers. When adding or subtracting numbers in this form it is quite possible that the result (unless the original operands are appropriately scaled in the computer program) will go out of range, for example become greater than one. Consequently, it is necessary to include special logic with the arithmetic circuits to detect when this occurs. The output of the detecting logic is used to set a control bistable (or flag), the state of which may be examined under program control using a conditional jump type instruction.

Similarly, when registers overflow, due for example to their contents being shifted left by too many places, an indication must be given. A simple way of detecting overflow in a serial adder is to examine the final condition of the carry bistable (which is always reset to zero at the start of an arithmetic operation). In the case of a parallel adder (and in general for all registers) an extra, overflow stage must be included at the most significant end of the register.

To determine a method for detecting when arithmetic circuits go out of range we will consider a 5-bit number with the range $1 > x \geqslant -1$ and 2's complement notation for negative numbers. Now we have:

(a) maximum positive number is 0·1111 equivalent to 0·9375 or 15×2^{-4},
(b) smallest positive number is 0·0000, that is, zero,
(c) maximum negative number is 1·0000, that is, −1, and
(d) smallest negative number is 1·1111 equivalent to −0·0625 or -1×2^{-4}.

Let us now consider some arithmetic examples:

(a)	0·0110	0·3750	(b)	0·0110	0·3750
	0·1000	0·5000		0·1111	0·9375
	0 ← 0·1110	0·8750		0 ← 1·0101	1·3125
(c)	1·1111	−0·0625	(d)	1·0000	−1·0000
	1·1100	−0·2500		1·1110	−0·1250
	1 ← 1·1011	−0·3125		1 ← 0·1110	−1·1250
(e)	1·1111	−0·0625	(f)	1·0000	−1·0000
	0·1111	0·9375		0·1111	0·9375
	1 ← 0·1110	0·8750		0 ← 1·1111	−0·0625

In examples (a), (c), (e), and (f) the correct answer is obtained; that is, the result is within the number range of the machine representation. Example (b) is incorrect, however, since the number has gone out of range resulting in a *negative* number instead of positive; similarly, in example (d) the answer has gone *positive*.

From the examples above, and the complete truth-table shown in Table 5.8, we may state the following rule:

The number *overflows*, that is, goes negative, if both operands are positive and the sum is negative and the number *underflows*, that is, goes positive, if both operands are negative and the sum is positive.

This may be expressed algebraically as:

$$\text{Out of range} = \text{overflow} + \text{underflow}$$
$$= \overline{A}\,\overline{B}S + AB\overline{S}$$

where A and B are the sign digits of the operands and S the sign digit of the sum.

In some cases this circuit cannot be implemented because the sign digits of the operands are lost in the arithmetic process, for example, when transferring the sum back into a register which originally held one

Table 5.8
Out-of-range circuits

Sign digits of operands		*Carry*	*Sign digit sum*	*Overflow*	*Underflow*
A	B	C	S	O	U
0	0	0	0	0	0
0	0	0	1	1	0
0	0	1	0	×	×
0	0	1	1	×	×
0	1	0	0	×	×
0	1	0	1	0	0
0	1	1	0	0	0
0	1	1	1	×	×
1	0	0	0	×	×
1	0	0	1	0	0
1	0	1	0	0	0
1	0	1	1	×	×
1	1	0	0	×	×
1	1	0	1	×	×
1	1	1	0	0	1
1	1	1	1	0	0

Table 5.9
Truth table for 5421 adder

	B_5	B_4	B_2	B_1
	0	0	0	0
	0	0	0	1
	0	0	1	0
	0	0	1	1
	0	1	0	0
	1	0	0	0
	1	0	0	1
	1	0	1	0
	1	0	1	1
	1	1	0	0
Don't-care terms	0	1	0	1
	0	1	1	0
	0	1	1	1
	1	1	0	1
	1	1	1	0
	1	1	1	1

	A_5	A_4	A_2	A_1	C
	0	0	0	0	0
	0	0	0	0	1
	0	0	0	1	0
	0	0	0	1	1
	0	0	1	0	0
	0	0	1	0	1
	0	0	1	1	0
	0	0	1	1	1
	0	1	0	0	0
	0	1	0	0	1
	1	0	0	0	0
	1	0	0	0	1
	1	0	0	1	0
	1	0	0	1	1
	1	0	1	0	0
	1	0	1	0	1
	1	0	1	1	0
	1	0	1	1	1
	1	1	0	0	0
	1	1	0	0	1
Don't-care terms	0	1	0	1	0
	0	1	0	1	1
	0	1	1	0	0
	0	1	1	0	1
	0	1	1	1	0
	0	1	1	1	1
	1	1	0	1	0
	1	1	0	1	1
	1	1	1	0	0
	1	1	1	0	1
	1	1	1	1	0
	1	1	1	1	1

S_5	S_4	S_2	S_1	C′
0	0	0	0	0
0	0	1	0	0
0	0	1	1	0
1	0	0	0	0
1	0	0	1	0
1	0	1	1	0
1	1	0	0	0
0	0	0	1	1
0	0	1	0	1
0	1	0	0	1

etc.

of the operands. One method of solving this problem would be to store the sign digits beforehand, but it is also possible to get the out-of-range information by examining the overflow carry digit and the carry into the sign-digit position. The rule may be restated as follows:

Overflow occurs if both operands are positive and there is a carry into the sign-digit position, and underflow occurs if both operands are negative and there is no carry into the sign-digit position.

Algebraically we have:

$$\text{Out of range} = \overline{C_P}C + C_P\overline{C}$$

where C_P is the carry into the sign-digit position, and C the overflow carry. The implementation of this circuit, for example in a serial arithmetic unit, would entail providing an additional carry bistable.

5.10 Binary-coded decimal adders

So far we have restricted our discussion to pure binary arithmetic processes. However, a small minority of computers (and some special purpose machines) work in the binary-coded decimal (BCD) notation.[9,10] Consequently, it is sometimes required to design BCD adders that will accept two 4-bit BCD numbers as inputs, and produce the sum in BCD notation. A block diagram of a single BCD stage is shown in Fig. 5.18(a); note that there are nine inputs (two BCD numbers and a carry) and four outputs plus a carry to the next decade.

Let us now consider a BCD adder in more detail and to this end we shall design a parallel adder operating in the 5421 code. The adder could be designed theoretically as a combinational circuit using formal logic design methods, but the amount of computation required for the minimization would be excessive unless a digital computer was used. A shortened form of the truth-table is shown in Table 5.9, and it is necessary to combine each B term with all the A terms (that is, 0000 with 00000 through to 11001, etc.) to obtain the full truth-table which contains 512 entries. Note that it is a nine-variable five-function problem with 200

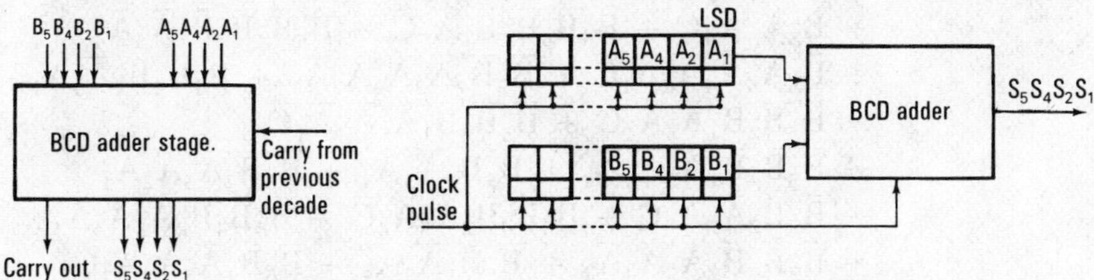

(a) Parallel 5421 adder stage. (b) Serial 5421 adder.

Figure 5.18 Binary coded decimal adders.

specified terms and 312 don't-care terms. The switching equations for the 5421 adder circuit computed using an ICL 1907[11] computer, are given below:

$$\begin{aligned}
C' = {} & B_5B_4A_4 + B_5B_4A_2 + B_5B_4A_1 + B_5B_4C + B_5B_2B_1A_2 \\
& + B_5B_2B_1A_1C + B_5B_2A_4 + B_5B_2A_2A_1 + B_5B_2A_2C \\
& + B_5B_1A_4 + B_5B_1A_2A_1C + B_5A_5 + B_5A_4C \\
& + B_4A_5A_4 + B_4A_5A_2 + B_4A_5A_1 + B_4A_5C \\
& + B_2B_1A_5A_2 + B_2B_1A_5A_1C + B_2A_5A_4 + B_2A_5A_2A_1 \\
& + B_2A_5A_2C + B_1A_5A_4 + B_1A_5A_2A_1C + A_5A_4C
\end{aligned}$$

$$\begin{aligned}
S_1 = {} & B_4A_4\overline{C} + B_4A_2\overline{A}_1\overline{C} + B_4A_1C + B_2B_1A_2A_1\overline{C} \\
& + B_2B_1\overline{A}_1C + B_1\overline{A}_4\overline{A}_2\overline{A}_1\overline{C} + B_2\overline{B}_1A_4\overline{C} \\
& + B_2\overline{B}_1A_2A_1C + \overline{B}_4B_1\overline{A}_2A_1\overline{C} + \overline{B}_4B_1\overline{A}_4\overline{A}_2\overline{A}_1C \\
& + B_1A_4C + \overline{B}_2B_1\overline{A}_2A_1C + \overline{B}_2B_1\overline{A}_4\overline{A}_1\overline{C} \\
& + \overline{B}_4\overline{B}_2\overline{B}_1A_1\overline{C} + \overline{B}_4\overline{B}_2\overline{B}_1\overline{A}_4\overline{A}_1C
\end{aligned}$$

$$\begin{aligned}
S_2 = {} & B_4A_4\overline{C} + B_4A_2A_1 + B_4A_2C + B_2B_1A_2A_1C \\
& + B_2B_1\overline{A}_2\overline{A}_1C + B_2A_4C + B_2\overline{B}_1\overline{A}_4\overline{A}_2\overline{C} \\
& + B_2\overline{B}_1\overline{A}_4\overline{A}_2\overline{A}_1 + \overline{B}_4\overline{B}_2A_2A_1\overline{C} + \overline{B}_2B_1\overline{A}_2A_1 \\
& + \overline{B}_2B_1\overline{A}_4\overline{A}_2C + \overline{B}_4\overline{B}_2\overline{B}_1A_2\overline{C} + \overline{B}_4\overline{B}_2\overline{B}_1A_2\overline{A}_1 \\
& + \overline{B}_4\overline{B}_2\overline{A}_2A_1C
\end{aligned}$$

$$\begin{aligned}
S_4 = {} & B_4A_4C + B_4\overline{A}_4\overline{A}_2\overline{A}_1\overline{C} + B_2B_1\overline{A}_2A_1\overline{C} + B_2B_1\overline{A}_4\overline{A}_2\overline{A}_1C \\
& + B_2\overline{B}_1A_2\overline{A}_1\overline{C} + B_2\overline{B}_1\overline{B}_2A_1C + \overline{B}_2B_1A_2A_1\overline{C} \\
& + \overline{B}_2B_1A_2\overline{A}_1C + \overline{B}_4\overline{B}_2\overline{B}_1A_4\overline{C} + \overline{B}_4\overline{B}_2\overline{B}_1A_2A_1C
\end{aligned}$$

$$\begin{aligned}
S_5 = {} & B_5B_4A_5A_4 + B_5B_4A_5A_2 + B_5B_4A_5A_1 + B_5B_4A_5C \\
& + B_5\overline{A}_5\overline{A}_4\overline{A}_2\overline{A}_1\overline{C} + B_5B_2B_1A_5A_2 + B_5B_2B_1A_5A_1C \\
& + B_5\overline{B}_4\overline{A}_5\overline{A}_4\overline{A}_2\overline{C} + B_5\overline{B}_4\overline{A}_5\overline{A}_4\overline{A}_2\overline{A}_1 + B_5B_2A_5A_4 \\
& + B_5B_2A_5A_2A_1 + B_5B_2A_5A_2C + B_5\overline{B}_4\overline{B}_1\overline{A}_5\overline{A}_4\overline{A}_1\overline{C} \\
& + B_5\overline{B}_4\overline{B}_1\overline{A}_5\overline{A}_4\overline{A}_2 + B_5B_1A_5A_4 + B_5B_1A_5A_2A_1C \\
& + B_5\overline{B}_4\overline{B}_2\overline{A}_4\overline{C} + B_5\overline{B}_4\overline{B}_2\overline{A}_5\overline{A}_4\overline{A}_1 + B_5\overline{B}_4\overline{B}_2\overline{A}_5\overline{A}_4\overline{A}_2 \\
& + B_5A_5A_4C + B_5\overline{B}_4\overline{B}_2\overline{B}_1\overline{A}_5\overline{C} + B_5\overline{B}_4\overline{B}_2\overline{B}_1\overline{A}_5\overline{A}_4 \\
& + \overline{B}_5A_5\overline{A}_4\overline{A}_2\overline{A}_1\overline{C} + \overline{B}_5\overline{B}_4A_5\overline{A}_4\overline{A}_2\overline{A}_1 + \overline{B}_5B_2B_1\overline{A}_5A_2 \\
& + \overline{B}_5B_2B_1\overline{A}_5A_1C + \overline{B}_5\overline{B}_4\overline{B}_1A_5\overline{A}_4\overline{A}_1\overline{C} \\
& + \overline{B}_5\overline{B}_4\overline{B}_1A_5\overline{A}_4A_2 + \overline{B}_5B_2\overline{A}_5A_4 + \overline{B}_5B_2\overline{A}_5A_2A_1 \\
& + \overline{B}_5B_2\overline{A}_5A_2C + \overline{B}_5\overline{B}_4\overline{B}_2A_5\overline{A}_4\overline{C} + \overline{B}_5\overline{B}_4\overline{B}_2A_5\overline{A}_4\overline{A}_1 \\
& + \overline{B}_5\overline{B}_4\overline{B}_2A_5\overline{A}_4\overline{A}_2 + \overline{B}_5B_1\overline{A}_5A_4 + \overline{B}_5B_1\overline{A}_5A_2A_1C \\
& + \overline{B}_5\overline{B}_4\overline{B}_2\overline{B}_1A_5\overline{C} + \overline{B}_5\overline{B}_4\overline{B}_2\overline{B}_1A_5\overline{A}_4 + \overline{B}_5\overline{A}_5A_4C \\
& + \overline{B}_5B_4\overline{A}_5A_4 + \overline{B}_5B_4\overline{A}_5A_2 + \overline{B}_5B_4\overline{A}_5A_1 + \overline{B}_5B_4\overline{A}_5C \\
& + \overline{B}_5\overline{B}_4A_5\overline{A}_4\overline{A}_2\overline{C}
\end{aligned}$$

Table 5.10
Input combinations for 5421 parallel adder

Decimal number	*Input combination* (A + B)	BCD	*Error sum*	*Correction*
0	(0 + 0)	0000		
1	(1 + 0)	0001		
2	(2 + 0)(1 + 1)	0010		
3	(3 + 0)(1 + 2)	0011		
4	(4 + 0)(1 + 3)(2 + 2)	0100		
5	(5 + 0)(1 + 4)*(2 + 3)*	1000	0101	+3
6	(6 + 0)(1 + 5)(2 + 4)*(3 + 3)*	1001	0110	+3
7	(7 + 0)(2 + 5)(1 + 6)(3 + 4)*	1010	0111	+3
8	(8 + 0)(1 + 7)(2 + 6)(3 + 5)(4 + 4)*	1011	1000	+3
9	(9 + 0)(1 + 8)(2 + 7)(3 + 6)(4 + 5)	1100		
10	(2 + 8)*(1 + 9)*(3 + 7)*(4 + 6)*(5 + 5)	10000	1101	+3
11	(2 + 9)*(3 + 8)*(4 + 7)*(5 + 6)	10001	1110	+3
12	(3 + 9)*(4 + 8)*(5 + 7)(6 + 6)	10010	1111	+3
13	(4 + 9)*(5 + 8)(6 + 7)	10011	10000	+3
14	(5 + 9)(6 + 8)(7 + 7)	10100		
15	(6 + 9)*(7 + 8)*	11000	10101	+3
16	(7 + 9)*(8 + 8)*	11001	10110	+3
17	(8 + 9)*	11010	10111	+3
18	(9 + 9)*	11011	11000	+3
19	[(9 + 9) + 1]*	11100	11001	+3

* Sums which require correction.

The solution is not optimum as the algorithm is partly heuristic; the minimization routine required some 50 seconds of central processor time. The equations may be implemented in the usual way, using two-level logic or by factorizing to a multi-level circuit.

The circuit may of course be designed intuitively and we shall investigate this approach. Table 5.10 lists all possible combinations that can arise in the addition of two parallel BCD numbers (A and B) in the 5421 code. It soon becomes apparent on examination of the table that a correction factor of +3 needs to be added in certain cases. This is conceivable since we are counting:

$$0000 - 0001 - 0010 - 0011 - 0100 \overset{+3}{-} 1000$$

in the 5421 code. Thus when, as a result of adding two 5421 BCD numbers together in a parallel binary full-adder, the following sums are produced:

0101 0110 0111 1101 1110 1111 1000
10000 10101 10110 10111 11000 11001

they must be corrected by adding +3. This correction may be performed using a second parallel full-adder stage, with a controlled +3 wired-in input.

The first six terms in the list above may be minimized to

$$\text{Correction} = S_4'S_2 + S_4'S_1$$

where $S_1'S_2'$ and S_4' are the intermediate sums produced in the first full-adder (see Fig. 5.19). The other terms, including the term 1000 brought

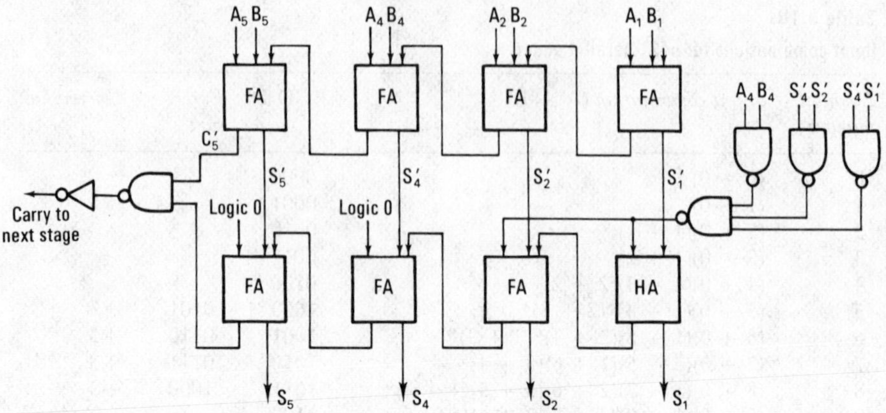

Figure 5.19 BCD 5421 parallel adder stage.

about by the addition of two fours, may be corrected by A_4B_4, the primary inputs to the adder. Note (from Table 5.10) that the inputs A_4 and B_4 are only present for these terms. For example, consider the generation of sum output 13; this is produced by the input combination (5 + 8) or (6 + 7) or (4 + 9). Now (4 + 9) = 0100 + 1100 = 10000 which is in error and both the A_4 and B_4 digits are present. The other input combination, that is, (5 + 8) = 1000 + 1011 = 10011, does not have these digits present and generates the correct result. Thus we may derive the final correction factor:

$$\text{Correction} = S'_4S'_2 + S'_4S'_1 + A_4B_4$$

The full circuit diagram is shown in Fig. 5.19; note that the output of the correction network is used to introduce a wired-in +3 to the second adder; a schematic is shown in Fig. 5.18(b). In this case it is not possible

This is just one example of an intuitive design for this circuit and there is no way of telling, except perhaps by trial and error, if this is a minimal solution. Furthermore, in practice the circuit may well need to be simulated (see Chapter 8) to ensure that it functions correctly under all conditions.

It is also possible to design a serial version of the 5421 BCD adder; a schematic is shown in Fig. 5.18(b). In this case it is not possible to derive the sum output bit-by-bit as in the case of the binary serial adder, since all four bits need to be examined before the final sum can be produced. This means that there will be no output from the circuit for the first word-time, that is, four digit-times, and the circuit must be clocked a further word-time (during this time the next BCD number can be entered) to obtain the serial sum. Thus the addition of N 4-bit BCD numbers takes $N + 1$ word-times, and requires a 4-bit register in the adder circuit.

As in the case of the parallel adder, this circuit is best designed theoretically using switching theory to express the adder as a synchronous finite-state machine. As before, however, this approach becomes laborious able 5.11

m outputs for 5421 serial adder

Outputs after third clock pulse						Required bistable outputs					
Carry circuit	*Carry bistable*	*Sum circuit*	*Bistable outputs*			*Carry bistable*	*Sum bistables*				*Comments*
C_p	C	S_5	S_4	S_2	S_1	C	Q_5	Q_4	Q_2	Q_1	
0	0	0	0	0	0	0	0	0	0	0	Correct results
0	0	0	0	0	1	0	0	0	0	1	
0	0	0	0	1	0	0	0	0	1	0	
0	0	0	0	1	1	0	0	0	1	1	
0	0	0	1	0	0	0	0	1	0	0	
0	0	1	0	0	0	0	1	0	0	0	
0	0	1	0	0	1	0	1	0	0	1	
0	0	1	0	1	0	0	1	0	1	0	
0	0	1	1	0	0	0	1	1	0	0	
0	1	1	0	0	0	0	1	0	1	1	
0	0	0	1	0	1	0	1	0	0	0	
0	0	0	1	1	0	0	1	0	0	1	
0	0	0	1	1	1	0	1	0	1	0	
0	0	1	1	0	1	1	0	0	0	0	Set carry bistable
0	0	1	1	1	0	1	0	0	0	1	
0	0	1	1	1	1	1	0	0	1	0	
1	1	0	0	0	0	1	0	0	1	1	Correct result
1	0	0	1	0	0	1	0	1	0	0	
1	0	0	1	0	1	1	1	0	0	0	
1	0	0	1	1	0	1	1	0	0	1	
1	0	0	1	1	1	1	1	0	1	0	
1	0	1	0	0	0	1	1	0	1	1	Use C_p in correction
1	0	1	0	0	1	1	1	1	0	0	
0/1	1	0	0	0	1						Don't-care terms
0/1	1	0	0	1	0						
0/1	1	0	0	1	1						
0/1	1	0	1	0	0						
0/1	1	0	1	0	1						
0/1	1	0	1	1	0						
0/1	1	0	1	1	1						
0/1	1	1	0	0	1						
0/1	1	1	0	1	0						
0/1	1	1	0	1	1						
0/1	1	1	1	0	0						
0/1	1	1	1	0	1						
0/1	1	1	1	1	0						
0/1	1	1	1	1	1						
1	1	0	1	0	0						
1	1	0	1	0	1						
1	1	0	1	1	0						
1	1	0	1	1	1						
1	1	1	0	0	1						
1	1	1	0	0	0						
1	0	0	0	0	0						
1	0	0	0	0	1						
1	0	0	0	1	0						
1	0	0	0	1	1						
1	0	1	0	1	0						
1	0	1	0	1	1						

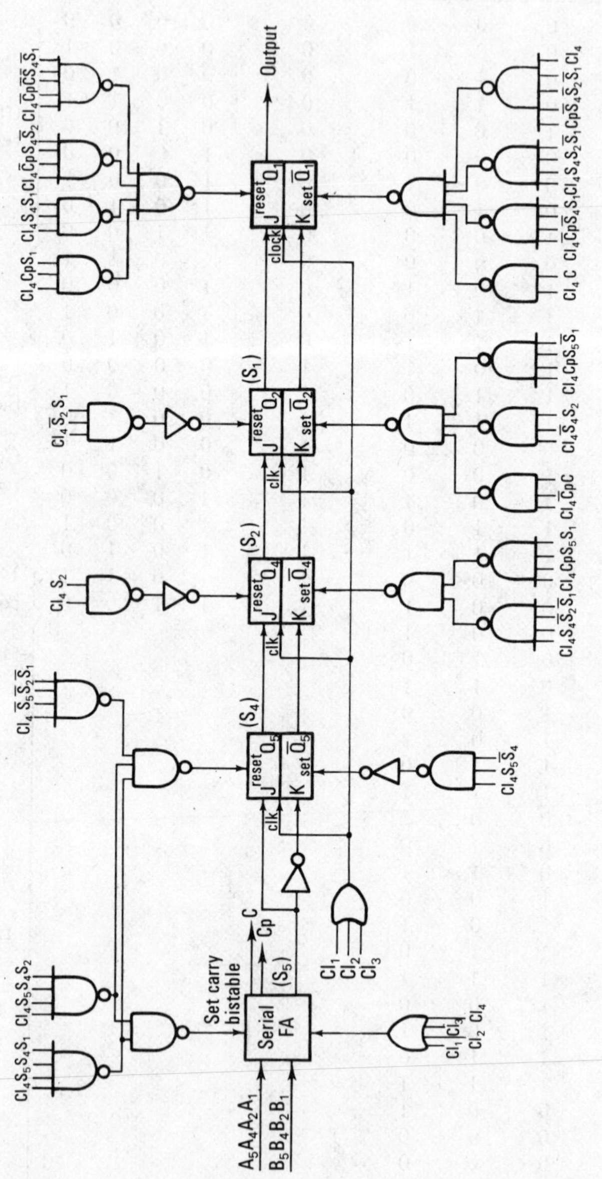

Figure 5.20 BCD 5421 serial adder.

without computer aids and we will once again use a part intuitive approach. The first step is to produce the pure binary sum of the two BCD inputs using a conventional serial full-adder circuit, the result being clocked into a shift register. The next step in the procedure is to correct the sum output and then finally to clock out the corrected serial BCD sum.

Figure 5.20 shows a logic diagram for the serial BCD adder. The operation commences with the BCD numbers being clocked into the full-adder and JK bistable shift-register circuits. After the first three clock pulses a situation exists such that S_5 appears at the output of the combinational adder logic, and S_4, S_2, and S_1 are stored in the first three stages (Q_5, Q_4 and Q_2) of the shift-register. At this point it is necessary to set the shift-register stages (Q_5, Q_4, Q_2 and Q_1) to the required sum outputs, correcting where necessary. This correction procedure occurs in clock pulse 4, after which the process is repeated with the next pair of numbers being entered to the adder simultaneously with the corrected sum result being clocked out to the answer register.

Table 5.11 shows a list of possible sum outputs for the serial BCD adder; the table must be read in conjunction with Table 5.10. The bistable equations (for the set and clear inputs of the JK bistables) are obtained by examining the required bistable outputs and noting the corresponding circuit conditions (C_P, C, S_5, S_4, S_2, and S_1) in clock pulse 3. These conditions may then be used to set the bistables in clock pulse 4, for example, Q_1 must be set to 0 for $\overline{C}_P\overline{C}\,\overline{S}_5S_4\overline{S}_2\overline{S}_1$ and to 1 for $\overline{C}_P\,\overline{C}\,\overline{S}_5\,\overline{S}_4\,\overline{S}_2\,S_1$, etc. The terms extracted in this way may be plotted on a K-map and minimized in the usual way, note that the don't-care (can't-happen) terms may also be included. Furthermore it is also possible to compare the present and next outputs of the shift register stages, that is, S_1Q_2, S_2Q_4, and S_4Q_5 for the usual don't-care conditions that exist when extracting input equations for set–reset bistables,[12] that is, the transitions $0 \rightarrow 0$ and $1 \rightarrow 1$. Note also that in the case of the outputs 001101, 001110, and 001111 the carry bistables must be set to 1. The terms 101000 and 101001 require special attention and the state of the carry circuit output (C_P) must be used to determine the correct set and reset conditions. The complete set of design equations are:

$$\begin{aligned}
\text{Set } Q_1 &= \text{CL4}\,[C + \overline{C}_P\overline{S}_4S_1 + S_4S_2\overline{S}_1 \\
&\qquad + C_P\overline{S}_4\overline{S}_2\overline{S}_1] \\
\text{Reset } Q_1 &= \text{CL4}\,[C_PS_1 + S_4S_1 + \overline{C}_PS_4\overline{S}_2 \\
&\qquad + \overline{C}_P\overline{C}\,\overline{S}_4\overline{S}_1] \\
\text{Set } Q_2 &= \text{CL4}\,[\overline{C}_PC + \overline{S}_4S_2 + C_PS_5\overline{S}_1] \\
\text{Reset } Q_2 &= \text{CL4}\,[\overline{S}_2S_1] \\
\text{Set } Q_4 &= \text{CL4}\,[S_4\overline{S}_2\overline{S}_1 + C_PS_5S_1] \\
\text{Reset } Q_4 &= \text{CL4}\,[S_2] \\
\text{Set } Q_5 &= \text{CL4}\,[S_5\overline{S}_4] \\
\text{Reset } Q_5 &= \text{CL4}\,[\overline{S}_5\overline{S}_2\overline{S}_1 + S_5S_4S_1 + S_5S_4S_2] \\
\text{Set carry bistable} &= \text{CL4}\,[S_5S_4S_1 + S_5S_4S_2]
\end{aligned}$$

What we have designed above is just one possible circuit for the 5421 BCD adder, and there is no justification whatever for assuming it may be minimal. The approach used is basically that used by the practising design engineer – a mixture of intuitive and formal methods.

It will be obvious that the design of serial BCD adders is rather a complicated and costly process compared to binary adders, moreover the complexity will depend on the choice of code. Few computers operate in this mode but it is possible that special purpose computers, and certainly calculators, would utilize these concepts. It is also possible to obtain LSI chips which perform BCD addition and subtraction.

5.11 Binary multiplication

This is normally performed in digital computers by repeated addition in an analogous manner to the normal 'paper and pencil' method. Table 5.12 shows the binary multiplication table, and the equivalent 'paper and pencil' and machine method of binary multiplication. The 'paper and pencil' method (Table 5.12(b)) involves inspecting the multiplier digits one at a time and writing down the multiplicand (shifted one place left) when they are equal to 1. The final step in the procedure is the addition of individual partial products, taking account of the protracted carries produced in the process. Note that multiplying together two n-bit numbers yields a $2n$-bit result, that is, a double-length product is produced.

Binary adders, as we have seen, generally add two numbers at a time (except for specially designed multiple-input adders) because of the difficulties involved in handling protracted carries. Consequently, it is more convenient to add partial products as they are formed, shifting the partial product sum one place right after each step of the operation to ensure the multiplicand is added in to the right order. Alternatively, the

Table 5.12
Binary multiplication

(a) Multiplication table

×	0	1
0	0	0
1	0	1

(b) 'Paper and pencil' method

Multiplicand (X)	1111	
Multiplier (Y)	1101	
	1111	Partial products
	0000	
	1111	
	1111	
Product	11000011	

(c) Machine method

	Over-flow	*Double length register*	
Multiplicand (X)		1111	
Multiplier (Y)		1101	
Add m'plicand		1111	0000
Shift right		0111	1000
Shift right		0011	1100
Add m'plicand	1	0010	1100
Shift right		1001	0110
Add m'plicand	1	1000	0110
Shift right		1100	0011
End		1100	0011

multiplicand may be shifted one place left before each addition. Table 5.12(c) shows an example of the steps involved in the machine multiplication of binary numbers.

Thus one method of performing binary multiplication in a digital computer is by a suitable configuration of the following basic units:

(a) A binary adder that adds the multiplicand to the partial product sum if, and only if, the current multiplier digit is 1.

(b) A shifting circuit that shifts the partial product sum one place right after each multiplier digit is sensed; the same result may be obtained by shifting the multiplicand one place left.

(c) A shifting circuit that shifts the multiplier one place right after each operation to allow the next most significant digit to be in the correct position for sensing.

The multiplication algorithm may be implemented using the above units in a variety of ways; Fig. 5.21 shows in schematic form some typical circuits. In Fig. 5.21(a) the operation is as follows: The 8-bit accumulator (double-length) is initially cleared and the 4-bit multiplicand placed in the right-hand half of the double-length multiplicand register. The least significant digit of the multiplier (held in a 4-bit register) is sensed and if equal to 1 the contents of the multiplicand register are added to the accumulator, using either a parallel or serial adder circuit. The multiplicand register is then shifted one place left and the multiplier register one place right and the operation repeated. The complete multiplication operation requires n additions and shifts, where n is the number of bits in the word; a control unit, embodying a counter circuit, will also be required to supervise the operations. An alternative approach is shown in Fig. 5.21(b). In this case the multiplicand is stored in a single-length register and added in to the left-hand half of the double-length accumulator; both the accumulator and the multiplier registers are shifted right. Finally, in Fig. 5.21(c) we have the most economical circuit which requires only three single-length registers. This is identical in operation to the circuit shown in Fig. 5.21(b) except that the multiplier register is now used to store the least-significant half of the double-length product. Note that in this case the multiplier is lost since it is now shifted out of the register rather than shifted end-around as in the previous circuits. The choice of a suitable structure for a multiplication circuit depends mainly on the availability and type of registers. For example, in a digital computer, registers may already be provided for other machine operations and to use these existing registers could economize in the hardware necessary to implement the algorithm. In special purpose multipliers, however, the major design criterion would be economy.

Let us now consider multiplier circuits in more detail, starting with a full serial multiplier employing the same algorithm as described above. Serial computers normally employ clocked end-around static shift-registers to store the operands, though earlier first generation computers used delay-line type registers (see Chapter 6). In both cases the basic mode of operation is identical and timing pulses (derived from the basic clock signal) must be available for each digit-time. For example, timing pulse T_0 would synchronize with the least significant digit of the serial operand (and the start of the computer word, since operands are circulated least significant digit first in serial systems). This would be followed by T_1, T_2, etc., up to the most significant digit when the pulses are repeated; this is sometimes called a **minor cycle**. In many cases it is necessary to include an extra digit pulse (called a **gap pulse**) to allow for the overflow of registers and circuit timing. The need for this will be obvious from Table 5.12(c), where the most significant digit, which is

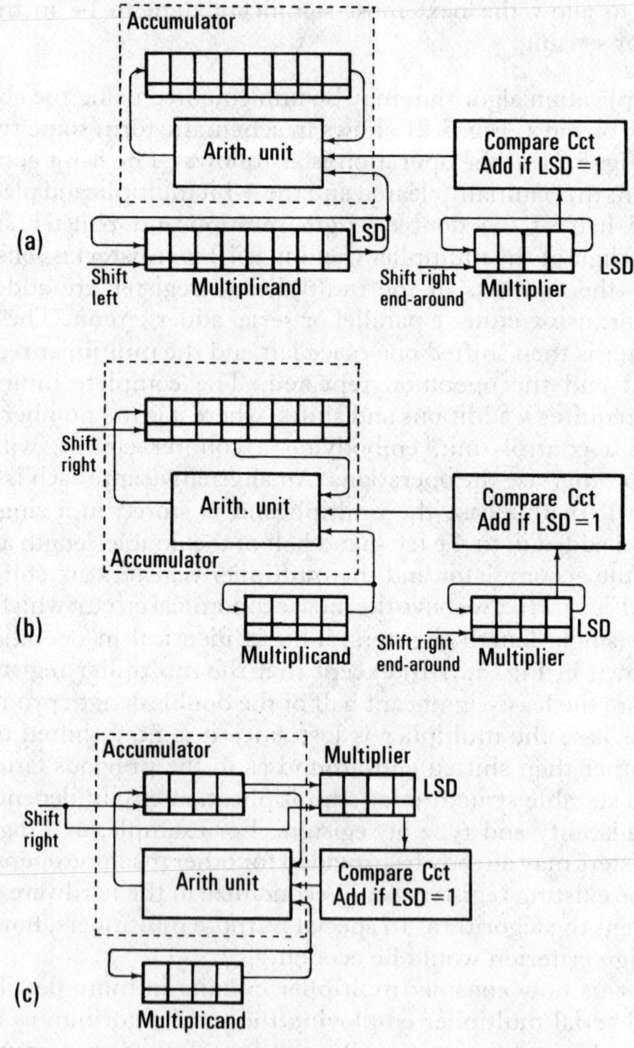

Figure 5.21 Alternative forms of multiplier circuit.

shifted out of the double-length register, must be retained and shifted back to get the correct result. These timing pulses would be generated from a counter and decoding circuit as described in the last chapter, or more probably from a micro-program unit.

A logic schematic for a serial multiplier circuit is shown in Fig. 5.22. Note that the multiplicand and multiplier are stored in circulating (end-around) registers, and the product registers are shifting one place right each word-time. The product is accumulated in a double-length shifting register, which must be cleared at the start of each multiplication, and a conventional serial full-adder is used to add the multiplicand to the partial products. From the timing diagram, Fig. 5.23, the operation of the circuit is as follows. The timing of the multiplication procedure is controlled by the repetitive clock pulses T_0, T_1, T_2, T_3, and T_4 (the gap digit) occurring every word-time (5 bit-times). We will assume that the multiplier and multiplicand registers have been filled with the appropriate operands, and that the double-length product register has been cleared to all zeros before the commencement of the actual multiplication operations. The multiplication operations are initiated by a control waveform M, which goes to 1 in time T_0 and stays up for the duration of the multiplication instruction, that is, four word-times. At time T_0 the least significant digit of the multiplier is examined, and if it is a 1 the sense bistable is set. The output of the sense bistable allows the contents of the multiplicand register to be added to the most significant half of the product register (MS), the partial sum being shifted back into the same register. This addition takes 4 bit-times (T_0–T_3 inclusive) and an extra bit-time (T_4) is allowed for carry overflow and shifting right. In bit-time T_4 the entire contents of the double-length product register are shifted one place right, thus the least significant digit of the top half (MS) becomes the most significant digit of the bottom half (LS). During this time the MS output (as well as the multiplicand output) is inhibited from entering the adder by clock pulse $\overline{T}_4$. Note, however, that in this time the serial adder outputs the contents of the carry bistable (that is, the overflow digit). The operation then repeats as before, commencing on the next occurrence of timing pulse T_0; from the timing diagram it is clear that at least four word-times are required to execute the multiplication instruction.

A fully parallel multiplier may be designed on the same principles; in this case a parallel full-adder is required to add the multiplicand to the partial sum. As before, the least significant digit of the multiplier is sensed to control the entry of the multiplicand to the adder, input gates now being required for each bit of the operands. The shifting operations may be performed in parallel using up and down gating or, alternatively, shift-registers may be used. A logic schematic for a parallel multiplier is shown in Fig. 5.24; the operation is as follows:

(a) The cycle counter in the control unit and the operand, sum, and input registers are first cleared to zero. Then the multiplicand and multiplier are entered (in parallel) to their appropriate registers.

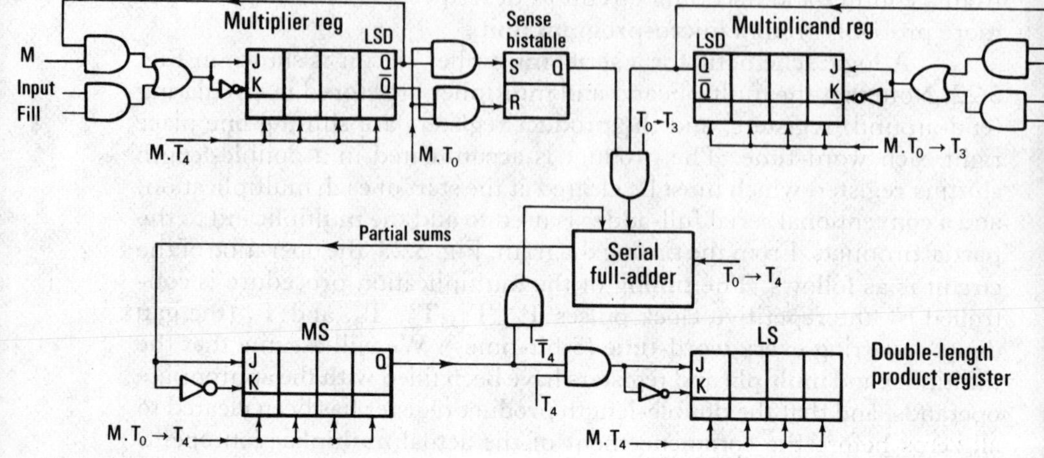

Figure 5.22 Schematic for a serial multiplier.

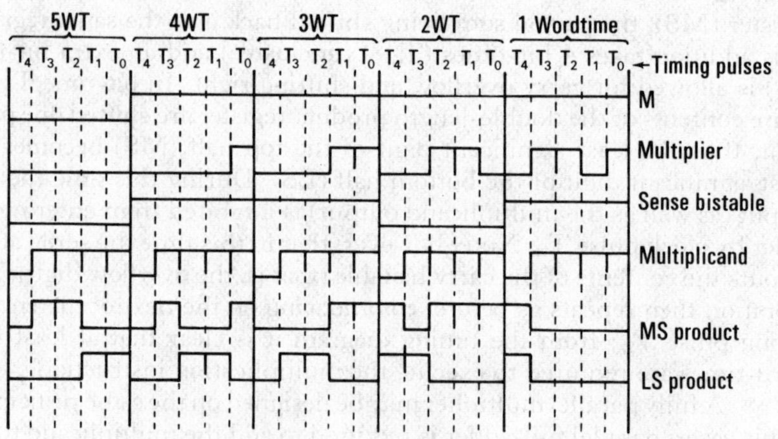

Figure 5.23 Serial multiplier timing.

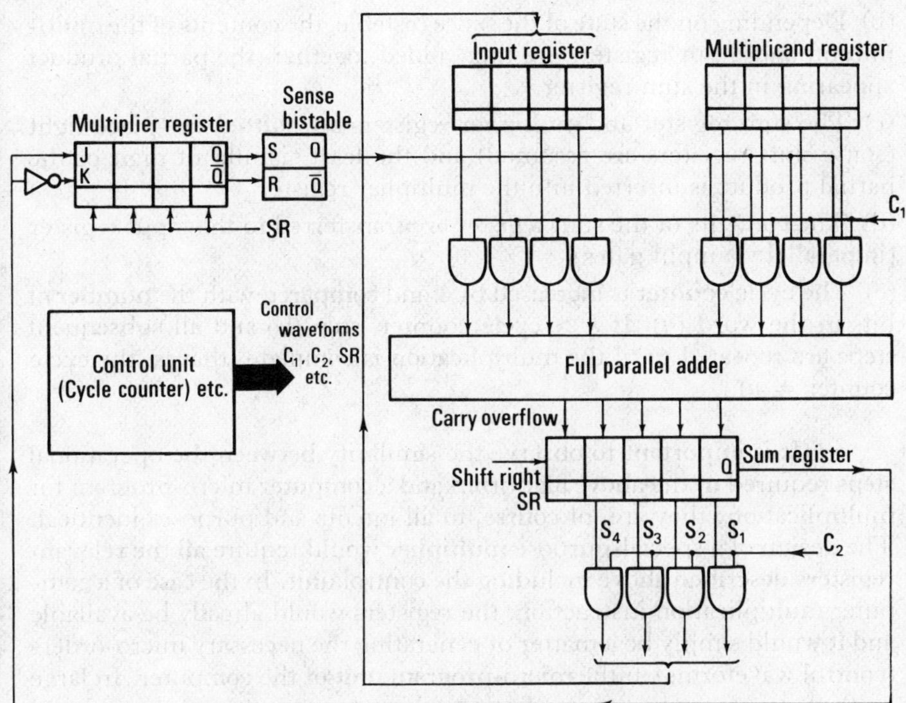

Figure 5.24 Schematic for parallel multiplier.

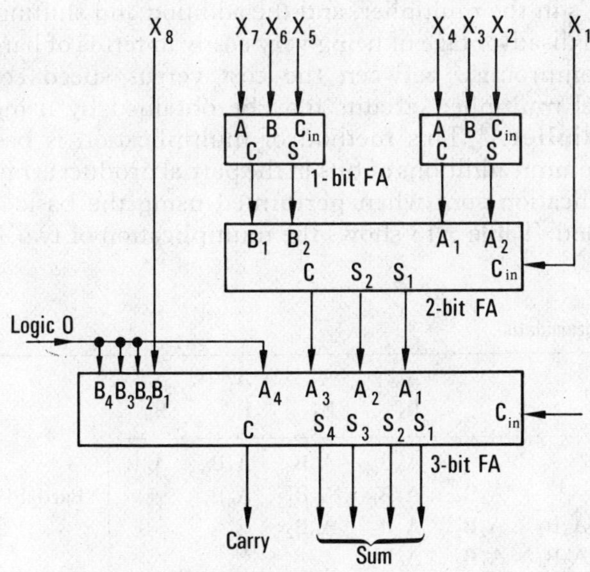

Figure 5.25 Combinatorial parallel counter.

(b) Depending on the state of the sense bistable, the contents of the multiplicand and input registers are then added together, the partial product appearing in the sum register.
(c) The sum register and multiplier registers are shifted one place right (static shift registers are assumed) and the least significant digit of the partial product is inserted into the multiplier register.
(d) The contents of the sum register are transferred to the input register (in parallel via input gates).
(e) The cycle counter is increased by 1 and compared with the number of bits in the word (n). If $n >$ cycle counter, step (b) and all subsequent steps are repeated until the multiplication is complete (that is, the cycle counter $= n$).

It is important to observe the similarity between the operational steps required in the above algorithm and a computer micro-program for multiplication; they are, of course, to all intents and purposes identical. The design of a special-purpose multiplier would require all the relevant registers described above including the control unit. In the case of a computer multiplication instruction, the registers would already be available and it would simply be a matter of generating the necessary micro-orders (control waveforms) in the micro-program unit of the computer. In large computer systems, however, when speed is at a premium wired-in control logic is often employed for the multiplication (and division) instructions.

The parallel version of the multiplier is approximately n times as fast as the serial version, assuming the worst case when n additions are required. The actual multiplication time will depend primarily on the number of 1's in the multiplier, and the addition and shifting speeds; it has the usual disadvantage of being very costly in terms of hardware.

A compromise between the cost versus speed equation for serial/parallel multiplier circuits may be obtained by using a **quasi-serial multiplier.**[13] This method of multiplication is based on the column by column additions of bits in the partial product array generated by a multiplication sum when performed using the basic 'paper and pencil' method. Table 5.13 shows the multiplication of two 4-bit words

Table 5.13
Partial product accumulation

				A_4	A_2	A_1	A_0	
				B_4	B_2	B_1	B_0	
				A_4B_0	A_2B_0	A_1B_0	A_0B_0	Partial Products
			A_4B_1	A_2B_1	A_1B_1	A_0B_1		
		A_4B_2	A_2B_2	A_1B_2	A_0B_2			
	A_4B_4	A_2B_4	A_1B_4	A_0B_4				
S_{64}	S_{32}	S_{16}	S_8	S_4	S_2	S_1	S_0	Partial Sums

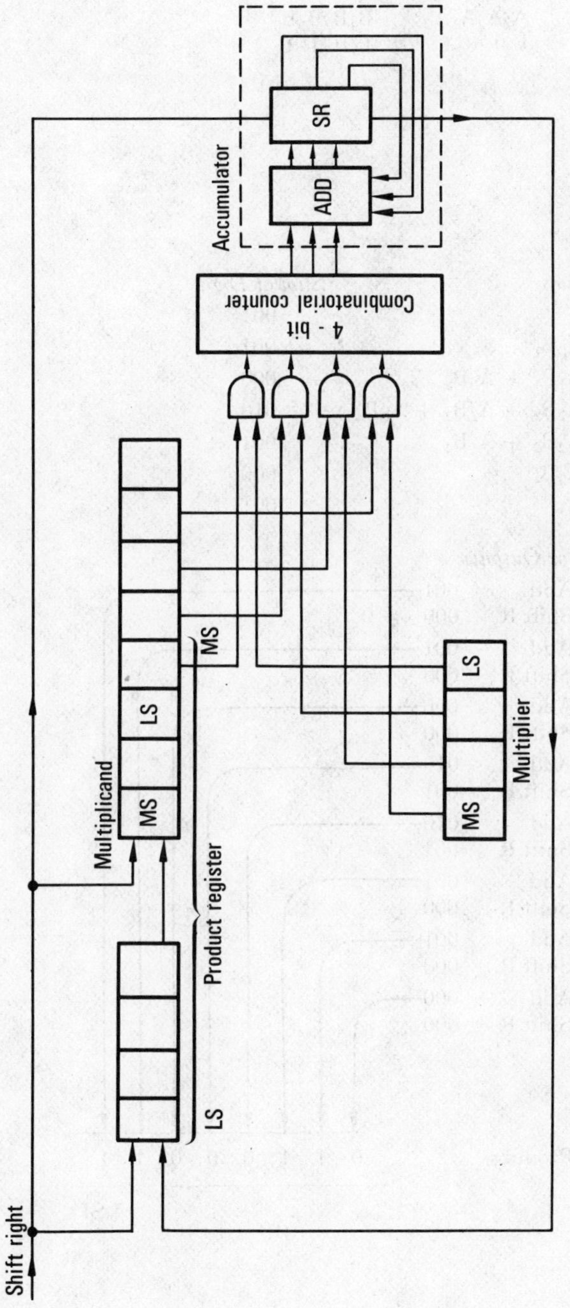

Figure 5.26 Quasi-serial multiplier.

Table 5.14
Quasi serial multiplication

a)

```
  1001      A4A2A1A0    B4B2B1B0
  1011      1 0 0 1     1 0 1 1
  ----
  1001
 1001
1001
--------
01100011
--------
```

$A_4A_2A_1A_0$ = 1 0 0 1 $B_4B_2B_1B_0$ = 1 0 1 1

b)

Partial Sums	*Sum of Digits*
A_0B_0	001
$A_1B_0 + A_0B_1$	001
$A_2B_0 + A_1B_1 + A_0B_2$	000
$A_4B_0 + A_2B_1 + A_1B_2 + A_0B_4$	010
$A_4B_1 + A_2B_2 + A_1B_4$	001
$A_4B_2 + A_2B_4$	000
A_4B_4	001

c) *Accumulator Outputs*

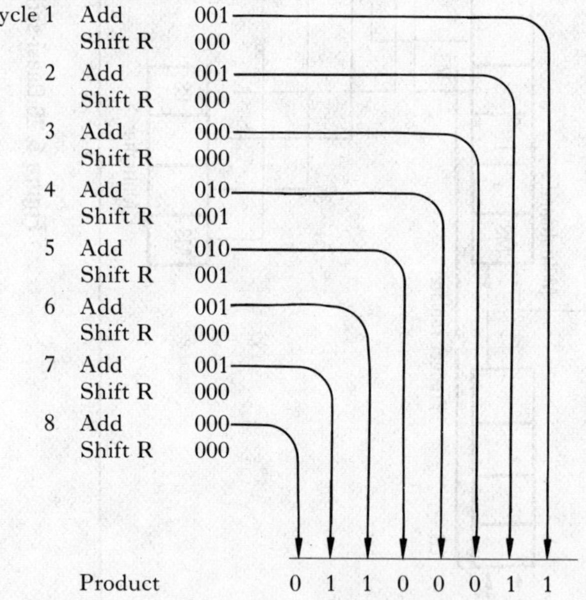

A and B and the array of partial products and sums formed in the process.

Fast column addition of the partial sums is executed in the quasi-serial multiplier by using a parallel mode **combinatorial counter** which forms the binary sum of the number of one's presented simultaneously to its inputs. A block diagram for an 8-bit combinatorial counter is shown in Fig. 5.25. Note that the first level consists of 1-bit full-adder units (without carry propagation) which generate the sum of the one's presented to its inputs (this becomes obvious when the sum and carry outputs shown in Table 5.2(a) are interchanged, with the sum as the least significant digit, and then interpreted as a 2-bit number). The two 2-bit numbers, together with X_1, are next added in a 2-bit adder to give a 3-bit output, this is then added to the X_8 input in a 3-bit adder to give the final sum output. For example, if the input vector is 10111011, then the output from the 1-bit adders is 10 and 10 respectively, this gives 101 as the output from the 2-bit adder with the final sum, 110, being obtained from the output of the 3-bit adder. The block diagram for a 4-bit word-length quasi-serial multiplier is shown in Fig. 5.26; note that this consists of a double length shift register to hold the multiplicand (and the LS half of the double length product) a single length multiplier register, a single length shift register to hold the MS half of the product and a combinatorial counter and accumulator circuit. The multiplication operation commences with the multiplicand left justified in the double length shift register and the multiplier in the single length register. During the first cycle the least significant digits of the multiplier and multiplicand are compared (in the AND gates) and if they are both one's the counter will generate a count of 001 which is added to the previous contents (initially zero) of the accumulator, the LSD of which is used to form the LSD of the product. The contents of the product and multiplicand registers are then shifted one place to the right and the process repeats on the next cycle until eight cycles ($2n$, where n is the number of bits in the word) have been completed. Note that the multiplicand is shifted through the double length register during the multiplication operation; the process is illustrated in Table 5.14 taking as an example the multiplication of two 4-bit numbers.

The multiplier can easily be constructed from standard LSI components and is readily interfaced to a mini- or microcomputer; the hardware costs are relatively low. Assuming a 10 MHz clock rate is employed two 8-bit words can be multiplied in 1·6 μs – that is in 16 clock cycles which is comparable to the time required for an ADD operation in most mini- or microcomputers.

5.12 Fast multiplier circuits

Multiplication times may be decreased for the parallel system described above by utilizing high speed components or employing fast adders with carry look-ahead techniques, and so on. There are, however, a number of

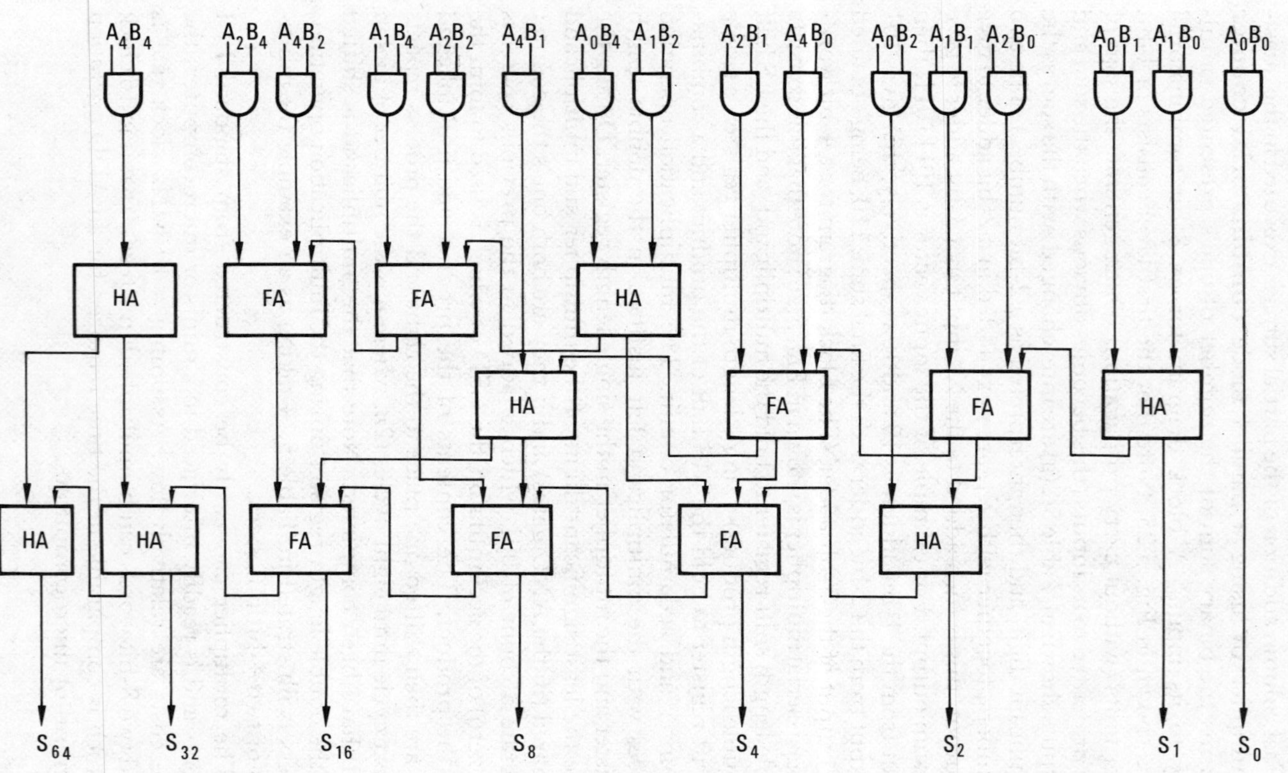

Figure 5.27 Simultaneous multiplier circuit.

other alternative approaches, such as the simultaneous multiplier, and techniques which have as their objective the reduction of the number of additions required for a multiplication operation.

One form of simultaneous multiplier may be designed as a combinational circuit, accepting the parallel operands and producing an instantaneous (ignoring gate delays) product. Table 5.15 shows a typical design, in this case the truth-table for a 4-bit combinational multiplier circuit. The parallel inputs A and B are multiplied together to produce the product P. The output functions may be simplified, using K-maps, to give the following circuit equations:

$$P_1 = A_1B_1$$
$$P_2 = \overline{A}_2A_1B_2 + A_1B_2\overline{B}_1 + A_2\overline{B}_2B_1 + A_2\overline{A}_1B_1$$
$$P_3 = A_2\overline{A}_1B_2 + A_2B_2\overline{B}_1$$
$$P_4 = A_2A_1B_2B_1$$

An alternative approach employs 1-bit adder stages to add together the individual terms of the partial products on a column by column basis. Table 5.13 shows a typical multiplication sum for 4-bit operands, and Fig. 5.27 shows the logic diagram for a multiplier based on this scheme. For example, the product terms A_1B_0 and A_0B_1 are added together to give S_1 and the carry taken to the next stage where A_1B_1 and A_2B_0 are added together. The sum output of this stage is then added to A_0B_2 to give S_2; the other product terms are obtained in a similar manner.

The time required to perform multiplication in a simultaneous multiplier depends on the number of gates (and adders) in the trans-

Table 5.15
4-bit combinational multiplier

Inputs				*Product*			
A_2	A_1	B_2	B_1	P_4	P_3	P_2	P_1
0	0	0	0	0	0	0	0
0	0	0	1	0	0	0	0
0	0	1	0	0	0	0	0
0	0	1	1	0	0	0	0
0	1	0	0	0	0	0	0
0	1	0	1	0	0	0	1
0	1	1	0	0	0	1	0
0	1	1	1	0	0	1	1
1	0	0	0	0	0	0	0
1	0	0	1	0	0	1	0
1	0	1	0	0	1	0	0
1	0	1	1	0	1	1	0
1	1	0	0	0	0	0	0
1	1	0	1	0	0	1	1
1	1	1	0	0	1	1	0
1	1	1	1	1	0	0	1

mission path from input to output; the worst case time may be estimated by considering the maximum signal path length. The high speed obtained with these multipliers is only achieved at the cost of employing a great number of components. Consequently, the methods are economically feasible for small numbers but quickly become prohibitive for larger wordlengths.

A different method of obtaining fast execution times is to reduce the number of addition cycles required in the multiplication operation. One obvious way of doing this is to omit the addition operation whenever a zero occurs in the multiplier and immediately initiating a shift cycle. This method is known as **shifting across zeros** (a simple example is the parallel multiplier described above) and it is used in practically all parallel computers. Using this technique the execution time for the multiplication instruction varies according to the number of 1's in the multiplier. The technique may be elaborated by arranging that shifting may occur over an arbitrary number of binary positions within one clock cycle. In this case it would be possible for a single shift to shift across a whole string of zeros.

Another technique for reducing the number of additions (and also dispensing with a variable shift) is called **multiplication by uniform multiple shifts**. In this method the multiplier bits are inspected two at a time, the multiplier being shifted right two places on each operation. However, in contrast to the methods described above, the partial product register is not shifted (the multiplicand being added at the least significant end) and the multiplicand register is shifted two places left on each operation. The appropriate action that is taken on each execution step of the multiplication instruction is shown detailed in Table 5.16;

Table 5.16
Multiplication by uniform multiple shifts
(a)

Multiplier bits	*Action*
00	Do nothing
01	Add multiplicand
10	Add multiplicand shifted left 1 place
11	Add multiplicand shifted left 1 place, add multiplicand

(b)

Multiplicand register	00001111	Multiplier register	1101
	00111100		0011
Partial product register	00000000		
	00001111	Add multiplicand	
	10000111	Add 2 × multiplicand	
	11000011	Add multiplicand	

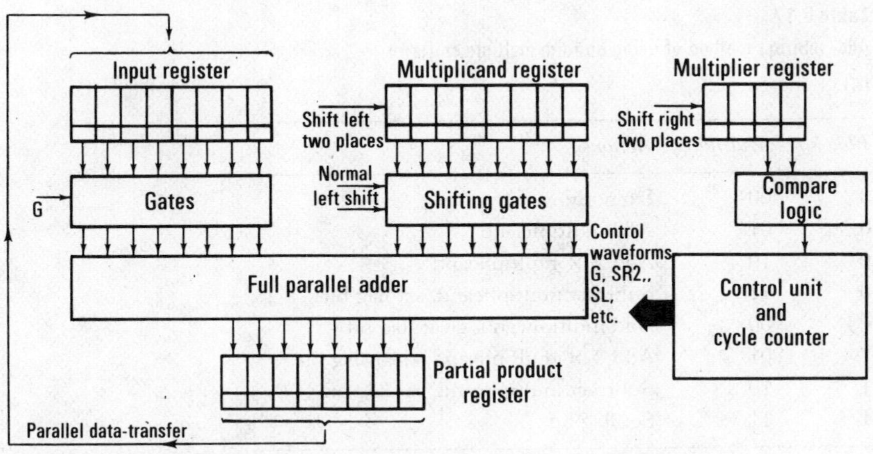

Figure 5.28 Uniform shift parallel multiplier.

note that the example is for a uniform shift of two bits. The first two operations in the table are easy to implement, the third operation, adding two times the multiplicand, requires the multiplicand to be shifted left one place before addition. This can easily be accomplished by interposing shifting gates between the multiplicand register and the parallel adder (see Fig. 5.28). The fourth operation is a combination of the second and third steps and must be performed using two addition cycles. It is obvious that the number of addition cycles depends on the number of 1's in the multiplier.

This technique requires a considerable amount of extra logic compared with the other methods since the multiplicand and input registers need to be double-length as does the parallel adder (it is also possible to devise a similar algorithm employing right-shifted partial products and an extended arithmetic unit). However, if the algorithm was being implemented as a machine-code instruction in a central processor, most of the registers would already be available in the arithmetic unit and it would only be necessary to provide a suitable micro-program.

A variation of this method is shown in Table 5.17. As before, the multiplier bits are examined in groups of two and the multiplicand shifted two places left on each operation. If the multiplier bits are zero, one, or two and the flag bit is zero, the correct multiple of the multiplicand (multiplied by four after each operation) is added to the partial product register. If the multiplier bits are three, with the flag bit zero, the multiplicand is subtracted and the flag bit set to one, and so on. The state of the flag bit must be examined together with the next higher order bits of the multiplier on each operation; the process proceeding as detailed in Table 5.17. Note that the operation must continue until the flag bit is cleared to zero.

Table 5.17
Add–subtract method of using uniform multiple shifts

(a)

Flag bit	*Multiplier*	*Action*
0	00	Do nothing
0	01	Add multiplicand
0	10	Add 2 × multiplicand
0	11	Subtract multiplicand, set flag bit
1	00	Add multiplicand, clear flag bit
1	01	Add 2 × multiplicand, clear flag bit
1	10	Subtract multiplicand, set flag bit
1	11	Set flag bit

(b)

Multiplicand register	00001111	Multiplier register	1101	Flag bit	0
	00111100		0011		0
	11110000		00		1

Partial product register	00000000	
	00001111	Add 1 × multiplicand
	11010011	Subtract 4 × multiplicand
	11000011	Add 16 × multiplicand

Table 5.18
Quatenary multiplication algorithm

Multiplier	*Action*
000	Do nothing
001	Add multiplicand
010	Add multiplicand
011	Add 2 × multiplicand
100	Subtract 2 × multiplicand
101	Subtract multiplicand
110	Subtract multiplicand
111	Do nothing

				Extra stage
Multiplicand register	00001111	Multiplier register	1101	0
	00111100		0011	01
	11110000		0000	1101

Partial product register	00000000	
	00001111	Add 1 × multiplicand
	11010011	Subtract 4 × multiplicand
	11000011	Add 16 × multiplicand

This method is based on the fact that, for example:

$$1111 \times 1101 = (1111 \times 00001) - (1111 \times 00100) + (1111 \times 10000)$$

$$15 \times 13 = (15 \times 1) - (15 \times 4) + (15 \times 16)$$

Note that this method only uses single add or subtract operations (compared to the double add operation in the last method) but requires an adder/subtractor unit and extra storage for the flag bit. In general, the method will reduce the number of additions required in the multiplication function, for example:

$$(\text{Multiplicand} \times 011111) = (\text{Multiplicand} \times 100000) - (\text{Multiplicand} \times 000001)$$

which requires one addition and a subtraction instead of five additions. With this method no arithmetic operations are performed if a long string of 'zeros' or 'ones' is encountered in the multiplier.

The scheme described above has the slight disadvantage that it requires a separate flag bit bistable to be set and reset. A quarternary algorithm, due to Kilburn and Aspinall,[14] in which the multiplier digits are inspected in groups of three, overcomes this disadvantage. Table 5.18 shows the algorithm; as in the method above the multiplicand is shifted two places left on each operation, but the partial product register is not shifted. An extra stage is required at the least significant end of the multiplier register (which is shifted two places right) but there is now no necessity to set or reset this stage. It is obvious that the two methods produce identical results.

In a serial multiplier the partial products are added successively using a single serial adder circuit. The **serial–parallel multiplier**[15] speeds up the multiplication time by using $n - 1$ serial adders and delays in cascade, where n is the number of bits in the serial multiplier word (see Fig. 5.29). The multiplier is entered to the circuit (least significant digit first) and the parallel multiplicand is taken directly to the inputs of the gates B_1, B_2, B_3, and B_4, the product appears at the output, least significant digit first. The choice of which input is the multiplier or multiplicand is purely arbitrary, as the circuit functions correctly in either mode.

The function of the circuit is such that if the multiplier digit is 1 (taken in sequence, least significant digit first) the input gates allow the multiplicand digits to enter the adder stages. In the first clock pulse the outputs of the delay stages will all be zero and the output of the circuit is the least significant digit of the product. Meanwhile the least significant digit of the multiplier is entered to the input of the delaying stages. On the next and subsequent clock pulses, the delay stages are shifted and the next digit of the output is produced. Effectively, the circuit is adding the multiplicand to the partial product (stored in the adding–shifting circuit) the requisite number of times, as determined by the multiplier, with the necessary shifting operations being performed by the delay elements. The

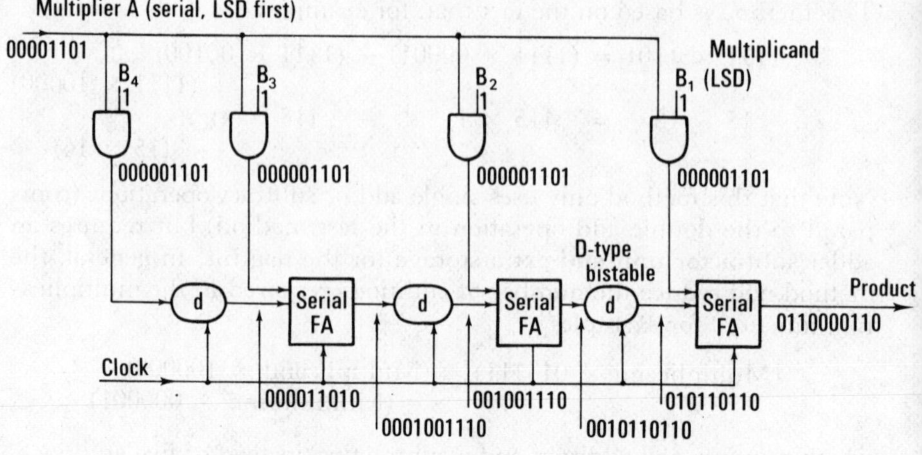

Figure 5.29 Serial-parallel multiplier.

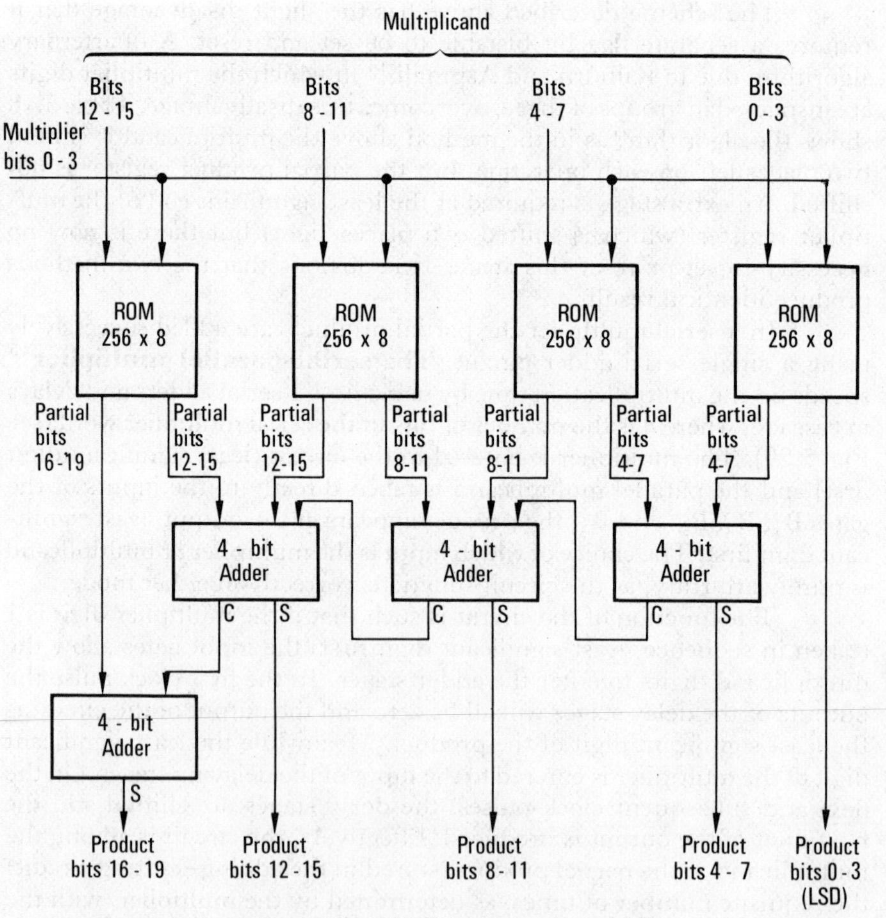

Figure 5.30 Multiplication using ROM's.

action of the circuit is similar to the modulo-2 multiplier circuits described in linear switching theory.[16]

The execution time for the multiplication circuit is $2n$ bit times (assuming equal length multiplicand and multiplier words) and the serial input must be followed by n zeros. Since the circuit has a symmetrical cascaded structure, it would be ideally suited for implementation in MSI circuit form, longer wordlengths being accommodated simply by adding extra cells consisting of adder, delay, and input gate. The 1-bit delay circuits are easily produced using clocked D-type or JK bistables.

An alternative approach to multiplication is to use a table look-up procedure employing read only memory modules (ROM) which are described in Chapter 6. However, to put a complete multiplication table, say for two 8-bit words, into a ROM can be prohibitively expensive and a compromise solution is normally employed using ROM in conjunction with adder modules.

For example, to multiply a 16-bit number by a 4-bit number the multiplicand is partitioned into 4-bit groups and a ROM table is used to obtain the 8-bit partial products (note that for a 4 × 4 multiplication table, 256 8-bit words are required). The four least significant digits of the multiplicand are used together with the 4-bit multiplier to address the ROM and generate the first partial product. The second partial product is produced by using the four next most significant digits of the multiplicand and so on until all the multiplicand bits have been used in sequence (note however that the addressing takes place in parallel). The partial products so formed are then added together using 4-bit adders, with carry propagation between stages; this process is shown in Fig. 5.30 and Table 5.19. Note that the operational speed of the circuit is determined by the ROM access time and the speed and propagation times for the adders. For larger operands the table look-up approach requires a large amount of hardware, for instance, a 16 × 16 multiplier would consist of 16 ROM's and 28 adders.

Table 5.19
ROM Multiplier

	Bits 16–19	12–15	8–11	4–7	0–3
Multiplicand		1011	0111	1110	0011
Multiplier					1010
Partial Product 1				0001	1110
Partial Product 2			1000	1100	
Partial Product 3		0100	0110	add	
Partial Product 4	0110	1110	add		
		add			
	carry				
Product	0111	0010	1110	1101	1110

Table 5.20
Binary division

(a)

```
        quotient                     1010
divisor | dividend         1110 | 10001100
                                   1110
                                   ----
                                    1110
                                    1110
                                    ----
                                    0000
```

(b)

Divisor	*Dividend–quotient*	*Action*
00001110	00000000 10001100	Initial left shift
	00000001 00011000	Compare, divisor < dividend, sub. 0
	00000010 00110000	Compare, divisor < dividend, sub. 0
	00000100 01100000	Compare, divisor < dividend, sub. 0
	00001000 11000000	Compare, divisor < dividend, sub. 0
	00010001 10000001	Compare, divisor ⩾ dividend, sub. divisor
	00000011 10000001	
	00000111 00000010	Compare, divisor < dividend, sub. 0
	00001110 00000101	Compare, divisor ⩾ dividend, sub. divisor
	00000000 00000101	
	00000000 00001010	Compare, divisor < dividend, sub. 0

Underlined digits are the inserted quotient.

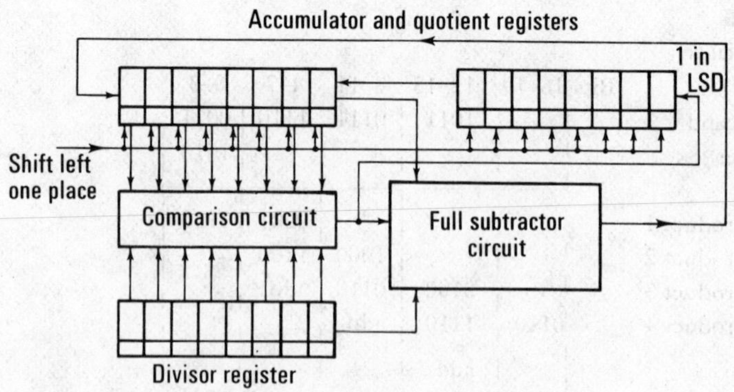

Figure 5.31 Block diagram of binary division circuit.

Many of the multiplier circuits described above are available as LSI chips. For example, a complete 16 × 16-bit parallel multiplier can be obtained on a single chip; this has a 300 ns multiplication time and costs in the order of £100. Table look-up and serial multiplier circuits are also available.

5.13 Binary division circuits

In the same way that multiplication can be implemented as a series of additions and shifts, so the division process can be reduced to a succession of subtractions and shifts. Table 5.20 shows the normal 'paper and pencil' method of division; the process can be expressed as a number of formal steps, these are shown below:

1 The divisor and dividend are compared, starting with one bit, then two bits, and so on, shifting the dividend one place left each time.

(a) If the divisor $\leqslant$ dividend a 1 is entered in the quotient and the divisor is subtracted from the dividend.

(b) If the divisor $>$ dividend subtraction is not possible and a 0 is entered in the quotient.

2 The process repeats from 1 above with the modified or partial dividend.

The procedure terminates when the correct number of places has been generated in the quotient; this, of course, is equal to n, where n is the number of bits in the operand word. Thus in the example (Table 5.20) it is necessary to perform eight 'subtractions' (including zero subtractions) and eight left shifts of the dividend with respect to the divisor.

The algorithm is shown implemented in hardware form in Fig. 5.31. The divisor is held in an n-bit register, and the dividend in the least significant half of the combined double-length accumulator and quotient register. The division commences by shifting the contents of the accumulator and quotient registers one place left, and then the accumulator is compared with the divisor. If the partial dividend is greater than the divisor, the divisor is subtracted from the accumulator, and a 1 entered into the least significant end of the quotient register.

In the next step the contents of both accumulator and quotient registers are shifted one place left, and the partial dividend and divisor compared as before. The procedure is repeated until all the quotient bits have been derived, the remainder being left in the accumulator.

The inclusion of a special comparator circuit (see Chapter 4) in division logic can be expensive, consequently many computers perform the comparison operation by subtracting the divisor from the dividend. If the difference is negative the divisor is larger than the dividend, and the original dividend must be restored by the subsequent addition of the divisor. This method is called **restoring division**, and obviously results in longer multiplication times than those obtained using the basic technique. A flow-chart describing the algorithm (from which a control micro-program may be deduced) is shown in Fig. 5.32.

Table 5.21
Non-restoring division

	Accumulator register	*Quotient register*	
	0 0 0 0 0 0 0 0	1 0 0 0 1 1 0 0	
Shift	0 0 0 0 0 0 0 1	0 0 0 1 1 0 0 0	
Subtract	0 0 0 0 1 1 1 0		
Negative	1 1 1 1 0 0 1 1	0 0 0 1 1 0 0 0	
Shift	1 1 1 0 0 1 1 0	0 0 1 1 0 0 0 0	
Add	0 0 0 0 1 1 1 0		
Negative	1 1 1 1 0 1 0 0	0 0 1 1 0 0 0 0	
Shift	1 1 1 0 1 0 0 0	0 1 1 0 0 0 0 0	
Add	0 0 0 0 1 1 1 0		
Negative	1 1 1 1 0 1 1 0	0 1 1 0 0 0 0 0	
Shift	1 1 1 0 1 1 0 0	1 1 0 0 0 0 0 0	
Add	0 0 0 0 1 1 1 0		
Negative	1 1 1 1 1 0 1 0	1 1 0 0 0 0 0 0	
Shift	1 1 1 1 0 1 0 1	1 0 0 0 0 0 0 0	
Add	0 0 0 0 1 1 1 0		
Positive	0 0 0 0 0 0 1 1	1 0 0 0 0 0 0 1	
Shift	0 0 0 0 0 1 1 1	0 0 0 0 0 0 1 0	
Subtract	0 0 0 0 1 1 1 0		
Negative	1 1 1 1 1 0 0 1	0 0 0 0 0 0 1 0	
Shift	1 1 1 1 0 0 1 0	0 0 0 0 0 1 0 0	
Add	0 0 0 0 1 1 1 0		
Positive	0 0 0 0 0 0 0 0	0 0 0 0 0 1 0 1	
Shift	0 0 0 0 0 0 0 0	0 0 0 0 1 0 1 0	
Subtract	0 0 0 0 1 1 1 0		
Negative	1 1 1 1 0 0 1 0	0 0 0 0 1 0 1 0	END

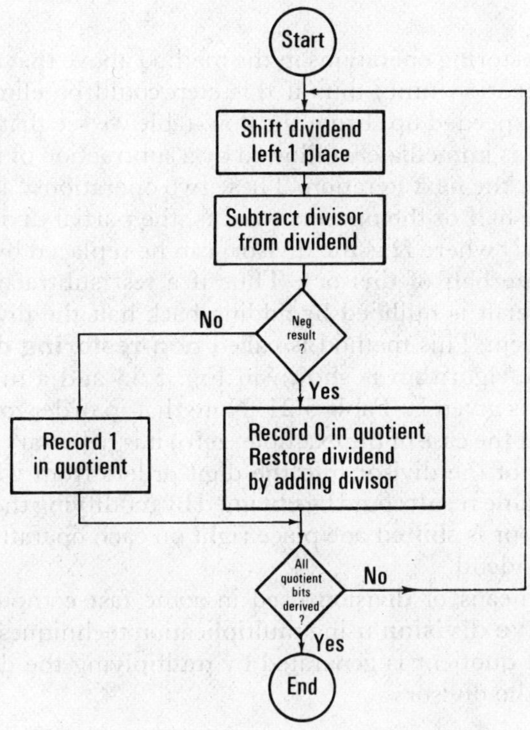

Figure 5.32 Flow chart for restoring division.

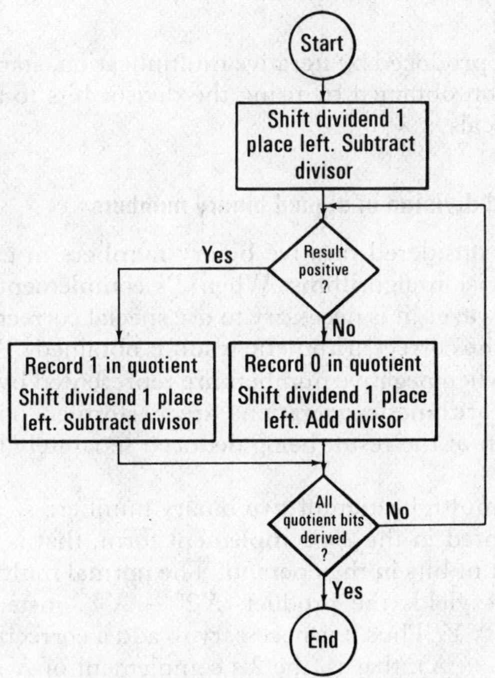

Figure 5.33 Flow chart for non-restoring division.

It is the restoring operations in the method above that account for the increased execution time, thus if this step could be eliminated the process would be speeded up. From the flow-table we see that each addition of the divisor is immediately followed by a subtraction of the divisor, divided by two, in the next iteration. These two operations 'add divisor' and 'subtract one half of the divisor' (that is, the partial dividend is increased by $D + \frac{1}{2}D$ where D is the divisor) can be replaced by the single operation 'add one half of divisor'. Thus if a test subtraction gives a negative remainder it is nullified by adding back half the divisor in the next arithmetic step. This method is called **non-restoring division**; a flow-chart for the algorithm is shown in Fig. 5.33 and a fully worked division example is given in Table 5.21. Note that in order to obtain the final remainder (in the case of the example, zero) it is necessary to perform an extra addition of the divisor into the digit orders from which it was subtracted. The same result could be obtained by modifying the algorithm such that the divisor is shifted one place right on each operation, instead of shifting the dividend.

A novel means of division used in some fast computers is the method of **iterative division** using multiplication techniques only.[17] In this approach the quotient is generated by multiplying the dividend by the reciprocal of the divisor:

$$Q = \frac{1}{D_v} \cdot D_d$$

The reciprocal $1/D_v$ is produced by iterative multiplication, starting with an initial approximation obtained by using the divisor bits to address a stored table of reciprocals.

5.14 Multiplication and division of signed binary numbers

So far we have only considered positive binary numbers in explaining multiplication and division algorithms. When 2's complemented numbers are employed, however, it is necessary to use special correction techniques to ensure that the correct arithmetic result is obtained. This is not necessary, of course, when negative numbers are represented by sign and magnitude since the arithmetic operations are performed on positive numbers only, the sign of the result being deduced separately by logical means.

Consider the multiplication of two binary numbers X and $-Y$, where $-Y$ is represented in the 2's complement form, that is, $2^n - Y$, where n is the number of bits in the operand. The normal multiplication of these two numbers yields the product, $X2^n - XY$, instead of the required result $2^{2n} - XY$. Thus it is necessary to add a correction factor of $2^{2n} - X2^n = 2^n(2^n - X)$, that is, the 2's complement of X scaled by 2^n; Table 5.22(a) shows an example of a corrected multiplication sum.

When X is negative and Y positive a correction factor of

Table 5.22
Multiplication correction technique

(a)

$2^4 2^3 2^2 2^1 2^0$	$n = 5$
0 1 1 1 1	$X = +15$
1 0 0 1 1	$Y = -13$
0 1 1 1 1	
0 1 1 1 1	
0 1 1 1 1	
0 1 0 0 0 1 1 1 0 1	
1 0 0 0 1	Correction factor $(2^n - X)2^n = 10001 \times 2^5$
1 1 0 0 1 1 1 1 0 1	$XY = -195$

(b)

$2^4 2^3 2^2 2^1 2^0$	$n = 5$
1 0 0 0 1	$X = -15$
0 1 1 0 1	$Y = +13$
0 0 0 0 1	
1 0 0 0 0	Sign digit of X
0 0 0 0 1	
0 0 0 0 1	
1 0 0 0 0	
1 0 0 1 0 1 1 0 1	
1	1×2^4
1 | 1 0 0 1 1 1 1 0 1	Sign digit inserted in 2^9 position

(c)

$2^4 2^3 2^2 2^1 2^0$	$n = 5$
1 0 0 0 1	$X = -15$
1 0 0 1 1	$Y = -13$
0 0 0 0 1	
0 0 0 0 1	
1 0 0 0 0	
1 0 0 0 0	
1 0 0 0 1	
0 1 1 1 0 1 0 0 1 1	
0 1 1 1 1	add $X \times 2^5$
1 0 1 0 1 1 0 0 1 1	
1	add 1×2^4
0 | 0 1 1 0 0 0 0 1 1	Sign digit inserted in 2^9 position

$2^n(2^n - Y)$ must be added, but in some cases the multiplier digits are discarded during the multiplication operation (see Section 5.11) and a different method must be used. It can be shown empirically that if the sign digit of X is added to the partial product each time the multiplier digit is zero this is equivalent to adding in the 1's complement (that is, the inverse) of Y in the 2^n position. An additional +1 must be added to the partial product (in the 2^{n-1} digit position) during the final operation to give the required 2's complement correction term. Furthermore, the sign digit of the multiplicand must be ignored during the add cycles of the multiplication procedure; the sign digit of the final product is automatically, and separately, determined from the fact that the product is negative when one of the operands is negative. Table 5.22(b) shows an example of this correction process.

When both the multiplier and multiplicand are negative the multiplication will yield the product:

$$(2^n - X)(2^n - Y) = 2^{2n} - X2^n + XY - Y2^n$$

In this case the correction factors $X2^n$ and $Y2^n$ must be added in to the partial product, the term 2^{2n} may be ignored as this causes the product register to overflow. The required correction may be performed by a combination of the two previous methods, see Table 5.22(c). Note that all digits of the multiplier are used, but the sign digit of the multiplicand is ignored in the addition operations. The sign digit must also be determined logically, as before, by inspecting the operands.

These correction methods are rather empirical and result in somewhat complicated logic correction circuits, though they were in fact used in first generation serial computers. A better technique is to implement **Booth's algorithm**[18] which is shown below.

(a) If the multiplier digit is 1 and the next lower order multiplier digit is 0, subtract the multiplicand from the partial product.
(b) If the multiplier digit is 0 and the next lower order multiplier digit is 1, add the multiplicand to the partial product.
(c) If the multiplier digit is the same as the next lower order digit, do nothing.

The procedure is implemented in hardware using a double-length product register with the arithmetic operations being performed on the most significant (single-length) half of the register only. The entire product register is shifted one place right after each examination of the multiplier digits but the sign digit is retained, as in a normal arithmetic shift. Table 5.23 shows how the algorithm functions for all possible combinations of the operand sign digits.

A slightly simpler variation of the above methods is shown in Table 5.24, in this case the sign digits of the operands determine the appropriate correction procedures. The actual multiplication operations

Table 5.23
Booth's algorithm

(a)

	Double length register	
Multiplicand	01111	
Multiplier	01101	
	00000	00000
Subtract m'plicand	10001	00000
Shift right	11000	10000
Add m'plicand	00111	10000
Shift right	00011	11000
Subtract m'plicand	10100	11000
Shift right	11010	01100
Shift right	11101	00110
Add m'plicand	01100	00110
Shift right	00110	00011

(b)

	Double length register	
Multiplicand	01111	
Multiplier	10011	
	00000	00000
Subtract m'plicand	10001	00000
Shift right	11000	10000
Shift right	11100	01000
Add m'plicand	01011	01000
Shift right	00101	10100
Shift right	00010	11010
Subtract m'plicand	10011	11010
Shift right	11001	11101

(c)

	Double length register	
Multiplicand	10001	
Multiplier	01101	
	00000	00000
Subtract m'plicand	01111	00000
Shift right	00111	10000
Add m'plicand	11000	10000
Shift right	11100	01000
Subtract m'plicand	01011	01000
Shift right	00101	10100
Shift right	00010	11010
Add m'plicand	10011	11010
Shift right	11001	11101

(d)

	Double length register	
Multiplicand	10001	
Multiplier	10011	
	00000	00000
Subtract m'plicand	01111	00000
Shift right	00111	10000
Shift right	00011	11000
Add m'plicand	10100	11000
Shift right	11010	01100
Shift right	11101	00110
Subtract m'plicand	01100	00110
Shift right	00110	00011

Table 5.24
Multiplication algorithm for signed operands

(a)

Multiplicand	*Multiplier*	*Correction procedure*
+	−	Add 2's complement of the multiplicand to result
−	+	Use arithmetic shift and retain sign digit
−	−	Use both correction procedures
+	+	No correction required

(b)

Multiplicand	01111	
Multiplier	10011	
	00000	00000
add	01111	00000
shift	00111	10000
add	10110	10000
shift	01011	01000
shift	00101	10100
shift	00010	11010
add	10001	11010
shift	01000	11101
Add 2's complement	11001	11101

(c)

	Overflow		
Multiplicand		10001	
Multiplier		01101	
		00000	00000
add		10001	00000
shift		11000	10000
shift		11100	01000
add	1	01101	01000
shift		10110	10100
add	1	00111	10100
shift		10011	11010
shift		11001	11101

(d)

	Overflow		
Multiplicand		10001	
Multiplier		10011	
		00000	00000
add		10001	00000
shift		11000	10000
add	1	01001	10000
shift		10100	11000
shift		11010	01100
shift		11101	00110
add	1	01110	00110
shift		10111	00011
add 2's complement	1	00110	00011

Table 5.25
Division of signed numbers

(a) $\frac{1}{2} \div \frac{3}{4} = \frac{2}{3}$

```
              0·1010...
       ─────────────
01100  0100000000
       0110
       ────
        10000
        01100
        ─────
          100  recurring
          ───
```

(b)

Divisor	*Dividend*	*Quotient*
0·1100	0·1000	
sign digits same, shift	1·0000	1
subtract	0·1100	
	0·0100	
sign digits same, shift	0·1000	1
subtract	0·1100	
	1·1100	
sign digits different, shift	1·1000	0
add	0·1100	
	0·0100	
signs same, shift	0·1000	1
subtract	0·1100	
	1·1100	End
pseudo-quotient	1·1010	
correction	1·0001	
	0·1011	

(c)

Divisor	*Dividend*	*Quotient*
0·1100	1·1000	
signs different, shift	1·0000	0
add	0·1100	
	1·1100	
signs different, shift	1·1000	0
add	0·1100	
	0·0100	
signs same, shift	0·1000	1
subtract	0·1100	
	1·1100	
signs different, shift	1·1000	0
add	0·1100	
	0·0100	End
pseudo-quotient	0·0100	
correction	1·0001	
	1·0101	

themselves are identical with those described in Section 5.12, for example, the serial–parallel multiplier (Fig. 5.29) would be corrected by adding the 2's complement of the parallel input to the final result and employing an arithmetic shift on the serial input. Note that in the examples (Table 5.24) when overflow occurs it should be retained and treated as the sign digit, that is, shifted right, into the most significant position of the product register.

The division of signed operands presents the same type of problems as we encountered above with multiplication. Moreover, in this case the problem is complicated further due to the fact that we are primarily dealing with fractional numbers. Even with integer numbers there are certain difficulties; for example, any finite number divided by zero yields infinity, while zero divided by any non-zero number is zero. It is generally the responsibility of the programmer, however, to provide safeguards against these conditions occurring by suitable problem scaling and indigenous program tests. Logical circuits also are sometimes included to indicate errors arising from overlooking these restrictions. In the case of machines using fractional representation it is essential to ensure that the quotient is within the prescribed number range ($-1 \leqslant x < 1$). This means that logical checks must be provided to determine if the divisor is greater than the dividend.

If non-restoring division is used with signed operands the following algorithm[15] will give the correct quotient for all cases:

(a) Compare the sign digits of the divisor and the partial dividend (that is, the remainder).

(b) If the sign digits are the same write a 1 into the appropriate quotient position, shift the partial dividend one place left and subtract the divisor from the remainder.

(c) If the signs are different write a 0 into the quotient, shift the remainder one place left and add the divisor to the remainder.

The quotient derived in this way is not the true quotient since in fact 0's have been written in place of 1's. The true quotient is obtained by adding $2^{-n} + 1$ to the final result, note that the pseudo-quotient must be calculated to $2^{-(n-1)}$ bits, where n is the number of bits in the operand. Table 5.25 shows examples of binary division using this algorithm. The accuracy of the result will obviously depend on the number of places to which the quotient is calculated; for example, in Table 5.25(b) and (c) if six places are used (correction factor of 1·00001) the answers become 0·10101 and 1·01011 respectively. This is due to the round-off action of the $2^{-n} + 1$ correction term (see next section).

5.15 Round-off of binary numbers

Both the division and multiplication processes generate more digits in the result than are generally required; for example, the multiplication of two 4-bit numbers produces an 8-bit product. Similarly, in division the quotient may be recurring, comprising an infinite series of digits regardless of the size of the operands.

In the majority of cases an n-digit approximation to a $2n$ digit number is required, that is, the single-length representation of a double-length result. There are a number of round-off procedures that exist for binary numbers, these may be employed either in software routines or incorporated into the arithmetic hardware.

The simplest technique is always to make the lowest order bit of the number to be retained a 1, irrespective of the value of subsequent lower order bits. This is a very convenient method to employ for hardware correction since it does not require the registers to propagate a carry. This technique is in fact used indirectly in the corrected non-restoring algorithm for division, described above. The method may be modified slightly by adding a 1 to the least-significant digit to be retained when that digit is a 1, but doing nothing if it is a 0. In this case, however, the product or quotient registers must have add or carry propagate facilities.

An alternative approach is to add a 1 to the highest digit order of that portion of the number that is to be ignored. This is equivalent to adding 1 to the least significant digit of the number to be retained if the highest order digit to be dropped is a 1. For example, the double-length product $00101\underline{1}0101$ would round-off single-length to 00110, and similarly $01011\underline{0}1101$ to 01011. When a 2's complement number is rounded-off it is first necessary to convert the number to its true representation, otherwise the round-off will occur in the opposite direction to that required. As before, the implementation is expensive if add or carry propagation facilities are not available. Furthermore, it is necessary to retain the double-length result (or at least the highest order digit of the least-significant half) throughout the arithmetic processes.

5.16 Floating-point binary arithmetic[19,20]

We have already seen in Chapter 2 how floating-point number representation eliminates the need for scaling (required in fixed-point notation) and increases the number range of the computer. However, the hardware implementation of floating-point arithmetic is complicated by the fact that it is necessary to perform arithmetic operations on both the exponent and mantissa parts of the numbers.

In the addition or subtraction of floating-point numbers it is necessary to align the mantissas before performing the arithmetic operations so that the numbers are added (or subtracted) into the correct orders. The alignment is achieved by equalizing the exponents (a, b) of the floating-point operands so that the scaling factors are the same, and at the same time shifting the mantissas (A or B) accordingly. The operation is performed by comparing the exponents and then proceeding according to the following rules:

(a) If $a > b$, mantissa B is shifted right K places, where K is the difference between the exponents.
(b) If $a < b$, mantissa A is shifted right K places.
(c) If $a = b$, no action takes place.

At the same time the corresponding exponent register must be incremented by $+1$ for each right shift of its mantissa; Fig. 5.34 shows a detailed flow-diagram of the operation, and Tables 5.26(b) and (c) show some numerical examples of floating-point addition and subtraction. Note that after the floating-point numbers have been aligned, the addition (or subtraction) of the mantissa proceeds in the normal manner.

One method of implementing the floating-point addition/subtraction algorithm is shown in Fig. 5.35. In this case separate registers are used to store the exponent and mantissa parts of the operands, the exponent registers are connected as 6-bit counters, and the mantissa registers as standard shift-registers. The parallel outputs of the exponent registers are taken to an asynchronous iterative comparator circuit (see Fig. 5.36) which has two outputs, X and Y, coded as follows:

If $a > b$ then $X = 1, Y = 0$
$a < b$ then $X = 0, Y = 1$
$a = b$ then $X = 0, Y = 0$

The operation of the comparator and its control unit is to compare the exponents, a and b, the resulting outputs Z_1 and Z_2 being used to shift the appropriate mantissa one place right and at the same time to increment its exponent by $+1$ in the least significant digit position. This procedure continues until the exponents are equal; the waveform Z_3 is a control signal which is used to inhibit the comparator output while the contents of the exponent registers are changing.[21] After the equalization has been completed (indicated by the waveform Z_4) the addition or subtraction is performed using normal parallel-adder techniques as described above.

When equalizing small-magnitude operands the least-significant digits may be lost (they are shifted out of the register) unless double-length registers are used. In order to economize on register stages it is common practice to use a single double-length register for the smallest operand (which is required to be shifted in the alignment process) and to determine this by an initial subtraction of the operands. The alignment procedure is usually terminated automatically after a limited number of shift operations (determined by the wordlength) in which case both the mantissa and exponent are forced to zero.

As we have seen earlier, floating-point numbers are usually distributed and stored in the computer in a normalized form. The arithmetic

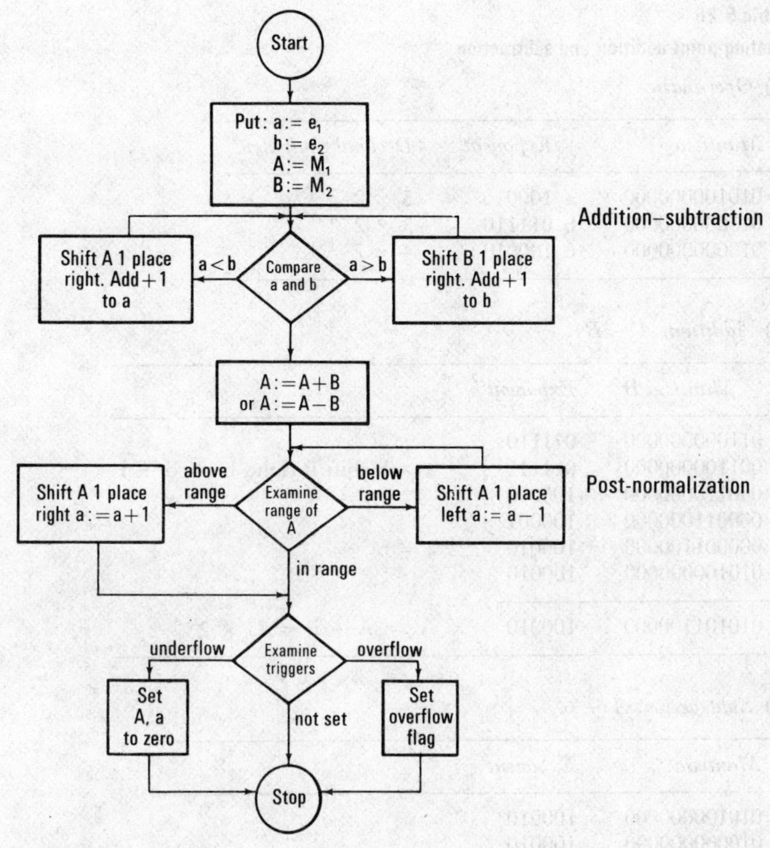

Figure 5.34 Flow chart for floating-point addition.

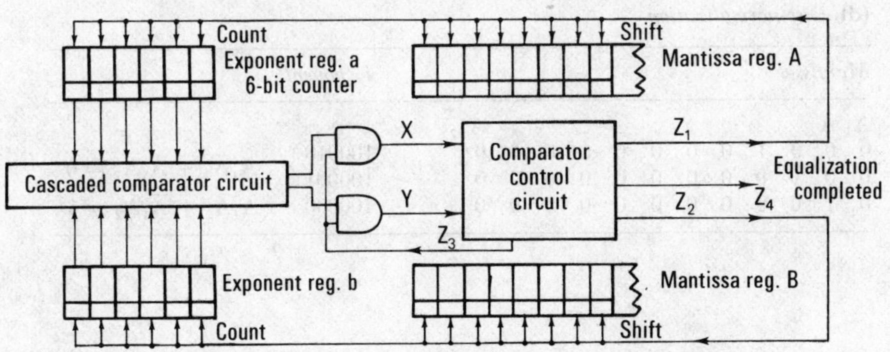

Figure 5.35 Floating-point addition/subtraction exponent equalisation.

Table 5.26
Floating-point addition and subtraction

(a) *Operands*

	Mantissa		Exponent	Decimal equivalent
A	010100000000	a	100010	5×2^{-1}
B	011000000000	b	011110	3×2^{-4}
C	010000000000	c	100010	4×2^{-1}

(b) *Addition, $A + B$*

	Mantissa B	Exponent b	
	011000000000	011110	
	001100000000	011111	a > b shift B right, b := b + 1
	000110000000	100000	
	000011000000	100001	
B	000001100000	100010	a = b
A	010100000000	100010	
	010101100000	100010	$A := A + B = 43 \times 2^{-4}$

(c) *Subtraction, $A - C$*

	Mantissa	Exponent	
A	010100000000	100010	
C	010000000000	100010	
	000100000000	100010	$A := A - C = 1 \times 2^{-1}$

(d) *Post-normalization*

Mantissa	Exponent	
A_1 A_2 . . . . A_{12}		
0 0 0 1 0 0 0 0 0 0 0 0	100010	$A_1\overline{A}_2 + \overline{A}_1A_2 = 0$
0 0 1 0 0 0 0 0 0 0 0 0	100001	
0 1 0 0 0 0 0 0 0 0 0 0	100000	$A_1\overline{A}_2 + \overline{A}_1A_2 = 1$

process, however, often results in a non-normalized form of number and consequently it is necessary to incorporate a **post-normalization** routine into the arithmetic logic. For example, in the subtraction of mantissas of similar magnitude the resulting difference could be very small, and the number must be brought back into the correct range.

In practice post-normalization consists simply of shifting the result of an arithmetic operation left (or right) until the number comes into range. One method of detecting when the number is in range is to arrange that the mantissa register includes an extra stage at the most significant end, for example in the case of mantissa A, this would be A_0, where A_1 is the most significant digit. The detection and normalization procedure may then be stated as follows:

(a) If the number is *below range* the mantissa is shifted one place left, and the exponent decremented by $+1$, until the condition $A_1\overline{A}_2 + \overline{A}_1A_2 = 1$ is satisfied, that is, below range is indicated by $A_1\overline{A}_2 + \overline{A}_1A_2 = 0$
(b) If the number is *over range*, indicated by $A_0\overline{A}_1 + \overline{A}_0A_1 = 1$, the mantissa is shifted one place right and the exponent increased by $+1$.

The post-normalization procedure is illustrated in Fig. 5.34 and Table 5.26(d). The same technique may, of course, be used to perform the normalization of any floating-point number, and some computers include an instruction of this type in their order repertoire.

The multiplication and division of floating-point operands are easier operations to perform, since in this case the exponents are only required to be added (or subtracted) and normal multiplication or division performed on the mantissas. In many computers the two arithmetic operations (on the exponent and mantissa) are performed simultaneously. For example, since the exponent normally has a short wordlength (typically 9-bits) the addition or subtraction can be performed serially at the same time as the parallel multiplication and division. As with floating-point addition and subtraction it is also necessary to perform a post-normalization routine after the main arithmetic operations.

Hardware floating-point is seldom incorporated in the basic central processor of small real-time control computers, but is generally available as a separate option, or as a software routine. The larger business and scientific machines would of course include floating-point arithmetic as a basic function.

5.17 Software implementation

Arithmetic functions such as multiplication and division and floating-point operation are not normally included in the instruction set of presently available microcomputers (and some minicomputers). Consequently if these operations are required they must be programmed as a software routine or provided as an add-on hardware unit. In bit-slice microcomputers these functions could of course be implemented at micro-code level.

Programmed sub-routines are usually slow in operation, and if speed is an important factor it is essential to use special purpose hardware. For example, to implement Booth's multiplication algorithm for 16-bit operands on an Intel 8080/MC 6800 microprocessor system would require some 100 bytes of storage and entail an operation time in the order of 1000 μs. In contrast, as we have seen, the multiplication operation can be performed in 3·2 μs using a quasi-serial hardware multiplier.

Most of the procedures presented in the above sections can be programmed as software routines.[22] Note however, that it is normally required to use multi-byte operation in microcomputers (8 or 16-bit words in general do not give the required accuracy, though this does of course depend on application) which will complicate and slow down the software processes. Because the arithmetic instructions of a microcomputer operate on a single byte any routine requiring addition or subtraction of multi-byte numbers must be able to handle the carries occurring between bytes (larger machines and some minicomputers, such as the DEC PDP11, have variable byte operation which enables multi-byte operands to be handled directly by the instruction set). In most microcomputers a special 'ADD with carry' instruction is provided which adds the contents of the Accumulator, plus the addressed operand, plus a carry bit, placing the sum back into the Accumulator and generating an external carry if required. With this instruction the addition of each byte of a binary number becomes a simple routine involving the 'ADD with carry', plus instructions to move data to and from the store. Similarly a 'SUB with borrow' instruction is provided to allow subtraction with multi-byte numbers. Numerical routines are further facilitated, especially multiplication and division, if the microcomputer has two or more registers which can be used as accumulators and suitable shift and rotation instructions.

5.18 Error-detecting arithmetic logic

Digital computers frequently include special hardware circuits (and software routines also) to detect errors in the logical and arithmetic operations of the machine. For this purpose an error may be defined as any logical output other than the normal output of the correctly functioning circuit. There are two main actions that can be performed by the computer system after detecting an error, they are:

(a) inform the user (programmer or maintenance engineer) that an error has occurred, and
(b) initiate the restart of an aborted operation.

The simplest method of indicating when an error has occurred is to use error lamps situated on the control console. A more useful approach, though, is to include in the order repertoire of the machine a conditional jump type instruction which transfers program control to an error routine if an error has occurred. This method places the responsibility of detecting when an error has occurred (and any subsequent action) with the programmer. Moreover it has the disadvantage that the user must continually check for errors by inserting the error jump instruction at suitable points in the main program. A better technique is to allow all errors to automatically interrupt (see Chapter 3) the main program immediately they occur, and to use an interrupt routine to initiate any error procedure. In all cases it must be possible to prohibit or cancel the error indication.

In order to check if an error has occurred during a logical operation it is necessary to know what the correct results should be. With binary operations it is virtually impossible to predict and store the correct outputs (for comparison) owing to the vast number of possible input combinations involved. The only practical alternative is to use identical circuits in parallel and to compare the outputs on a voting basis, for example two-out-of-three majority circuits (see Chapter 8). Since in general these methods are far too expensive to use in most commercial computers a compromise is made by using error-detecting codes.[23, 24, 25] In general these codes can detect single errors, but not necessarily multiple errors. Thus undetected errors are always possible, but the proper choice and use of code can make the probability of undetected errors occurring very small.

The basic idea of error-detecting codes is that the binary data is encoded in such a way (for example by including additional information such as a parity bit) so that errors may be detected by examination of the coded form after a logical operation has taken place. The two main types of code used in error-detection circuits are the **parity** and **residue** codes.

Parity checking codes, as we have already seen in Chapter 3, are obtained by including an extra digit (or digits in the case of Hamming codes) with the information bits such that the total decimal sum of 1's in the number is either odd or even. For example if the number of 1's in the data is even, the additional parity digit to be included with the data would be 1 for odd parity and 0 for even parity. Checking of parity coded binary numbers is easily performed by the modulo 2 addition of the individual digits in the word; modulo 2 addition is simply normal binary addition without carries, that is, the exclusive OR function. Figure 5.37 shows typical parity checking circuits for serial and parallel numbers.

Residue codes are based on modulo arithmetic; for example, if we divide a number N by another number p (called the **modulus**) we obtain a quotient and a remainder, or **residue** R, that is:

$$R(N) = N \bmod p$$

The residues of a number uniquely identify it, and they can be used as codes to represent the numbers; Table 5.27 shows the numbers 0–20 and

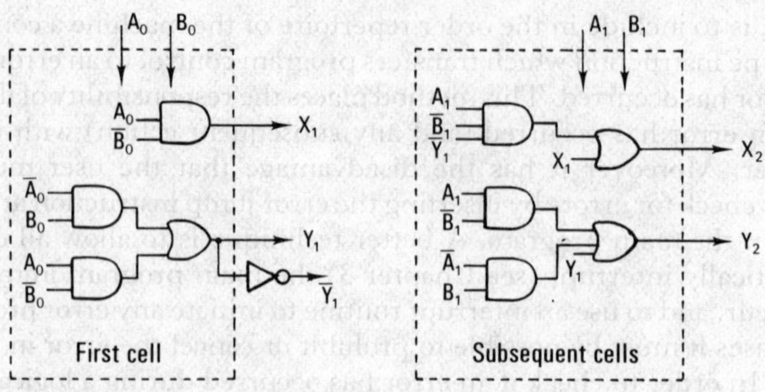

Figure 5.36 Cascaded comparator circuit.

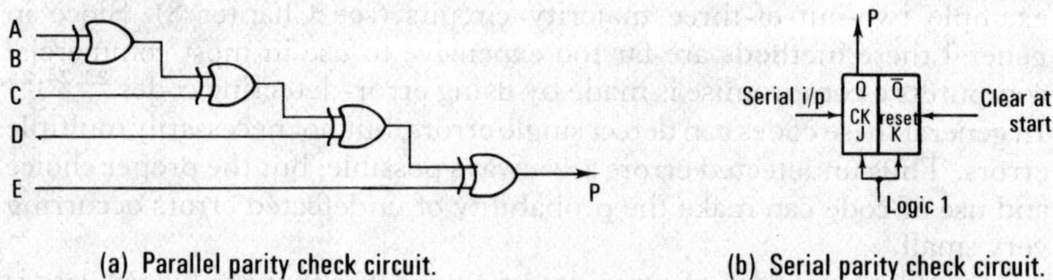

(a) Parallel parity check circuit.

(b) Serial parity check circuit.

Figure 5.37 Parity check circuits.

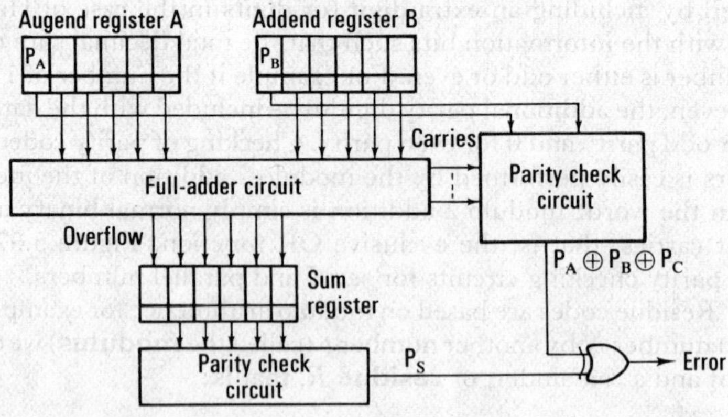

Figure 5.38 Parity checked adder circuit.

Table 5.27
Residue numbers

Number N	2	3	5	7	*Number N*	2	3	5	7
	Residue to modulo					*Residue to modulo*			
0	0	0	0	0	11	1	2	1	4
1	1	1	1	1	12	0	0	2	5
2	0	2	2	2	13	1	1	3	6
3	1	0	3	3	14	0	2	4	0
4	0	1	4	4	15	1	0	0	1
5	1	2	0	5	16	0	1	1	2
6	0	0	1	6	17	1	2	2	3
7	1	1	2	0	18	0	0	3	4
8	0	2	3	1	19	1	1	4	5
9	1	0	4	2	20	0	2	0	6
10	0	1	0	3					

their residues modulo 2, 3, 5, and 7. It is also possible to perform arithmetic operations on the residue numbers; for instance, addition requires only the addition of the corresponding residues in their respective number system (that is, addition modulo p), carries being ignored. Thus:

$$\begin{array}{rl} & 2\ 3\ 5\ 7 \\ 13 \equiv & 1\ 1\ 3\ 6 \\ 7 \equiv & 1\ 1\ 2\ 0 \\ \hline 13+7=20 \equiv & 0\ 2\ 0\ 6 \end{array} \qquad \begin{array}{rl} & 2\ 3\ 5\ 7 \\ 9 \equiv & 1\ 0\ 4\ 2 \\ 8 \equiv & 0\ 2\ 3\ 1 \\ \hline 9+8=17 \equiv & 1\ 2\ 2\ 3 \end{array}$$

Multiplication of residue numbers is performed very simply by the multiplication (modulo p) of corresponding residues, again ignoring carries, for example:

$$\begin{array}{rl} & 2\ 3\ 5\ 7 \\ 3 \equiv & 1\ 0\ 3\ 3 \\ 6 \equiv & 0\ 0\ 1\ 6 \\ \hline 3\times 6=18 \equiv & 0\ 0\ 3\ 4 \end{array} \qquad \begin{array}{rl} & 2\ 3\ 5\ 7 \\ 5 \equiv & 1\ 2\ 0\ 5 \\ 4 \equiv & 0\ 1\ 4\ 4 \\ \hline 5\times 4=20 \equiv & 0\ 2\ 0\ 6 \end{array}$$

Because the multiplication of residue numbers is extremely fast, the technique has been used as the basis of a fast hardware multiplier for special applications.[26]

These ideas also form the basis of the 'casting out nines' arithmetic check, often described in many elementary arithmetic books. For example:

$$\begin{array}{rlrl} N_1 = & 256 & R(N_1) = & 256 \bmod 9 = 4 \\ N_2 = & 345 & R(N_2) = & 345 \bmod 9 = 3 \\ \hline N_1 + N_2 = & 601 & R(N_1 + N_2) = & 601 \bmod 9 = 7 \end{array}$$

Thus we may check the arithmetic operations by performing the corresponding modulus arithmetic on the residue numbers. It can be shown[27] that since we are using the modulus $p = r - 1$, where r is the radix of the number system, the calculation of the residues may be greatly simplified, in this case we have:

$$R(N_1) = (2 + 5 + 6) \bmod 9 = (13) \bmod 9 = 4$$
$$R(N_2) = (3 + 4 + 5) \bmod 9 = (12) \bmod 9 = 3$$

Note also that if the modulus of the number had been chosen equal to the radix, the residue would be equal to the last digit of the number. These techniques of residue checking are commonly incorporated in hardware for the purpose of detecting errors in arithmetic operations.

A residue code may also be used for the general checking of data and there are two main methods in general use. The first one consists of finding the residue of a number using a suitable modulus, converting it to check digits, and then appending it to the original number. For example:

$$N = 101110 \quad R(N) = N \bmod 3 = 46 \bmod 3 = 01$$
thus, $\text{codeword} = 10111001$

The second method is to multiply the number to be encoded by a chosen modulus m, the resulting number mN will then have a residue equal to zero. For example:

$$N = 101110, \qquad m = 3$$
thus, $\text{codeword} = 10001010$

Note that the first method allows the check digits and numbers to be handled separately, whereas the second method includes self-checking properties.

We will now discuss the way in which arithmetic operations performed with coded operands, using the above error-detecting codes, may be checked with hardware to ensure that the logic circuits are functioning correctly. In the case of the **parity checked adder** the general technique is first to form the normal binary sum of the operands, ignoring the parity digits. Then the actual parity digit corresponding to the sum output is derived and compared with the predicted parity. The predicted parity of the sum output is given by:

$$\begin{aligned} P_S &= S_n \oplus S_{n-1} \oplus \cdots \oplus S_0 \\ &= (A_n \oplus B_n \oplus C_{n-1}) \oplus (A_{n-1} \oplus B_{n-1} \oplus C_{n-2}) \oplus \cdots \\ &\qquad (A_0 \oplus B_0 \oplus C_{1N}) \\ &= (A_n \oplus A_{n-1} \oplus \cdots \oplus A_0) \oplus (B_n \oplus B_{n-1} \oplus \cdots \\ &\qquad \oplus B_0) \oplus (C_{n-1} \oplus C_{n-2} \oplus \cdots \oplus C_{1N}) \end{aligned}$$

Thus, $P_S = P_A \oplus P_B \oplus P_C$

Where P_A and P_B are the parity check digits of the input operands, and

P_C the parity of the internally generated carries. Examples of this procedure are shown below:

		P.B.
A	0 1 1 1 0 1	0
B	0 1 1 0 1 0	1
Sum	1 1 0 1 1 1	

$$\text{Actual } P_S = 1 \oplus 1 \oplus 0 \oplus 1 \oplus 1 \oplus 1 = 1$$
$$\text{Predicted } P_S = 0 \oplus 1 \oplus 1 \oplus 1 = 1$$

		P.B.
A	0 1 1 1 0 1	0
B	0 1 1 0 1 0	1
Sum	1 1 0 1 1 0 (↑ error)	

$$\text{Actual } P_S = 1 \oplus 1 \oplus 0 \oplus 1 \oplus 1 \oplus 0 = 0$$
$$\text{Predicted } P_S = 0 \oplus 1 \oplus 1 \oplus 1 = 1$$

Note that even parity is used in the examples, but if odd parity is required it is simply a matter of inverting the outputs of the checking circuits. A block diagram of the parity checked adder is shown in Fig. 5.38.

Unfortunately this circuit has the disadvantage that carry errors are undetected. The reason for this is that carry errors always produce an equivalent sum digit error, and parity checked binary numbers do not detect even numbers of errors. In practice, this situation can be alleviated if the full-adder is designed using half-adder circuits (see Section 5.2) and the half-adder sum equation factorized (and implemented) in the form:

$$S = (A + B)\overline{AB}$$

where AB is, of course, the carry generate term, and A + B the carry transmit. In this way the same circuitry is shared by both sum and carry logic thus permitting an error in the carry to cause a detectable error in the sum. Note that this technique does not give a complete check on all possible fault conditions, since it is still possible for an error in (A + B) or $(\overline{AB})$ to produce a carry error without causing S to be in error.

The standard method of detecting carry errors is to generate a duplicate carry, for each carry digit, using separate hardware. For example, consider the carry equation for a ripple full-adder (Section 5.2), that is,

$$C' = G + TC$$

now a duplicate carry would be generated for each stage:

$$C_d' = G_d + T_dC$$

An error in G or T cannot cause an error in C'_d because it is independent of G and T; errors in C, however, can cause both C' and C'_d to be in error. All these carry errors can be detected by comparing C' and C'_d using the exclusive OR function, that is, carry errors are given by $C' \oplus C'_d$.

Combining the carry error detection circuit with the sum parity checking circuit described earlier, gives a complete error detection circuit; this is shown in Fig. 5.39. These techniques and ideas can, of course, be extended to carry look-ahead type adders, but in the majority of cases the large amount of additional logic involved prohibits its use in most commercial computer applications.

It is also possible to use logic circuits to check arithmetic operations using the residue system, the hardware equivalent of 'casting out nines'. Because of the increased cost and loss of speed, plus the added complexity of the control unit, residue checking is not an attractive proposition at the present time. However, at least one commercial computer company (UNIVAC) uses a mod 3 residue check on its binary arithmetic operations.

The block diagram of a residue checked adder is shown in Fig. 5.40; note that the overall scheme closely follows the worked examples of residue checking described earlier. The binary operands to be added together, A and B, are entered to a conventional adder and also to the mod m residue circuits. The residues, a and b, are then added mod m and taken to a comparator circuit where they are compared with the residue mod m of the sum $A + B$. If an error occurs in the binary adder the residue of the sum will differ from the sum of the operand residues and an error will be indicated. Note that only the input and output of the binary adder circuit is used in this type of checking, and consequently the technique may be used with any form of adder.

After addition the next most fundamental arithmetic operation is that of shifting, and here again it is possible to check errors using either the parity or residue systems. As we have seen earlier (Chapter 4) there are basically three ways of performing the shift operation – the logical shift, the arithmetic shift, and the end-around shift. Furthermore, we must also take into consideration the transfer of data from one shift register to another. In all these cases the parity or residue of the shifted operands can be computed from the bits shifted out, the bits shifted in, and in the case of residue the number of shifts.

In the case of parity, the parity of the shifted number P_S is given by:

$$P_S = P_N \oplus P_O \oplus P_I$$

where P_N is the parity of the original number, P_O the parity of the bit shifted out and P_I the parity of the bit shifted in.

As one would expect, the checking computation involved when using the residue system is rather more complex. The residue of the shifted number R_S is given by:

$$R_S = [2^s . R(N) - R_O + R_I] \bmod m$$

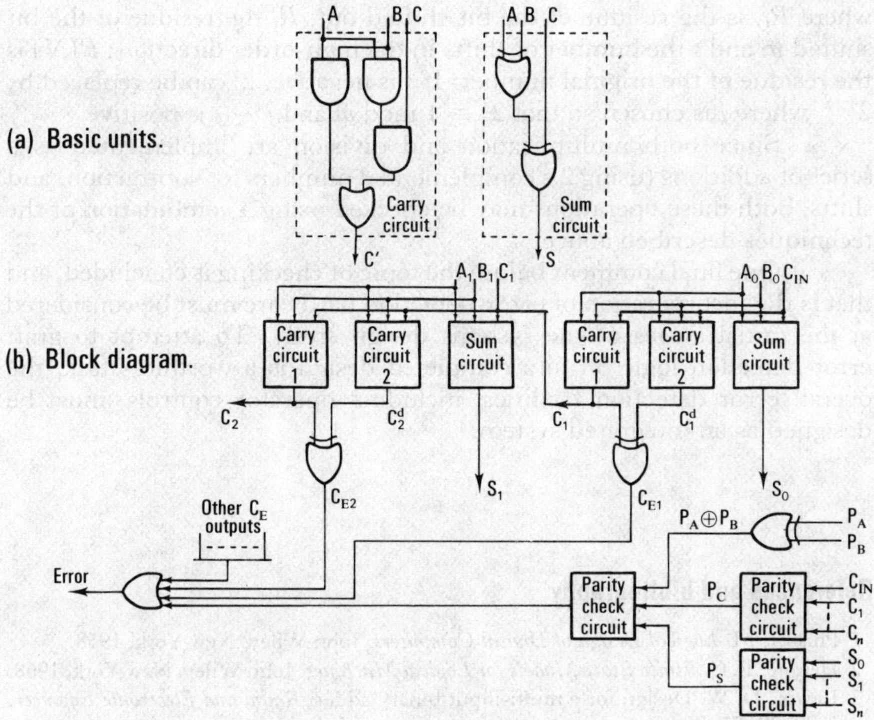

Figure 5.39 Completely checked adder circuit.

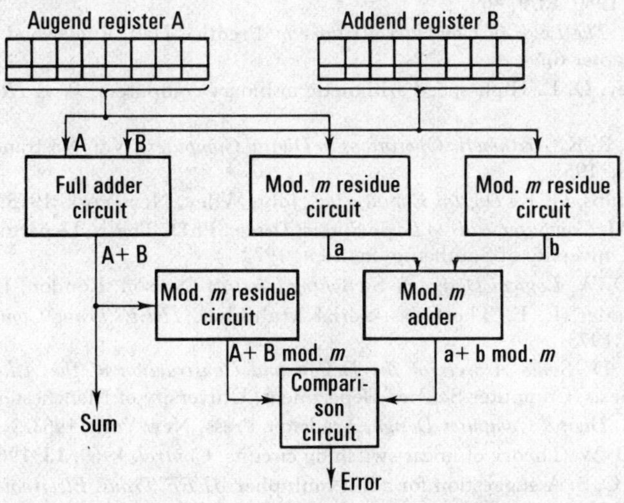

Figure 5.40 Residue checked adder.

where R_O is the residue of the bit shifted out, R_I the residue of the bit shifted in and s the number of shifts in the high-order direction; $R(N)$ is the residue of the original number. If s is negative, 2^s can be replaced by 2^{j-s}, where j is chosen so that $2^j \equiv 1 \bmod m$ and $j - s$ is positive.

Since both multiplication and division are implemented as a series of additions (using 2's complemented numbers for subtraction) and shifts, both these operations may be checked using a combination of the techniques described above.

One final comment before the topic of checking is concluded, and that is the incorporation of error-detection hardware must be considered at the initial stages of the systems design study. To attempt to graft error-detection logic on to a completed design is a worthless task, the overall error detection facilities, including operator controls, must be designed as an integrated system.

References and bibliography

1 Phistor, M. *Logical Design of Digital Computers*. John Wiley, New York, 1958.
2 Hennie, F. C. *Finite State Models for Logical Machines*. John Wiley, New York, 1968.
3 Lewin, D. W. Design for a multi-input binary adder. *Radio and Electronic Engineer*, 1970, **39**, 77–80.
4 Burks, A. W., Goldstone, H. H., and Von Neumann, J. *Preliminary Discussion of the Logical Design of an Electronic Computing Instrument*. Princeton, N.J., Institute for Advanced Study, June, 1946.
5 Reitwiesner, G. W. The determination of carry propagation length for binary addition. *IRE Trans. Electron. Comput.*, 1960, **EC9**, 35–38.
6 Hendrickson, H. C. Fast high-accuracy binary parallel addition. *IRE Trans Electron. Comput.*, 1960, **EC9**, 469.
7 Flores, I. *The Logic of Computer Arithmetic*. Prentice-Hall, Englewood Cliffs, N.J., 1963 (Chapter 6).
8 MacSorley, O. L. High-speed arithmetic in binary computers. *Proc. IRE*, 1961, **49**, 67–91.
9 Richards, R. K. *Arithmetic Operations in Digital Computers*. Van Nostrand–Reinhold, New York, 1955.
10 Kostopoulos, G. K. *Digital Engineering*. John Wiley, New York, 1975.
11 Waters, M. *Computer Aids to Logic System Design*. PhD. Thesis, Department of Electronics, University of Southampton, Sept. 1972.
12 Lewin, D. W. *Logical Design of Switching Circuits*. Nelson, London, 1974.
13 Swartzlander, E. E. The Quasi-Serial Multiplier, *IEEE Trans Computers*, **C22**, 317–321, 1973.
14 Aspinall, D. *Some Aspects of the Design and Construction of the Atlas Computer*. PhD. Thesis, Computer Science Department, University of Manchester, Oct. 1961.
15 Braun, E. *Digital Computer Design*. Academic Press, New York, 1963.
16 Lewin, D. W. Theory of linear switching circuits. *Control*, 1969, **13**, 196–203.
17 Wallace, C. S. A suggestion for a fast multiplier. *IEEE Trans. Electronic Computers*, 1964, **EC13**, 14–17.
18 Booth, A. D. A signed binary multiplication technique. *Q.J. Mech. Appl. Math.*, 1951, **4**, 2, 236–240.

19 Sweeney, D. W. An Analysis of Floating Point Addition, *IBM Syst J.*, **4**, 31–42, 1965.

20 Sterbenz, P. H. *Floating Point Computation.* Prentice Hall, Englewood Cliffs, N.J., 1974.

21 Lewin, D. W. Delay generation in asynchronous logic systems. *Electronic Engineering*, 1969, **41**, 356–360.

22 Peatman, J. B., *Microcomputer-Based Design.* McGraw Hill, New York, 1977.

23 Hamming, R. W. Error detecting and error correcting codes. *Bell Syst. Tech. J.*, 1950, **29**, 147–160.

24 Peterson, W. W. and Weldon, E. J., *Error Correcting Codes.* MIT Press, Cambridge, Mass., 1972.

25 Garner, H. L. The residue number system. *IRE Trans. Electron. Comput.*, 1959, **EC8**, 140–147.

26 Flores, I. *The Logic of Computer Arithmetic.* Prentice-Hall, Englewood Cliffs, N.J., 1963 (Chapter 18).

27 Sellers, F., Hsiao, M., and Bearnson, L. *Error Detecting Logic for Digital Computers.* McGraw-Hill, New York, 1968 (Chapter 3).

Tutorial problems

5.1* Design a parallel arithmetic unit for the addition and subtraction of binary numbers represented in (a) the sign and modulus system and (b) 1's complement notation.

5.2 Design a parallel arithmetic logic circuit for adding and subtracting binary-coded decimal integers represented in the 8421 code.

5.3* Construct the addition table for modulus 3 binary arithmetic; using this table devise the logic circuits for a mod 3 residue adder.

5.4* Describe how the arithmetic function 'Add half the contents of register A to register B, putting the 2's complement of the result in register C' may be performed. Give a logic diagram of the system and a suitable control micro-program; consider both parallel and serial implementation.

5.5 Repeat question 5.3 for a mod 3 residue multiplier circuit.

5.6* Write a software routine (give flow-charts only) to implement the corrected multiplication algorithm for signed operands shown in Table 5.24. Assuming the parallel computer structure described in Chapter 4, devise a suitable micro-program to implement the algorithm in hardware.

5.7 Repeat question 5.6 for corrected non-restoring binary division.

5.8* Devise a detailed logic design for the comparison and control circuits of the uniform shift parallel multiplier shown in Fig. 5.28.

5.9 Consider the problems involved in parity checked subtraction and then design a suitable logic circuit to implement the necessary checking operation.

5.10* Consider in detail how the following logical functions may be designed into the arithmetic unit of a computer.

(a) The collation function, $A := A.B$.

(b) The inclusive and exclusive OR functions,

$$A := A + B, A := \overline{A}B + A\overline{B}$$

(c) 2's complementation of operands, $A := 2^n - A$.

(d) The NAND function, $A := \overline{A.B}$.

6

Storage systems

6.1 Introduction

As we have seen in the earlier chapters, the digital computer is based on the stored program concept and consequently depends for its operation on the availability of a **memory** to store data and instructions. In fact, storage has been a continuing limitation in the design and cost of a computer in addition to playing a central role in characterizing its structure. The speed of computers is limited by the time required to store and retrieve information, while the costs of a system are largely determined by the data storage capacity of its memory. As a result considerable effort has been devoted to the development and improvement of memory devices and systems.

In particular, dramatic advances have been made as a result of the tremendous progress in semiconductor technology and LSI.[1] This has resulted in the development of very cheap semiconductor memories which could remove many of the constraints previously imposed on computer architecture. Since 1963, when the first semiconductor memories were produced, operating speed and reliability have increased by at least an order of magnitude as physical size, power consumption and cost per bit of storage, have been reduced by a similar factor. There is of course a physical limitation to this progress, but it is envisaged that comparable improvements can be achieved over the next decade.

The importance of these developments on computer architecture could be considerable, since memory will now play a dominant role rather than just acting as a simple system component.

In this chapter we shall first consider the fundamental properties and characteristics of storage devices, and then proceed to discuss in detail the techniques of using and organizing storage in computer systems. The emphasis will, in the main, be on the efficient and economical use of storage since, in most practical applications, the choice of storage capacity is often a compromise between what is desirable and what can be provided.

6.2 Basic characteristics of storage devices

A digital memory must contain a discrete physical storage cell, capable of being set by an external signal into one of two distinct states, for each bit of the computer word to be stored. The cell must remain in this set state indefinitely or until it is changed to the other state by another external signal. The two distinct states of a storage cell can be naturally occurring states which require no external energy sources to be maintained. This is true for ferro-magnetic, ferro-electric, and superconducting cells all of which have the property that the quantity defining the state (magnetic induction, electric polarization, or induced super current respectively) has stable remanent states corresponding to zero energization. It is also possible to use storage elements which require external energization to maintain the stored state. As well as the obvious bistable circuits employing semiconductor electronics, these include capacitive, optical, magnetic and electrostatic devices. Typical examples of current storage devices are ferrite cores, thin magnetic films, magnetic tapes and discs (ferromagnetic), magnetic bubbles, Schottky diodes, optical (laser and holographic) and LSI arrays (semiconductor electronics).[2,3,4,5,6]

The most significant feature of a storage device or system is the speed with which a word can be read out, or written into, the store. This characteristic is called the **access time**. The total time for reading and writing information is normally split into two parts:

(a) the time required for addressing and locating the required word, and
(b) the switching or operation time of the storage element.

In fast systems it is also necessary to take into account the propagation times for the address and information to travel to and from the CPU and memory via the common highway bus logic. Thus, true access time is best defined as the interval between the generation of a store address in the CPU and the receipt by a CPU register, for example the accumulator, of the information contained in the addressed location. In a **random access** store all words are equally accessible; that is, the time required for addressing and locating a word is constant. The information in a **cyclic access** store, however, circulates continuously in a repetitive loop and each word is accessible only as it passes the reading and writing stations. Thus, the access times for cyclic stores are inherently longer since, on the average, half a cycle must elapse before the desired word becomes available. **Serial access** systems, such as magnetic tape and bubble memories also exist where the data are read or written in a longitudinal serial fashion along a finite length of tape or shift register.

Storage devices may have either **destructive** or **non-destructive** read-out of the stored data. In the former case, extra logic must be provided to remember and then restore the data to their previous state (this is typical, for example, of most ferrite core stores). **Volatile** storage systems, such as bistables and most other types of semiconductor memory, require a continual or periodic application of power in order to prevent deterioration or a complete loss of the stored information. In some cases this last property may also be a characteristic of non-volatile stores – for example, unless suitable precautions are taken, transients produced by switching off the main power supply can cause loss of data in a ferrite core store.

Magnetic tape and disc stores, etc., which are non-volatile and retain their data after the removal of the power supply, are termed **permanent** stores.

Erasability is another important property of a storage device, particularly when used as the main store in the CPU. External storage systems, however, may employ non-erasable media such as punched cards and tape or photographic systems. The exception to this would, of course, be the use of a **read-only memory**[7] in the CPU micro-program unit, and the implementation of standard software routines in firmware (see Chapter 4). Storage systems may be further classified into **static** (for example integrated circuit arrays and ferrite core stores) and **dynamic** (magnetic tape and drums, shift registers, etc.) depending on whether they operate in the space or time domains respectively.

6.2.1 Storage hierarchy[8]

Since it is not economically nor as yet technically feasible to use one type of storage element for all the memory requirements in a digital computer system, the program and data storage is normally organized in a hierarchy of levels based on speed, capacity, and cost. In fact systems designers use a wide variety of storage devices in an effort to achieve the best performance and largest capacity at a reasonable cost. A typical storage structure for a large computing system is depicted in Fig. 6.1 with the characteristics of the major categories of storage that comprise the hierarchy in Tables 6.1 and 6.2.

The fastest semiconductor storage devices are used for tag registers, control and arithmetic operations, pointer registers, stacks etc. For example, modifier and accumulator registers used to store constants, intermediate results and so on would be realised using high speed bipolar RAM's (**scratch-pad memory**) since they are compatible with the CPU logic. As we have seen semiconductor MOS read-only memory is

Table 6.1
Major categories of computer storage

Category	*Capacity (bits)*	*Access time*	*Access*	*Technology*
Scratch pad and control	500–200k	50 ns–100 ns	Random	Semiconductor
Cache store	50k–200k	50 ns–500 ns	Random	Semiconductor
Main store	10k–1 M	0·3 μs–1 μs	Random	Core/Semiconductor
Paging or swapping	50k–400k	10 μs–50 μs	Random	Core/Semiconductor Bubbles/CCD
Backing Store	10 M–2000 M	500 μs–50 ms	Cyclic/Serial	Disks, drums
Mass memory	10^9–10^{12}	seconds	Serial	Tape/Optical

Characteristics of storage devices

Storage Device	Access Time and type	Capacity	Recording Density	Data Rate	Physical Characteristics	Main Application
Magnetic Drum	8·5–17 ms cycle access	200k–200 M bits	1600 BP1	2–8 M bits/sec	Permanent storage	Backing storage
Magnetic Disc	25–70 ms cyclic access	800k–500 M bits	1000–6000 BP1	800k–8 M bits/sec	Permanent storage Rigid disc Fixed or moving heads	Backing and file storage
Floppy Disc	100–400 ms cyclic access	2–5 M bits	3000 BP1	250k bits/sec	Permanent storage flexible disc	Cheap form of Disc Store; transportable packs
Magnetic Tape	Several minutes serial access	30–800 M bits	1600–6250 BP1	10^7 bits/sec	Permanent storage	Mass storage, on-line and archival
Cassette Tape	50–500 s serial access	1–5 M bits	800 BP1	8–20k bits/sec	Permanent storage	Cheap high capacity store; transportable packs
Magnetic Cartridge	10–50 s serial access	20–22 M bits	1600 BP1	48–160k bits/sec	Permanent storage	Cheap high capacity store; transportable packs
Magnetic Core	250–750 ns (cycle times 650 ns–1 μs) random access	4k–1 M bytes	—	—	Volatile unless protected; destructive read-out	Main store and backing store extension
Semiconductor						
Bipolar RAM	45–60 ns (cycle times 60–100 ns) random access	256–1k bits	—	—	Volatile	Registers, accumulators, buffers, push-down stores. Cache stores
CMOS RAM	50–100 ns (cycle times 500–600 ns) random access	1–4k bytes	—	—	Volatile	Cache stores Fast main memory
PMOS/NMOS RAM	150–400 ns (cycle times 350–800 ns) random access	1–16k bytes	—	—	Volatile	Main memory Buffer stores
MOS/Bipolar ROM/PROM	55–500 ns random access	1–64k bytes	—	—	Permanent/Volatile	Read Only Memory Micro-program control stores, control programs, table-look-up
CCD	50–500 ns serial access	4–64k bits	—	5–20 M bits/sec	Volatile	Backing Store
Magnetic bubbles	Several milliseconds serial access	16k–64k bits	—	100–300k bits/sec	Volatile	Backing store Mass memory

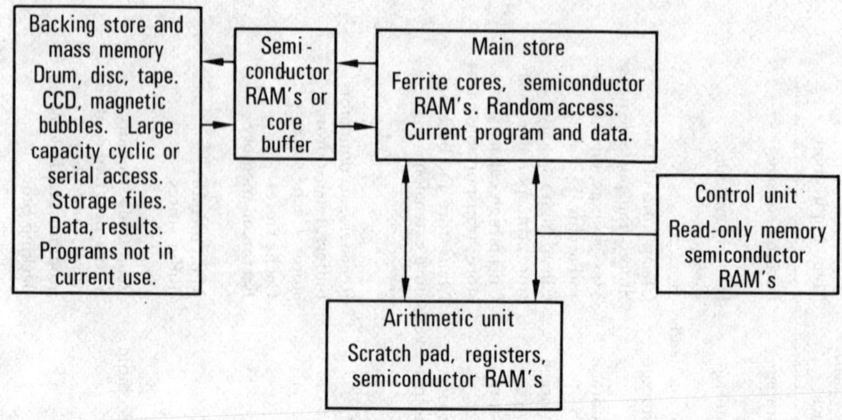

Figure 6.1 Storage hierarchy.

also used in the CPU for micro-program control, look-up tables, function generation, etc. The **main-store** which holds the current data and program being executed requires fast, high capacity random access storage. At the present time magnetic core and semiconductor RAM storage is used for this purpose; core storage, mainly because of its low cost, is still dominant in many computer designs though it is rapidly being replaced by semiconductor memory.[9] To speed up the overall processing rate of the machine a **cache store** is often used (see later) as an high speed buffer between the slower main memory and the CPU; fast semiconductor storage is normally used for this purpose.

Backing Storage is used to store program and data (such as files) not currently in use but which must be available in the main CPU memory at short notice. The requirement here is for medium speed, high capacity but low cost storage, and magnetic drums and discs (both rigid and flexible) are usually employed for this purpose. It is anticipated that in the future moving magnetic devices will be superseded by semi-conductor stores using CCD (charge coupled devices) or magnetic bubble technology.

In many large systems, for example when time sharing is required, core or semiconductor storage is used as a buffer between the main store and the backing-store to facilitate the paging and swapping techniques required to implement a virtual memory system (see later).

When vast quantities of data storage are required for example, archival information such as data-files and data-banks, magnetic tape is used to provide low speed but high capacity storage; this is known as **mass memory**. In smaller microcomputer based systems cassette tape and magnetic cartridges are used to provide low cost backing store and mass memory facilities.

A very important concept in backing and mass storage systems is that of modularity. A module is an independently operable block of memory and any given system will consist of a number of such modules.

For example, the magnetic disc store (or file) is organized into a number of distinct disc packs each of which may be removed from the main system and replaced with other packs as and when required. In this way data may be recorded on one disc pack which may then be removed and used as the data input on a completely different computer system, providing of course the disc file stores are interchangeable. The same advantages apply to magnetic tape where the data are stored on reels of tape; these may, of course, be removed and used on any other compatible tape handler. Tape cassettes, floppy-discs and magnetic cartridges also utilise transportable storage modules but in general the modules are not interchangeable.

6.2.2 Addressing random access stores

The addressing structure of a random access store, that is, the means whereby the contents of a particular store location may be isolated, can take the form of either a **word-organized** or a **bit-organized** system. In the first method (also known as the **two-dimensional** system) the storage elements are arranged in a rectangular array with each row corresponding to a word, and each column to a bit (see Fig. 6.2). To read or write a word into store the appropriate address must first be decoded in order to select and energize a particular wordline. The store address would be represented in the usual way by an N-bit binary number, which must be decoded to select one out of 2^N possible output lines, that is, one for each word. In the case of the 10-bit address used in our running example this would mean 2^{10}, that is, 1024 word-lines. To write a word into the store the bit-lines must be energized according to the pattern of 1's and 0's in the n-bit computer word (held in the store's input/output register) the corresponding wordline having first been selected by the decoder circuits. The same process of selecting and energizing a wordline also applies when reading a word from store; in this case, however, output signals corresponding to the binary contents of the word appear on the bit-lines. Note that these signals must be amplified and registered and, in the case of destructive read-out, written back into the store.

Many different forms of storage cell have been used in this type of random access store – for example, ferrite cores, semiconductor devices such as bipolar and MOS transistors and Schottky diodes. Moreover, it is also possible to design a simple read-only memory using this technique, for example, by using a diode–resistor matrix or MOS transistors to store the digit patterns.

In the method described above the address selection operation is normally completely separate from the basic storage function. If, however, the storage cells themselves were allowed to participate in address selection a considerable economy could be achieved in the decoder circuitry, this is the major advantage of the bit-organized or **three-dimensional** system of addressing. Consider the block diagrams shown in Fig. 6.3; in this case the storage elements are arranged in the form of separate planes each containing one bit of the stored word. Thus for an 18-bit word we would require 18 planes. The total number of words that can be held in the store is determined by the dimensions of the matrix; for a 10-bit address a 32 × 32 plane, giving 1024 words, would be required. The store is accessed by applying simultaneously two address signals, in the *X* and *Y* directions, to wires which are threaded through the same position in each plane of the store assembly (or stack). The response of the selected storage cells (one only per plane) to these signals is sensed to form the output word.

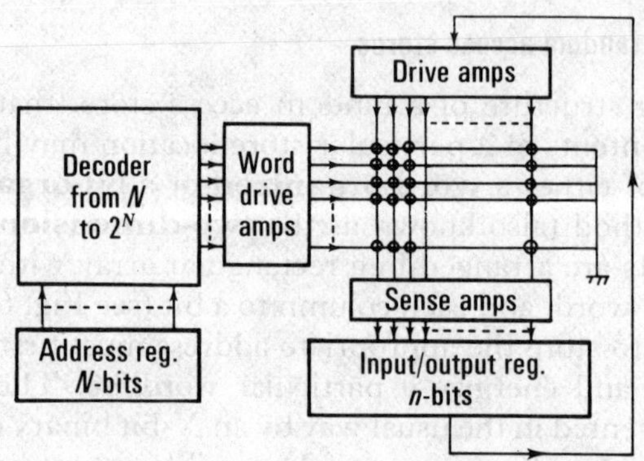

Figure 6.2 Word-organized selection.

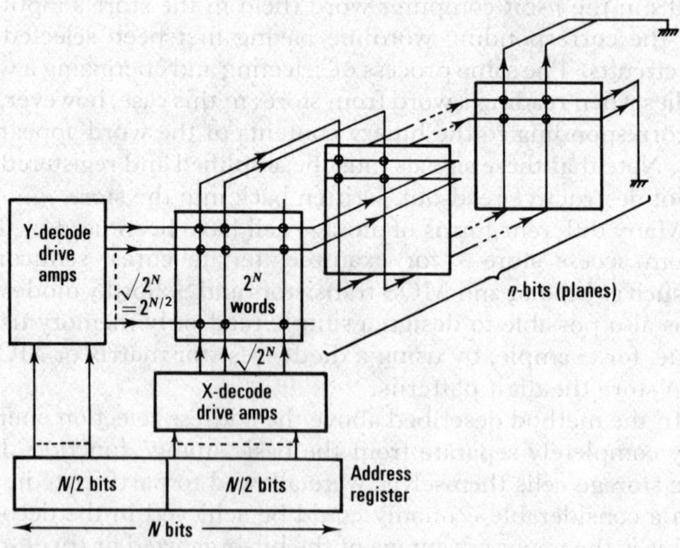

Figure 6.3 Bit-organized (coincident) selection.

The N-bit address digits are normally decoded in two equal parts, each corresponding to $N/2$ bits, to produce $\sqrt{2^N} = 2^{N/2}$ distinct X (or Y) address lines. For example, a 10-bit binary address would be decoded to select one out of 2^5, that is, 32 separate address lines. Thus, the decoding requirements are reduced from 2^N outputs for a word-organized system, to $2.\sqrt{2^N}$ for the bit-organized scheme, that is, for a 10-bit address, from 1024 to 64 address lines. In the general case, assuming that it is not possible to divide the address bits equally (which implies that the store planes must also have an unequal number of rows and columns), the N address bits are divided into N_1 and N_2 bits where $2^{N_1+N_2} = 2^N$, the total number of words in the store. In this case the decoder must select one address line out of 2^{N_1} (or 2^{N_2}) and the total number of address lines is $2^{N_1} + 2^{N_2}$.

The bit-organized method of addressing random access stores is particularly applicable when ferrite cores are used as the storage element, this form of storage, together with numerous others, is described in detail in the next sections.

6.3 Ferrite core stores

Magnetic core storage elements are based on the use of materials (usually ferrites moulded in a toroidal form) which exhibit a square-loop hysteresis characteristic (see Fig. 6.4(a)). The essential property of these materials (as far as the storage function is concerned) is that they can be put into one of two states of remanent magnetism ($\pm B_r$) by the application of a field H_m in the appropriate direction. Thus it is possible to represent the binary values of 0 and 1 by the polarity of the residual remanent magnetism of the core. However, since the stored information can only be read by sensing the flux change when a field H_m is applied (causing the core to traverse its hysteresis loop) the read-out is necessarily a destructive process.

The switching time of a ferrite core is given by the expression:

$$T = \frac{\tau H_c}{H_m - H_c} \text{ seconds}$$

where H_m is the applied field (in oersteads), H_c the coercive force of the material (the value of H where the hysteresis loop crosses the H-axis), and τ is a constant of the material. Since the applied force is usually limited to $2H_m$ by system considerations (such as using the **coincident current** technique, see later) and the product τH_c is largely independent of the composition of the material, the switching time can only be reduced by increasing H_c. This means in practice that, since proportionally higher drive currents must be used, the linear dimensions of the core must be reduced to keep the power requirements and drive currents at an acceptable level. Thus the switching speed of the core depends on its size (the switching current requirement decreases linearly with the diameter, since the flux path is along the circumference) with typical outside diameters varying between 14–30 mils, with 22 mils being the normal value.

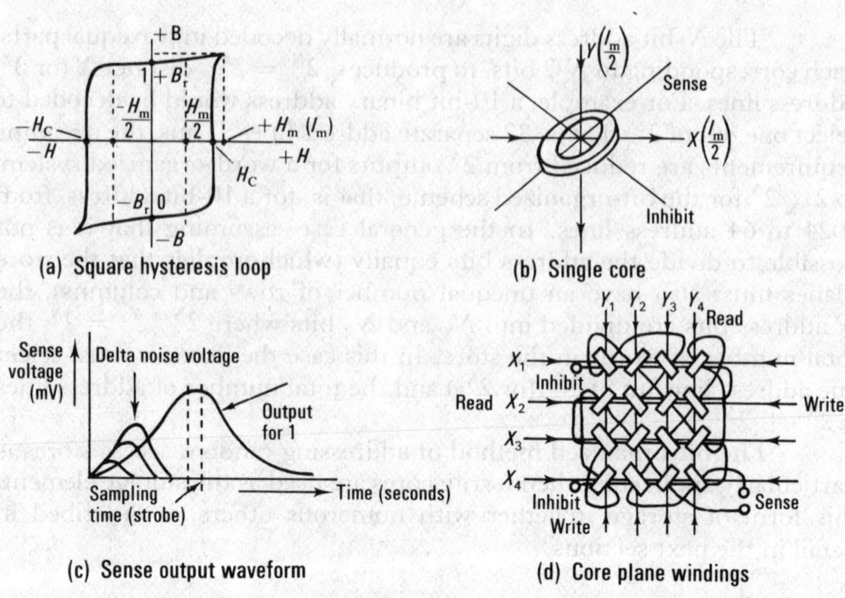

Figure 6.4 Coincident current core stores.

Ferrite cores are non-volatile in the sense that once a core has been set to a particular state it will remain in that state until a disturbing force of proper sign and magnitude is applied. However, if the power supply is cut off either intentionally or accidentally the resultant transients could be large enough to switch the cores, thus aborting the stored data.

The first core stores were three-dimensional arrays, with individual words being addressed on a bit-organized basis using a coincident current technique. The system requires one core per bit of data, and as described in Section 6.2.2, each plane of the stack is common to one bit in the word with the X and Y coordinate wires being connected in series as shown in Fig. 6.3. Assuming that information has been previously stored in the stack, the cores are interrogated by passing half-current ($I_m/2$) signals down selected X and Y address wires. These currents pass through all the cores in the chosen row and column of each plane, but only one core in each plane will experience two coincident half-current pulses. Since the currents pass through these cores in the same direction, they are additive, giving a full current pulse (I_m) which generates a magnetic field H_m. If the core was originally in the remanent state corresponding to a binary 1, the magnetic field will reverse the magnetization of the core to the 0 state, with a consequent change of flux. This change of flux links with the *sense* winding of the core (see Fig. 6.4(b)) and induces a voltage, in the order of millivolts, into the winding. The voltages from the sense winding of each plane must be amplified and registered to form the word output of the store. During the read operation all the cores in the selected row and column of the plane will have been subjected to half-current pulses and will partially switch, moving round a minor loop on the hysteresis curve. Since the sense wires thread all the cores in a plane, the final output will contain **delta-noise** (see Fig. 6.4(c)) which must in general be gated out (strobed) in the read amplifiers. Noise problems can be considerably reduced, however, by arranging that the cores in a plane are placed at right angles to each other (to minimize the coupling between drive and sense windings) and by using a diagonal wiring pattern (see Fig. 6.4(d)) such that the direction of the induced voltages due to partial switching is opposite in alternate cores, thereby cancelling out. Note that each of the wires used for address driving, sensing, and so on, effectively pass a single turn through each core; multiple turns, though reducing the current requirements, would increase the cost and complexity of the store as well as decreasing its operating speed.

The writing process is similar to the reading scheme, but in this case the XY address current drives are of opposite polarity; that is, the currents are reversed. A straightforward interrogation in this manner would result in all the selected cores being set to 1, so it is necessary to provide a fourth and final wire through the core, called the **inhibit winding**. This winding carries a half-current pulse in the reverse direction to the XY drives through all the cores in a plane and is used to cancel out half the switching field if a zero is required to be stored.

A block diagram for a typical ferrite core store system is shown in Fig. 6.5, with the corresponding timing waveforms in Fig. 6.6. Owing to the destructive read-out nature of the devices, the operation of the store is based on a read/write cycle. A typical read operation commences with the address of the required location being placed in the address register. When the control unit receives a start signal – the R/W control waveform – the address decoder outputs are allowed to generate the appropriate X and Y current drives to the stack. The drive signals have the effect of setting all the selected cores to the 'zero' state, and the resultant change of flux from cores which were initially in the 'one' state produces an output voltage on the sense lines. These voltages are amplified and strobed by the read amplifiers and the resultant logic level signals used to set and reset the bistables of the memory register.

On the write cycle those stages of the memory register which contain 0's are used to enable the gated outputs of the inhibit drive amplifiers. At the same time the XY drive currents are reversed (which, unless inhibited, will set all the selected cores to the 'one' state) with the result that the original data are written back into the same location. The procedure must be slightly modified when new data have to be written into the store. In this case the data to be written are transferred to the memory register and the address of the location placed in the address register. On receipt of the write-only signal, the control unit will initiate a read/write cycle in the usual way, but this time the outputs of the read amplifiers will be inhibited so that the contents of the memory register will not be overwritten.

A typical fast ferrite core store would require XY drive currents

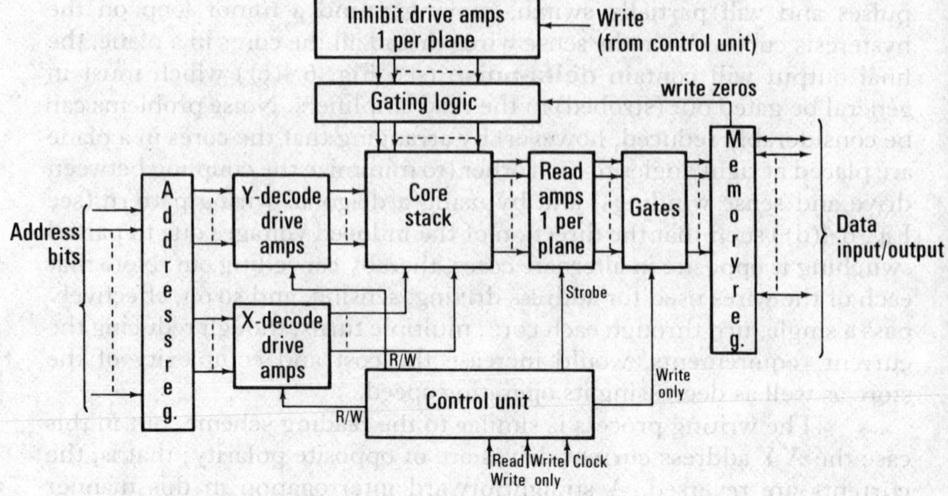

Figure 6.5 3D core store system.

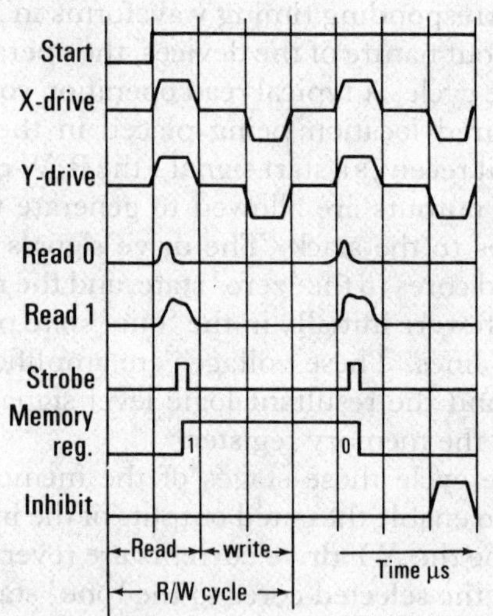

Figure 6.6 Timing diagram for core store.

in the order of 400 mA, and would generate a read out signal of some 20 mV. The switching time of the cores would be in the region of 200 ns with a read/write cycle time of 500 ns. For some applications, however, (in particular as buffer or bulk storage) cheap slow-speed stores with read/write cycle times in the order of 1–10 μs are perfectly adequate. The practical limitation on the speed (and cost) of core stores is the physical size of the cores, and the difficulties encountered in the necessary hand-assembly techniques. For this reason it is unlikely that read/write cycle times of less than 250 ns will ever be obtained in practice.

The cost per bit of a ferrite core store depends more on the external circuitry than on the cost of the actual ferrite material itself. Each plane requires its own read and inhibit drive amplifiers, so that the length of the stored word will determine the amount of logic (and core planes) needed for a particular sized store. This is another good reason for using a byte organized system (see Section 2.4) since, while the computing wordlength may be composed of a number of basic bytes, the actual stored wordlength would be, for example, a constant two bytes long. The number of words stored in the stack will determine the size of the plane and hence the number of ferrite cores as well as the XY drive circuitry. The use of a coincident current method of addressing reduces the amount of selection and drive circuitry, but there are two further factors that limit the storage capacity that may be obtained using a single stack. The first one is that the unwanted delta noise generated by the half selected cores increases with the number of cores in a plane, and secondly, the number of cores that can be driven for a given read/write cycle time is limited owing to the back e.m.f. produced by the stack. Various techniques have been tried to reduce these effects, but their success is obtained at the expense of increased complexity of core threading and electronic circuitry.

One way of overcoming the problem of half-switched cores is to use the two-dimensional word organized system of addressing,[10] as described earlier. Using this method, reading is performed by applying a full current pulse (I_m) to the selected wordline. The switched cores (those previously containing a 1) will induce a voltage into the bit lines which will be sensed and amplified in the usual way; note that the reading process is still destructive and provision must be made to rewrite the word back into the store. Writing into the store is achieved by simultaneously applying half-current pulses to the selected wordlines and the bit lines corresponding to those inputs which are binary 1. It will be obvious that only one core per bit line will be switched during the read operation, thereby eliminating the delta noise problem. The main disadvantage of the system is the large amount of electronics required in the selection process, that is, 2^N drive amplifiers compared to $2\sqrt{2^N}$ for the three-dimensional technique.

The cycle time of the word organized store can be reduced by overdriving the cores (by increasing the current amplitude) with the effect of decreasing the switching time; however, this technique can only be applied in reading. Because use is made of coincident current selection

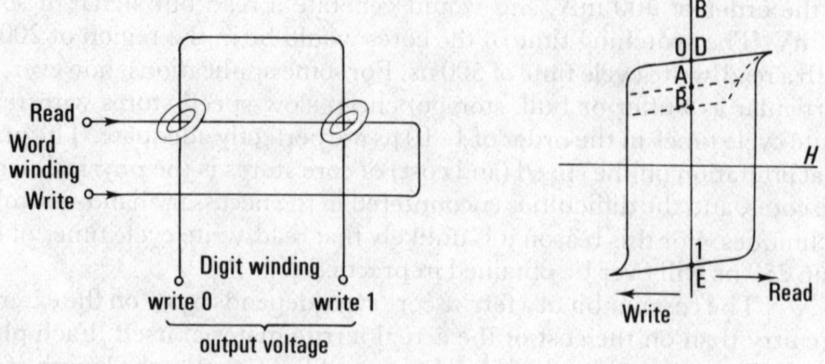

Figure 6.7 Two-cores per bit system.

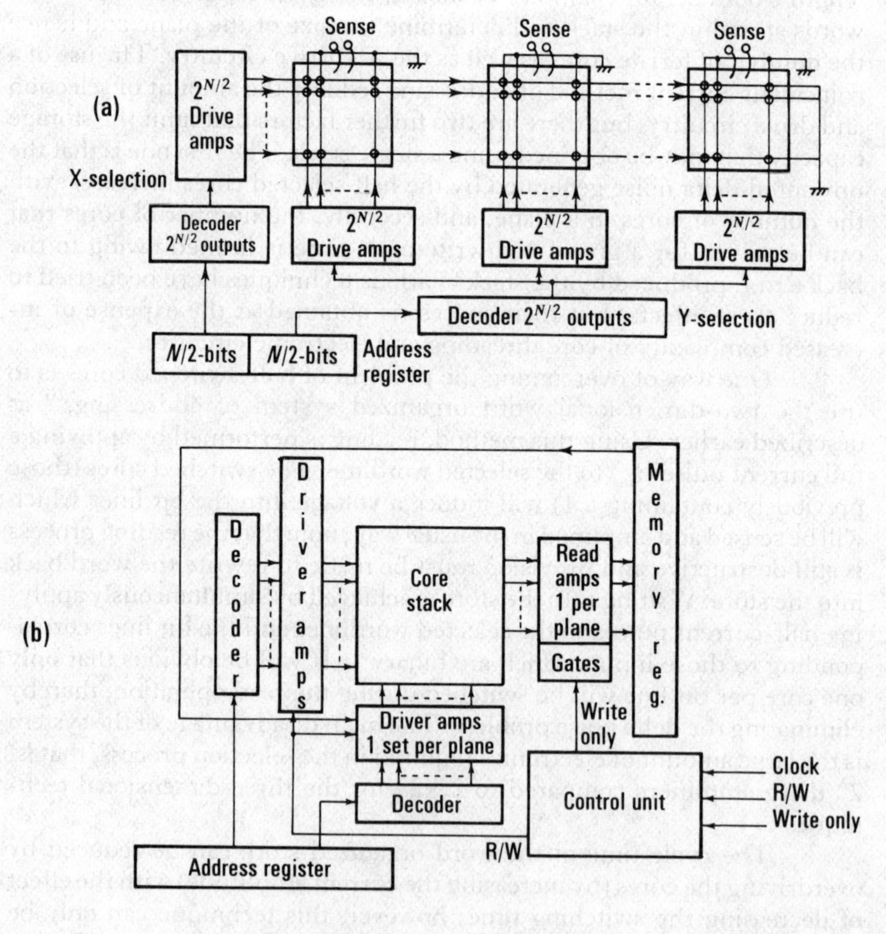

Figure 6.8 $2\frac{1}{2}$D core store system.

during the writing process, the selected cores will begin to switch if the amplitude of the currents is increased beyond a certain level.

The use of two cores per bit[11] in a word organized system allows overdriving techniques to be used without any deleterious effects. Consider the diagram in Fig. 6.7; let us assume that both the cores, A′ and B′, have been initially set by a read current to point D on the hysteresis curve. The write current (in the opposite direction) will tend to switch both cores towards the other stable remanent position, point E on the curve. However, if a simultaneous digit current is applied in the write 'zero' direction it will aid the switching action in core B′ but will oppose it in core A′, thus the cores will now set at points A and B on the curve respectively. When a read current is applied under these conditions opposing e.m.f.'s are generated, corresponding to the flux changes induced by the switched cores in the digit windings. Similarly a write 1 digit current will reverse the flux settings of the cores and hence the polarity of the output voltage during the read cycle. Thus the data are stored as the difference in switched flux in the two *partially* switched cores. The partial switching of the cores allows a large current pulse (overdriving the cores) but of short duration, thereby permitting a reduction in the store cycle time. In addition to the obvious disadvantages of increased costs and complexity due to the use of two cores per bit, the specification and control of the core parameters is more critical. Moreover, since it is a word organized system, the amount (and complexity) of the associated electronic circuitry is also increased.

With the greatly reduced costs of electronics as a result of the introduction of integrated circuits the balance between core matrix and circuitry costs has been drastically changed, with a consequent effect on systems cost and the design philosophy. For example, two-dimensional core stores are now considered economically viable for the main storage of large general purpose machines. This has led to the adoption for most fast modern stores of the **$2\frac{1}{2}$-dimensional array**,[12] which is a compromise between the complexity of electronics and the number of cores driven by one wire. The $2\frac{1}{2}$D system enables a faster, and in some applications a cheaper, store to be built than the 3D system. The $2\frac{1}{2}$D system is basically very simple; if the Y drive wires in a conventional 3D system (which thread *all* planes in series) are replaced by separate Y wires and drives for each plane we have the $2\frac{1}{2}$D system, as shown in Fig. 6.8.

Information may be written into the core array by energizing the selected X wire and the appropriate Y wire for each of those bits where 1's are to be written. The information may be read out by reversing the direction of the X currents and selecting the appropriate Y wire for each bit. Note that individual sense lines (one per plane) are still required, the resulting change of flux being amplified in the usual way. With this technique the number of cores on each drive line is less than in the 3D system but since the selection circuits are duplicated for each Y plane the hardware is considerably increased.

The main advantage of this method is that the inhibit winding is no longer required, thus the core sizes may be smaller (only three wires need to be threaded through each core) and a source of heat has been eliminated. Furthermore, the cores may now be arranged parallel to each other giving a higher packing density which minimizes the wiring losses and allows thinner wires to be used. Since the use of smaller cores enables high-speed cycle times to be obtained, the $2\frac{1}{2}$D store can be operated at approximately twice the rate of a conventional 3D store.

Ferrite core stores are a well-established and reliable component in computer systems design, particularly when ultra high-speed operation is not a requirement, for example as a buffer for the main CPU store. It seems likely that they will continue to dominate the storage scene, at least for the immediate future.

6.4 Semiconductor storage[13,14]

Large scale integration (LSI) techniques have brought about the possibility of manufacturing large arrays of storage elements on a single chip of silicon. These storage cells, using either bipolar or MOS technology, consist of multi-component circuits in a conventional bistable configuration. Confident predictions have been made about the low cost of these active element memories, and their ultimate replacement of other forms of storage as the main high speed store in the CPU.

Unfortunately, storage of this type is inherently volatile since the bistable elements require a constant source of power to maintain the stored information. This can be a severe disadvantage in some applications where it is essential that the data should not be irretrievably lost due to a power failure. However, this disadvantage may be overcome by incorporating power standby circuits which detect when the main power supply is about to fail and switches over to a battery source; in this way the contents of the store can be retained. Though little is known about the long term reliability of semiconductor memories it would appear that they are inherently more reliable than other forms of storage, moreover, it is also comparatively easy to include error detection and correction circuits in the basic modules.

The immediate advantages of semiconductor storage are the high packing density, low drive requirements and high speed of operation (especially if bipolar devices are used) coupled with the promise of plummeting prices. The maximum capacity of this type of store has been estimated at 600k bits/chip, with cycle times (read or write) ranging from 500 ns to 0.5 μs depending on capacity and on whether bipolar or MOS devices are used. A chip is a piece of silicon on which the memory cells are manufactured, each cell occupies about 1000 μm^2 in area and the overall size of a 256 bit memory would be some 2·54 × 2.54 mm; for convenience in use the chip would be mounted in a 16 pin dual-in-line package.

The MOS device has been particularly exploited in this application since, though slower in operation than bipolar devices they take up less substrate area (thereby increasing the packing density) require less power and fewer process steps. However, the logic circuits employed in the computer are normally constructed using bipolar technology which are incompatible with MOS circuits unless suitable interface circuits are provided. Most semiconductor memories are word-organized and a typical commercial RAM store would have a capacity of 16k bits/chip with a read/write cycle time of 500 ns.

One of the attractive future possibilities of semiconductor storage arrays is the change it could bring about in computer systems architecture. At the basic level it is now possible for the main store to consist entirely of semiconductor storage with all the attendant advantages. However, since the store itself is composed essentially of logic elements, and is thus perfectly compatible in speed and power levels with computer logic, there is no longer any reason to separate the storage and logic functions. Consequently, the computer of the future could evolve as an homogenous distributed logic and memory system (see Chapter 9).

6.4.1 Static semiconductor RAM's

The basis for a semiconductor storage cell is the simple transistor bistable circuit shown in Fig. 6.9. Note that both bipolar transistors and metal oxide semiconductor (MOS) field effect transistors[15] can be used for this purpose. The operation of the circuit is such that only one transistor can be conducting at any one time. For example, when T_1 is turned ON current flows through T_1 to ground, putting node A (the collector in this case) also to ground. This in turn puts the base of T_2 to ground potential and prevents any flow of current through T_2. When T_2 is non-conducting node B assumes the level of voltage Vc which also goes to the base of T_1 thus holding the transistor ON. Consequently we have two stable static states, T_1–ON and T_2–OFF (which can arbitrarily be called logic 0) and T_1–OFF and T_2–ON (logic 1). To change the stored states nodes A and B must both be brought to either Vc or 0 volts, for instance, to write logic 1 when the bistable is storing logic 0 node B must be brought down to 0 volts and node A to Vc volts. This action can easily be performed using additional transistor gates connected to nodes A and B. Note that for an efficient memory organization (such as 2D or $2\frac{1}{2}$D addressing) the requirement is for A and B to be selected coincidentally. The basic circuit for a MOS memory cell using coincident $2\frac{1}{2}$D addressing is shown in Fig. 6.9(b). Coincident signals on the wordline, and either one of the two bit/sense lines, results in a change of bistable state thus effecting a write 1 or 0 operation. To write a logic 0, regardless of the initial state of the cell, a positive voltage is applied to the wordline in coincidence with a pulse bringing the bit/sense write/read 0 line down to ground potential. The word pulse on the gate input of T_3 turns the MOS transistor on and since its input source is at ground level node A is also brought to ground (note that the action of the MOS transistor gate is analogous to a relay contact). If the memory cell was already in the logic

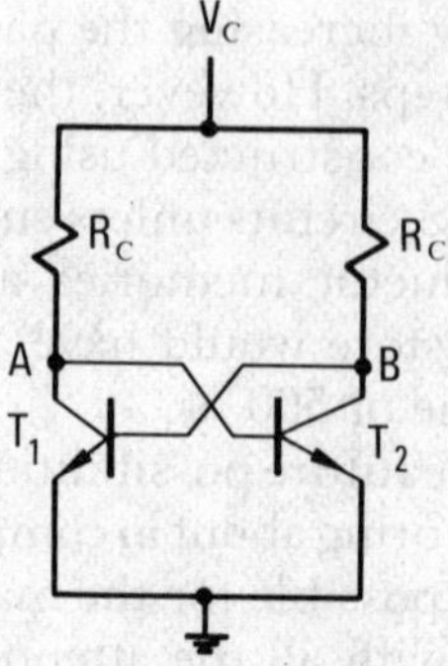

(a) Bipolar memory cell.

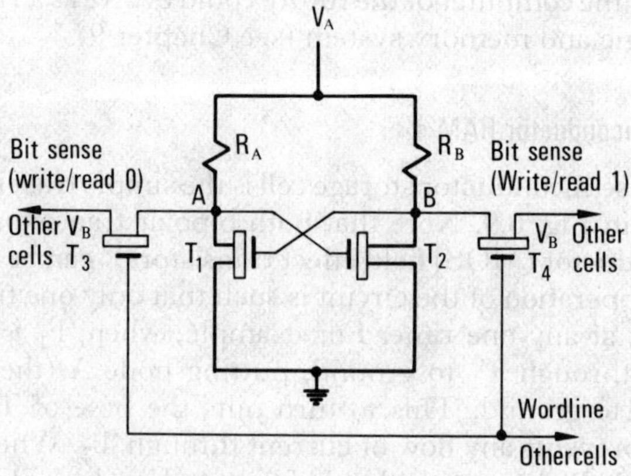

(b) MOS memory cell with selection gates.

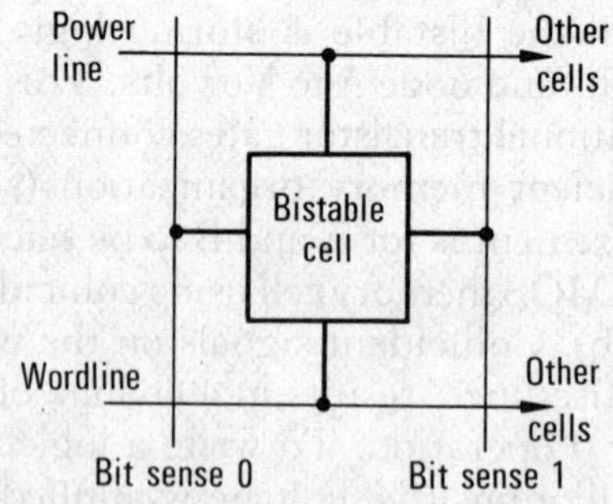

(c) Representation for single cell.

Figure 6.9 Bistable memory cells.

0 state node A would be at 0 volts so nothing would happen. However, if the cell was initially in the logic 1 state, the effect of bringing node A to ground (from V_A volts) is to turn off T_2 whereupon node B would increase to V_A volts (from 0 volt) causing T_1 to conduct and switching the cell to the logic 0 state. An analogous sequence of operation would apply if it were required to write a logic 1, except that in this case the bit/sense write/read 1 line would be put to ground.

To read the contents of the memory cells a smaller positive pulse is applied to the wordline but in this case leaving the bit/sense lines unactivated. The effect of this word pulse is to cause both T_3 and T_4 to conduct slightly and whichever node is at 0 volts will cause a small current to flow through its corresponding transistor thus giving rise to a drop in the normally high bit/sense line voltage. This negative going voltage pulse must be amplified to bring it to standard logic levels before it can be used in the system. Note that the reading process is non-destructive and that the contents of all the cells connected to a particular wordline are read-out simultaneously. Fig. 6.10(a) shows a schematic of a 4×4 MOS memory array together with the corresponding read/write waveforms in Fig. 6.10(b).

In a practical storage system data are written and read one word at a time, consequently an input/output buffer register is required together with suitable decoding logic for word addressing; a block diagram of the store organization is shown in Fig. 6.11(a). In practice the decoder and the drive and sense circuits are normally mounted on the same LSI chip as the memory array. Moreover, in many cases a hybrid technology is used, utilizing bipolar circuits for decoding and I/O circuits (giving high speed operation and logic compatibility) and MOS for the memory array where low cost and high packing density is required; all these circuits would be mounted on the same substrate and packaged as a single module.

In many cases the RAM store module is produced as an array of single bit memory cells rather than *n*-bit words, for example a module containing 1024 bits per chip. This gives the advantage that words of any desired length can be constructed by simply paralleling the required number of chips; note that the number of possible words will be equivalent to the number of bits/chip. However, the addressing function becomes slightly more complicated and in order to optimize this operation each chip is provided with its own decoding and selection logic for the bit/sense and word lines. A block diagram for a 1024 8-bit word store is shown in Fig. 6.11(b); note that the decoding logic selects 1 bit of the word from each chip using coincident addressing and that the address lines are common to each chip.

RAM memory modules normally have provision for an ENABLE input which can be used to inhibit the operation of the decoder and sense circuits. The READ and WRITE control inputs are normally single bit lines which are used to activate or deactivate the ENABLE input of the decoders.

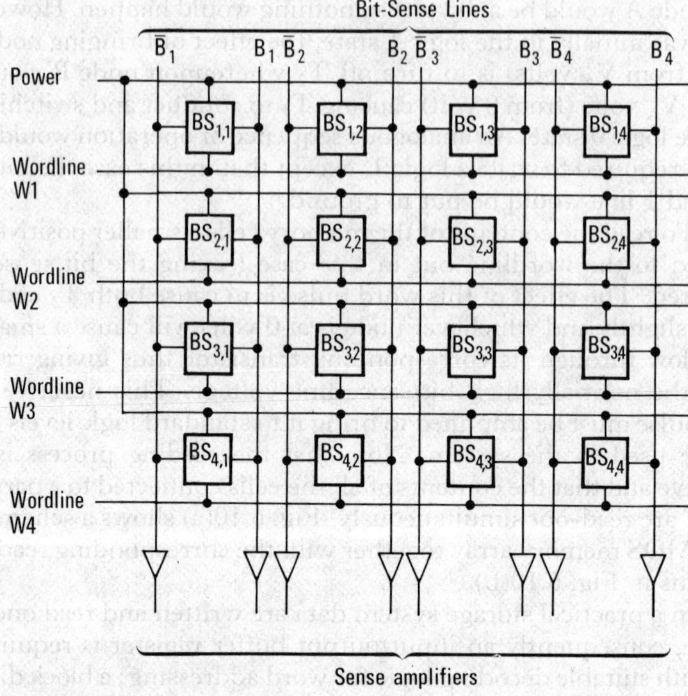

(a) MOS 4 x 4 Memory cell array.

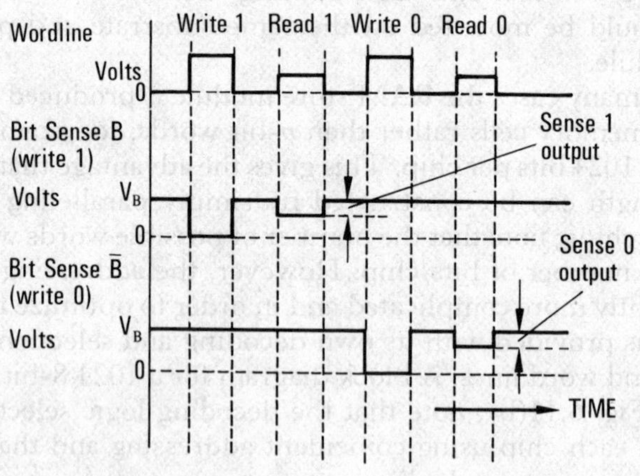

(b) Timing waveforms for read/write operations.

Figure 6.10 MOS memory array.

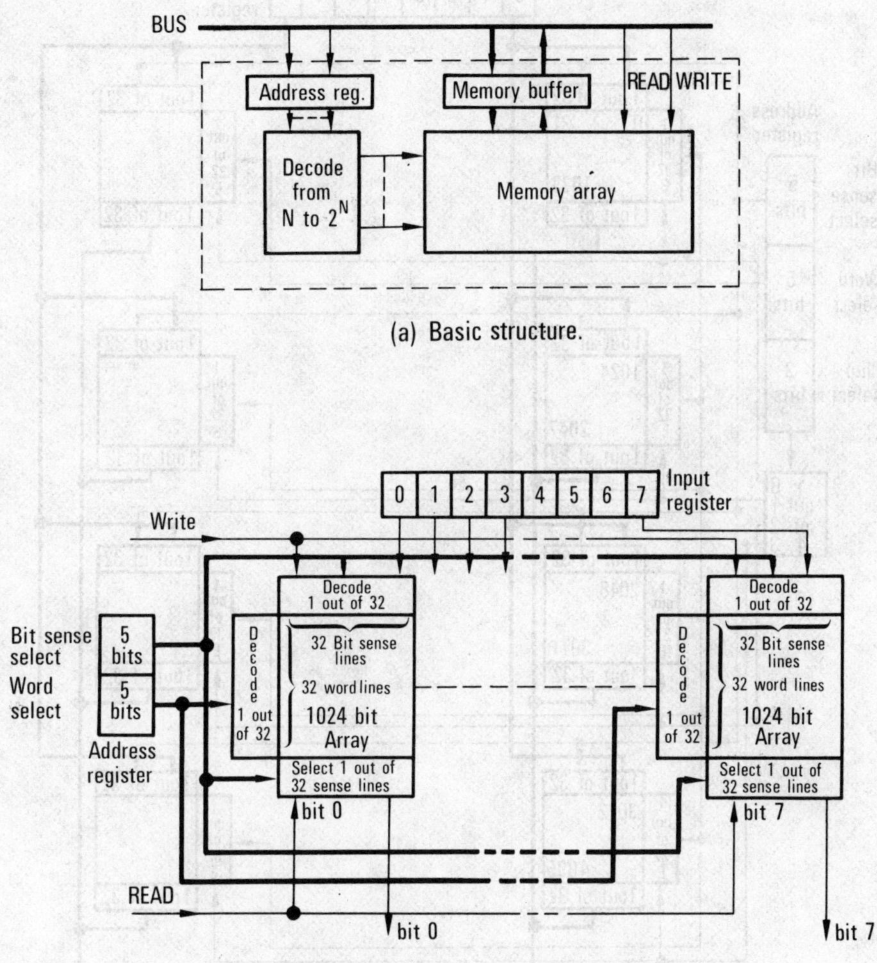

Figure 6.11 Organization of semiconductor RAM stores.

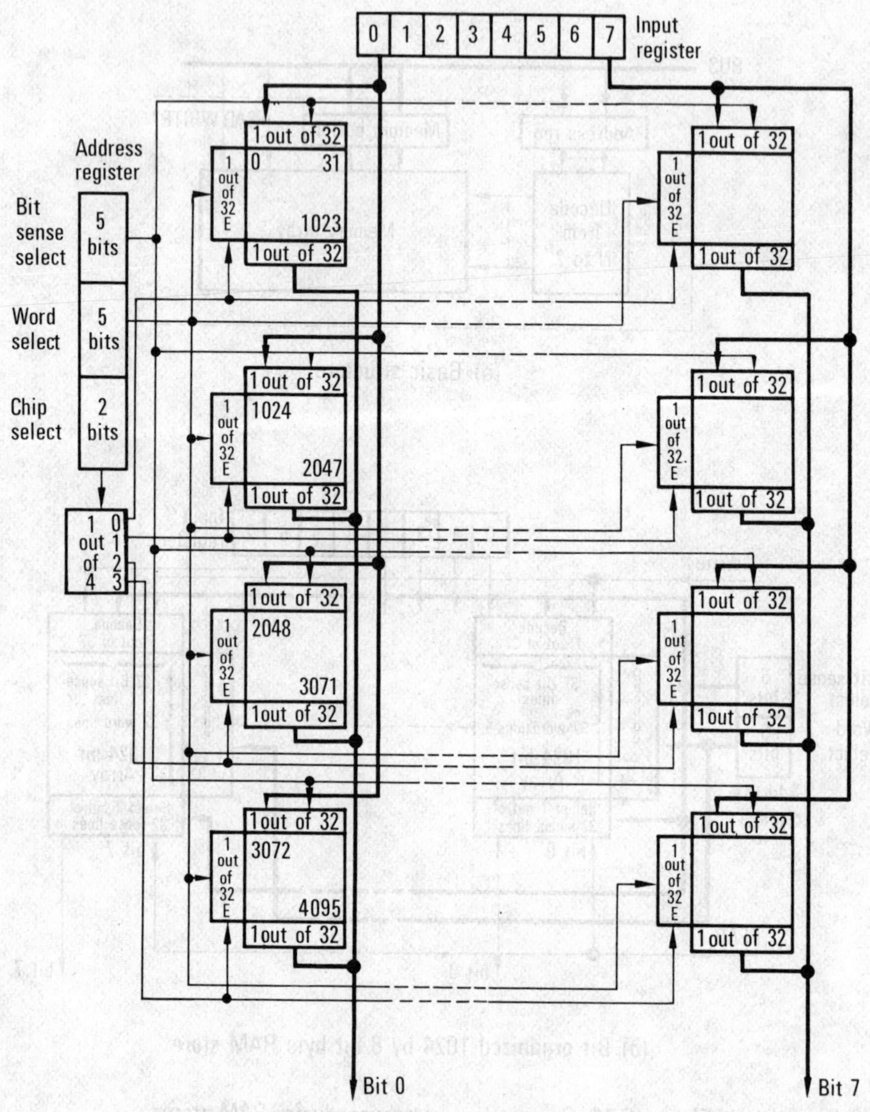

Figure 6.12 Decoding using enable input.

The ENABLE input can also be used as an extra level of decoding, for example, when expanding to a store size larger than the number of available bits on a chip. For instance, suppose it was required to build a 4096 8-bit word store using 1024 bit RAM modules. In order to achieve this size it is necessary to use a 2-dimensional array, consisting of 4 × 8 RAM modules, instead of the 1-dimensional scheme shown in Fig. 6.11(b). In this case the address would be 12-bits in length which is greater than the 10-bit decoders available to select a bit in the 1024 bit array and the additional 2-bits must be decoded externally. Fig. 6.12 shows how this may be done using the ENABLE input to select the appropriate row of the RAM array; note that each row is equivalent to 1024 words and the last two bits of the 12-bit address are effectively counting module 1024.

6.4.2 Dynamic MOS memories

In order to reduce the cost of semiconductor memory it is necessary to increase the number of memory cells on the chip. The major restrictions on the chip density when using static bistables are due to the number of devices per bit (usually 4–6 transistors) and the power dissipation which prevents close packing. The storage cell area on a chip can be reduced by using MOS transistors in a dynamic circuit configuration based on charge storage techniques. Data is stored in the form of charge on the gate capacitance of the transistors; since this charge can leak off through the circuit connections to the previous device it is necessary to continuously regenerate the information using clocking or refresh techniques.

Figure 6.13(a) shows the circuit diagram for the basic three transistor MOS charge storage cell. The MOS transistor T_1 stores the charge, the magnitude of which is used to represent a logic 0 or 1, on its gate input (the storage capacitor is symbolized by C_1 consisting of the gate and stray circuit capacitances). A high voltage on the write enable input causes T_3 to conduct thereby allowing the signal on the write bus to charge or discharge the gate capacitance of transistor T_1. When the write enable voltage is low, transistor T_3 is non-conducting and the charge is retained on T_1 irrespective of the voltage on the write bus. The cell may be interrogated for the presence of a charge (that is to establish whether the cell contains a 0 or a 1) by applying a read enable pulse to the gate of T_2 which connects the output of T_1 to the read bus via transistor T_2.

The MOS memory cell may be used in a 2D addressed M row × N column array similar to that described earlier for the bistable cell; the basic cell module is shown in Fig. 6.13(b). Note that a separate refresh amplifier is required for each column (bit output) to periodically restore the charge on transistor T_1. The effect of activating any one of the M read enable row lines is to write the contents of the corresponding N storage cells into their respective refresh amplifiers. This read operation is followed by a write enable signal on the same row which restores the contents of the memory cells in that row. Note that this is analogous

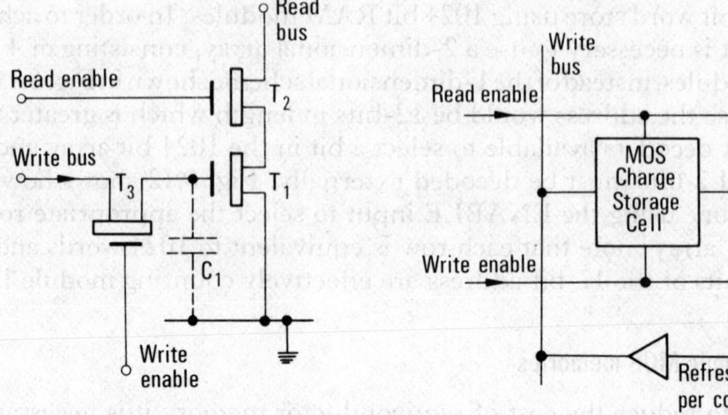

(a) Basic storage element. (b) Single cell representation.

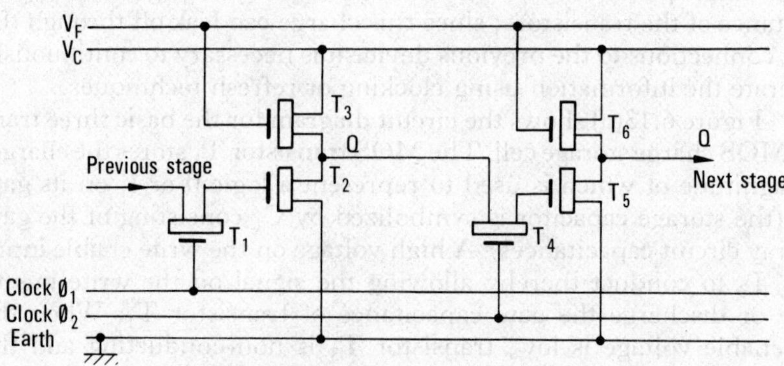

(c) Single bit stage of dynamic MOS shift.

Figure 6.13 Dynamic MOS memories.

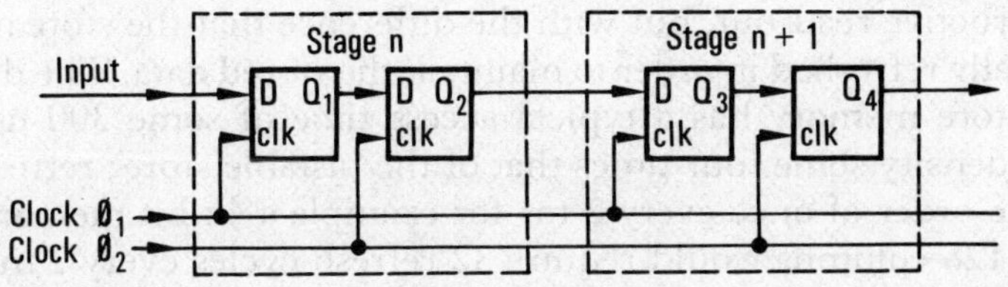

(a) Dynamic shift register.

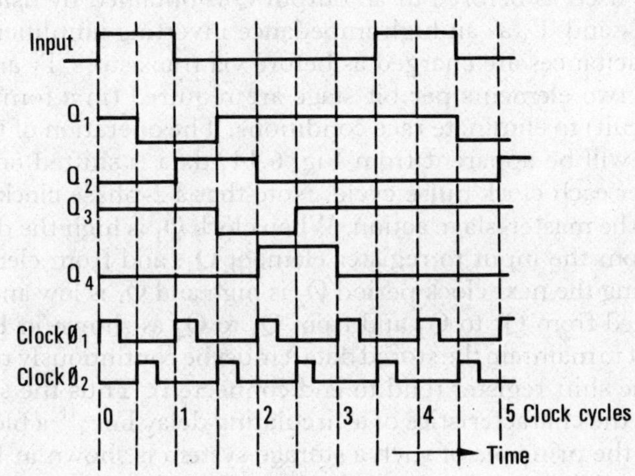

(b) Timing diagram.

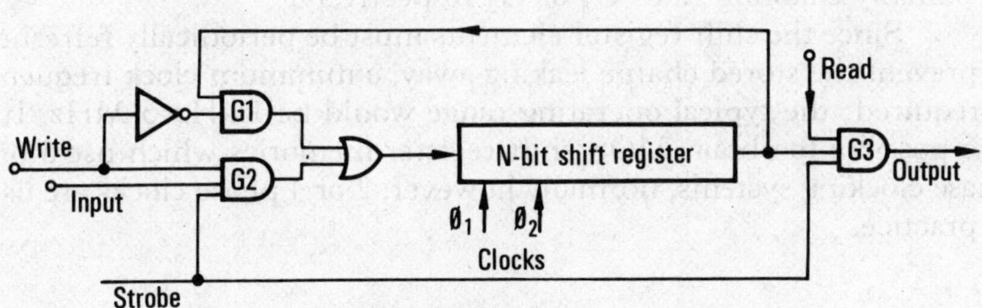

(c) Block diagram of circulating store.

Figure 6.14 MOS shift register memory.

to a destructive read-out, but with the difference that the store must be periodically refreshed in order to maintain the stored data. The dynamic charge store memory has a typical access time of some 300 ns and a packing density some four times that of the bistable store; refresh rates are of the order of once every 2 ms for example a 4k bit memory of 32 rows by 128 columns would require 32 refresh cycles every 2 ms.

An alternative way of implementing MOS dynamic memory is to use a circulating shift register configuration, the circuit for a single register stage is shown in Fig. 6.13(c). In this case the same basic MOS circuit is used as before but an output Q is obtained by using the transistors T_2 and T_3 as an high impedance inverting amplifier; the MOS gate capacitances are charged as before via transistors T_1 and T_4. Note also that two elements per bit stage are required (in a form of master-slave circuit) to eliminate race conditions. The operation of the dynamic memory will be apparent from Fig. 6.14; data is shifted one bit to the right after each clock pulse cycle. Note that a 2-phase clock is required to effect the master-slave action. When clock $\emptyset_1$ is high the data is transferred from the input to register element Q_1, and from elements Q_2 to Q_3. During the next clock period $\emptyset_2$ is high and $\emptyset_1$ is low and the data is transferred from Q_1 to Q_2 and from Q_3 to Q_4 as shown in Fig. 6.14(b). Note that to maintain the stored data it must be continuously re-circulated round the shift register (end to end connected). Thus the shift register store has the characteristics of a circulating delay line;[16] a block diagram showing the principle of such a storage system is shown in Fig. 6.14(c). In the normal circulating mode both the Read and Write control signals are low thereby inhibiting gates G_2 and G_3; G_1 is enabled which establishes the feedback path. Data may be read from or written into the store by suitably enabling gates G_3 or G_2 respectively.

Since the shift register elements must be periodically refreshed, to prevent the stored charge leaking away, a minimum clock frequency is required; the typical operating range would be 1 kHz–5 MHz. It is also possible to obtain MOS shift register memories which use 3 or 4 phase clocking systems, normally however, 2 or 4 phase clocks are used in practice.

6.4.3 Charge coupled memory systems[17,18]

The charge coupled device (CCD) is basically a serial shift register which propagates data bit by bit (represented by the presence or absence of a charge) at a rate established by an external clock. The CCD is an outgrowth of *n*-channel MOS technology and is based on the capacitive coupling that exists between the MOS gate electrode and the substrate material. A voltage applied to the gate charges the capacitance, and the charge regulates the current flow in the substrate; as we have seen in section 6.4.2 this capacitance is the storage site for charge in MOS dynamic memories.

Physically the CCD is a linear array of closely spaced MOS capacitors or gates, with 'potential wells' (depletion regions) situated beneath the gates at or near the surface of the silicon substrate. The device operates by storing and transferring charge between these potential wells, which correspond to the unit memory cell of the CCD. The wells are formed and controlled by the closely spaced MOS capacitors and a phased voltage (clocks) applied to the gates. Thus charge coupling is the process of transferring the mobile electric charge within a well to an adjacent well when a periodic clock is applied to the gates.

The process is illustrated in Fig. 6.15. The input voltage V_{in}

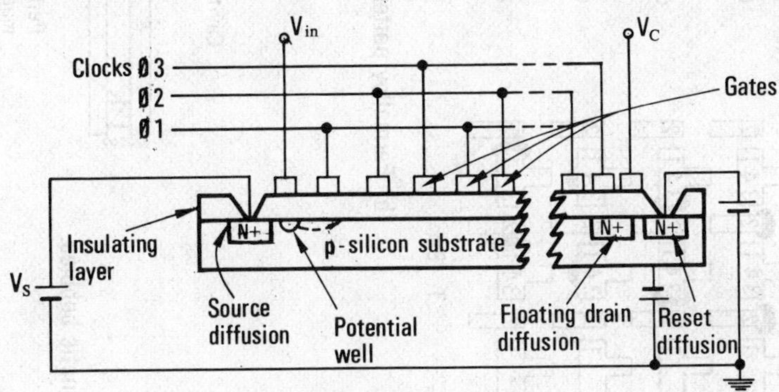

Figure 6.15 CCD shift register.

controls the injection of electrons into the CCD channel, when injected the first potential well is filled to the level controlled by the source voltage V_s. Subsequently these charge packets are moved down the register by applying the series of clock pulses $\emptyset_1$, $\emptyset_2$ and $\emptyset_3$ and eventually reach the floating diffusion at the end of the shift register. That in turn controls a high impedance gate which provides an external output. The control voltage V_c periodically gates the charge in the floating diffusion to the drain diffusion (a reset operation) so that the floating diffusion can accept the next charge. Thus the basic operations involved are charge injection, charge movement and finally charge detection and regeneration.

As well as the basic CCD shift register various other circuits are required on the chip to produce a functional memory (some of these are shown in Fig. 6.15). The circuits required are:

(a) **Charge injection circuits** – this is performed by gating a bias line connected to the N_+ diffusion region which allows charge to be injected as packets of minority carriers.

(b) **Charge detection and regeneration circuits** – charge is detected at the output of the CCD as a change in surface potential and capacitance. A floating diffusion (not biased) adapts its potential to the surface depletion potential (the potential well) and this change in potential can be detected, and if required, connected to the injection control gate of the same CCD or other CCD devices, thus regenerating the charge

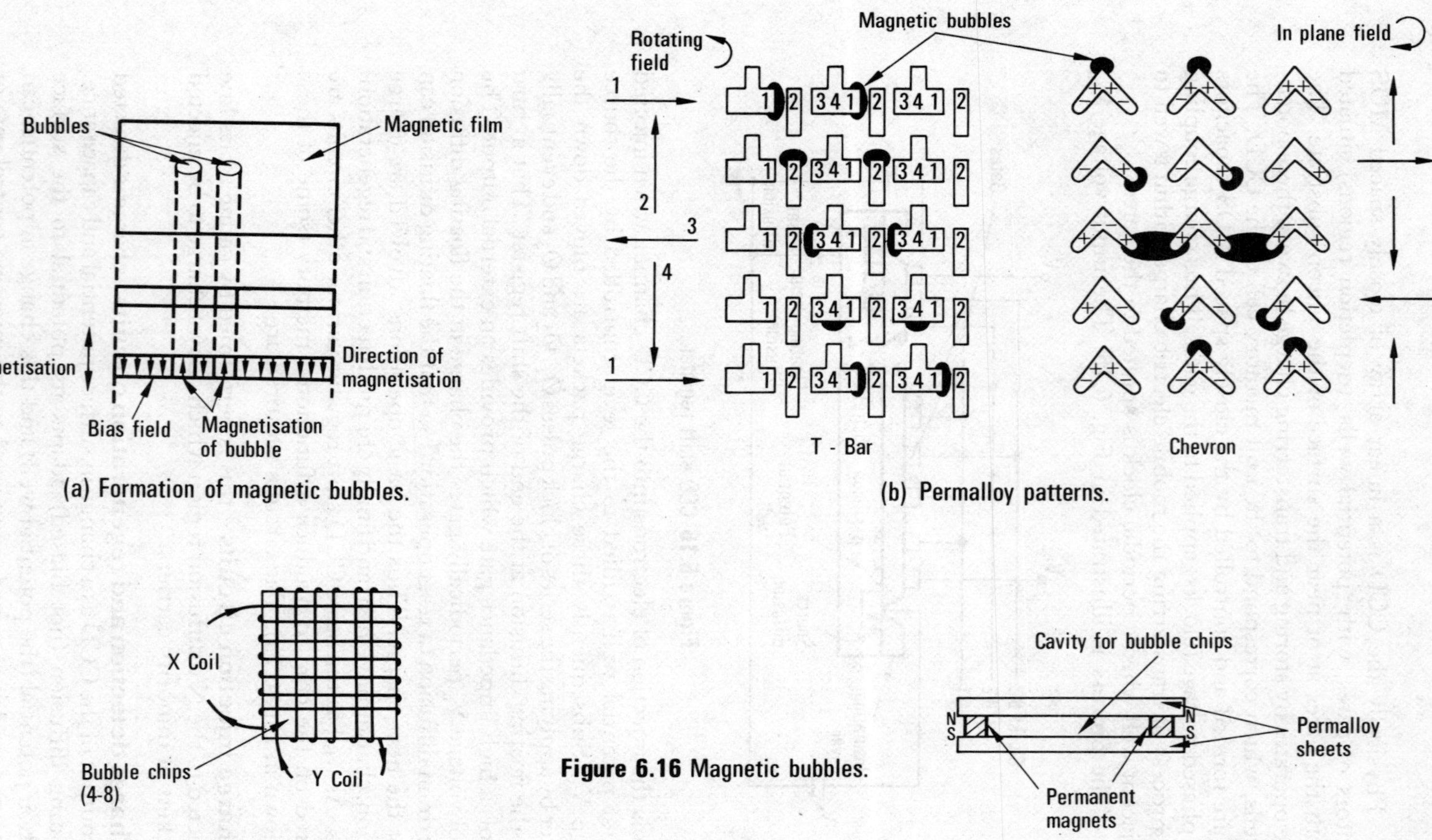

(a) Formation of magnetic bubbles.

(b) Permalloy patterns.

(c) Drive field structure.

(d) Bias field structure.

Figure 6.16 Magnetic bubbles.

stream. Note that each regeneration results in a signal inversion.

(c) **Read/Write Circuits** – when the CCD is used as a serial memory system read/write logic circuits are required. These would in essence fulfil the same functions as those shown in Fig. 6.14(c) for the MOS dynamic store.

(d) **Data Buffers and Decoders** – in order to interface the CCD memory with the rest of the system TTL compatible data buffers are required at the input and output of the chip. Decoder circuits, the exact form depending on the storage configuration used, will also be required.

(e) **Clock Drivers** – these are normally included on the chip for ease of system design. When high current drivers are required (as for example in high speed operation) these would normally be external to the chip due to the problems of power dissipation.

The basic parameters of the CCD memory and limiting constraints on its operation are the **transfer efficiency** and **storage time**. The transfer efficiency is dependent on the amount of charge lost, due to charge being trapped in the local well areas, while shifting. The upper frequency of operation of CCD memories is determined by the transfer efficiency. Storage time refers to the maximum time data can be stored in the register without leaking away (in the order of 2–10 ms). Transfer efficiencies of 10^{-3} to 10^{-5} have been reported, which will allow shift register chips of some 10–1000 bits in length to be built without the need for integral refresh amplifiers.

6.4.4 Magnetic bubble memories[19,20]

Magnetic bubble memories can be considered as a solid-state version of rotating memory such as discs and drums, since they all store data in the form of magnetized regions and are cyclic in operation. In the integrated circuit version the magnetized regions are cylindrical magnetic domains contained within a thin layer of magnetic material of opposite magnetization, hence the term **magnetic bubble**. Binary data is represented by the presence or absence of these bubbles at specific locations in the structure. The data bits can be made accessible by moving the domains, under the influence of a magnetic field, within the solid layer to some reading device.

Magnetic bubble domains can be produced in a film of magnetic material grown epitaxially on a non-magnetic substrate; the magnetic material is anisotropic having the easy axis of magnetization in a direction perpendicular to its plane. In the normal state, without any bias field, the vertical magnetization domains occupy equal up and down areas. If a vertical bias field (in the down direction) of sufficient strength is applied the up areas decrease up to a point where they become isolated cylinders – this is shown in Fig. 6.16.

Magnetic bubbles can be propagated in the plane of the film by applying weaker magnetic fields at right angles to the principal field. These fields can be generated by depositing a suitable permalloy (that is a soft magnetic material) pattern on the magnetic film and using an in-phase rotating field.[21] Permalloy patterns can take various forms, two of the more useful and commonly used patterns are the **chevron** and the **T-bar** shown in Fig. 6.16(b). The magnetic polarities of the isolated permalloy shapes shift around in a regular manner in synchronism with the rotating drive field, thus steering the bubbles from pattern to pattern. Each rotation of the drive field makes the bubbles move one complete pattern; the method employed to obtain the drive field is shown in Fig. 6.16(c). If a bubble is assumed to indicate a logic '1' and its absence a logic '0' then the resulting action resembles that of a shift register.

The storage property is maintained using a bias field set up by permanent magnets. Without the bias field the magnetic bubbles would spread out and coalesce; the permalloy sheets produce a near uniform magnetic field throughout the cavity. A schematic of the bias field structure is shown in Fig. 6.16(d), the cavity usually contains several drive-coil structures. Note that since the magnets do not require a power source the bubble store is non-volatile.

In order to realize a shift register memory, in addition to the basic storage property it is also necessary to be able to write, read, erase and select data. The write and erase operations are obtained by locally varying the bias field using separate conductor loops. A bubble can be generated (equivalent to writing) by reducing the bias current using a hairpin shaped conductor loop which is energized with a current pulse; a magnetic bubble is formed inside the end of the hairpin. The erase operation is performed in a similar manner, in this case however the procedure is reversed with the objective of locally increasing the bias field.

The read operation uses the magneto-resistance of the permalloy strips. The bubble is expanded by increasing the number of parallel chevron paths (in a particular section of the propagation circuit) which leaves the bubble size unchanged in the direction of propagation but stretches the bubble into a wide strip in the transverse direction. This causes a distinct change in the magneto-resistance of the permalloy which can be detected to generate a sense signal of several millivolts.

Another important function is that of replication which allows a non-destructive read operation to take place. This is achieved by a similar method to that employed for reading (the bubble is stretched but is then cut into two) and consists of reading the information (which is erased afterwards) whilst allowing the original data to be retained in memory.

In order to minimize costs the drive and bias circuitry must be shared by the maximum number of bits. Since however large drive coils are difficult to drive fast there is a basic trade-off between speed and cost which could prove to be a fundamental engineering limitation. Typical drive frequencies range between 50 to 300 kHz, with 1 MHz being considered a potential maximum.

Magnetic bubbles range in size from about 160 to 240 μinches and the distance between centres is in the order of 1 mil. These dimensions give a bit density of about 2×10^6 bubbles per square inch, not counting the essential attendant requirements of read, write, erase circuitry, terminations etc.; bubble chips have been produced with 16–100k bits/chip and packaged in a DIL pack. Magnetic bubble memories are normally organized in one of two ways, the simplest being as a straightforward shift register store. This structure resembles one track of a magnetic tape unit and has all the associated disadvantages of serial access time etc. Note that for a 100k bit/chip capacity and a 100 kHz shift rate it would take the bubble memory 1 second to circulate once, giving an average access time of 0.5 s.

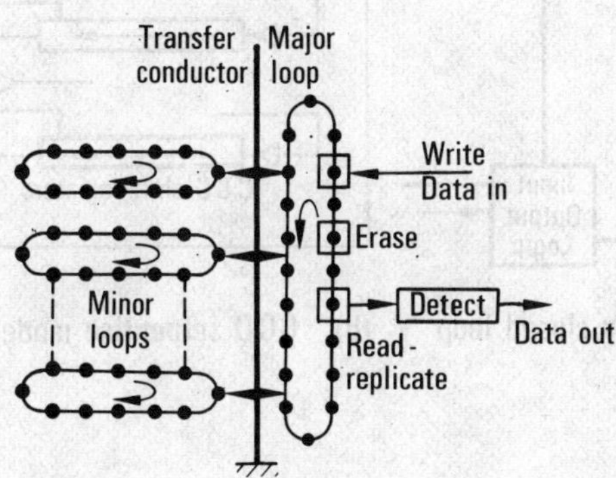

Figure 6.17 Bubble store organization.

For applications requiring a shorter access time the **major-minor loop** configuration is employed, this is shown in Fig. 6.17. In this structure multiple parallel shift registers are used to reduce the access times, for example, the 100k bits capacity could be distributed in, say, 200 minor cycles each one containing 500 bits. At a 100 kHz shift rate the minor loop would then cycle in 5 ms or an average access time of $2\frac{1}{2}$ ms. In order to dispense with the requirement for separate read/write circuitry for each minor loop, the major loop is used to accept (and transmit) data from the minor loops. This operation is performed in parallel (one bit at a time from each minor loop) using a coupling conductor energized by a suitable current pulse. The block of data (200 bits) is then read serially from the major loop; to write data into the store the block is first introduced serially into the major loop and then transferred in parallel to the minor loop. Note that though fairly complicated addressing and timing logic is required this is less complex than that necessary for rotating mass memories. The address logic takes the form of two external counters both operated at the bubble shift rate. One counter selects the required position in the minor loop; the other counter is started when the block is transferred to the major loop and indicates major loop cycles which are also used to control read/write operations.

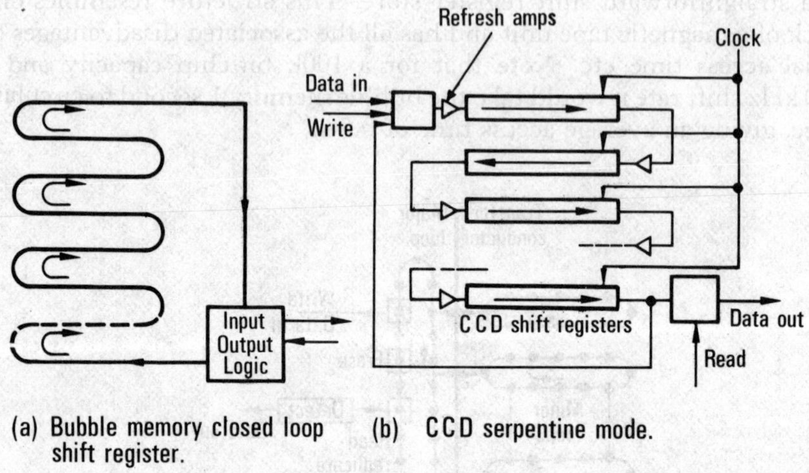

(a) Bubble memory closed loop shift register. (b) CCD serpentine mode.

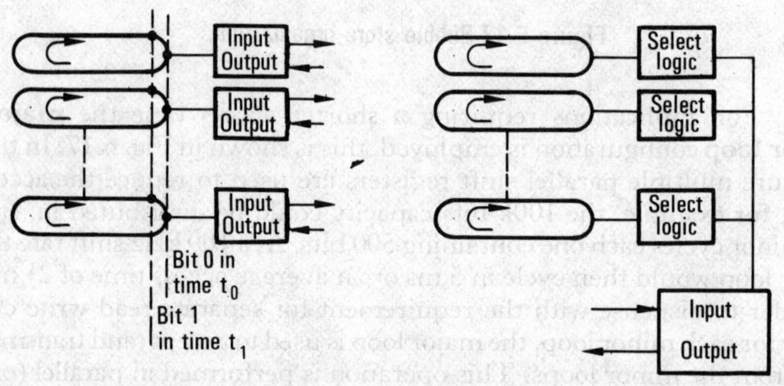

(c) Bubble memory using multiple shift registers. (d) Decoded shift registers.

Figure 6.18 Serial memory systems.

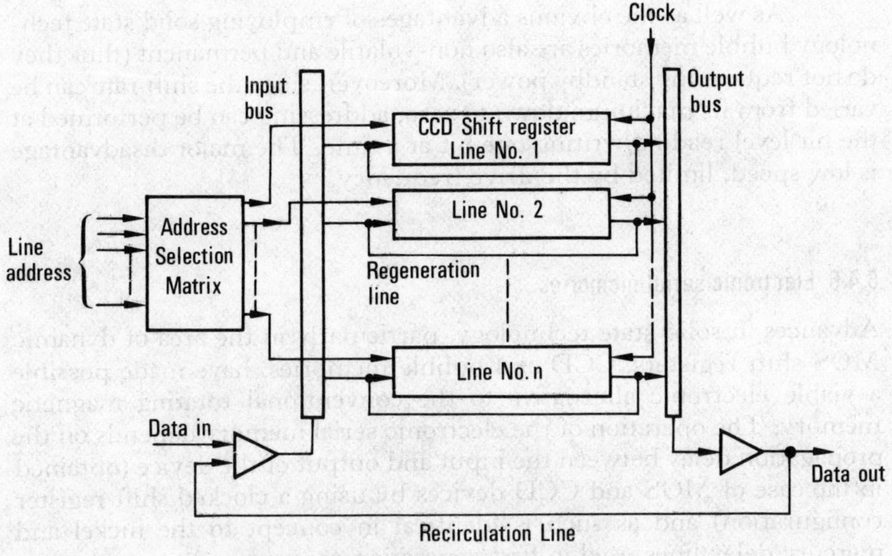

(e) CCD memory - LARAM organization.

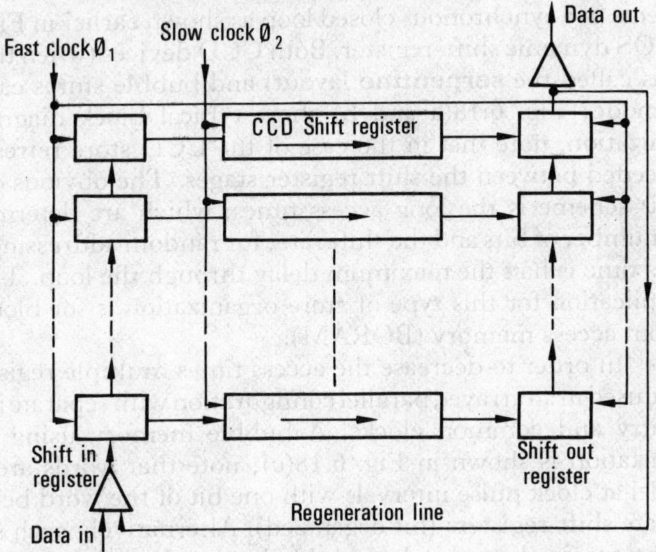

(f) Serial - parallel serial memory.

Figure 6.18 contd. Serial memory systems.

As well as the obvious advantages of employing solid state technology bubble memories are also non-volatile and permanent (thus they do not require any standby power). Moreover, since the shift rate can be varied from its maximum down to zero, addressing can be performed at the bit level reading/writing one bit at a time. The major disadvantage is low speed, limited by the drive frequency.

6.4.5 Electronic serial memories

Advances in solid state technology, particularly in the area of dynamic MOS shift registers, CCD and bubble memories, have made possible a viable electronic alternative to the conventional rotating magnetic memory. The operation of the electronic serial memory depends on the propagation delay between the input and output of the device (obtained in the case of MOS and CCD devices by using a clocked shift register configuration) and as such is identical in concept to the nickel and mercury delay lines used in first generation computers.[22]

The simplest organization for a serial memory is to connect the elements in a synchronous closed loop as shown earlier in Fig. 6.14(c) for an MOS dynamic shift-register. Both CCD devices (when the configuration is called the **serpentine** layout) and bubble stores can be used in this mode; Fig. 6.18(a and b) show typical block diagrams for this organization, note that in the case of the CCD store refresh amplifiers are needed between the shift register stages. The obvious disadvantage of this scheme is the long access times, which are determined by the total number of bits and the shift rate; for random addressing the average access time is half the maximum delay through the loop. The main area of application for this type of store organization is for block organized random access memory (BORAM).

In order to decrease the access times multiple register elements can be used in an arrayed parallel configuration with separate input/output circuitry and common clocks. A bubble memory using this type of organization is shown in Fig. 6.18(c); note that words are read out in parallel at clock pulse intervals with one bit of the word being stored in separate shift registers (bit organized). Alternatively each shift register may be accessed separately and the data read out in serial mode (word organized). In practice this type of organization is limited by the number of input/output pins that can be provided on a chip. More important this structure uses a large amount of device area and hence limits the storage density. These difficulties can be overcome by using the major-minor loop structure as described earlier.

An alternative method (known as **line addressable random access memory** – LARAM) is to select the data circulating in a shift register loop by direct decoding, thereby allowing the data bits to appear in serial at the common input/output port; this is shown schematically in Fig. 6.18(d).

The LARAM configuration can also be used for CCD memories, as shown in Fig. 6.18(e); the address decoding matrix would be implemented on the chip in MOS technology. Each line can be addressed randomly and only the particular line selected is clocked, all other lines remain in an idle or slow refresh mode; note that data can only be read from, or written into, the addressed line. Moreover since only one line is operative at any one time the power dissipation is minimized for this mode of operation. The access time, as before, depends on the number of bits per line and the clock rate; the store can also be structured on either a word or bit organized basis. The main application for a LARAM is as a fast, low power, block organized random access memory and would be used for example as a main memory or paging store. One special application of BORAM would be as a micro-program control store when the long words required in horizontal micro-programming could be stored in a CCD shift register. When very low power dissipation and high packing density is required a serial-parallel-serial (SPS) organization is employed, this is shown in Fig. 6.18(f).

In the SPS mode two single channel shift registers of n bits service one large multi-channel shift register consisting of n serial registers operating in synchronism. A serial bit stream is fed into the input shift register at a fast clock rate, transferred in parallel using a slow clock rate into the multi-channel register and then gated out serially with a fast clock. The slow clock rate is $\frac{1}{n}$ of the fast clock, and since most data bits are transferred at this slower rate the power dissipation is low. The main disadvantage of this type of organization is the long access times and the need for two different clock frequencies. The technique does have the advantage however of a high packing density, and hence is chiefly used for serial mass memory and large data bases.

CCD memories typically use clock frequencies of 1–5 MHz, though low frequencies of 10–100 kHz are also used to conserve power under static conditions. A typical CCD memory chip would have a bit capacity of 16k bits with access times (at 5 MHz) varying between 25–800 μs depending on storage organization and data structure. It is anticipated that CCD memory chips could easily achieve a storage capacity of some 65–120k bits in the near future.

The simplest method of addressing information in a circulating store is to count the system clock pulses in an address counter; the contents of this counter are continuously compared with the required address (also held in a register) using a comparitor circuit. When the address counter equals the specified address a gating pulse is generated which initiates the read/write operations. Note that individual words or blocks of words can be addressed in this way. For example, assuming a memory containing 8192 (8k) 16 bit words arranged in blocks of 64 words, then individual words can be addressed by counting modulo 16, and blocks by counting modulo 1024. If we also assume a clock rate of 10 MHz the maximum access time will be $10^{-7} \times 8192 = 819{\cdot}2$ μs. Note that this is considerably faster than many drum or disc access times.

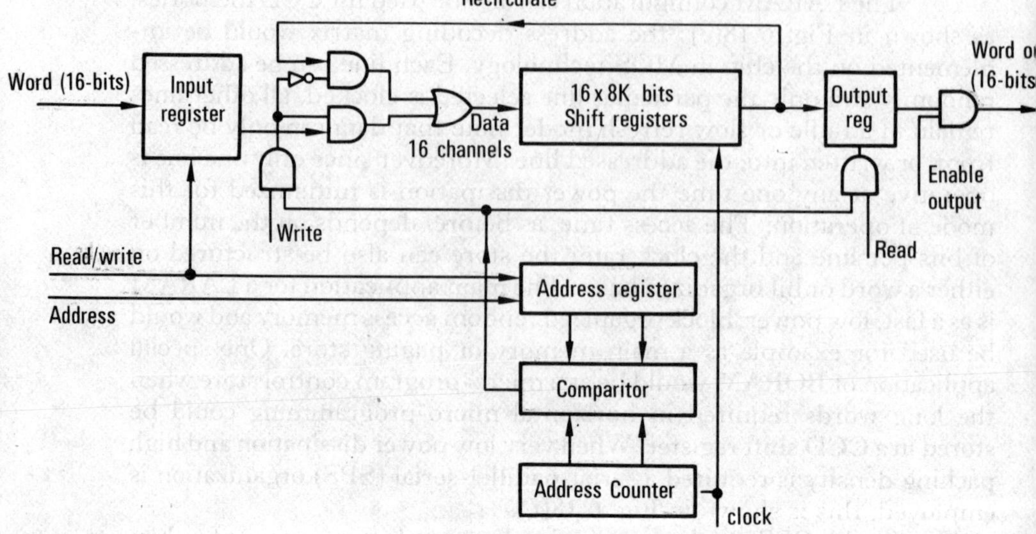

Figure 6.19 Addressing serial memory.

Arrayed or multi-channel shift register memories may also be accessed in the same way. Using a parallel configuration in which the shift registers are circulating in synchronism, with each line containing one bit of a particular word, a parallel word output can be obtained. Alternatively, each line could be accessed independently (the LARAM mode) in which case the line address must also be specified and suitable selection logic provided. A block diagram for a 16-channel memory system is shown in Fig. 6.19; note that logic for one channel only is shown in the diagram.

The inherent disadvantage of a dynamic serial memory is that data must be cycled continuously and this causes difficulty in achieving both a large storage capacity and a short access time. In a cyclic memory the access to a randomly selected item increases linearly with the size of the memory.

Since electronic serial memories are normally constructed from an array of interconnected shift registers there is no theoretical reason why the normal linear cascaded configuration should be used. Recent research[23, 24, 25] has indicated that the use of non-linear interconnection patterns for shift register memories can be extremely effective in reducing access times. One such interconnection pattern is called the **perfect-shuffle** which was originally suggested by Stone for parallel processing[26]; the perfect-shuffle permutation pattern for an n-element vector is shown in Fig. 6.20. The pattern takes its name from the fact that it is analogous to shuffling cards in a deck of playing cards. The memory cells on the left of Fig. 6.20 are shuffled by dividing the cells into two

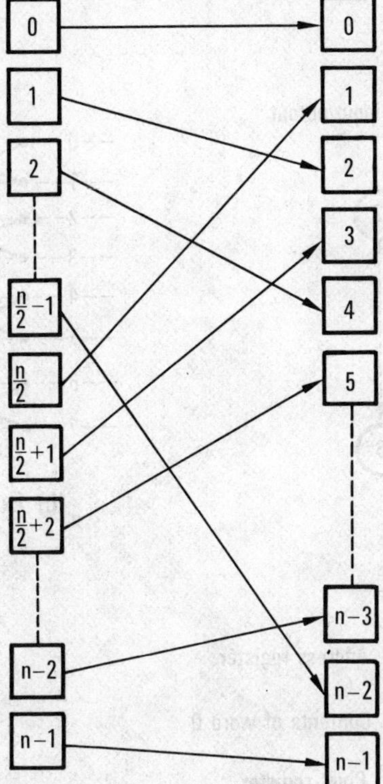

Figure 6.20 The perfect-shuffle.

groups with a 'cut' in the middle of the memory. The memory cells on the right receive data from the memory cells on the left by interlacing data from the two groups just as the two halves of a deck of cards are interlaced when shuffled.

In Fig. 6.20 the cell indices on the left are mapped into the indices on the right according to the permutation P such that

$$P_i = 2i \qquad 0 \leqslant i \leqslant \frac{n}{2} - 1$$

$$= 2i + 1 - n \qquad \frac{n}{2} \leqslant i \leqslant n - 1$$

An alternative, and perhaps simpler, view of the shuffle can be obtained by considering the binary representation of the indices of the vector elements as shown in Table 6.3. The permutation mapping is obtained by simply performing an end-around right shift on the binary indices. Note also that after m shuffles, where m is the number of binary variables, the elements will assume their original values. Furthermore, if the elements are paired, as shown in Table 6.3, we see that adjacent cells differ by only one variable; this difference is retained after shuffling with the difference variable being shifted one place right.

The properties of the perfect-shuffle can be used to construct a dynamic random access shift register memory containing n words in which any word can be accessed in no more than $\log_2 n$ clock periods. The

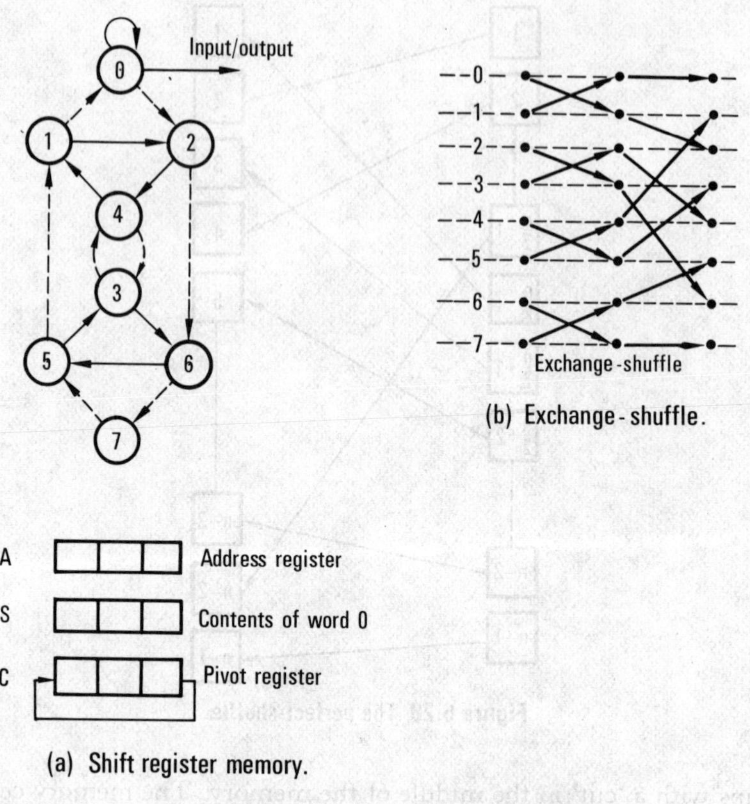

(a) Shift register memory.

(b) Exchange-shuffle.

Figure 6.21 Perfect-shuffle shift registers.

Table 6.3
Perfect-shuffle

Index	$2^2 2^1 2^0$		$2^2 2^1 2^0$		$2^2 2^1 2^0$		$2^2 2^1 2^0$
0	0 0 0		0 0 0		0 0 0		0 0 0
1	0 0 1		1 0 0		0 1 0		0 0 1
2	0 1 0		0 0 1		1 0 0		0 1 0
3	0 1 1	⟶	1 0 1	⟶	1 1 0	⟶	0 1 1
4	1 0 0	Shuffle	0 1 0	Shuffle	0 0 1	Shuffle	1 0 0
5	1 0 1		1 1 0		0 1 1		1 0 1
6	1 1 0		0 1 1		1 0 1		1 1 0
7	1 1 1		1 1 1		1 1 1		1 1 1
Difference between Paired indices	↑ 2^0		↑ 2^2		↑ 2^1		↑ 2^0

state-diagram for the shift-register memory is shown in Fig. 6.21(a); note that the nodes represent one word registers which are uniquely identified by the address contained within the node, and that each word-register is connected to two other registers within the memory. The solid lines show the perfect-shuffle connection, while the dotted lines show an alternative connection, related to the shuffle, called the **exchange-shuffle** which is shown in Fig. 6.21(b). The shift register memory operates normally in the perfect-shuffle configuration, however when it receives a control signal it goes into the exchange-shuffle mode for one clock period. The clock period is determined by the number of bits in the word registers, since complete words must be exchanged during a permutation cycle. Thus, for a basic clock rate of 10 MHz and assuming 16-bit words the clock period would be equal to $10^{-7} \times 16 = 1{\cdot}6$ μs.

The three registers A, S and C are part of the addressing logic, and will be of length $\log_2 n$ bits for a memory with n words of storage; in our example with $n = 8$ the registers will be 3 bits long. Input and output to the memory is via word-register 0, which acts as an interface to the external system. The A register is an address register which holds the address of the required word in memory, S contains the address of the word that is currently in word-register 0 and C is a cyclic shift register containing a single bit. As we have seen earlier the addresses of the even-odd word-registers differ by one and the same variable in this binary representation. The register C is used to indicate the position of this variable (called the pivot point) in relation to the contents of the other registers. After each shuffle operation the C register is updated by performing an end-around shift, thus the S and C registers together describe precisely how the words are permuted in memory at any given instant of time. Note that the contents of word-register 0 are unchanged by a perfect-shuffle, but are modified by an exchange-shuffle. Note also that the pivot bit indicates exactly where the addresses associated with words 0 and 1 (which are always interchanged during the exchange operation) differ so that the updating operation for the S register after an exchange-shuffle is given by $S \leftarrow S \oplus C$.

The addressing algorithm for the memory is shown in Fig. 6.22. To access an item it has to be placed in word-register 0, and thus its address must appear in the S register. Consequently the first operation is to check whether $A = S$, if this is the case then the required item is immediately available and the addressing operation is concluded. If the address is not already in the S register then the memory sequence can be organized to place it there by a succession of appropriate perfect-shuffle and exchange-shuffle operations; this procedure is determined by testing the inequality $A \cap C \neq S \cap C$. If the required address and the current address of the item in word-register 0 agree in the pivot position then a perfect-shuffle is performed, otherwise an exchange-shuffle takes place. This process is iterated, as shown in Fig. 6.22, until the condition $A = S$ is obtained; the addressing operation is illustrated in Table 6.4 where the required word is accessed in $\log_2 n = 3$ clock periods.

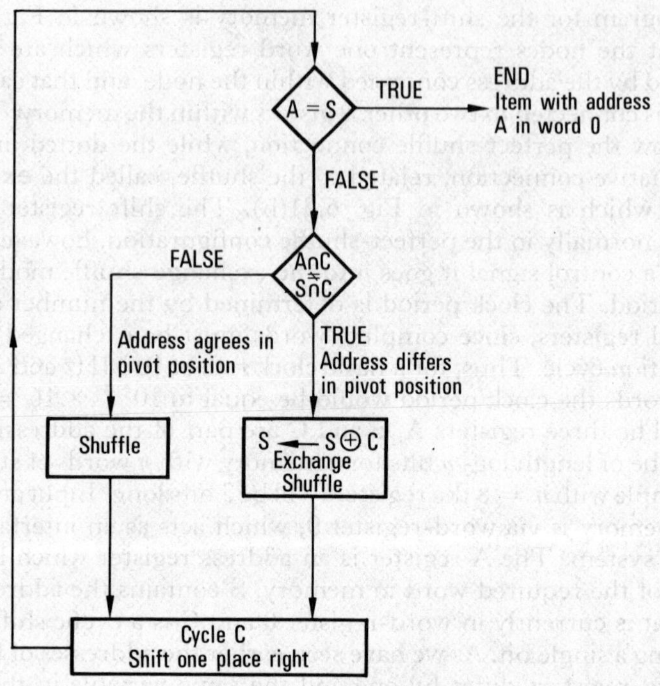

Figure 6.22 Addressing algorithm.

Table 6.4
Memory access in perfect-shuffle SR

Word Register	*Memory Contents*			
0	0	2	3	7
1	2	3	7	5
2	4	0	2	3
3	6	1	6	1
4	1	6	1	6
5	3	7	5	4
6	5	4	0	2
7	7	5	4	0
Register A	111	111	111	111
Register S	000	010	011	111
Register C	010	001	100	010
Comment:	Initial state on access	After Exchange-Shuffle	After Exchange-Shuffle	After Exchange-Shuffle

The shuffle interconnection patterns can be implemented using current technologies, but difficulties could arise due to the requirement for planar interconnections. For memories of $n > 8$ the connection patterns become non-planar and can give rise to implementation problems. The memory described above with $n = 8$ can easily be realized with two layers of interconnection. A modification of this method has been reported by Aho and Ullman[24] who have proposed a memory addressing system which gives a random access of $\log_2 n$ clock periods to the first two words and a sequential access thereafter of 1 clock period. Note that this corresponds to a latency of $2 \log_2 n$ and a transfer rate of one word per unit time.

Electronic serial memories are quasi-serial or pseudo-random in any practical implementation and consequently their operation is similar to either a serial access head-per-track disc system or a random access semiconductor memory. Magnetic bubble memory will certainly compete, both in cost and performance, with discs as a backing store and mass storage media.[27,28] Moreover, there is the added advantage that bubble memory controllers are considerably simpler than their disc drive counterpart, thereby considerably reducing the cost of a total system. Another important characteristic is that unlike discs, bubble memories do not depend on large capacities to achieve a low cost per bit. Thus it is conceivable that small low cost units could be developed for use with mini- and microcomputer systems. However, it is important to remember that CCD and magnetic bubble memories are not just direct replacements for existing storage devices but should be considered as new system components that can be exploited for novel applications, for example in associative and parallel processing.

6.4.6 Semiconductor ROM's

The **read-only memory** (ROM) is a permanent non-volatile store, addressed in exactly the same way as the RAM on a n-word by m-bit basis. Thus the contents of the store are arranged in m-bit words which are accessed by selecting one out of 2^n. The words are prewritten either at time of manufacture, or as we shall see later by the user, and remains permanently in the store – it is not possible to write new data into the store using normal programming operations. The ROM is a random-access memory with the read-time being the time taken between supplying the address input and the appearance of the m-bit word. MOS and bipolar technologies are mainly used for ROM's; commercial devices can range from 1k bit (256 4-bit words) up to 64k bits (8k 8-bit words) with an average read-time in the order of 50–300 ns.

The main application of ROM's are in code conversion, table-look-up and as a store for special purpose control programs (such as micro-code and calculator programs). Note that the ROM performs essentially as a combinational circuit (and can obviously be used as a

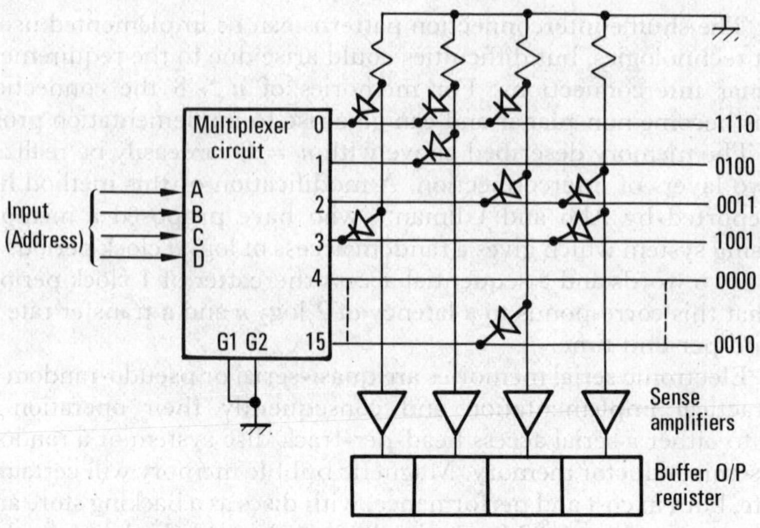

(a) ROM store using a diode matrix.

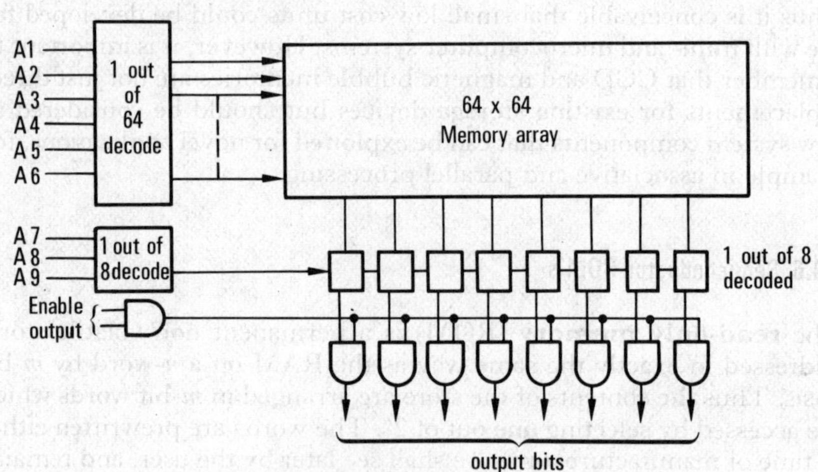

(b) 4096 bit bipolar ROM chip.

Figure 6.23 ROM stores.

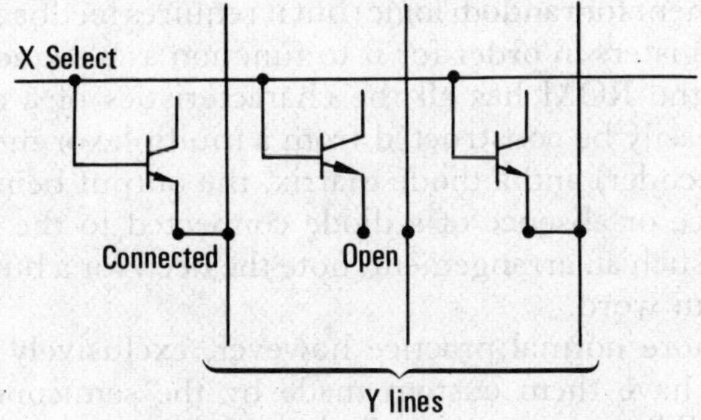

(c) ROM using bipolar transistors.

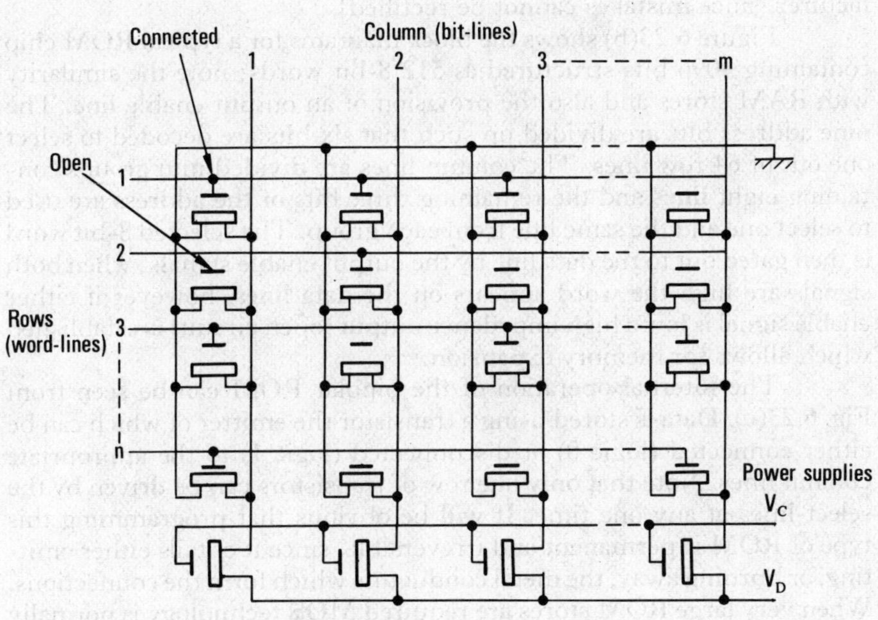

(d) ROM using MOS transistors.

Figure 6.23 contd. ROM stores.

direct replacement for random logic) but it requires feedback connections via external registers in order for it to function as a sequential circuit.

Since the ROM has all the characteristics of a combinational circuit it can easily be constructed from a multiplexor module (used as the address decoder) and a diode matrix, the output being determined by the presence or absence of a diode connected to the bit lines. Fig. 6.23(a) shows such an arrangement, note the need for a buffer register to store the output word.

The more normal practice however, exclusively so with large ROM's, is to have them custom made by the semiconductor device manufacturer. The user must supply the manufacturer with the required coding for each word (from which a mask is made) and then during the final fabrication stages the specified pattern is permanently entered into the ROM. This can be an expensive process initially but once the mask is made identical ROM's can be produced very cheaply. It is very important that the correct ROM pattern should be supplied to the manufacturer, since mistakes cannot be rectified!

Figure 6.23(b) shows the block diagrams for a typical ROM chip containing 4096 bits structured as 512 8-bit words; note the similarity with RAM stores and also the provision of an output enable line. The nine address bits are divided up such that six bits are decoded to select one out of 64 row lines. The column lines are divided into groups containing eight lines and the remaining three bits of the address are used to select one and the same line from each group. The selected 8-bit word is then gated out to the data line by the output enable signals; when both signals are high the word appears on the data lines, however if either enable signal is low a high impedance output (open circuit) is established which allows for memory expansion.

The internal operation of the bipolar ROM can be seen from Fig. 6.23(c). Data is stored using a transistor the emitter of which can be either connected (logic 0) or disconnected (logic 1) to the appropriate column lines. Note that only one row of transistors can be driven by the select-lines at any one time. It will be obvious that programming this type of ROM is permanent and irreversible, since it entails either omitting, or burning away, the metal conductors which form the connections. When very large ROM stores are required MOS technology is normally used. In this case the bipolar transistor is replaced by a MOS transistor, as shown in Fig. 6.23(d), which functions in an analogous manner.

Certain types of ROM can be electrically programmed by the user and are called programmable ROM's or **PROM's**. PROM's are normally supplied by the manufacturer with all the storage bits set to zero; the storage density can range up to 4k bits. In the bipolar PROM store a logic 1 can be electrically programmed at any of the bit locations by selecting, and then physically severing, a fusible metal connection using a high current source to melt the fuse link. Note that a zero error can always be changed into a one, but not vice versa! An alternative approach is to create a short-circuit, using a method known as avalanche induced migration, which involves electrically 'blowing' a diode junction into a short-circuit.

It is also possible using MOS devices to produce an electrically alterable or reprogrammable ROM, called an **EAROM** or **EPROM**. In this case an electric field or electromagnetic radiation is used to inject charge into an oxide trap above the gate region of a connecting MOS transistor; note that the MOS transistor memory cell has essentially isolated 'floating-gate'. When the gate is charged a permanent drain-source channel is established which allows the representation of a logic 1 at that point. Current EAROM devices are reprogrammed electrically using microsecond wide pulses from 25–75 volts with a 98 % duty cycle; erasure is accomplished using ultra-violet light, with exposure times of up to 30 minutes. Reprogramming can be repeated as often as required; note that the contents of the *entire* memory must be erased. The chief application of PROM's and EAROM's are in the development of small special purpose microprocessor and microcomputer systems where the cost of producing a commercial ROM mask can be prohibitive. In this type of system the control program (for example to regulate a real-time process) would eventually be stored in ROM.

During the development stages of a microcomputer system it is essential to be able to modify and change the control program to ensure that the final ROM program is error free. Though this can be done using an EAROM a much better technique is to use RAM, in place of ROM, in the first instance, and then once the program is fully developed a permanent ROM (or PROM) store can be substituted. Alternatively, simulation programs can be used to prove and debug control programs before being used in the actual system.

6.5 Magnetic drum and disc storage systems[29]

In the dynamic type of magnetic storage system, data are recorded by inducing magnetic dipoles in a moving magnetic surface. The relative motion of these magnetized areas of the storage medium past a read-out transducer induces an output voltage which may be sensed electronically in the usual way. Digital recording differs from analogue recording in that it requires only two levels of magnetization (based, as with magnetic cores, on the hysteresis phenomenon) the magnetic surface being saturated into one or other of its two remanent states to represent the digital values of 0 and 1.

Magnetic heads, comprising a high permeability split-ring core wound with a fine wire coil, are used for both the reading and writing functions (see Fig. 6.24). When a current is passed through the coil a magnetic field is generated around the core gap which instantaneously magnetizes the area of magnetic medium immediately beneath the gap. The magnetization will be in one of two senses, depending on the direction of current flow through the coil, with its direction in the plane of the surface and parallel to the direction of surface motion. Thus the pattern

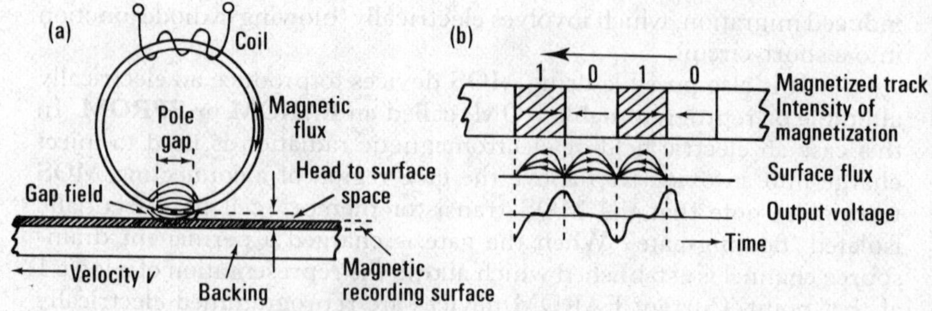

Figure 6.24 Magnetic recording principles.

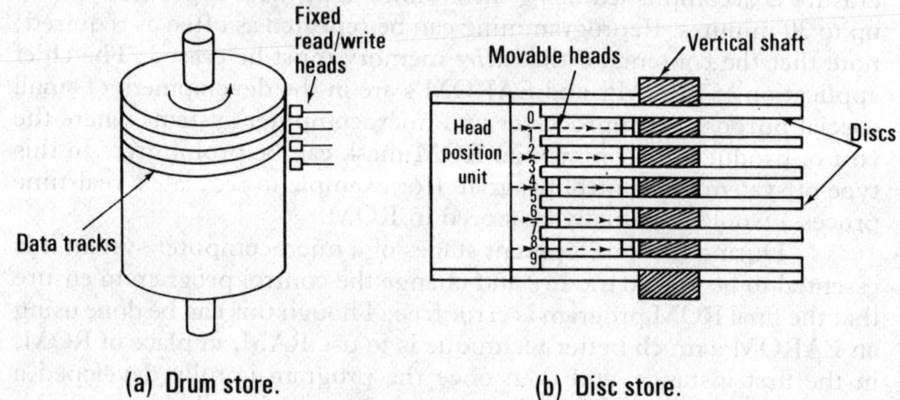

(a) Drum store. (b) Disc store.

Figure 6.25 Drum and disc mechanisms.

of magnetization in time, and therefore spatially along a track, represents the digital data. The surface magnetic flux radiating from the magnetized cells induces a voltage across the reading head the polarity of which depends upon the direction of the flux reversals.

The magnetic drum (see Fig. 6.25(a)) consists of a precision machined and balanced cylinder or drum of non-magnetic material, the surface of which is coated with a thin magnetic layer. The drum is rotated at high speeds, in the order of 120–3600 rev/min, causing a narrow band or track (in the order of 0·013 in) to pass circumferentially beneath the magnetic recording head. The axial length of the drum may be divided into a number of such data tracks, the space between them being determined by the distance between the heads (the head pitch), typical values being 0·005–0·025 in. The information storage capacity is determined by the number of bits per inch (bpi) along a data track (bit density) and the number of tracks per inch (tpi) along the drum axis (track density). The product of these two factors is the data storage density per square inch of the recording surface. Digit packing densities can range from 50–2000 bpi and the track density from 10–150 tpi but the normal values encountered in practice with drum stores are 100 bpi and 40 tpi respectively.

Magnetic head cores are usually made of ferrite materials in order to achieve a high permeability magnetic path and hence an intense magnetic field, which is required during the writing process. To ensure that the flux generated when writing does not unduly interfere with the magnetic flux of an existing cell, the width of the pole gap and the head to surface gap, both of which affect the spread of flux, must be maintained as small as possible. Moreover, the head to surface gap needs to be very small at high packing densities so as to be able to discriminate between individual magnetized cells. Air gaps of the order of 0·0005–0·001 in are used in practice, with the same distance approximately from head to surface. Read/write heads may be either **fixed**, in which case one head per track will be required, or of the **scanning-head** type where one head is shared between several tracks along the axial length of the drum. The fixed head mounting is a simple and economical solution but requires a relatively large head to surface gap to prevent the head accidentally touching the drum surface; this limits the packing density to about 100–150 bpi. A further development of the stationary head system is to float the head on a thin film of air, forming a boundary layer, which is carried round on the surface of the drum. In this method the head to surface space is determined by the ratio of the head lift, created by the pressure of the air layer, and the opposing pressure of a spring forcing the head towards the surface. The pressure generated by the air film opposes the direct contact of the head onto the surface and it is possible with this system to operate with a head to surface gap in the order of 0·0001 in.

The scanning (or flying) head is mounted on an extendable arm which moves across the tracks until the head is directly in line with a selected track; this form of head is particularly favoured with magnetic disc stores. Though the system is economical in the use of heads, there are disadvantages in the time required to position the head, and the complexity of the head control system. The head operating mechanism is required to position the read/write heads directly over any one of a number of tracks (with a track span of several inches) within milliseconds and with an accuracy in the order of 0·0001 in.

The disc store shown in Fig. 6.25(b) consists of a number of aluminium discs coated on both sides with a ferric oxide material which are mounted on a vertical shaft and rotated at high speed. Data are recorded serially, along each of several concentric tracks, on both sides of the discs; each side is scanned in a horizontal direction by a separate read/write head. All the heads are positioned together so that one track per disc side is available to the read/write head at any given position. Such a set of tracks is called a **cylinder**. On some large capacity disc stores a single scanning head is used, which not only scans the tracks, but can also be positioned vertically to select a particular disc. Typical high speed, high density disc units would have bit densities of 1000–5000 bpi with 100–400 tracks per inch radially.

An alternative and cheap form of disc store, widely used in mini- and microcomputer systems, is the **floppy disc** originally developed by

IBM. In this system the disc is fabricated from a flexible plastic material (usually Mylar) some 8 inches in diameter with an oxide coating on one side only; the disc is permanently packaged in a cheap 8 × 8 inch plastic envelope. In operation the floppy disc is inserted into the disc drive unit which clamps the centre of the disc and rotates it inside the stationary jacket. A slot in the jacket allows the read/write heads access to the circular tracks; during the read/write operations the heads are in contact with the disc surface but are retracted when the data transfers cease to minimize wear. A further hole is provided in the envelope to allow optical sensing of an index mark indicating the starting point for each track.

The floppy disc typically rotates at 100–400 rpm with an average access time to any track of half a revolution time; some 6–10 ms is required to move from one track to the next. Floppy discs normally have 70 tracks and 48 tpi, with a bit density of 1000–3000 bpi; data can be transferred at 250–500k bits/sec with a total capacity in the order of $\frac{1}{2}$ million characters.

The floppy disc is significantly less reliable than floating head disc drive units, due mainly to the head to disc contact wear, and has a disc life in the order of 10^6 passes per track (equivalent to 2–3 years head life). It is also necessary to incorporate some form of error detection (such as a Hamming code check character for each record) in order to reduce the error rate (which is in the order of 1 error for 10^9 bits unchecked). However the floppy disc is considerably more reliable than paper-tape, cassette-tape or cartridge-tape units.

6.6 Digital recording techniques[30]

One of the major problems that is inherent in any form of digital data recording is that of **pulse crowding**. For a current pulse input to the read/write coil the voltage output signal will be a dipulse with each individual pulse of width n (see Fig. 6.26). This arises because a single current pulse records a single discrete cell on the surface, having two closely spaced but oppositely directed magnetization transitions at its edges owing to the trailing and leading leg fields of the magnetic heads. In Fig. 6.26, let h define the length of individual magnetic cells, that is, the minimum spacing between current reversals. The number of cells, or pulses per inch (ppi) is equal to $1/h$, thus if the individual identity of each pulse is to be preserved:

$$h \geqslant n \quad \text{and} \quad \text{ppi (max)} = 1/n$$

During the writing process the binary digits 0 and 1 must be converted into their respective states of magnetic surface saturation, that is, the binary input data must be converted to a sequence of recording current signals of a suitable form to activate a recording head. There are basically six different methods that may be used to encode the input current to the head coil and we shall now describe each one of these in detail; Fig. 6.27 shows the encoding of the binary input 101101 for each technique.

In the first method, known as the **return to zero** (RZ) technique, the magnetic surface is magnetized to one state of remanence for a binary 1, but left unmagnetized for a binary 0. That is, a dipole is recorded for a 1 and nothing for a 0, for this reason the technique is also known as the **dipole** method. An alternative form of the RZ technique, known as the **bipolar** method, is to magnetize the surface in opposite directions for binary 0 and 1, the pulses being separated by a period when no current flows and hence the surface is left unmagnetized. Note that in both the return to zero methods the flux density is always returned to the reference level after each digit is recorded. The technique allows single bits to be altered without rewriting the entire pattern; that is, a discontinuous writing mode is possible. However, it is not possible to overwrite recorded information in this system, and it is necessary to erase previously recorded data (using a separate erase head) before recording. In order to distinguish between actual zero's and areas of no magnetism during the reading process, a separate clock track must be recorded on the surface in synchronism with the recorded data.

Method C in Fig. 6.27 is known as the **non-return to zero** (NRZ) technique. In this form of recording the magnetic coating is always magnetized to one or other of the remanent states, throughout the length of the track, depending on whether a 1 or a 0 is to be recorded. The flux changes only occur when the digital input data change from $0 \rightarrow 1$ or $1 \rightarrow 0$, thus consecutive 1's and 0's will remain at the same level. With this continuous writing system the magnetic coating is always magnetized in one direction or the other, thus the erasure of previously recorded data is not required.

In the **non-return to zero one** (NRZ1) method, which is a modified NRZ technique, the current is reversed producing a flux change every time a 1 is recorded, but this time no flux change is generated for a 0; that is, the system changes polarity each time a 1 is recorded. Note that a 1 is indicated by either a positive or negative output pulse and a 0 by the absence of a pulse. The advantage of NRZ1 encoding is that if a bit is misread only that bit is in error; with NRZ, if a bit is in error all succeeding bits will be corrupted until the next signal pulse is encountered.

The NRZ method is the most efficient way of encoding binary data, and requires at most one saturation reversal (or output pulse) per bit, the maximum occurring for an alternating sequence of 1's and 0's. Thus with the NRZ method a response of from 0 (d.c.) to the digit frequency is required from the read amplifiers. Because only the binary 1's are identified by an output pulse, accurate clocking is necessary to interpret and recover the recorded data correctly. In fact, the chief limitation of NRZ coding is that self-clocking is not possible, this is essential in some systems to counteract the problems of interhead timing-skew when recording data in parallel along several tracks. For the NRZ technique,

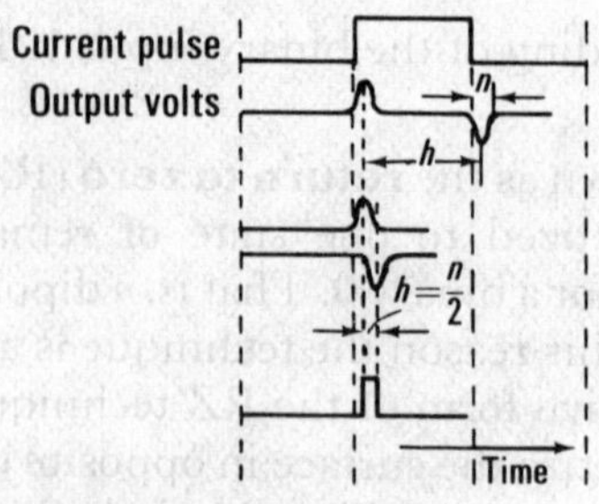

Figure 6.26 Pulse crowding in digital recording.

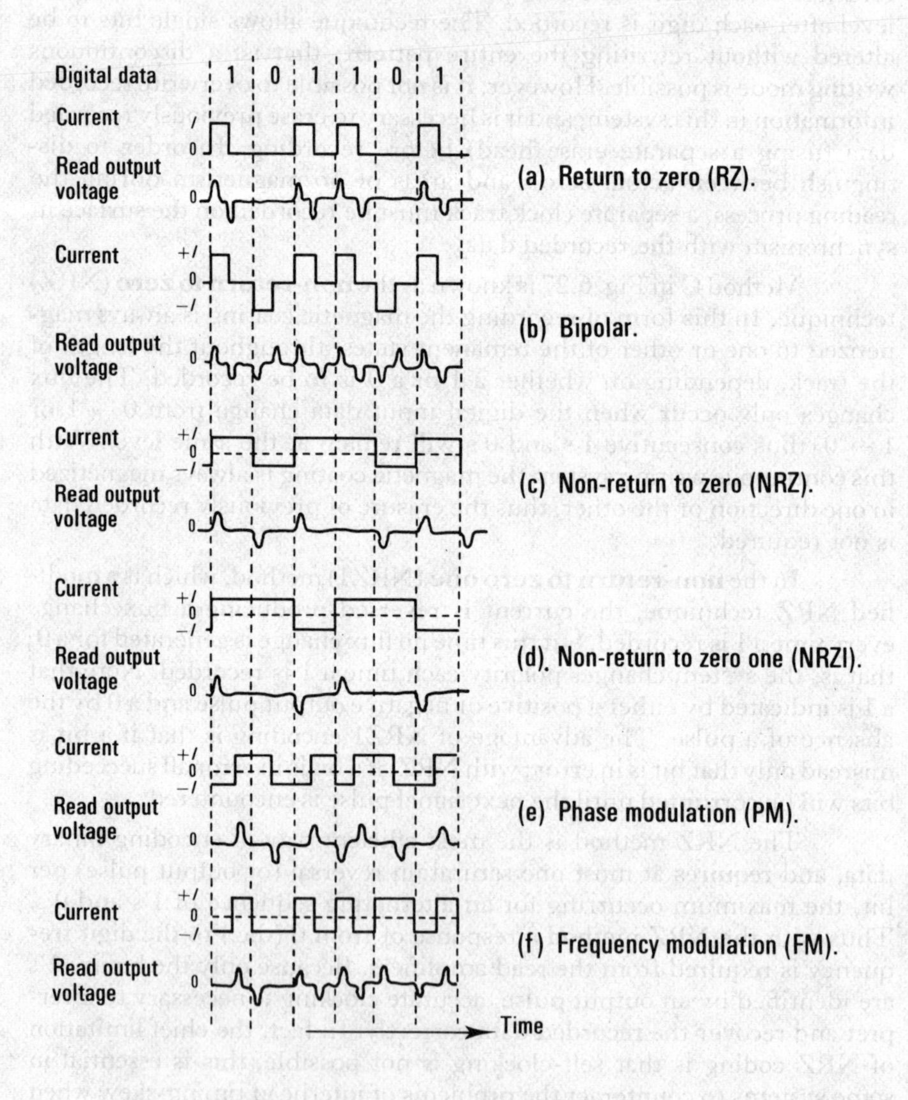

Figure 6.27 Digital recording techniques.

bpi = $1/h$, and the maximum value of the output pulse density (ppi) is just equal to the recorded bit density.

Phase modulation (PM) shown in Fig. 6.27(e), is a form of NRZ recording, first described by Williams *et al.*,[31] in which half of the bit cell is magnetized in opposite directions. A 1, for example, may be written by a positive change in saturation current and a 0 by a negative change at data clock rates. In this way a positive read pulse is produced for every recorded 1, and a negative pulse for each 0. Thus with phase modulation it is possible to discriminate between a 1 and a 0 by simply detecting the polarity of the output pulses. Moreover, since both binary 1 and 0 provide output pulses, there is at least one output pulse per bit period, thereby allowing a clocking signal to be continuously generated from the output, that is, the system is self-clocking. Because phase modulation recording gives a bit density equal to $1/2h$ bpi, bit crowding will occur at a lower bit density than with the corresponding NRZ technique. However, the technique does have the advantage of requiring a smaller frequency bandwidth, varying between digit frequency to half the digit frequency.

In the **frequency modulation** (FM) method, shown in Fig. 6.27(f), a flux change always occurs at the boundary of the magnetic cells, with a binary 1 causing a flux change in the mid-cell position; no change is produced for a binary 0. Consequently, the presence of a voltage output pulse in the mid-cell position indicates a 1, and the absence of a pulse indicates a 0. This technique is very similar to phase modulation, and except for slightly simpler output detection, has the same advantages and disadvantages as the PM technique.

A modified form of frequency modulation (called MFM or **delay modulation**[32]) has been used by IBM in some disc file systems. In MFM, as in FM, a 0 and 1 correspond to the presence or absence, respectively, of a transition in the centre of the corresponding bit cell. However in this case transients at the cell boundaries only occur between bit cells that contain consecutive zeros. Thus the method allows self-clocking but also achieves the maximum bit density obtainable with NRZ1 methods. The technique requires more complex data detection circuitry than the conventional FM mode and in some cases, depending on the data patterns, large d.c. components can be generated.

Because phase modulation techniques have been used extensively (and still are) in drum and disc systems (NRZ1 is preferred for magnetic tape), we shall consider the PM method in slightly more detail. The initial binary input may be encoded into the PM form using the logic circuit shown in Fig. 6.28(a). This is a simple inverted exclusive OR circuit (non-coincidence circuit) performing the function $W = \bar{c}\bar{d} + cd$, where c is the basic clock and d the serial data input. A complete timing diagram for the PM system is shown in Fig. 6.28(b). The output of the PM encoder is applied to the input of the write amplifier which then generates the required currents in the coil of the read/write head. Signals from the read/write head are amplified, clipped, and then differentiated. Since the

Clock c

Serial data d

To write amplifier w

(a) Write encoding.

Data d

1 0 1 1 0 1 0

Clock c

Write wfm W

Readout voltage

Amplified and clipped

Differentiated (if necessary)

Strobe (differentiated clock)

Ones

Zeros

Note: 1s and 0s are ½ digit-time late compared with original data.

(b) Read decoding.

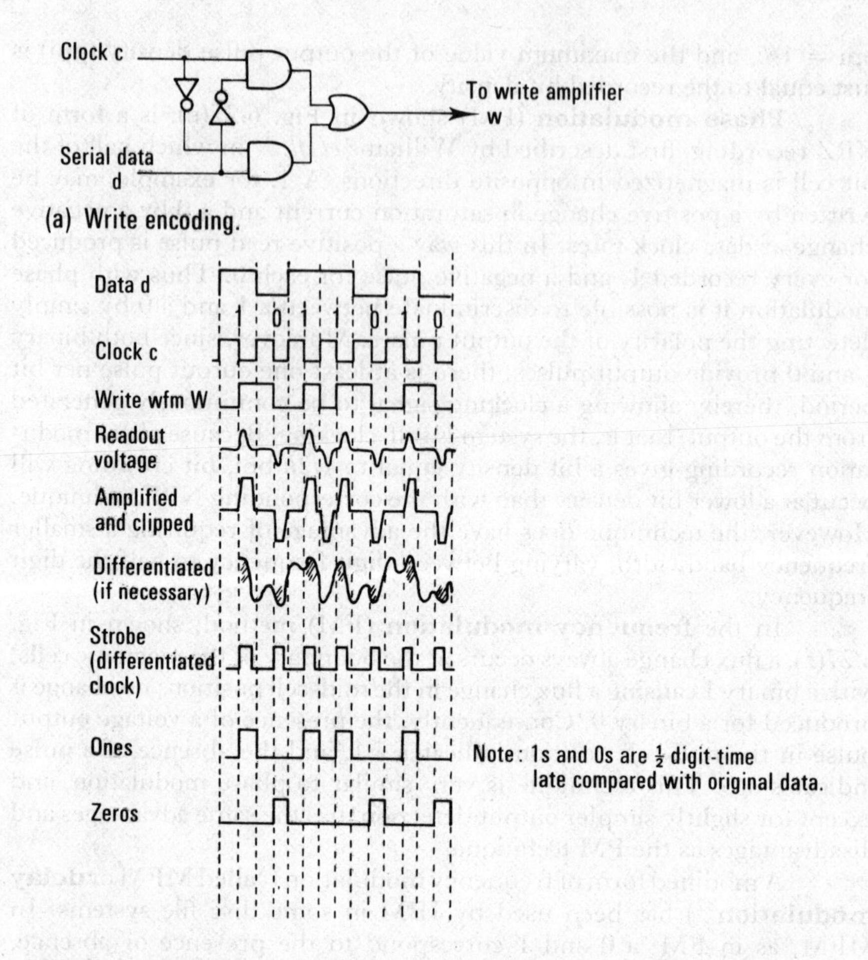

Figure 6.28 Read and write techniques for drum store.

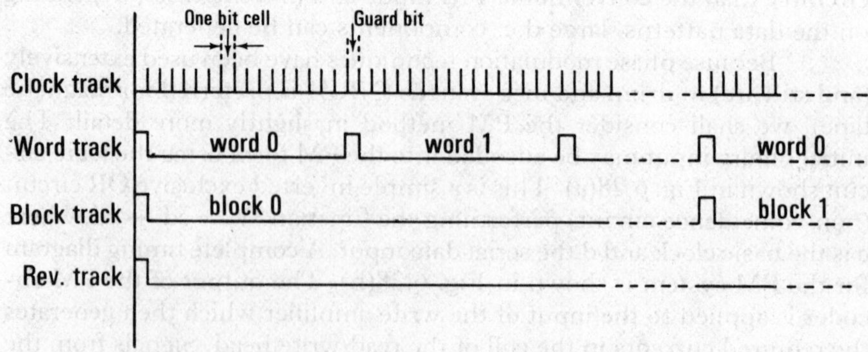

Figure 6.29 Drum store clock and address tracks.

mid section of the recorded pulse rises or falls according to whether it represents a 1 or a 0, differentiating will generate a positive pulse from the rising edge and a negative pulse from the falling edge. This waveform is gated with a strobe pulse (derived from clock) and the generated 0's and 1's used to set the input bistable stage of a shift register. Note that the identification of the 1's or 0's take place half a digit-time after the start of the original serial input data due to clock timing.

The basic clock pulses, used for the overall timing of the storage system, are supplied from a special track permanently written on the drum or disc, usually at the time of manufacture. Other timing pulses, or *address tracks*, indicating the position of words, blocks, sectors, etc., recorded on the drum, are also permanently written on separate tracks.

A recently developed method of magnetic recording which combines two advantages not found together in previous techniques – absence of d.c. components in the output signal and a high recording efficiency – has been described by IBM.[33] The method called **zero modulation** (ZM) also has the advantages of being self-clocking, operates within a narrow bandwidth and provides easy and reliable error checking due to the code structure employed.

The method of coding is to sequentially map every bit in the data stream into two binary code digits such that any two consecutive ones are separated by at least one, and at the most three, zeros. This encoded sequence is then recorded using the NRZ1 technique. Since the narrowest pulse in the ZM waveform spans two digits in the coded sequence this is the width of the data cell.

The encoding algorithm is somewhat complicated but can be described in terms of the data bit to be encoded together with its preceding and following data bits, and the two **coded** digits corresponding to the preceding data bit. In addition two parity functions are used; **look-ahead parity** P(A) is the modulo 2 addition of the ones in the data stream beginning with the data bit to be encoded and counting forward to the next zero bit while **look-back parity** P(B) is the modulo 2 sum of the zeros in the data stream from the beginning up to the data bit under consideration. Table 6.5 shows the coding and decoding functions for ZM using these parameters. Note that d is the data bit and a,b the encoded digits, the subscripts $-1, 0, +1$ represent proceeding, current and following bits respectively.

The non-existent bit preceding the first data-bit is taken to be a 1 and its look-back parity 0, similarly the non-existent bit following the last data bit is taken to be 0. Table 6.6 shows a typical example of ZM encoding; note that the parity summations include the current data digit.

In implementing a ZM encoder the look-back parity function can easily be generated during the encoding process by updating a single bit bistable each time a zero occurs in the bit stream (i.e. the conventional method of parity checking serial data). Unfortunately the look-ahead parity function is not so simple since it depends on the length of a string of ones in the *following* sequences of the data stream. Since the algorithm

Table 6.5
Zero modulation

a) Encoding Algorithm

Condition	*Mapping* $d_0 \rightarrow a_0b_0$
$d_{-1} = 0$	$0 \rightarrow 10$
$d_{-1} = 1$ and $a_{-1}b_{-1} = 00$	$0 \rightarrow 10$
$d_{-1} = 1$ and $a_{-1}b_{-1} \neq 00$	$0 \rightarrow 00$
$d_{-1} = 0$ and $P(A) = 0$ and $P(B) = 1$	$1 \rightarrow 10$
$d_{-1} = 1$ and $a_{-1}b_{-1} = 00$	$1 \rightarrow 10$
$d_{-1} = 1$ and $a_{-1}b_{-1} = 10$	$1 \rightarrow 00$
Otherwise	$1 \rightarrow 01$

b) Decoding Algorithm

Condition	*Mapping* $a_0b_0 \rightarrow d_0$
$a_{+1}b_{+1} = 00$	$10 \rightarrow 1$
$a_{+1}b_{+1} \neq 00$	$10 \rightarrow 0$
$a_{-1}b_{-1} = 10$	$00 \rightarrow 1$
$a_{-1}b_{-1} \neq 10$	$00 \rightarrow 0$
None	$01 \rightarrow 1$

Table 6.6
ZM encoding

Data stream d	0	1	0	0	1	1	1	1	0	1
Look-ahead Parity	0	1	0	0	0	1	0	1	0	1
Look-back Parity	1	1	0	1	1	1	1	1	0	0
ZM encoding	00	01	00	10	10	00	10	00	10	01

Waveform

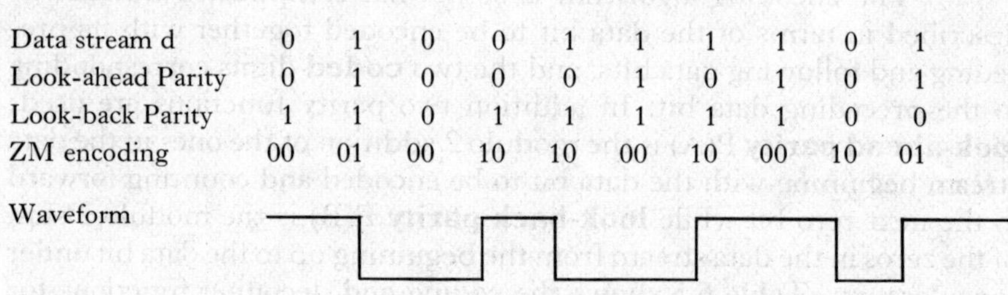

imposes no limits on the length of this stream the memory requirements are unlimited! Consequently in practice a modified ZM encoding algorithm is used which limits the memory requirement for computing P(A) to a specified number of bits. This is done by inserting an extra parity bit with the value of P(B) in the data stream at specified intervals of k bits (there are no theoretical limits to *k* but in practice values of 100 or more are used). With this constraint the computation of P(A) proceeds in the usual manner but now extends only to the following $k - 1$ data bits.

6.7 Organization of drum and disc storage systems

The surface of the drum is normally organized into tracks (typically in the order of 32) which are sub-divided into **sectors** or **blocks** each holding a certain number of words. For example, each track might consist of 128 blocks each containing 32 16-bit words, giving a track capacity of 65,536 bits, and a total storage capacity in the order of 2 megabits. To save time on data transfers, information held in backing stores is normally transferred in blocks (in this case consisting of 32 16-bit words).

Owing to its mode of storage (the recorded data pass continuously under the read/write heads, in a serial fashion) the drum is a cyclic access system, analogous in operation to the delay line store. Thus the access time (also called the **latency** time) for a randomly addressed block is, on the average, half the time for one revolution of the drum, but could in the worst case extend up to one revolution time. That is, assuming a typical speed of 3600 rev/min, we have $P = 1/60$ second/revolution, and a worst case access time of approximately 17 ms. This applies to fixed read/write heads only, when scanning-heads are employed the access time will be increased by the time required to move the head into position over the selected track.

To locate data on the drum (before a read or write operation) it is necessary to specify the relevant track and sector number, this information must be included in the address portion of the corresponding input/output instruction from the computer. The timing and organization of the addressing system is controlled by the special clock and timing tracks recorded on the drum (see Fig. 6.29). Note that the start of the clock track, and the word and block markers are synchronized by a REV pulse which occurs once per revolution of the drum. In actual fact the word and block markers are not essential to the system, since this information may be generated externally using counting logic. Though the provision of these additional marker pulses reduces the amount of control logic that is required, the basic cost of the drum is increased accordingly. An alternative addressing technique is to record a binary-coded address for each block on a separate drum track, these addresses are read out in sequence and compared with the required block address (held in a circulating register) using a serial comparator circuit.

A block diagram for a serially organized drum store is shown in Fig. 6.30(a). The basic clock track is counted in a $\div 16$ bit-counter, with the carry output providing the input to a $\div 32$ word-counter, which in turn feeds the block counter; alternatively the block markers (if provided) could be counted directly. The contents of the block counter are compared with the specified address and when a coincidence is obtained a coincidence pulse (lasting for the duration of one block-time) is generated and used either to gate out the stored data into the buffer store, or to write from the buffer store onto the drum track.

The method employed to locate a specific track depends on whether the drum is of the fixed or scanning head type. In the case of the

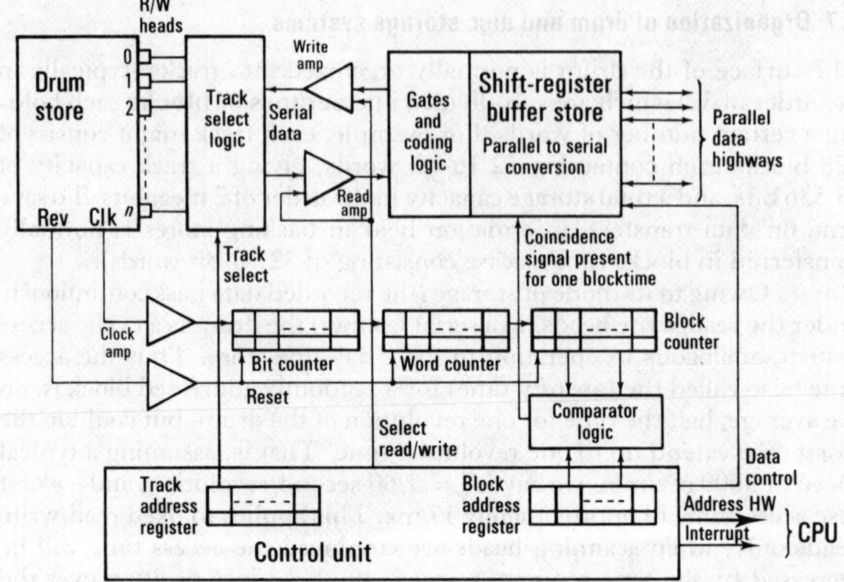

(a) Serial drum system.

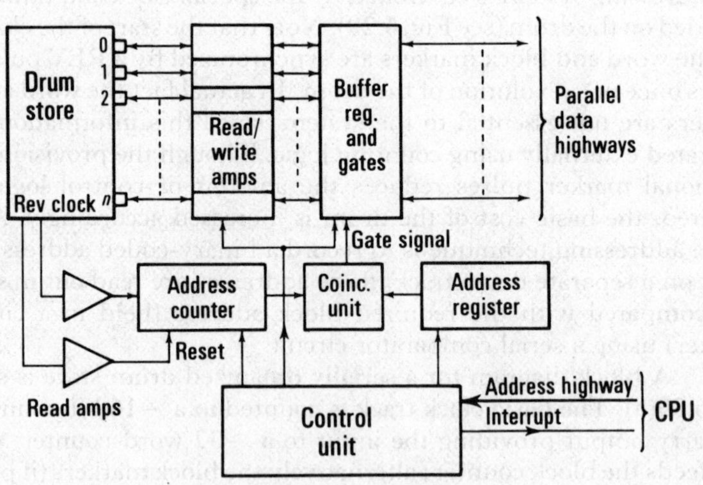

(b) Parallel drum system.

Figure 6.30 Drum store logic systems.

scanning-head unit, the decoded instruction is used to select and position the arm holding the read/write head into alignment with the required track. With fixed heads, however, there is no physical movement involved, and the track selection is achieved by electronic switching, using for example, reed relay devices. The basic problem in the fixed-head system is firstly to select the read/write head for the required track, and secondly to put the appropriate amplifier and heads into either the read or write mode. The track selection function may be performed by arranging that the read/write heads are permanently connected in a matrix format with output leads from both the rows and columns. Selection is performed on a coincident row and column basis with the drive signals being derived from the decoded track address. Note that only one read or write amplifier is required, the input (or output) being switched to the selected head; the clock track (and any other timing pulses) must of course have its own separate amplifier.

An alternative method of organizing drum storage is to lay out the tracks in parallel fashion (see Fig.6.30(b)). In this way the data transfer rate may be increased, but at the cost of increased read/write electronics since each head must now have its own amplifiers. The data are recorded in parallel on the tracks, so that all n bits of a word appear simultaneously on the output of the drum read channels. The main advantage of parallel operation is the increased speed of read-out, brought about by the fact that there are no longer any delays due to track switching or the serial assembly of stored data.

All the techniques described in the earlier sections for drum storage systems apply equally well to disc stores, particularly fixed-head disc stores, with the obvious difference that with the disc store the data are recorded on both sides of a flat magnetic disc. Owing to the geometry of the disc system the track length is progressively reduced (resulting in an increased packing density) as the track radius is decreased. If the same total number of bits is required for each track (which is usually the case) the electronics must be designed in such a way as to compensate for this non-uniform packing density. Alternatively, only a small band of tracks at the outer (or inner) perimeter of the disc would be employed in order to enable a uniform packing density to be obtained.

A typical moving-head disc system would consist of six discs, providing 10 recording surfaces each capable of holding 100 tracks. Data are recorded on the disc in blocks or **sectors**, containing in a typical case 128 words each of 24 bits (see Fig. 6.31) each track has eight sectors giving a total of 1024 words per track. Since the tracks on all 10 surfaces are accessed at the same time a complete *cylinder*, consisting of 10 tracks each of 1024 words, is available for each track position. The total access time to randomly addressed data (in the order of 100–250 ms) is governed by the head-positioning (**seek**) time, and the normal rotational delay, (**search** time), the latter being of the same order as for drum storage. The principal delay, however, is due to the head positioning (which varies according to the track selected) and on average would be in the order of

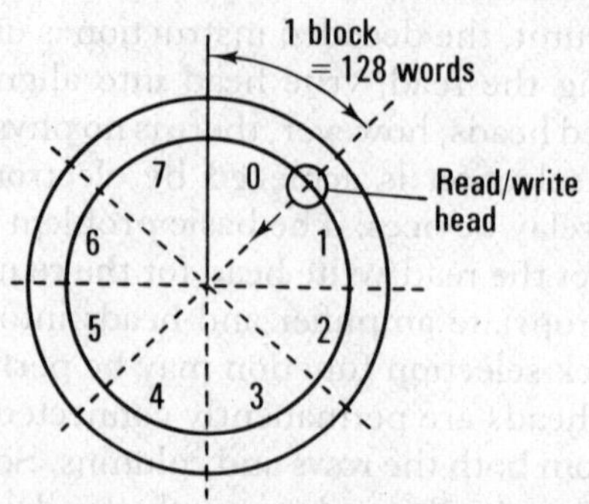

Figure 6.31 Disc store organization.

50–200 ms. The heads can be moved from track to track in 10–25 ms, including settling time. The data are normally recorded in a serial mode onto a selected sector of the track, the computer instruction specifying the track, disc surface, and sector for each data transfer. To minimize the access times, recording is normally performed on a complete cylinder at a time, the track is selected first and then the read/write heads are switched to access each disc surface in that track position.

Track addressing is normally achieved either electronically or mechanically but sector addressing necessitates a stored address, usually in the form of a pre-recorded sector marker at the start of each sector. In addition a stored clock or timing track is also required to facilitate the decoding operations.

Compatibility between different disc units is obtained by using a special bit pattern for track location which is pre-recorded during manufacture. The **servo** track is written on one disc surface in the disc pack, and then read back during operation and used to control the position of the heads on the other discs. The pattern is recorded in such a way that a reversal of magnetism occurs when moving transversely across the centre line of the data tracks on the other surfaces. A special head is used to read this track which gives a minimum signal when the head is aligned with the centre of the track, and varies positively or negatively as the heads move from side to side over the centre line of the track. The output from the head is used to control a linear motor drive which is used to line up the heads on the track. In this way any disc pack can be read or written on any disc drive under direct control of the servo system.

6.8 Magnetic tape stores[34,35]

Magnetic tape systems are the predominant means, at the present time, for cheap mass storage. The magnetic tape is usually stored in reels, containing some 2400–3600 ft, and is transferred from one reel to another in the process of reading and writing; Fig. 6.32 shows a typical unit. The tape is moved past the read/write heads (one per track with typically eight tracks) by means of driving rollers, called the **capstans**. These may be of the **pinch roller** type, which mechanically grip, or pinch, the tape between a free moving and a driven roller, or of the **vacuum** variety which 'suck' the tape onto a moving roller. Since it is necessary to be able to start, stop and reverse the tape very quickly, a tape buffer must also be provided, including a tape level servosystem, to control the slack tape caused by the inertia of the transport mechanism.

The digital data are recorded (typically at 200–1000 bpi) on the magnetic tape using the same techniques as described in Section 6.6, the NRZ1 method being preferred for high speed systems. The output voltage waveforms are amplified and decoded using a peak sensing technique; that is, a 1 is detected as either a positive- or negative-going peak, consequently there is no need to record a separate clock on the tape. Access to the data is sequential, with the information being recorded in a serial fashion along the length of the tape. The data rates that may be achieved with this type of system (once the tape unit is reading the selected data) are in the order of 150×10^3 to 2500×10^3 bits/second, with a start–stop time of between 3–10 milliseconds. It is difficult to quote an access time for this type of storage, a random search, for example, could cover the entire length of the tape (taking minutes). In normal practice, however, the tape is manually (or automatically) positioned to some predetermined point on the tape, whereupon reading or writing proceeds sequentially along the length of the tape.

Data are normally recorded on the magnetic tape one 6-bit character or byte at a time, in either pure binary or BCD form, using seven recording channels (though 9-track has also been used); Fig. 6.33(a) shows the typical format for IBM magnetic tape units. As each character is recorded, a parity count is performed, and a check digit written in the seventh channel. When the tape is read back the check digit is recomputed

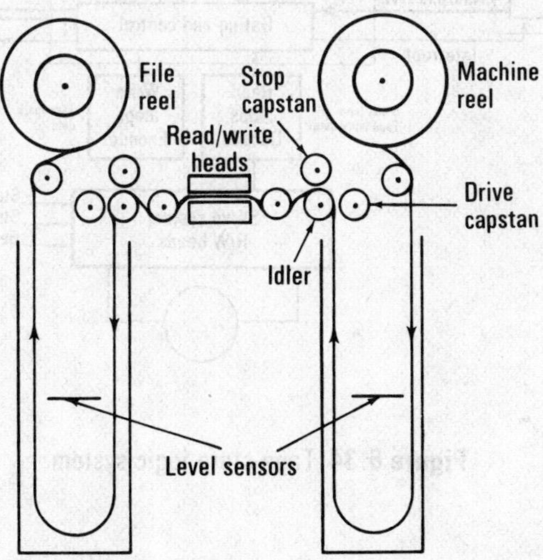

Figure 6.32 Magnetic tape mechanism.

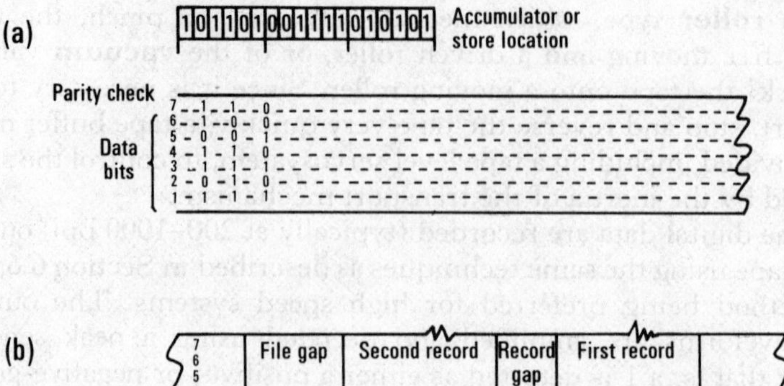

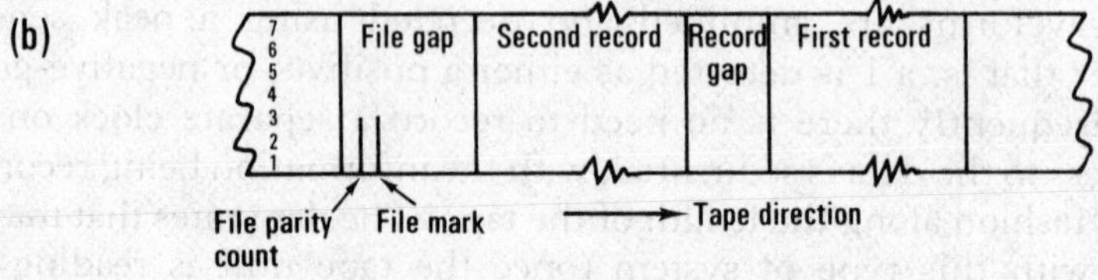

Figure 6.33 Magnetic tape data format.

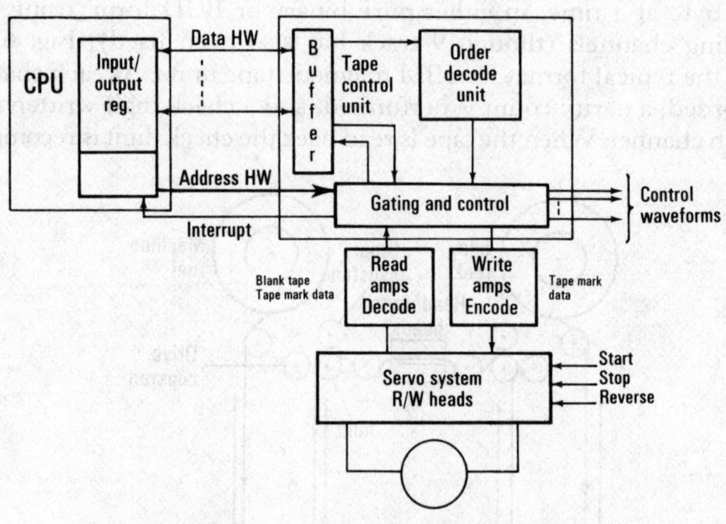

Figure 6.34 Tape store logic system.

Table 6.7

Typical magnetic tape instructions

1. Read BCD/binary, 3 chs/word	– Starts tape mechanism, inputs and assembles data into I/O register, checks parity; reading continues until a no-data gap is reached.
2. Write BCD/binary, 3 chs/word	– Starts tape mechanism, outputs contents of I/O register one word at a time, inserts parity. Record gap written at end of operation.
3. Write end of file	– Starts tape mechanism, after appropriate delay writes file mark and file parity count on the tape.
4. Space forward one file/one record	– Starts tape mechanism and searches for record gap/file mark.
5. Back space one file/one record	– Starts tape mechanism and searches in reverse direction for record gap/file mark.
6. Rewind	– Starts tape mechanism and rewinds.

and compared with the recorded check digit, any discrepancy being signalled as a tape error. In addition to the vertical count, a horizontal count is also performed and written, when requested, at the end of a block of information. A character, or group of characters, written on the tape is referred to as a **record**. There is no restriction on the length of a record, but if more than one record is placed on a tape a blank portion of tape known as a **record gap** must be placed between them. (The record gap is automatically inserted at the cessation of the write operation.) Records may be grouped together to form **files**, which must also be separated by blank tape called a **file gap**, and files are also identified by writing a special **tape mark** character at the end of the file. The tape mark character or gap (or both) is used to differentiate between the records and files written on the tape (see Fig. 6.33(b)). Tapes are written or read in the forward direction only but the tape may be back-spaced or forward-spaced by one block or file if so desired (control of the spacing operations being governed by the record and file gaps).

A typical set of instructions required to control the operations of the magnetic tape store is shown in Table 6.7. Each instruction must initially start the tape unit, assuming it is stationary, and thereafter the reading and writing proceeds continuously. However, because information is transferred to magnetic tape one word at a time it is necessary to repeat the read/write instructions to maintain continuous operation. After each read or write operation the tape comes to rest on a blank portion of the tape and this effective time lag can be used to ensure that the tape mechanism is up to full speed before the read/write operations commence.

Since none of these instructions requires a specified store address they may be represented by either the general INA or OTA instruction (suitably coded) or by using the spare zero-address instructions. Alternatively, the tape unit could be treated as a peripheral device (see Chapter 7) and instructions passed to the tape control unit as data words where they would be decoded and obeyed locally. A block diagram for a tape store system using this technique is shown in Fig. 6.34.

The location and addressing of particular blocks of stored data is the responsibility of the programmer. Thus each block or file must be preceded by a reference tag which must be read and examined under program control so that it can be identified. The layout and organization of magnetic tape files is a major problem in the software design of a systems application.

The magnetic tape units described above are referred to as reel-to-reel devices to distinguish them from the low cost Phillips type cassettes originally developed for audio use and later modified and enhanced for digital applications. The magnetic tape cassette is a small, convenient, low cost, re-usable, non-volatile medium that can easily store 100–550k bytes in a 2-track format, recorded in alternative directions of the tape; cassettes come in a number of tape lengths, the standards being 150, 300, and 450 feet of 0·15″ wide tape packaged in a plastic case 4″ × 2·5″ × 0·5″. Cassette devices provide a cheap storage medium for small, normally microcomputer based, systems such as intelligent terminals, data collection and data entry systems and small business systems. Moreover most LSI manufacturers provide a standard interface chip for cassette type devices, for use with microcomputer systems.

In contrast to many other forms of storage magnetic tape, and in particular cassettes, have been the subject of internationally agreed standards. The American National Standards Institute (ANSI) standard for cassette tape specifies 800 bits per inch and phase modulation encoding. Information is recorded in bit serial mode, least significant digits first, in the form of bytes; blocks of data can vary from 4 to 256 bytes long separated by an inter-record gap with a specified preamble and postamble for each record. NRZ1 encoding is also used but this necessitates a recorded clock track to overcome speed fluctuations, which restricts the system to single track data recording.

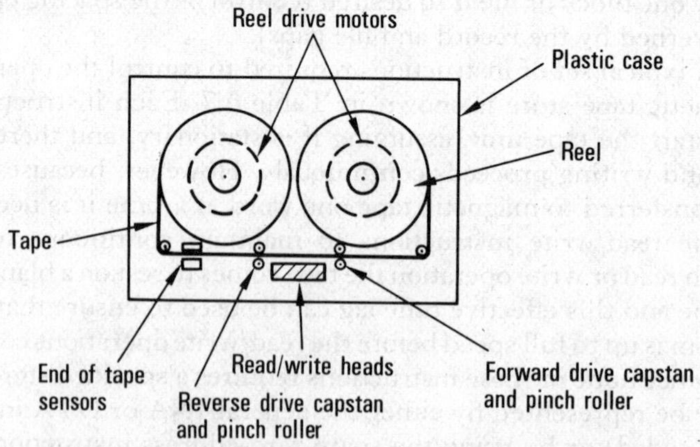

Figure 6.35 Cassette drive unit.

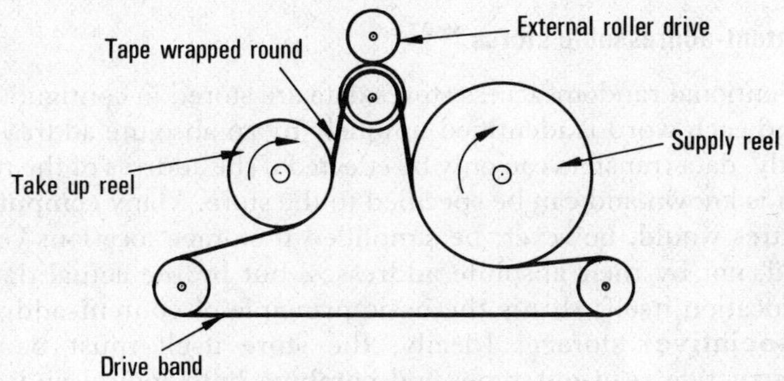

Figure 6.36 Cartridge drive unit.

The type of tape transport used is either reel-to-reel, using two reel drive motors, or capstan drive; capstan devices are usually more complex but have a faster response time. In the drive unit shown in Fig. 6.35 two contrarotating capstans are used driven by constant speed motors, in conjunction with two reel motors (single drive capstan units are also used). Each capstan supports a high inertia flywheel to reduce short term fluctuations and the constant capstan speed provides a linear tape velocity throughout the length of the tape. The reel-to-reel system employing two motors introduces a variation in tape speed which reduces the maximum possible storage density. This variation in speed can be compensated for by using a separate pre-recorded clock track (thereby sacrificing the possibility of using a second data track). During operation the clock track's output frequency is continuously compared with a reference frequency and the error signal used to control the speed of the drive motor. Note however that using this system limits the interchangeability of the cassettes, since two track recording is usually accepted as standard.

Data rates for tape cassettes are rather low, the drive speeds ranging from 4–20 inches per second with typical transfer rates in the order of 1–2k bytes per second. The access time may be considered as comprising a **response time** which is the time taken to respond to a command when the unit is in the required read/write position (in the order of 15–20 ms) and the **search time** which for random block distribution and assuming a 300 ft tape ranges from 25–100 ms.

An alternative magnetic tape store which falls between reel-to-reel and cassette type systems is the **cartridge** tape. The transport system for a cartridge tape unit, shown in Fig. 6.36, is considerably simpler than the cassette transport and uses an elastic belt drive system, which both drives and tensions the tape. Though the cartridge is more expensive than the cassette its storage capacity is some 4–8 times greater and is inherently faster in operation due to the tape being wider and in one long continuous loop. Typical operating speeds are 30 inches/sec, recording at 1600 bpi using a PM encoded bit serial format; the transfer rate is in the order of 48k bits/sec.

6.9 Content-addressable stores [36,37,38]

In conventional random-access stores data are stored in contiguous locations and each word is identified uniquely by an absolute address. Consequently, data transfers can only be effected if the address of the relevant location is known and can be specified to the store. Many computational procedures would, however, be simplified if storage locations could be accessed, not by their absolute addresses, but by the actual data held in the location itself; this is the basic principle of content-addressable (or **associative**) storage. Ideally, the store itself must be of the non-destructive read-out type, and combine both logical and storage functions. Early attempts at implementing associative stores utilized cryogenic and multi-aperture ferrite cores but today current technology would dictate semiconductor storage employing in the main MOS or bipolar devices,[39,40] though CCD devices have been used. Though the content addressable memory (CAM) is obtainable as a specialised chip from a number of manufacturers (for example: Solid State Scientific Inc. have produced a 64 bit array organized as eight 8-bit words with a 100 ns interrogate time) it is still not as commercially available as other semiconductor storage chips.

The mode of operation is such that a **descriptor** or **tag** specifying all or, as is more generally the case, some of the bits held in a required location would be presented to the store. The tag would be compared simultaneously (in a parallel fashion) with the corresponding bits in all the locations in the store, and the identified word (or its address) would be extracted and placed in a register. The write operation is performed in a similar manner with the exception that, if a location is unspecified, the information would be placed at random anywhere in the store, normally in the first vacant position. To write into a specific location, the identifying tag is first specified to the store and when the address of the required location is found a normal writing cycle is initiated; note that most content-addressable stores can also be used in the conventional manner with specified absolute addresses. (A block diagram for an associative store is given in the solution to problem 6.7 – Fig. S.14.)

One disadvantage of the system is the problem of **multiple-hits**, that is, when more than one location contains the data sought by the search operation. This problem is, of course, aggravated when only a selected number of bits is used for comparison. In this case there is no alternative but to refine the search procedure or to examine the contents of all the relevant locations, a suitable indication of this condition must of course be provided in the associated hardware.

The associative store is ideal for data-handling operations requiring bit-by-bit and symbolic manipulations, this function will be discussed further in Chapter 9. A small store of this type has also found application in *paging* systems, which will be considered in a later section. Unfortunately at the present time the cost of developing a very large content-addressable storage system would be prohibitive.[41]

6.10 Organization and structure of storage systems

Over the last few years CPU speeds have increased by a factor of more than 1000, and the increased ability to handle larger problems has resulted in a demand for significantly greater storage capacity. In fact, storage requirements have increased from 1k words (where k = 1024) to something in the order of 1000k words. Unfortunately the desire for large and fast storage systems to keep pace with CPU speeds are opposing goals, and the only way to resolve this conflict is to ensure that a properly designed storage hierarchy exists in the system.

The ideal solution would be, of course, to have single-level storage, with the entire store consisting of fast random-access storage. This unfortunately is not a practical or an economic proposition and an inevitable resort must be made to multi-level storage systems – for example, a small fast store of up to 32k words capacity (residing in the CPU) acting as a **slave** or **buffer** to a main memory system consisting of a slow speed ferrite core or semiconductor store of up to 10^7 words. The relative speeds (read/write cycle times) of the two storage systems would be in the order of 10:1, with the buffer store 1–3 times slower than the CPU cycle time.

The justification for such a system, (called a **cache** memory) which necessitates the continual transfer of data to and from the main memory, is based on the frequency with which information is used or modified in a computer program – known as the **information value**. For instance it has been stated that in any computer program only 20 % of the storage space contains program, the rest being data. This is a gross oversimplification but nevertheless has some truth, since it is obvious (if one compares program execution time and program writing time) that some of the instructions will be executed many times over owing to loops, etc., in the program. Moreover, the value of the information is not constant with respect to time, since the importance of a particular program (particularly in a time-shared environment) can change from one moment to the next. Thus, if we assume that the principle of information value is valid (which has in fact been proved in practice) then the possibility exists of matching the information and storage hierarchy such that the majority of accesses within a computer system will be to the fastest storage.

The term 'memory hierarchy' is somewhat ambiguous with todays technology since it has been superseded by the general concept of a single level store (also known as **virtual memory**). The two most significant types are the **cache-main memory** and the **main memory-disc** organizations. Note that the former bridges the speed gap between main memory and the processor while the latter overcomes the differences in storage capacities between main memory and back-up storage. While the main memory-disc system is generally referred to as virtual memory (see section 6.10.2) it is important to realize that both systems are designed to produce a single level homogeneous system to the user.

(a) Effective address word (**N** bits).

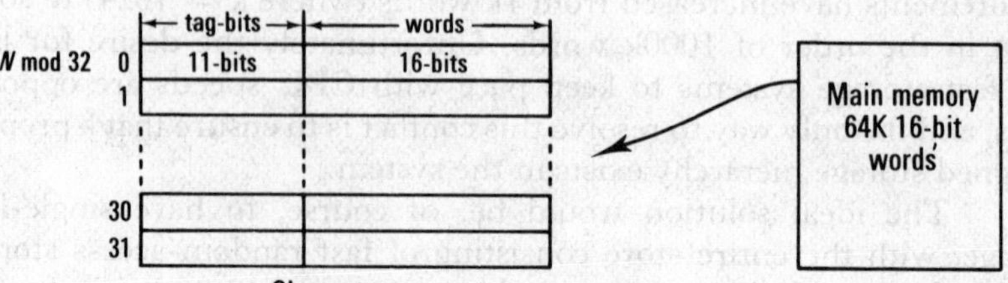

(b) Organization.

Figure 6.37 Slave store technique.

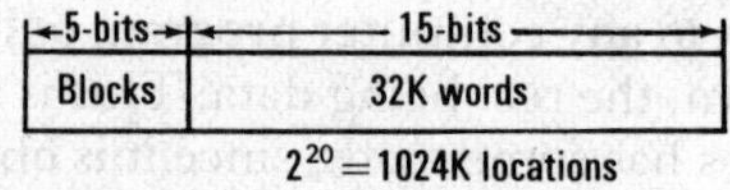

(a) Effective address word.

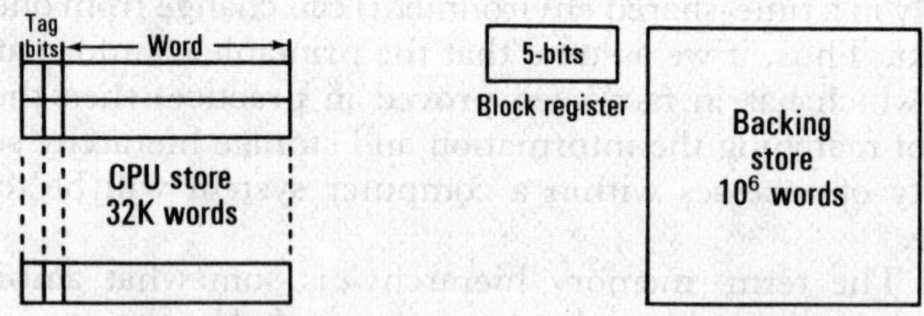

(b) Organization

Figure 6.38 Buffer store technique.

6.10.1 Cache memory systems

The basic ideas can be illustrated very well by considering the slave (or buffer) memory system proposed by Wilkes.[42] Suppose we wish to access *any word* in a large main memory say of 64k words, but wish to cut down the access time (and hence the computation rate) by using a very fast slave store capable of holding only the current instructions that are in use. To attempt a complete block transfer would be prohibitive both in time and organizational hardware, but using the ideas of information value we may obtain the same ends more economically.

This is achieved by making each word of the fast slave memory equal in length to that of the main memory, but with an additional eleven bits referred to as the **tag-bits**. Instructions are read down from the main store *only when required in the program*, and inserted into location N mod 32 (where N is the address in the main store) of the slave, at the same time the 11 most significant digits of N are copied into the tag-bits (see Fig. 6.37). The fast store is always searched first (by comparing the 11 most significant digits of N with the tag-bits) when an instruction is required. If the tag-bits are equal to N, the instruction is taken from the fast store, otherwise it is obtained from main memory and copied into the slave. On a writing operation the slave must be examined first to see if it contains the required location, if so *both* the main store and the slave are updated. Note that a variation of this technique could form an alternative hardware method (in place of software sectorization, see Section 2.5) of expanding the overall storage capacity of a system.

The technique described above is really only applicable to the fast control of a computer but the ideas can be easily extended to the overall control of a large backing store using a normal sized CPU core memory as a slave. Let us assume that the main CPU contains a 32k word store (the *slave*) with a 100 ns access time, and that it is backed by a large core store of 10^6 words with a 1 μs read/write cycle time. In the simplest case the programs would be partitioned into 32k word blocks, with a **block register** in the backing store holding the address of the current block (see Fig. 6.38). Each word of the fast store includes two additional tag-bits which are used to indicate firstly, when a word has been written into the slave store from the backing store, and secondly, if a change has been made to a stored location.

Whenever an instruction is called for in the program the slave store is examined first (under hardware control) to see if it contains the required instruction. This is done by accessing the location that might contain the instruction (using the least significant 15 bits of the effective 20-bit address, that is, the addresses of the words within a block) and examining the tag-bits. If the first tag-bit is set to 1, indicating that the required word is already available in the slave store, no reference is made to the backing store and the instruction is read out from the slave. If, however, the tag-bit equals zero the 15-bit address is added to the contents of the block-register (or the effective address is used) and access made to the backing store; the contents of this location are then written into the slave store and the first tag-bit set to 1. The same process is followed when writing, except that if a word in the slave memory is changed the second tag-bit is set to 1. Note that a word is only copied into the slave store when called for in the program and that, as time goes on, the slave memory will accumulate all the words of a program in active use. When the number in the block-register is about to be changed, so that a new program may become active, a scan of the slave memory is initiated and if the condition '11' is found in the tag-bits (that is, *both* set to 1) the word is copied into its appropriate place in the backing store. The tag-bits associated with the fast slave could be stored in the CPU hardware using active memory, this would require (in our example) 1024 words each of 64 bits, with an access time in the order of 100 ns.

The concept may be extended further so that the fast store acts as a slave to more than one 32k block in the backing store. This would necessitate providing a separate block register for each program block address, and increasing the number of tag-bits for each word in the fast store to enable the address of the block register to be defined. Suppose, for example, seven block registers are used, then four tag-bits will be required. The first three bits are used to specify the address (001 → 111) of the block register, with the all zero combination indicating that the required word must be obtained from the backing store. The fourth bit is used to note whether or not a word has been altered in the slave store.

At any given time one of the seven blocks of program is actively being used, and whenever an instruction is required the fast store is searched first. This is done by examining the tag-bits of the word in the corresponding position in the slave store. If the first three bits are all 0 the word is obtained from the backing store (using the address in the active block register) and written into the slave store. Non-zero tag-bits are compared with the active block register and if they agree the word is read down from the slave store. If the bits do not agree, the fourth tag-bit is examined and if this is 0 the reading proceeds as before, overwriting the previous contents of the slave store. If, however, the fourth bit is 1, the contents of the location in the slave store must be written into the backing store before performing the read operation. (Alternatively, a software procedure may be initiated each time the active block number is changed; the function of this is to scan the fast store and write into the backing store all those words belonging to the displaced block which have been altered, that is, those for which the fourth tag-bit is 1). A similar procedure is followed for the write operation, except in this case the fourth tag-bit is always written to 1 when a word in the fast store is overwritten.

These ideas, and others, have been successfully incorporated into the storage structure of the IBM 360/85 computer system.[43] In this case, called the sector-buffer technique, both the backing and buffer storage are logically divided into sectors consisting of 1k bytes. A sector from the backing storage can be mapped into *any* sector of the buffer store, which has a total capacity of 16 sectors. A 14-bit tag address is associated with each sector of the buffer in order to identify which sector of the backing store that it currently contains (see Fig. 6.39). The tag-bits are implemented in the same register technology as the CPU, thereby enabling a fast simultaneous comparison (that is, an associative look-up) to take place whenever a data transfer is required.

Transfers from the backing store to the sector buffers are performed a block at a time, on a demand basis, each block consisting of 64 bytes of information. Thus each sector buffer is further divided into 16 congruently mapped blocks, each with an additional tag-bit (the **validity** bit) to indicate when the block has been referenced and filled from the backing store. One, and only one, sector from backing storage can be mapped at any given time into each buffer sector. All transfers into the sector buffers are ordered using an activity list, with the sector at the top of the list being the one most recently referenced, and so on in order down the list. When another sector is referred to it is moved to the top of the activity list and all intervening ones are moved down one position; new sectors are assigned to sector buffers at the bottom of the activity list. Note that the operation of an activity list does not involve any physical reorganization of the contents of the sector buffers, it is purely a logical control mechanism.

The reading procedure is similar to that already described in the last section, the buffer store is always searched first using the 14-bit sector tag, followed by the block address (both specified in the effective word address). If the block is in the buffer store the required word is accessed and transferred to the CPU, otherwise the main backing store must be accessed and the appropriate block written into the sector buffer, setting the validity bit to 1. If, however, the initial sector search yields a blank, a new sector must be initiated in the buffer and the specified block transferred from the backing store to the sector buffer, marking the validity tag as before. To write into a location the same search procedure is followed, but in this case the backing store is always updated. The main disadvantage of this associative mapping approach is the amount of hardware involved and the time required to search through the sector tags. However, it is possible to speed up the searching routine by using associative storage techniques, implemented with fast active devices such as semiconductor storage.[44]

Other buffer methods have been considered for the IBM 360 system, notably the **direct mapping technique**,[45] described by Scarratt[46] and very similar in principle to the Wilkes scheme discussed above. In this method both the backing and buffer stores are divided into blocks, and each block has its own specific tag. In this scheme if there are k blocks in buffer storage, then every kth block from the backing store may be mapped into a specific block of the buffer. For example, from Fig. 6.40 blocks, 0, 128, 256, 384, etc. of backing store would all be mapped into the same block of the buffer store. To achieve this, the 7-bit block portion of the effective address is used to define specifically the *only* block in which the addressed block could reside in the buffer store. The 11 most

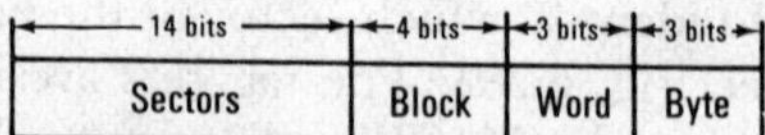

(a) Effective address word.

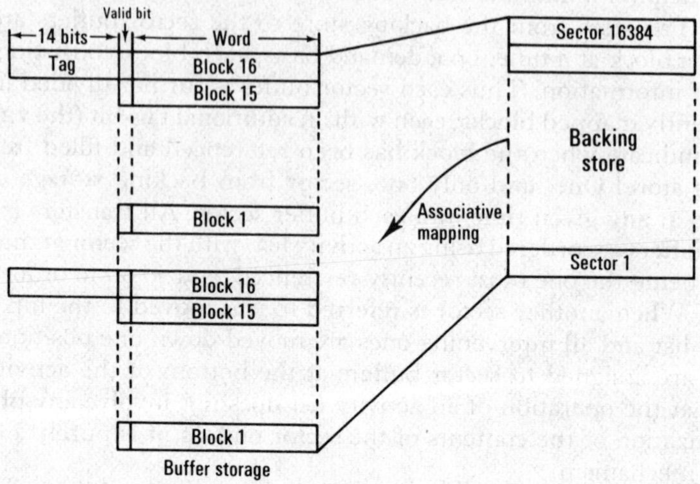

(b) Organization.

Figure 6.39 Sector buffer system.

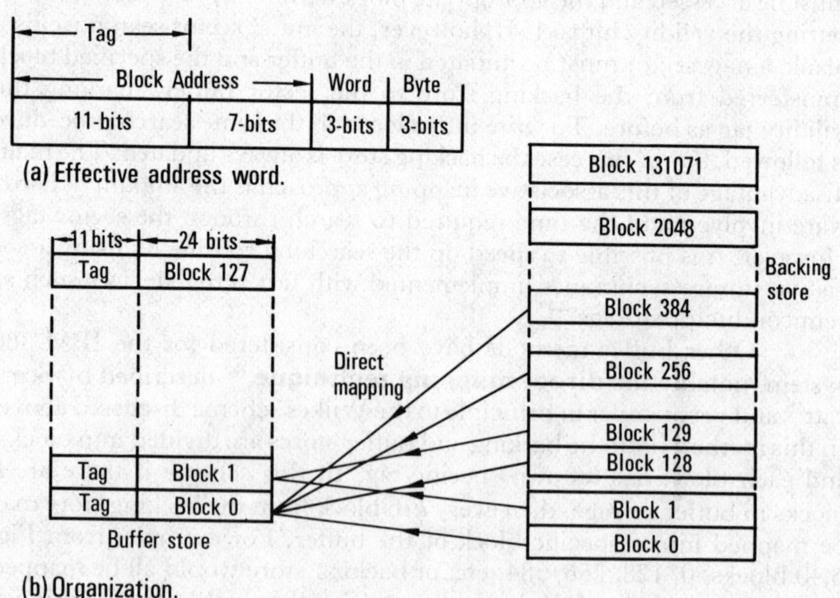

(a) Effective address word.

(b) Organization.

Figure 6.40 Direct mapping buffer.

significant bits are used as a tag for each block (they are in fact written at the time of transferring from the backing store to the buffer) and their function is to indicate whether or not the correct data are in the block. For example, when called for by the program, the block containing byte 10111011010 101 1010 110 101 would be transferred from the backing store, using the 18-bit block address, to block 90 of the buffer and the tag-bits written to 10111011010. The next time the program required these data the buffer store would be accessed first at block 90, and the tag-bits compared with the effective address to see if the required data were present. The main advantage of this method is that simultaneous access is made to the required data and tag, and no associative comparisons or searches need to be made. Note however that contention problems can arise if two blocks in the backing store which share the same blocks in the buffer are required on successive memory references.

6.10.2 Virtual memory systems

One of the most economical ways of operating a large computer system, while at the same time allowing direct user contact with the machine ('hands-on time') is in the **multi-programming** or **time-shared** mode.[47] With this technique the computing resources are switched from one user to another, in such a way that each user can interact with the computer from his own personal teletype terminal, under the impression that he has sole access to the machine. (With processing power and semiconductor storage now becoming available at very low cost the need for time-sharing systems could easily diminish, since small, personal desk-top computers could satisfy many user requirements.)

The design of a time-shared system places particular responsibility on the storage organization of the computer complex. For example, in the simplest case, it would be quite impractical to let the user write his programs using absolute memory addresses, since this would assume that it was known in advance which users were to be time-shared together. Also any one program mistake could corrupt all the others; these remarks incidentally also apply to *any* group software project. Thus, it is essential to have efficient **dynamic** allocation of the main memory, which means in practice some form of indirect addressing system.

Until recently the standard computer systems model viewed the arithmetic and logic unit as residing in the centre of the system, hence the use of the term central processing unit (CPU). This situation has now changed, and most modern systems design considers the storage unit as the central resource (see Fig. 6.41). In a computer system which allows several independent programs (or processors) to share the memory, actual storage locations can only be allocated during program running time, that is, dynamically. Thus it is necessary for programs to be coded (or compiled) in terms of a **virtual** store,[48] and for the computer system dynamically to map the virtual address space into the real, physical address space.

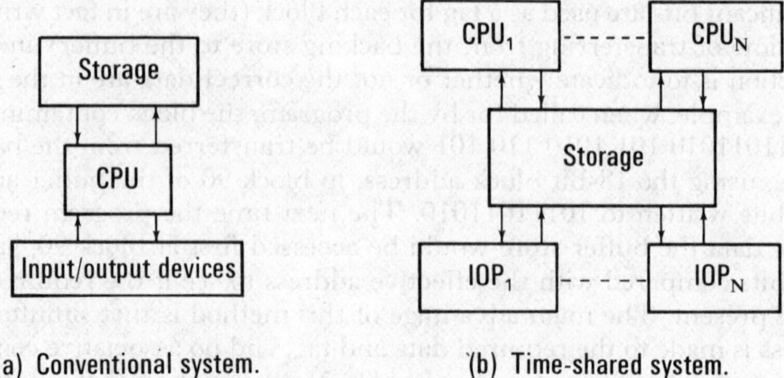

Figure 6.41 Computer systems structure.

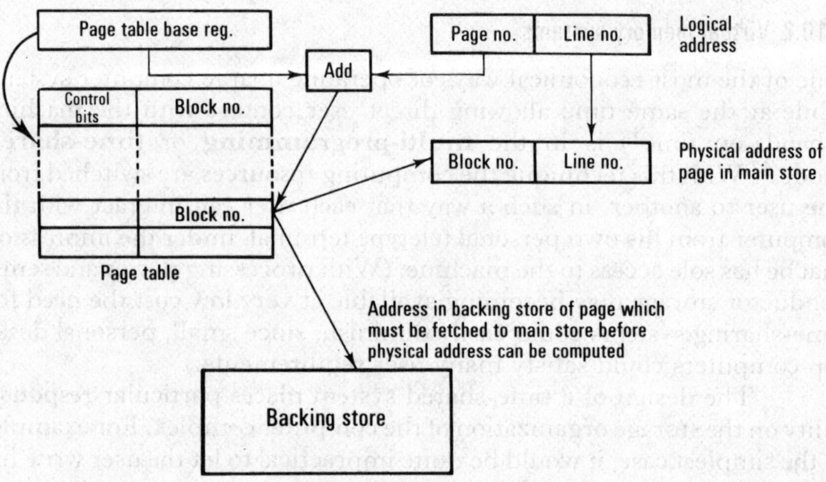

Figure 6.42 Paging system.

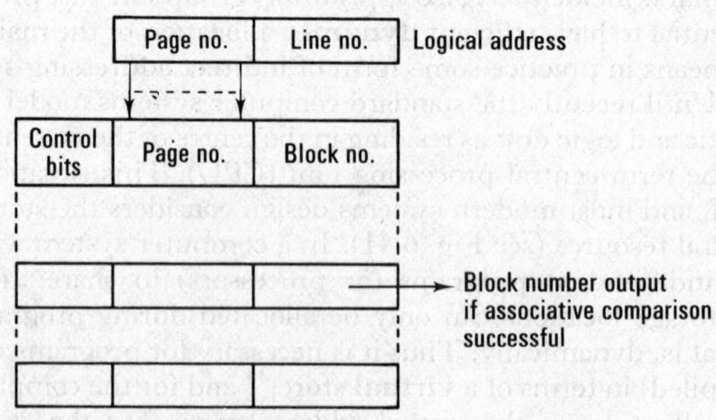

Figure 6.43 Associative page map.

This concept of address space needs clarification. **Addressing** is the means by which a process or program distinguishes between the storage locations in its address space. We may define the two address spaces which exist in the following way:

1 The **physical-address** space consists of the set of actual main memory locations which are *directly* addressable.

2 A **virtual** or **logical address** space consists of the set of abstract or logical locations addressed by a process.

The logical address space, the locations of which are usually identified by a number called the virtual address, consists of all the storage associated with the system, that is, main memory plus any auxilliary storage including drum, disc, core, etc. If the logical address space is larger than the physical address space, the term virtual memory is employed.

In order for a process to be run on a hardware processor at least some of the instructions and data must reside in the physical main memory. Thus the major design problem in time-shared memory systems is to provide a suitable logical storage space and a mechanism to translate logical address references into the actual addresses of the instructions and data in the main store. Moreover, it is also essential to provide memory protection, to ensure that each individual program is free from interference from other users. The translation and protection processes may be performed using either software or hardware techniques, but we shall concern ourselves primarily with the hardware implementation of the translation process.

There are three ways in which present computer systems may perform the mapping process, they are:

1 The procedure may be written as an operable program with all locations specified in the physical memory; the result is an **absolute** program which is always assigned to the same place in store.

2 When the program is loaded into the computer; this is known as **static relocation**.

3 When the program is being executed in the computer; known as **dynamic relocation**.

Programs using static relocation are usually assembled as if they were to be stored in location zero onwards, with succeeding instructions and data being placed in contiguous locations. During the assembly of the initial source program all instructions or data words with address references are marked by the assembler, and a relocatable object tape produced. When this object tape is loaded into the computer the location of the first word in the program, called the **base** (or **block**) address, is specified and the loader adjusts all the address references accordingly, before storing the instructions. Thus, using static relocation, a program can be initially loaded anywhere in the store. However, if the program is transferred to auxilliary storage and then retrieved (for example, swapping one user's program for another) it must always be returned to the same location in store. This leads to difficulties, since the same contiguous block of store may not always be available when the program is required, hence the need for a dynamic relocation scheme.

One of the simplest and most common methods of dynamic relocation employs a **base register** to modify the address of each store access cycle. Programs (written as though they commenced in location zero) can be loaded into any contiguous block of the store simply by changing the contents of the base register. One advantage of this technique is that simple program protection may be implemented by setting a limit on the maximum address that can be used in a program block. Thus the absolute program addresses must range from the base value up to base value plus the maximum address; it is a relatively simple task to check if an address exceeds the limit.

There are many variations of the base register approach; for instance, the flexibility of the system may be increased by using two registers. One such technique is to make instructions refer to their own program blocks via one base register, but to reference stored data through a second register. Control of the base registers may be performed using either software (as in the IBM 360 system where base registers can be directly accessed by program) or hardware techniques. The size of the logical address space using static or dynamic relocation with base registers is usually equal or less than the physical address space. In practice, a larger physical address space can often be simulated by *overlaying* a portion of program not immediately required with new program from auxilliary storage.

One of the problems that exists with the base register technique is the question of memory utilization. To transfer a block of program from backing storage to the main store a free contiguous block of storage must be available. If this is not the case, as often happens, the resident programs must be swapped or compacted together to make room for the new program, resulting in wasted time and storage space.

These disadvantages may be overcome by dividing the program and main storage into small blocks (**pages**) which can be located anywhere in the main memory. Moreover, as we shall see, **paging techniques** allow a logical memory space larger than the physical memory space to be implemented economically. In a paged system the physical memory is considered to be divided up into blocks of a fixed size, preferably 128, 256, or 512 words, with the term 'page' also referring to equal sized units in the logical memory. Addresses in this system can be represented by two numbers, a page address and a line (within a page) address. For a machine with an N-bit address field the most-significant M digits are considered to be the page address, and the remaining least-significant $N-M$ digits is the line address.

One of the important concepts in relocation systems is the idea of a **memory map**, which translates logical address space into physical address space. In the static relocation technique this function is performed by the loader program, while in the base register method the registers themselves are the map. In dynamic relocation using paging a **page map** or **table**, which can be considered as a set of multiple base registers (implemented in either hardware or software form), is used to perform this function. A typical paging system is shown in block diagram form in Fig. 6.42. Each program must have its own page-table, which is contained in the main memory, and a page-table base-register is used to point to the corresponding page-table for the active program. The block address in the main store corresponding to a given page is found by a table-look-up procedure, using the page number in the logical address (modified by the base-register) as the key. Control bits are used in the page-table to indicate whether the required page resides in the main memory or in the backing store. The address in the block portion of the page-table may refer either to the actual starting address in the main memory or to the location of the block in the backing store. In this way the logical address space can be made larger than the actual physical storage.

The major disadvantage of the system outlined above is that all store references require an additional store access in order to retrieve the relevant block number from the page-table. One method of reducing this limitation is to use a hardware associative slave store in conjunction with the page-table. The associative map (see Fig. 6.43) can be addressed by content rather than by an explicit address by comparing the page number of the logical address, simultaneously, with all the page numbers in the table. If the page number is found, the physical address is formed by concatenating the line and block numbers. If, however, the comparison is unsuccessful, reference must be made to the page-table in the main store, and the new page number-block number pair inserted in the associative map. This technique is identical in concept to the slave buffer idea working on the principle of information value, as discussed in the introduction to this section.

An alternative approach to dynamic relocation is the **segmentation** method[49, 50, 51] which employs a variable page size or **segment**. Though paging techniques are often used to implement this method there is however a fundamental difference between paging and segmentation. Paging is concerned with the allocation of physical storage space while segmentation is primarily concerned with the allocation of logical space. One of the major problems with simple paging systems using base registers is logical memory allocation, particularly if program and data are to be shared and data structures allowed to contract and grow at will without being explicitly reserved in the program.

Paged systems eliminate the constraints imposed by static relocation (that is once the program is loaded absolute physical addresses are entered in the program) by introducing mechanisms which allow physical addresses to be relative to a base number or block address the contents of which are set at *execution* time. However the particular base register value, for instance, is still required to be specified in the program and thus once loading is complete absolute *logical* addressing is implied. The main advantage of segmentation is that it enables *relative* addressing to take place within the logical memory space. Moreover, from the users point of view it allows a natural structuring of the code and data elements of a program with the allocation of corresponding code and data segments, and the use of conceptually independent areas of data which can vary dynamically during run-time. Thus, a segment is a self-contained logical entity of related information defined and named by the programmer, such as a procedure, data array, symbol-table or push-down stack. Inherent in this approach is the protection of data; since each segment contains items with similar attributes and requiring common processing the **segment description** can specify whether or not the segment can be read, written into, executed etc. The segment descriptor is generated by the operating system at the time when the user creates the segment, and from then on is under the control of the operating system.

Returning to the idea of relative addressing within the logical address space, a segment (S) can be considered as an ordered set of data elements (x), usually computer words. The symbolic name S is translated at *run-time* (by the operating system) into a base register number, and the symbolic data element name x into a relative location within the segment. The process of mapping symbolic segment address pairs (S/x) to physical locations requires a **descriptor base register** which points to a **segment-descriptor table**, and a **procedure base register** which contains the generalized segment address, as shown in Fig. 6.44. The procedure base register contains the segment number (a transformation of the segment name S obtained at execution time) and the location in the segment of the procedure about to be executed. The segment descriptor table is in effect an array of base registers, the contents of which are combined with the relative location held in the segment address to

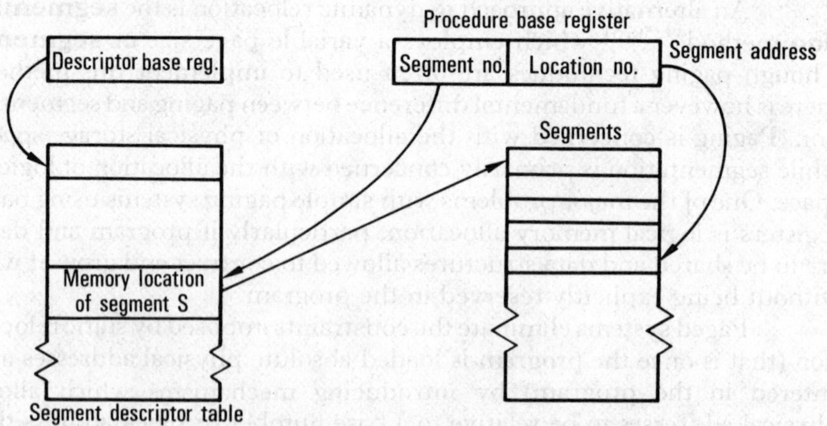

Figure 6.44 Segmentation - addressing of physical space.

yield the address in physical space. The use of a descriptor base register enables the segment table to be relocatable thus allowing the segment descriptor table to be handled like any other segment in the system.

A major disadvantage of segmentation is that it tends to be wasteful of physical memory space since a segment normally requires contiguous storage locations in main memory and the segments can all be of varying lengths. Consequently the transfer of a new segment into main memory depends on locating an empty and contiguous region of the correct size. Thus though there may be many empty areas available in the store they cannot be used. The fixed page organization is more efficient since though contiguous storage is still required, each page slot is of a fixed size which makes the task of finding empty space in the main store considerably easier.

The concept of virtual memory, or single level storage as it was known, was first introduced on the Manchester ATLAS computer. This was quickly followed by the Burroughs B5000 machine which also used segmentation, the ideas being considerably enhanced in the later B6000/B7000 series[52]. A segmented virtual store with paging was also employed in the MU5 and ICL 2980 computers.[53] The Plessey system 250[54] also uses a segmentation approach employing special segment descriptors called **capability registers** which provide sophisticated protection properties for use in real-time message switching systems.

The term virtual memory is often confused in the literature with memory mapped systems which allow computers, restricted by their architecture to some upper limit of direct addressing of main memory, to have extended addressing capability (and hence access to a larger physical memory). For instance, the PDP 11/40 machine has a basic 16-bit wordlength which allows direct addressing of up to 32k words. However, since the addressing logic has been designed to handle 18-bit words it is possible to extend the main memory up to 128k words. This is achieved using a **memory management unit** in which the 16-bit address is considered as a 'virtual address' and used to construct a new 18-bit physical address using a base register technique. Note that with this method the concept of a one level memory employing the *total* addressing capability of an architecture, to provide a logical address space larger than the physical main memory, does not apply, hence the system is not strictly speaking a virtual memory.

6.10.3 Real-time system design considerations[55, 56]

The type and capacity of the storage required for a given digital control system application is generally dictated by the operational specification but there are some basic guiding principles which can affect the final choice. Let us consider the question of program storage; this is all important in any real-time system application as the loss of a control program (due to storage faults such as dropping or gaining bits, or overwriting, etc.) can have disastrous consequences. Consequently, the use of any form of volatile store for program storage must be avoided. Program protection in this case is vital and can be accomplished by using permanent storage such as a magnetic drum or disc and ensuring that only the read circuits function (by inhibiting the writing circuits). With the availability of semiconductor memory however ROM's are being almost exclusively employed for the storage of control programs. It is essential of course that the programs should be properly debugged before committing them to ROM, and suitable software aids must be available for testing and simulation and the preparation of the detailed ROM coding. In many microcomputer based systems the demarcation between ROM and RAM becomes an essential design parameter and if a high level language is used (normally available as a cross-compiler on a separate computer) it must be able to handle this allocation problem.

In many systems, for example air traffic control, some form of permanent record storage is essential. This provides a continuous record of system operation which may be used, in case of accidents or process break-down, for post-mortem enquiries. The storage capacity required for this function depends on the system and the time which can safely elapse before the process or procedure becomes dangerous. Short-term storage can often be cheaply provided by a continuous loop of magnetic tape or magnetic cartridges.

A further use of storage in a control system environment (as we have already seen) is to provide buffering between equipment working at different speeds. In general, centralized random-access buffer storage with time sharing facilities provides an economical solution to most systems peripheral timing problems (see Chapter 7). Moreover, in many real-time applications it is possible to save time (which in some cases can mean that a particular operation becomes a feasible proposition!) by the inclusion of extra buffer storage. It is also essential that the main storage system be expandable in modules, thus providing a safety factor to allow for the contingency of the computers originally proposed for system control running out of program computation time.

The major problem in the design of real-time computer control systems is how to perform the required control functions in a specified time cycle, consequently processing speed is always at a premium. For this reason virtual memory systems, designed essentially for multiprogramming requirements, can be more noxious than advantageous. It is essential, for example, to be able to minimize the number of backing-store transfers, because of the long access times required, and to estimate the running time of processes – functions which are difficult to perform when virtual memory systems are employed. Consequently the tendency is to use extended main memory whenever possible and to write special purpose store management software.

Program and data protection is an essential requirement for real-time systems and consequently the storage system used must have some form of error detecting (and correction) property. In the simplest case this would involve hardware circuits to perform simple parity checks or more sophisticated testing using cyclic codes, such as the Hamming code, on the transfer of data to and from the storage devices. As we have seen protection can also be provided at the systems level using a segmentation approach employing special segment descriptors.

The current approach to real-time system design is to use distributed processing (see chapter 9) with dedicated processors and storage devices connected to a shared highway system. In this case both local and global storage is required, with a common shared store acting as a 'post-box' to effect communication and data transfers between processors. The problems of protection are even more important in systems of this type, for example contention can occur between processors requesting access to the same store, and some form of system protection is essential.

Storage devices may also be used in control systems design as a means of combining logical operations with the memory function – for example, the generation of complex functions, using table-look-up techniques implemented with ROM devices. Another example occurs in the case of input/output devices, which require both buffer storage and data-handling operations such as decoding and error-detecting, etc. In many cases, the necessary logical operations can be conveniently and cheaply performed by a simple extension of the storage device. For instance, serial to parallel conversion with parity checking of a teleprinter input could easily be combined with the necessary buffer storage.

The majority of storage problems encountered in the design of real-time computing systems are connected with the control and operation (including buffering) of the peripheral devices. This aspect of storage organization (for example, the control of backing stores) which is pertinent to all computer systems, will be dealt with in detail in the next chapter.

References and bibliography

1 *Scientific American*, Special issue on Microelectronics, September 1977.

2 Hodges, D. A. A Review and projection of Semiconductor Components for Digital Storage, *Proc. IEEE*, **63**, August 1975, 1136–1147.

3 Cohen, M. S. and Chang, H. The Frontiers of Magnetic Bubble Technology, *Proc. IEEE*, **63**, August 1975, 1196–1206.

4 Chen, D. and Zook, J. D. An Overview of Optical Data Storage Technology, *Proc. IEEE*, **63**, August 1975, 1207–1230.

5 Haughton, K. E. An Overview of Disk Storage Systems, *Proc. IEEE*, **63**, August 1975, 1148–1152.

6 Gillis, A. K., Hoffmann, G. E. and Nelson, R. H. Holographic memories – fantasy or reality?, *Proc. AFIPS NCC*, **44**, 1975, 535–539.

7 Kramme, F. Standard read-only memories simplify complex logic design. *Electronics*, **43**, January 1970, 89–95.

8 Rege, S. L. Cost, performance, and size tradeoffs for different levels in a memory hierarchy, *IEEE Computer*, **9**, April 1976, 43–51.

9 DeAtley, E. The Big Memory Battle: Semis take on Cores. *Electronic Design*, **18**, July 1970, 70–77, 113–114.
10 Renwick, W. A magnetic core matrix store with direct selection using a magnetic core switch matrix. *Proc. I.E.E.*, 1957, **104**, 436–444.
11 Rhodes, W. H., L. A. Russel, F. E. Sakalay, and R. M. Walen. A 0·7 microsecond ferrite core memory. *IBM J. Res. Dev.*, 1961, **5**, 174–182.
12 Gilligan, T. 2½D high speed memory system – past, present, and future. *IEEE Trans. Electron. Comput.*, 1966, **EC15**, 475–485.
13 Feth, G. C. Memories: Smaller, Faster and cheaper, *IEEE Spectrum*, **13**, June 1976, 36–43.
14 Allan, R. Semiconductor Memories, *IEEE Spectrum*, **12**, August 1975, 40–45.
15 Wallmark, J. and Carlstedt, L. *Field-Effect Transistors in Integrated Circuits*. Macmillan, London, 1974.
16 Mrazek, D. MOS delay lines. *Digital Systems Handbook*. NS Inter-National, 1969, 139–146.
17 Crouch, H. R., Cornett, J. B. and Eward, R. S. CCD's in Memory Systems Move into Sight, *Computer Design*, **15**, September 1976, 75–80.
18 Panigrahi, G. The Implications of Electronic Serial Memories, *IEEE Computer*, **10**, July 1977, 18–25.
19 Salzer, J. M. Bubble Memories – Where do we stand?, *IEEE Computer*, **9**, March 1976, 36–41.
20 Juliussen, J. E. Magnetic Bubble Systems Approach Practical Use, *Computer Design*, **15**, October 1976, 81–89.
21 Chen, Y. S. and Nelson, T. J. Maximum Frequency of Propagation of Magnetic Bubbles by Permalloy Overlay Circuits, *Journal of Applied Physics*, **44**, 1973, 3306–3309.
22 Renwick, W. and Cole, A. J. *Digital Storage Systems*. Chapman and Hall, London, 1971.
23 Stone, H. S. Dynamic Memories with enhanced data access, *IEEE Trans. Computers*, **C21**, 1972, 359–370.
24 Aho, A. and Ullman, J. Dynamic Memories with rapid random and sequential access, *IEEE Trans. Computers*, **C23**, 1974, 272–276.
25 Stone, H. S. Dynamic Memories with fast random and sequential access, *IEEE Trans. Computers*, **C24**, 1975, 1167–1174.
26 Stone, H. S. Parallel Processing with the Perfect Shuffle, *IEEE Trans. Computers*, **C20**, 1971, 153–161.
27 BhandarKar, D. and Juliussen, J. E. Tutorial: Computer System Advantages of Magnetic Bubble Memories, *IEEE Computers*, **8**, November 1975, 35–40.
28 Stone, H. S. The Organisation of Electronic Cyclic Memories, *IEEE Computer*, **9**, March 1976, 45–50.
29 Hoagland, A. S. *Digital Magnetic Recording*. John Wiley, New York, 1963.
30 Whitehouse, A., and L. Warburton. Information coding on magnetic drums and discs. *I.E.E. Conf. Computer Science and Tech.*, 1969, Pub. No. **55**, 237–245.
31 Williams, F. C., T. Kilburn, and G. E. Thomas. Universal high-speed digital computers: a magnetic store. *Proc. I.E.E.*, 1952, **99**, 95–106.
32 Hecht, M. and Guida, A. Delay Modulation, *Proc. IEEE*, **57**, July 1969, 1314–1316.
33 Patel, A. M. Zero Modulation Encoding in Magnetic Recording, *IBM Journal of Res. and Dev.*, **19**, 1975, 366–378.
34 Davis, S. Update on Magnetic Tape Memories, *Computer Design*, **13**, August 1974, 127–140.
35 Hobbs, L. C. Low Cost Tape Devices, *IEEE Computer*, **9**, March 1976, 21–29.
36 Hanlon, A. G. Content-addressable and associative memory systems – a survey. *IEEE Trans Electron. Comput.*, 1966, **EC15**, 509–521.

37 Minker, J. An Overview of Associative or content-addressable memory systems and a KW1C index to the literature 1956–1970, *Computing Reviews*, **12**, October 1971, 453–504.

38 Parhami, B. Associative Memories and Processors – an overview and selected bibliography, *Proc. IEEE*, **61**, June 1973, 722–730.

39 Carlstedt, G., Peterson, G. P. and Jeppson, K. O. A Content Addressable Memory Cell with m.n.o.s. transistors, *IEEE J. Solid-State Circuits*, **SC8**, 1973, 338–343.

40 Lea, R. M. Low Cost High Speed associative memory, *IEEE J. Solid-State Circuits*, **SC10**, 1975, 179–181.

41 Lea, R. M. The Comparative Cost of Associative Memory, *The Radio and Electronic Engineer*, **46**, 1976, 487–496.

42 Wilkes, M. V. Slave memories and dynamic storage allocation. *IEEE Trans. Electron. Comput.*, 1965, **EC14**, 270–271.

43 Liptay, J. S. Structural aspects of the system/360 model 85-II the cache (high speed buffer). *IBM Systems J.*, 1968, **7**, 15–21.

44 Edwards, D., D. Aspinall, and D. Kinniment. Associative memories in large computer systems. *Proc. IFIPS Congress, Hardware 1*, 1968, D86–90.

45 Conti, C. J. Concepts for buffer storage. *IEEE Computer Group News*, 1969, **2**, 9–13.

46 Scarrott, G. G. The efficient use of multi-level storage. *Proc. IFIPS Congress 1965*, Spartan Books, Washington, 137–141.

47 Watson, R. *Time Sharing System Design Concepts*. McGraw-Hill, New York, 1970.

48 Denning, P. Virtual memory. *Computing Surveys*, 1970, **2**, 153–189.

49 Morris, D. and G. Detlefsen. An implementation of a segmented virtual store. *IEE Conf. Computer Science and Tech.*, 1969, Pub. No. **55**, 63–71.

50 Dennis, J. R. Segmentation and the Design of Multiprogrammed Computer Systems, *JACM*, **12**, 1965, 589–602.

51 Hoare, C. A. R. and McKeag, R. M. Store Management Techniques, *Operating System Techniques* eds. C. A. R. Hoare and R. H. Perrott. Academic Press, New York, 1973.

52 Organick, E. I. *Computer Systems Organisation : The B5700/B6700 Series*. Academic Press, New York, 1973.

53 Morris, D., Detlefsen, G. D., Frank, G. R. and Sweeney, T. J. The Structure of the MU5 Operating System, *Computer Journal*, **15**, 1972, 113–116.

54 Fabry, R. S. Capability-Based Addressing, *Comm. ACM*, **17**, 1974, 403–412.

55 Freedman, A. L. and Lees, R. A. *Real-Time Computer Systems*. Crane Russak, New York, 1977.

56 Martin, J. *Design of Real-Time Computer Systems*. Prentice Hall, Englewood Cliffs, N.J., 1967.

Further reading

Wilkes, M. V. *Time-Sharing Computer Systems*. Macdonald, London, 1968.

Hawkins, J. K. *Circuit Design of Digital Computers*. John Wiley, New York, 1968.

Doran, R. W. Virtual Memory, *IEEE Computer*, **9**, October 1976, 27–37.

Tutorial problems

6.1 Design the X and Y decoder circuitry for a 2048 16-bit word ferrite core store employing the 3D system in conjunction with a 32 × 64 bit plane. Contrast the amount of hardware required, including drive and sense amplifiers, with the corresponding 2D and $2\frac{1}{2}$D methods.

6.2* (a) How many tracks are required on a magnetic drum if the bit spacing is 100 bpi, the wordlength is 40 bits, the drum is 8 inches in diameter and 4096 words are to be stored?

(b) What is the average access time if the drum rotates at 1800 rev/min?

(c) What is the clock rate?

6.3* Consider the use of a conventional 3D core store to replace a drum store in an existing special purpose equipment. The drum has 32 tracks, each capable of storing 256 8-bit words, and rotates at 3600 rev/min. Access is required to one drum track at a time, the input and output appearing in serial mode.

Devise a means of providing cyclic access to the core store and a suitable addressing scheme. Draw a detailed schematic of the circuits required for the logical control of the system.

6.4* Design a word-addressed read-only memory, using a diode network, which will read-down the 7-bit Hamming code version of any 4-bit pure binary address. Consider the extension of these ideas to the general problem of function generation.

6.5 Consider the basic problems involved in accessing a disc store system. Suggest a suitable address coding, and a means of employing the INA and OTA orders, so as to enable the transfer of data to and from the disc store. Devise an appropriate micro-program to perform these functions.

6.6 Design the logic for an associative storage system, using as a basis a word-addressed core-store containing 512 words each of 18-bits. The store must be searched for the existence of a specified tag, the address of the location containing the tag being placed in a register. Normal addressing and read/write operations must also be provided.

6.7* In a paging system one of the major problems encountered is the mapping of virtual addresses into physical addresses; this function is normally performed using some form of table-look-up procedure. Try to estimate the amount of storage and the execution times needed to locate an address in a core memory of 1024 18-bit words using the following techniques: (a) normal programming techniques; (b) using special table-look-up instructions (suggest suitable micro-programs); (c) using an associative store.

7

Input/output systems

7.1 Introduction

The purpose of the input/output (I/O) unit of a computer is to serve as a communications link, in the broadest sense of the term, between the computer and its associated peripheral equipment. Basically its task is to act as an interface between the main memory in the CPU and the input/output devices, under the direction of the control unit. Such devices would include, for example, card-readers and punches, magnetic tape units, disc files, and line printers. There are two main input/output functions associated with a computer system; they are:

(a) to effect data transfers between a high speed computer and the slower external data-processing equipment, and

(b) to prepare data, and the consequent code transformations, to enable communication between devices and computer.

The second function has already been discussed in Chapter 3, and the main object of this chapter is to consider in detail the hardware and software mechanisms involved in data transfer operations. In particular we shall study the concept of interrupt systems and the use of buffer storage to overcome the problems brought about by the disparity in speed between the CPU and the peripheral devices.

There are many different types of input/output equipment that may be used in conjunction with the CPU. Table 7.1 shows a selection of currently available peripheral devices, and Fig. 7.1 depicts a typical computer system. The choice of input/output equipment rests entirely on cost, and the use to which the computer system is to be put – for example, scientific and commercial computing or real-time on-line process control, and so on. However, all computers have a basic requirement for the input of programs and data, and to output or display their results. Small quantities of data and program (for example, operating instructions) can be entered directly into the store by means of switches situated on the computer console, or by using keyboard devices such as tele-typewriters. Large programs are usually loaded into the main memory using punched-paper tape or cards (suitably coded in BCD or some other binary code) or, when high speed is required, magnetic tape. Computer output is normally presented as a printed page (obtained with a directly connected line-printer) or in some cases in graphical form using an X–Y plotter or a

Table 7.1

Computer input/output equipment

(a) Input devices

Device	*Input form*	*Reading method*	*Capacity*	*Speed*	*Comments*
Punched tape readers	Punched 5–8 hole paper or mylar tape. Standard $\frac{11}{16}$ in, $\frac{7}{8}$ in, and 1 in tape	Sensing, mechanical, photo-electric, or dielectric. Electro-mechanical feed. Parallel output.	100 ft reels with spooler	15–1000 ch. groups per second	Cheapest and most commonly used form of input.
Punched card readers	Punched cards 80 or 40 columns 12 rows, round or rectangular holes	Photo-electric and electro-mechanical sensing. Serially column or row at a time.	Magazines holding 250–3500 cards	250–1000 cards per minute	Widely used, as many existing business installations use punched cards. Availability of off-line sorters, etc.
Keyboard (typewriter and teleprinter)	Standard 44 Key α/n keyboard. Special purpose keyboards also available	Photo-electric and capacitive sensing used, but mainly electromechanical. Generates parallel 7–8 bit codes, converted to serial mode for input.	—	10–30 chs/s maximum	Used for operator inputs particularly in real-time systems.
Analogue/digital converters (ANDI) Shaft position encoders, etc.	Analogue current and voltage signals, shaft positions, etc.	*Shaft position encoders:* Brush or photo-electric sensing, parallel reflected code, 7–19 bits. *ANDI:* electronic LSI switching 8–16 bits. Binary, BCD, parallel or serial output.	—	*ANDI:* 5–2000 μs per conversion to 8–16 bit number *Shaft position encoders:* 5–500 rev/min	Used in real-time process control systems, usually time-shared or multiplexed.
Magnetic tape unit	Magnetic tape $\frac{1}{2}$ in wide	Magnetized moving tape induces signal in stationary read/write heads.	1200–3600 ft reels. 800–1600 bits/inch	10k–1000k chs/s Function of tape speed and packing density	Tape produced on one computer, then used as input to another. Must be compatible system, otherwise identical to magnetic tape storage.
Magnetic tape cassette	Magnetic tape $\frac{1}{8}$ in wide in Phillips type cassette	Magnetized moving tape induces signal in stationary read/write heads.	150–600 ft cassettes 300–800 bits/in	10–25k bits/sec	Provides high storage in compact package at very low cost. Used in microprocessor systems.
Modem	Serial, bidirectional concurrent-full duplex consecutive-half duplex	Electronic modulator demodulator ccts.		300–9600 bits/sec	Used to transmit data between remote points over telephone lines.
Direct digital	Switches, push-buttons, etc.	Electromechanical/electronic.	—	Computer speeds	Used in on-line control computers.
Graphic display (Cathode ray tube)	Use of 'light pen' and push-buttons	Photo-electric scanning.	—	Normal manual writing and drawing speeds	Essential for interactive work, such as CAD.

(b) Output devices

Device	*Output form*	*Method of operation*	*Capacity*	*Speed*	*Comments*
Paper tape punch	Punched 5–8 hole paper tape (sometimes Mylar tape)	Electromechanical/electronic control, parallel input	1000 ft reels	10–300 chs/s	Cheapest and most commonly used form of output, off-line printers must be used to obtain page copy
Card punch	Punched 80 or 40 column cards	Electromechanical/electronic control. Parallel input, column or row at a time	250–3500 cards/hopper	100–400 cards/min	Same comments as for card readers. Printed page produced off-line
Teleprinter/ electric typewriter	Printed page, character at a time	Electromechanical/impact or ink and thermal non-impact electronic control, parallel or serial input	Normal continuous stationery up to 13 in wide	10–45 chs/s	Very slow but useful as an on-line monitor, can also be coupled as input device
Line printers	Printed page, line at a time	Electromechanical impact laser/electrostatic non-impact electronic control, parallel input	Continuous stationery up to 24 in wide	300–2000 lines/min 5000–20000 non-impact 60–160 chs/line speed depends on no. of chs/line	Expensive but very fast means of output. Commercial and scientific systems
Tabular display (Cathode ray tube)	Visual presentation of alpha/numeric data on CRT	Electronically generated symbols or line drawings Refresh or storage	80 chs/line 300–4000 chs	10k–50k chs/s Refresh rate 60 Hz	On-line control systems where data needs to be presented to an operator Interactive CAD systems
Digital plotters	Graphic or line presentation	Electromechanical/electrostatic electronic control	Continuous chart or sheet stationery	12000–25000 steps/min. 0·005–0·01 in steps. With back-up average 100–400 points/min	Scientific, CAD, business applications. Can also be used off-line with cards, tape etc.
Digital/analogue converters	Analogue voltages and current for control purposes	Electronic LSI modules	8–16 bit words	1–20 μs/word	Used in on-line computer control in conjunction with multiplexer unit
Direct digital	Annunciators, lamps and indicator tubes, etc., or direct lines to control switches	Electromechanical/electronic	—	Computer speeds Word at a time	Used in on-line computer control

CRT visual display unit. Alternatively, since there is no cheap solution to the problem of obtaining a fast printed page output, an intermediate output medium, such as punched cards, or magnetic tape, is often used, the printed page being produced off-line. A further advantage of this technique is that the output data can be fed back into the computer (or other computers) for reprocessing later. Slow speed tele-typewriters are usually connected on-line to the computer for the express purpose of controlling the operating system, that is, to input (and accept) commands and instructions concerning the organization and running of programs.

A computer incorporated into a real-time control system must have special I/O equipment capable of converting analogue data (for example, from measuring devices and transducers) into digital signals,[1] that is **analogue-to-digital converters**. The reverse function, **digital-to-analogue** conversion, must also be provided to enable the digital outputs to control and operate the system directly using conventional analogue devices (such as three-term controllers, valve actuators, etc.).

7.2 Control of input/output equipment

The major problems associated with the operation of any peripheral equipment are those associated with the disparity that exists between the speed of the central processor unit and the on-line equipment. Speed in this case is assessed in terms of rate of data transfer. In the majority of cases the peripheral equipment is many orders slower than the central processor, thus the operating rate of the computer is limited by the input/output devices. This is because the physical movement of the punched tape or card must be carried out with electromechanical devices with an inherent limitation on the response time. The actual sensing of the data, say a row of punched holes in a paper tape, can be performed at CPU logic speeds, in the order of microseconds. However, the movement of the tape itself, to bring the next character to the reading station, requires something like 1–2 ms.

It will be seen then that the operation of reading (and writing) data involves moving the input medium, in this case paper tape, as well as sensing or writing the character groups. In general there is no reason why the central processor should not continue obeying machine-code instructions during the time required to move the tape (using the interrupt facility, see Chapter 3 and later sections). But it is not possible to give another input/output instruction until the first one is finished, therefore some means of indicating that the input/output channel is occupied must be provided by the equipment. This normally takes the form of a 'busy' signal generated, for example, by the peripheral equipment when in motion. Thus if two input instructions, say, are given in succession the operation of the computer will be held up until the first input instruction has finished. In practice the busy signal would act as a condition on the input/output microprograms causing the microprogram to wait (by looping) until the equipment was free.

The use of buffer storage, normally a random access core store or LSI registers, is one way of overcoming the difference in data transfer rates. Data to be transferred into or out of the computer are first assembled into blocks of information; for example, 8-bit characters from a magnetic tape unit can be assembled into a computer word (or words) before transfer to the central processor. In this way the programming time can be optimized since fewer delays are incurred waiting for the equipment to become free. Furthermore, it is also possible to replace some input assembly programs by hardware – for example, the parity checking and assembly of 8-bit character groups from a punched-paper tape reader into computer words (see Section 7.4.2).

Magnetic tape and disc file present a further problem (with which we are already familiar) – that of access time. The tape, for example, must be brought up to speed (it is normally stationary) and then searched for the correct information block before any transfers can be made. A busy line and interrupt technique can be used for this together with computer instructions which can start the mechanism and initiate a search, the data can then be placed in buffer storage and the current program interrupted; program instructions can of course be obeyed during the search period. An alternative and simpler method with magnetic tape is to find the start of the data block and then to read all the following data in sequence, at the maximum transfer rate.

As we have seen, the major problems in the design of input/output systems are concerned with the interface between the CPU and the peripheral equipment. As well as timing and control problems there are circuitry requirements, such as logic level changers, power amplifiers for line driving, and so on, which must also be considered.

7.3 Punched-paper tape equipment[2]

Before discussing the organization of the input/output system in detail let us first consider a particular equipment for handling punched-paper tape to appreciate fully the problems and requirements of the system. Punched-paper tape was used in communications teleprinter equipment long before computers as we know them today were built. Consequently, punched-paper tape equipment, readers, and punches were easily adapted to form a cheap, simple, and reliable means of communication with the computer. As we have already seen, paper tape can have 5, 6, 7 or 8 bit positions per row across the tape, each row being referred to as a character. In addition each row includes a guide, or sprocket hole, which is used to locate the characters and to provide tape transport in mechanical readers. The tapewidth varies from 11/16 in for 5-hole tape to 1 in for 8-hole tape, with the character rows normally spaced 10 to the inch.

Punched-paper tape readers read a complete character at a time using photo-electric, dielectric and, in earlier models, mechanical sensing. Data transfer rates vary from between 100–1000 characters per second

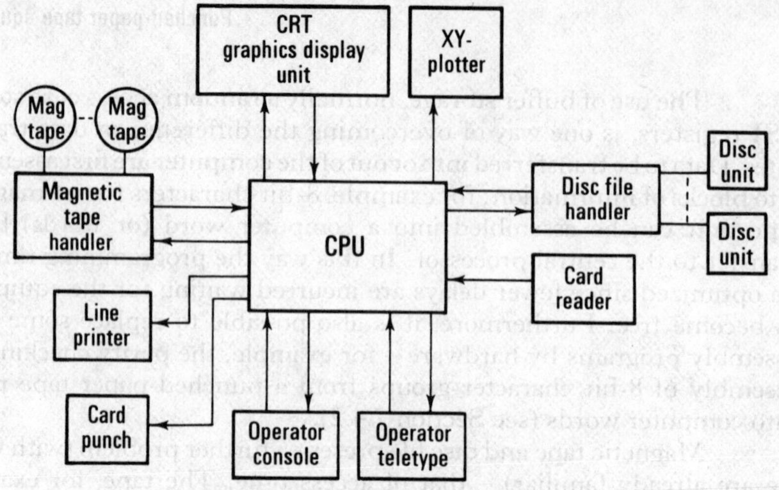

Figure 7.1 Typical computer system.

Lamp
Cylindrical lens
Brake
Guide rollers
Clutch driven by motor
Guide rollers
Perspex light guides
Driving roller
Pinch rollers
Spring return
Photo-transistors

(a) Mechanism.

RDTR
RDOP bistable
Current amplifiers
Clutch
Brake
Control unit
Busy
S Q
R $\overline{Q}$
Pulse generator
Sprocket sensor
Digit lines (logic levels)
Gates
Level changer
Digit 1 sensor
Input/output register
Level changer
Digit 2 sensor
Level changer
Digit 8 sensor
CPU
Sense amplifiers
Tape reader interface unit
Tape reader

(b) Logical control unit.

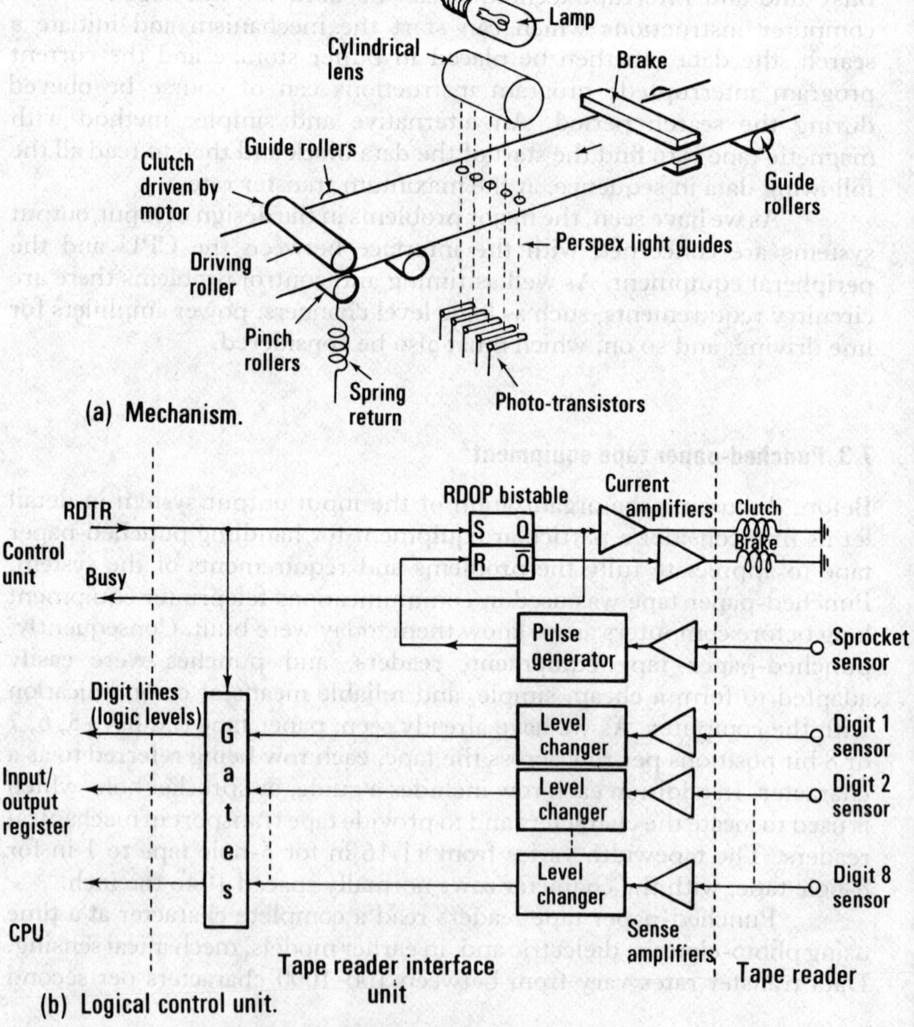

Figure 7.2 Punched-paper tape reader.

using electrical sensing, to 20–60 characters per second for mechanical readers. Figure 7.2(a) shows a typical paper tape-reader mechanism; there are four essential components – the clutch, brake and guide rollers, and the photo-sensing head. The basic operations performed by the reader are sensing and advancing the tape – the tape feed mechanism advances the tape one character at a time (or in some cases continuously) past the reading station where each character is sensed. The complete sequence of operations is as follows:

(a) In the normal quiescent condition the character under the reading head is the next character to be read. The brake coil is energized and the pinch rollers are out of contact with the tape; that is, the clutch is inoperative. Since the tape is ready to input data, the busy signal to the computer is down: $\overline{\text{busy}}$.
(b) When the instruction to input a character is about to be executed the tape reader is first checked to see if it is busy; if it is free the RDTR signal is generated (see Fig. 7.2(b)) and this gates the digit outputs to the input/output register. At the same time the RDOP bistable is set which causes the clutch to be engaged and the brake released, thus setting the tape in motion and generating the busy signal back to the computer. If the tape reader is occupied the computer waits (by looping in the microprogram) until the reader becomes available.
(c) When the tape (and punched characters) move forward, light is cut off from the photocells. The first photo-output to be affected is the sprocket output (since the sprocket hole is slightly smaller than the digit holes), then the digit outputs are affected. When the sprocket hole is *next* illuminated the signal from the sprocket sensor is used to generate a pulse which resets the RDOP bistable; this in turn switches off the clutch and energizes the brake; the reader is now ready to input the next character as indicated by the busy line being down.

Note that the actual reading is performed *during* the time needed by the electro-mechanical system to respond to the RDTR signal, and *before* starting to drive the tape. The tape must stop in such a position that the character holes are fully illuminated; this requires careful design of the clutch and brake mechanism. The maximum allowable stopping distance is three quarters of the diameter of the sprocket hole itself – some 0·033 inches. Thus, at 1000 characters per second the tape is travelling at 100 in/s and the total time available for stopping the tape is 670 μs. In order to achieve this it is necessary to ensure that the brake armature has a minimal distance to travel; in practice there is virtually no movement since the brake is always under a light spring pressure and the braking force is approximately three times that of the clutch.

Punched-paper tape equipment operates at data transfer rates varying between 10–30 characters per second, with the majority operating at some 100 characters per second. The basic functions of a paper tape punch are concerned with the punching and tape feed operations; a typical mechanism is shown in Fig. 7.3(a). Synchronization of the internal

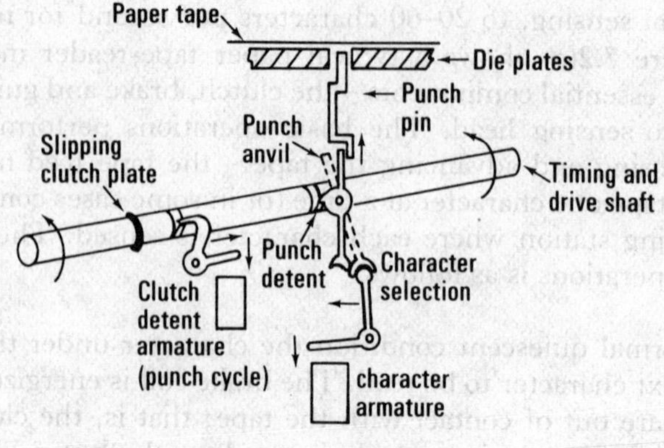

(a) Mechanism

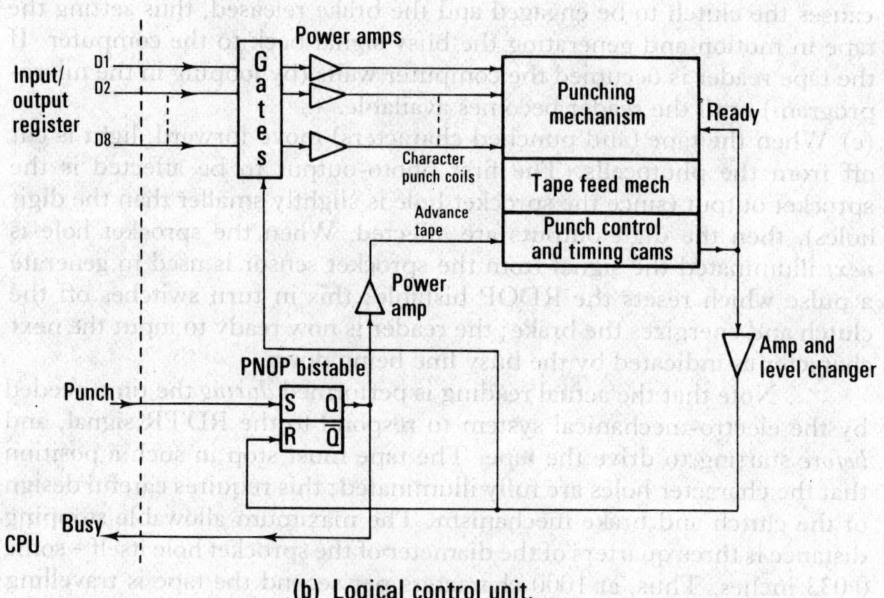

(b) Logical control unit.

Figure 7.3 Paper tape punch.

mechanical components of the punch is obtained by built-in electromechanical and mechanical devices such as eccentric cams, microswitches, and contacts, all of which are operated by a main timing and drive shaft which is continuously rotating. These internal interlocks ensure that the punch operations are performed in the proper sequence and with an adequate time interval between them. Thus the tape cannot be advanced until the punch pins have been completely withdrawn, but must of course be sufficiently advanced before the next character is punched. The input data must also be synchronized with the punch so that the electromagnets which select the punching mechanisms are only energized during the appropriate point in the punching cycle. In Fig. 7.3(b) the punch signal from the computer sets the PNOP bistable, the Q output of which is used to initiate a punch cycle. As before, the Q output is fed back to the computer as a busy line, where it is treated in the usual way. The tape advance signal energizes the clutch detent armature which causes the drive and timing shaft to rotate for one complete revolution. When the internal synchronizing signal (ready) occurs the amplified output is used to gate the input data from the input/output register to the punch unit – causing the character armatures to be energized and selecting the required punch pins. As the drive shaft rotates the selected punches are raised and the tape is punched with the appropriate character code and a sprocket hole. The ready signal is also used to reset the PNOP bistable, indicating that the equipment is free to accept and punch another character.

The input/output equipments described above have the characteristic of working directly connected to the computer; that is, the input/output instructions must refer exclusively to a particular equipment at any one time, and that instruction must be completed before any other instructions can be executed. Table 7.2 shows a typical micro-program for the instruction 'Input a character from punched paper tape to the accumulator'; note that there is a need to invent a new micro-order, RDTR, which effectively identifies the external equipment as well as initiating its operation. The rest of the micro-program is quite straightforward, with the contents of the input/output register being transferred to the M-register to allow the character to be OR'd with the contents of the accumulator, using the $\bar{I}_{M}$ micro-order.

It is obvious that this direct on-line technique of controlling peripheral equipment is very inefficient, since the effective computation speed is governed by the input/output devices which are inherently slow. For example, the initial input order can be executed in 2 μs (assuming the tape reader is free) but if another input order follows immediately after, the computer must wait until TRBUSY = 0, which could take up to 1 ms. A much better technique is to use the interrupt method of operation, discussed briefly in Chapter 3, which not only allows autonomous operation of the peripheral equipment but also allows for the simultaneous control of any number of devices. These techniques will be described in detail in the next section.

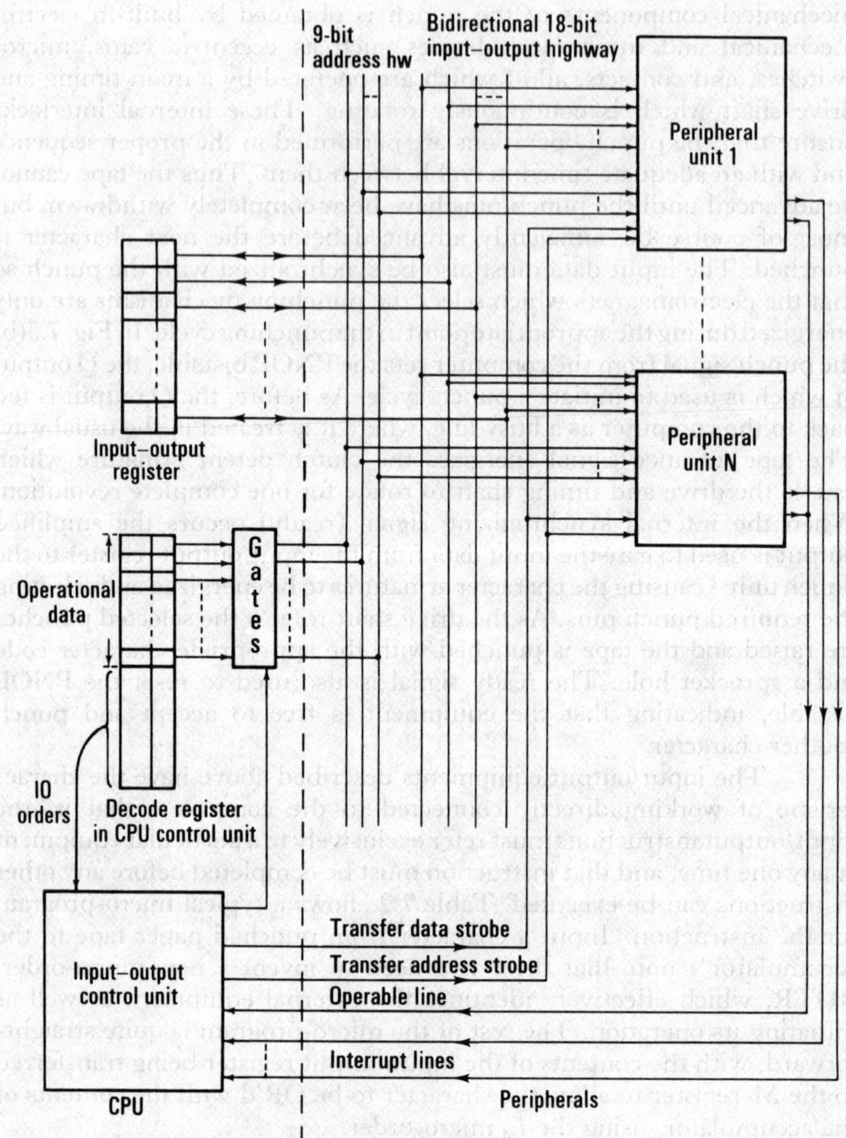

(a) Block diagram

Figure 7.4 Input–output organization.

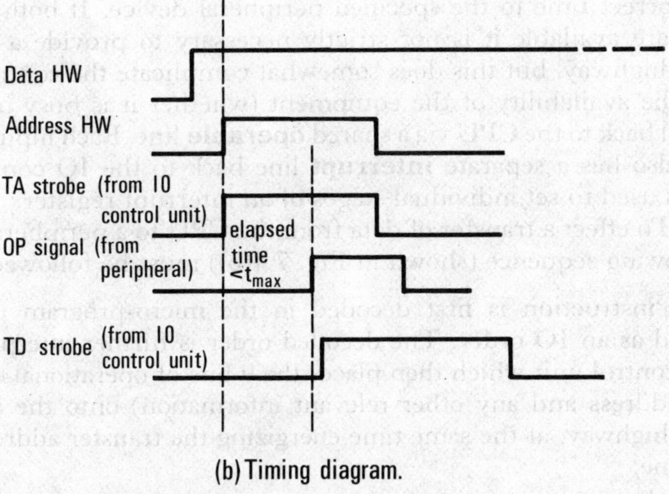

(b) Timing diagram.

Figure 7.4 cont. Input-output organization.

7.4 Organization of input/output systems

A block diagram of a typical input/output system is shown in Fig. 7.4; note that any number (or type) of peripheral devices may be connected to the CPU via a shared bidirectional **input/output highway** or **bus**. The data highway is used to transfer parallel information, one word at a time, between the CPU and the peripheral equipment. Some systems, however, have separate input and output highways, with suitable switching to the IO register. This radial, or **star** method of connection was adopted for early computer systems but proved to be uneconomical because of the large number of interconnections involved and the requirement in the CPU for individual hardware channels for each device. Note however that this approach does minimize the hardware needed by the device to handle data transfers since each device has its own dedicated highways. The method is still sometimes used for small systems, particularly those employing microprocessors where it is common practice to have separate input-output logic (called **ports**) for each device. However, most systems utilize the shared bus approach and effectively locate the input-output port in the corresponding device circuitry.

In the shared highway approach each peripheral is identified by a unique address which appears in parallel on the common **address highway** (9 bits are used for this function in our running example – that is, the IOS order bits). **Transfer strobe** lines (shared by all the IO devices) are associated with the data and address highways. These one-way control channels ensure the transfer of the appropriate information at the correct time to the specified peripheral device. If both transfer strobes are available it is not strictly necessary to provide a separate address highway, but this does somewhat complicate the control problems. The availability of the equipment (whether it is busy or not) is signalled back to the CPU via a shared **operable** line. Each input/output device also has a separate **interrupt** line back to the IO control unit which is used to set individual stages of an interrupt register.

To effect a transfer of data from the CPU to a peripheral device the following sequence (shown in Fig. 7.4(b)) must be followed:

(a) The instruction is first decoded in the micro-program unit and identified as an IO order. The decoded order is further interpreted by the IO control unit which then places the 9 bits of operational data (the device address and any other relevant information) onto the common address highway, at the same time energizing the transfer address (TA) strobe line.

(b) When the peripheral devices receive the TA strobe signal the information on the address highway is decoded and, if the addressed device is available, an operable signal (OP) is passed back to the IO control unit. There is normally a time limit t_{max} placed on the receipt of an operable signal, so that if an operable signal is not received by the IO control unit within some maximum time after sending the TA strobe, the computer regards the device as busy, turns off the TA strobe, and the IO instruction is assumed abortive.

(c) When the operable signal is received by the IO control unit a transfer data (TD) strobe is generated, which allows the addressed peripheral to accept the information currently on the data highway, that is, the contents of the IO register. The CPU normally assumes that the data have been successfully accepted by the peripheral device and the instruction is concluded.

A similar procedure is followed for inputting data from a peripheral device to the IO register in the CPU. In this case, however, the addressed device puts the data on the highway (thus setting the IO register) at the same time as it energizes the operable line. The information is maintained on the data highway until a TD strobe is received from the CPU indicating that the data have been accepted.

This method of organizing the IO system allows a number of peripheral devices to be multiplexed over a common highway or interface. Moreover, using this technique results in the design of the peripheral interface inputs (see Section 7.4.4) having the same basic structure for any IO device. Note that the timing of the interface operations assumes that the specified IO device has responded to a data transfer request from the CPU. This is governed by the transfer and operable lines (see above) – for example, the operable signal must occur a specified time after the TA strobe if a data transfer is to proceed. Note that the interface procedure described above is asynchronous, since the timing of the transfers is independent of the CPU master clock and is determined by synchronizing signals transmitted with the address and data information. This kind of communication protocol in which every signal transmission from a master or 'talker' (usually the CPU) is acknowledged by a return signal from the slave or 'listener' (normally a peripheral device) which is then used to determine whether or not the next phase of the transfer can proceed is known as **handshaking**. Note also that it is not essential to have separate address and data highways, and a common bus can be utilized for both purposes (as for example in the DEC PDP 11 Unibus system).

The alternative synchronous system, such as that described in section 7.3, is controlled entirely by CPU generated signals. This method has serious limitations when more complex peripherals are used, especially those which have an indigenous cycle-time such as the disc store, since unless the device responds within a prescribed time interval the transfer is corrupted.

Though the procedure described above allows any number or type of peripheral device to be operated, the technique is still restricted to single word data transfers to specified equipment. The basic micro-program (shown in Table 7.2) is still applicable, though it must be slightly modified to allow for the general IO instruction. For example, the common operable signal replaces the TRBUSY, but in this case the micro-program is concluded if the condition OP = 0 occurs (signifying that the device is busy). Note also that it is still necessary to transfer the contents of the IO register to the accumulator. Although with this system the problem of speed disparity between the CPU and the peripheral equipment still exists, there is now no longer any need for the computer to get irretrievably 'hung-up' on an IO instruction. To avoid delays of this type the device busy test may be taken out of the micro-program and replaced by a separate skip instruction of the form 'Skip next instruction if OP = 1 for specified device' (or alternatively if OP = 0). This means that each IO instruction must be preceded by a skip instruction which can either cause the computer to wait, or to carry on with an alternative program (see Table 7.3). With this type of skip instruction it is necessary to allow the operable signal to set a special-conditions bistable, which must be reset each time its output state is tested.

In this way the CPU time that is wasted waiting for equipment to become available can to some extent be reduced. However, there still remains the very difficult software problem of integrating useful alternative programs with the IO operations. For example, in Table 7.3(b) the alternative program must make provision for the IO instruction to be periodically repeated until the equipment is available and the order can be executed; this is a difficult function to program economically.

Table 7.2
Micro-program for the instruction 'Input character from paper tape reader to acc'

Timing (100 ns)	*Condition*	*Micro-order*
1		Read cycle including
⋮		modification and
17		interrupt
18	TRBUSY = 1	Rep 18
	TRBUSY = 0	IO_c, RDTR
19		M_i IO_o
20		A_i $\bar{I}_M$, END

Table 7.3
Programming IO instructions

(a) Computer waits until device free

Label	*Instruction*	*Comment*
BACK	SKP OP = 1 TR	Skip next instruction if device ready
	JMP BACK	Jump back to SKP instruction
	INA TR	Input ch. from tape reader

(b) Computer carries on if device busy

Instruction	*Comment*
SKP OP = 1 TR	Skip next instruction if device ready
JMP ALT	Jump to alternative program
INA TR	Input ch. from tape reader

Table 7.4
Input/output instructions

(a) IO Instructions

Order in octal	*Mnemonic*	*Comment*
000 011	ENAB	Enable general interrupts
000 012	INHB	Inhibit general interrupts
000 013	FTIN	Fetch contents of interrupt register to accumulator
205	INA	Input to accumulator from data highway
206	OTA	Output from accumulator to data highway
207	MASK	Mask interrupt register
210	SKP0	Skip next instruction if OP = 0
211	SKP1	Skip next instruction if OP = 1

(b) *OTA and INA orders*

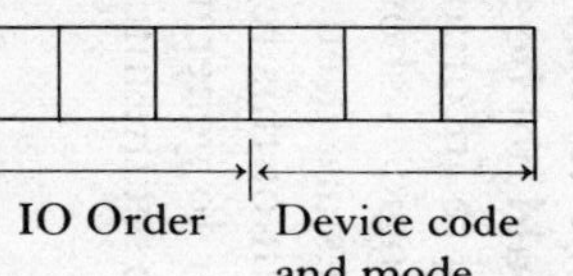

Octal code	*Device and mode*
000	Paper tape reader
010	Paper tape punch
001	Card reader
011	Card punch
002	Teletype (input from keys)
012	Teletype (output to page)
013	Line printer
004	Disc file (input ch.)
014	Disc file (output ch.)
005	Magnetic tape (input ch.)
015	Magnetic tape (output ch.)
016	Graphics (output ch.)
007	Keyboard (input word)
017	Keyboard (output word)
114	Disc file—output mode address word
115	Magnetic tape—output mode address word
116	Graphics—output address word

(c) *MASK orders*

			987	654	321
← MASK →			← Device →		

Bit	*Device*
1	Paper tape reader
2	Paper tape punch
3	Card reader
4	Card punch
5	Teletype
6	Line printer
7	Disc file
8	Magnetic tape
9	Graphics

7.4.1 Interrupt systems

The solution to this problem is, of course, to incorporate an **interrupt facility**[3,4] whereby a current computer program may be temporarily interrupted by a data transfer request from the external equipment. After satisfying the requirements of the peripheral devices the current program is resumed at the point where it was interrupted. How this function is performed has already been explained to some extent in Chapters 3 and 4, in particular the organization of software procedures (Section 3.3) and the micro-program implementation (Section 4.5). All that remains is to show how the interrupt facility is incorporated into the total input/output system.

Each peripheral device has a separate interrupt line back to the IO control unit, as shown in Fig. 7.4. An interrupt signal appears as soon as an equipment is ready to accept (or deliver) data to the CPU and it is maintained constant until the equipment is addressed by an IO instruction, when the signal is removed as the equipment now becomes busy. Thus, it is necessary to inhibit all interrupts (especially if the simple IO procedures described above are to be performed) until the interrupt mode of operation is required. This general prohibition on interrupts takes place automatically whenever program is loaded into the computer and must be overridden by software means when the interrupt facility is required. Consequently, machine-code instructions must be provided to enable (or prohibit) the interrupt action. It is also necessary to inhibit interrupts from particular peripheral equipment; this is done by a 'masking' instruction which only allows specified devices to interrupt (providing of course the general interrupt has previously been enabled). These orders, which do not require a store address, can be coded using the spare IOS or zero address instructions. Table 7.4 shows a possible encoding of the IO instruction set; note that in the case of the OTA, INA, and SKP instructions the two least significant octal digits are used to identify the equipment (a 1 in the second octal digit indicates a data output), with the most significant digit identifying a mode-address word (used in buffered systems, see next section). The mask instruction uses a separate bit to specify each device; if the bit is set to 1 the device interrupt is enabled, otherwise a zero digit prohibits the interrupt. Note that in this case the interrupt masks for the peripheral equipment must all be set up simultaneously. There is no reason, however, why masks should not be set up individually, provided the corresponding instructions are incorporated in the machine. A schematic of the interrupt control logic is shown in Fig. 7.5. Incoming interrupt signals from the IO devices are used to set up individual stages of the interrupt register. The output of this register is gated together with the contents of the mask register to generate the desired interrupt signals which are then OR'd together to set the interrupt trigger bistable.

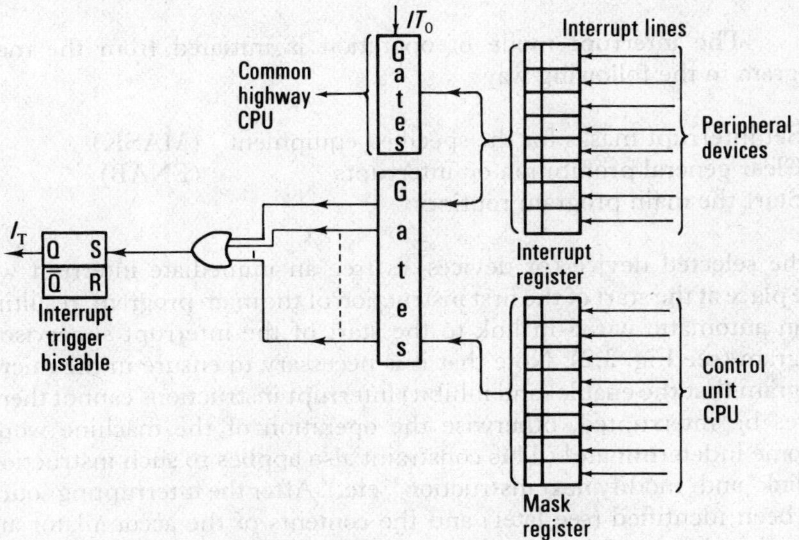

Figure 7.5 Interrupt control logic.

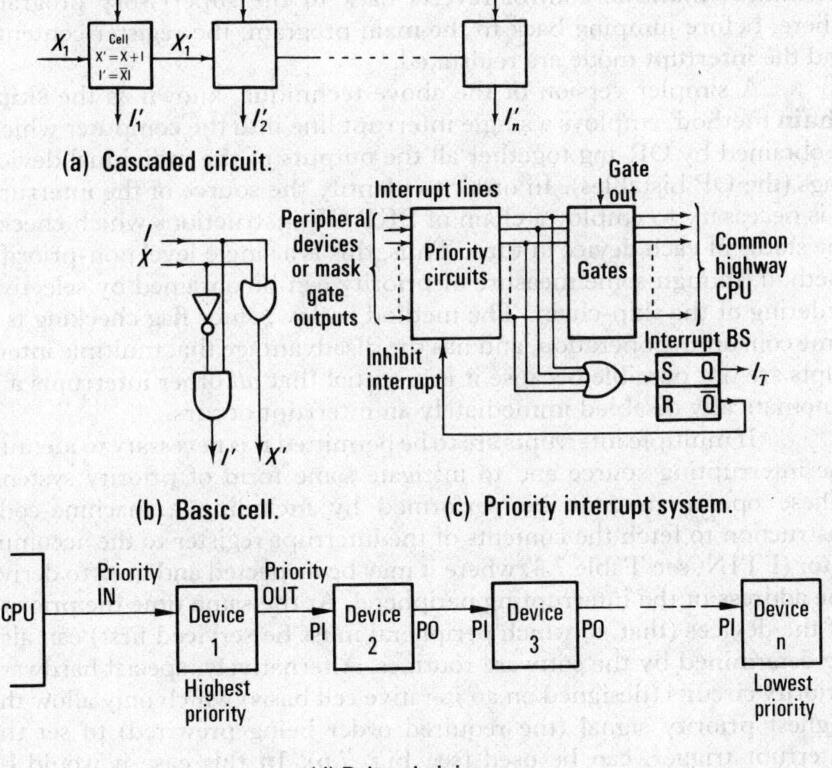

(a) Cascaded circuit.

(b) Basic cell.

(c) Priority interrupt system.

(d) Daisy chaining.

Figure 7.6 Hardware priority circuits.

The interrupt mode of operation is initiated from the main program in the following way:

1. Set interrupt masks for the specified equipment (MASK)
2. Clear general prohibition on interrupts (ENAB)
3. Start the main program routines

If the selected device (or devices) is free an immediate interrupt will take place at the start of the first instruction of the main program, resulting in an automatic wired-in link to the start of the interrupt supervisory program (see Fig. 3.2). Note that it is necessary to ensure in the micro-program that the enable (and inhibit) interrupt instructions cannot themselves be interrupted, otherwise the operation of the machine would become indeterminate! (This constraint also applies to such instructions as 'link' and 'modify next instruction', etc.). After the interrupting source has been identified (see later) and the contents of the accumulator and control registers safely preserved, the supervisory routine effects a jump to the appropriate IO routine. If it is not required (or feasible) to 'nest' interrupt requests, it will be necessary to prohibit any further interrupts at the start of the supervisory routine. When the IO procedure has been concluded, machine control reverts back to the supervisory program where, before jumping back to the main program, the register contents and the interrupt mode are reinstated.

A simpler version of the above technique, known as the **skip-chain** method, employs a single interrupt line into the computer which is obtained by OR-ing together all the outputs of the individual device flags (the OP bistables). In order to identify the source of the interrupt it is necessary to employ a chain of SKP OP instructions which checks the status of each device in turn. Thus, this is a single level non-priority method, though some measure of priority can be obtained by selective ordering of the skip-chain. The method is slow, since flag checking is a time consuming operation, and has the disadvantage that multiple interrupts are not possible because it is essential that *all* other interrupts are automatically disabled immediately an interrupt occurs.

If multiple interrupts are to be permitted it is necessary to identify the interrupting source and to instigate some form of priority system. These operations may be performed by including a machine-code instruction to fetch the contents of the interrupt register to the accumulator (FTIN, see Table 7.4) where it may be inspected and used to derive the address of the interrupting peripheral. At the same time the priority of the devices (that is, which peripheral must be serviced first) can also be determined by the software routines. Alternatively, special hardware priority circuits (designed on an iterative cell basis) which only allow the highest priority signal (the required order being prewired) to set the interrupt trigger, can be used (see Fig. 7.6). In this case it would be necessary to incorporate a machine instruction to fetch the outputs of the priority network to the accumulator to determine the address of the interrupting source. The input to the priority unit can be either the actual device interrupt lines or the result of gating these lines with the contents of the mask register.

An alternative procedure is to use the **daisy-chain** method of control, shown in Fig. 7.6(d), in which a priority line is connected into and out of each device in a simple chain with the highest priority device (usually power failure detection, parity check etc.) going direct to the processor. When the processor receives an interrupt (or interrupts) the priority line is interrogated, whereupon each device, in sequence, examines the signal and responds if, and only if, it requires servicing, otherwise the signal is passed to the next device in the chain. The responding device replies by transmitting its address to the CPU to initiate the appropriate interrupt routine. Note that requests from high priority devices block out those from devices with lower priority further down the chain. More sophisticated methods for arbitrating between concurrent requests from peripheral devices are discussed in section 9.4.1.

The procedure described above for multiple interrupts suffers from the disadvantage that a fast device of high priority may be locked out while a slower device of lower priority is being serviced. Ideally requests from a high priority device should be granted immediately, providing the device currently being serviced has a lower priority. This leads to the requirement to nest interrupts, that is an interrupt routine may itself be interrupted by a request from a higher priority device. The major difficulties encountered in nesting interrupts occurs in the storage of return addresses (such that the return path, via interrupt service routines, to the main program is clearly defined) and the preservation of active registers and flags in the interrupted routines. These difficulties can easily be overcome by using stack storage as discussed in section 3.9.

Using stack storage the servicing of an interrupt commences with the storage of the contents of the Program Counter and scratch pad registers etc. (called the **state** of the processor) on to the top of the stack, the interrupt service routine is then entered in the usual way. At the end of the interrupt routine a return to the main program is initiated by simply popping the top most words in the stack and using them to reinstate the Program Counter and register contents. Should another device of higher priority interrupt during the servicing of an interrupt routine the same procedure is followed. Thus the order of returning to the main program is stored inherently in the push-down stack thereby allowing interrupts to be nested indefinitely depending on the size of the stack.

This facility is becoming available in many machines, for example the DEC PDP11 has stack storage which is used in conjunction with a Processor Status Register (indicating the complete machine state of the program immediately prior to the occurrence of the interrupt) to simplify the use of nested interrupts (only two words, the Program Counter and the Processor Status Register, need be placed on the stack). Also most microprocessors, such as the Intel 8080 or Motorola MC6800 have stack facilities, normally implemented in RAM, which allow multiple interrupts to be nested with comparative ease and to any depth.

The interrupt mode of operation can also be used for the indication and remedial treatment of machine errors and power supply failures. In the last case an automatic unmaskable interrupt occurs if the mains power or HT supply to the system breaks down. If, as is generally the case, the mains supply is obtained from a motor generator set, the inertia of the flywheel system after cut-off will maintain the supply long enough (in the order of milliseconds) to allow an interrupt procedure to store away important data and results.

Another important benefit provided by program interruption is the **real-time clock** or internal timer. This facility allows a computer control system some indication of the passage of real-time (as distinct from machine-time) which is essential if process requirements are to be satisfied at the correct time intervals (for example, data sampling and correction procedures). A computer word is provided in a store location (or special CPU register) which can be set under program control to any desired value. At regular intervals (say 100 μs) the contents of the word are automatically decremented by +1 and tested for zero. If the contents are non-zero the operation of the computer is allowed to proceed as normal, otherwise the current program is interrupted. In this way it can be arranged for the computer to be interrupted at preset time intervals and program control switched to an appropriate service routine.

7.4.2 Buffered input/output systems

In all the IO systems described so far the transfer of data has proceeded one character, or one word, at a time. The direct on-line method involves performing the IO instruction (with attendant operating delays) many times over, and hence is very time consuming. This disadvantage still applies (to a lesser extent) even when the interrupt mode is used, since it is necessary to interrupt the current program and go through the supervisory and IO interrupt routines for each individual word transfer.

Considerable time can be saved when carrying out IO operations if a **buffer store** is included as an integral part of the peripheral control unit. In this mode of operation an initial IO instruction is used to select the device and initiate the transfer of a **block of data** (for instance, a complete disc sector) to or from the buffer store. The buffer store may consist of shift registers for small quantities of data but generally a semiconductor RAM store is used for this purpose. With some peripheral equipment, such as drum or disc stores, it is highly desirable to transfer one complete block of data at a time, since incremental operation is only possible in special cases.

Using this technique to read a sector from the disc store it is first necessary to output (using, for instance, instruction OTA 114) a **mode-address** word to the disc store control unit. This word is used to specify either the read or write function, and the address (track and sector numbers, etc.) of the required data; in some systems separate disc IO instructions would be used for this purpose. On receiving and decoding the mode-address word the disc control unit selects the appropriate read-head and then starts to read the complete sector into the associated buffer store. When the store is full (that is, the sector has been read down) the control unit generates an interrupt signal back to the CPU. The corresponding disc interrupt procedure contains a looped input routine which transfers the complete sector (one word at a time but at maximum transfer rates) directly into the CPU store. This procedure is shown programmed in Table 7.5.

Table 7.5
Input program for buffered disc system

(a) Initial call

Label	*Instruction*		*Comment*
	FET	MOAD	Fetch mode-address word to accumulator
BACK	SKPI	014	Test OP line
	JMP	BACK	
	OTA	114	Output accumulator to disc file unit
	Main program		

(b) Interrupt input routine

Label	*Instruction*		*Comment*
	FETM	SENO	Fetch constant (no. of words in sector) to modifier
LOOP	SKP1	004	Test OP line
	JMP	LOOP	
	INA	004	Input ch. from disc buffer
	STR	DIDA/M	Store ch. in CPU disc data block
	JMPM	FIN	Test modifier register
	INCM		Add +1 to modifier register
	JMP	LOOP	
SENO	(	)	Number of words in sector

Output instructions are performed in a similar way, in this case the peripheral buffer store must first be filled using a looped OTA instruction operating at maximum transfer rates. Note that this implies buffer store sequencing and addressing logic within the disc control unit. When the transfer is complete a mode-address word is outputted to the peripheral, which initiates the necessary addressing and write operations. When the write operation is complete, an interrupt signal is generated which indicates to the CPU that the peripheral is ready to accept more data into its buffer store.

In this way the speed disparity between the CPU and the peripherals can be largely overcome by the use of buffer storage in association with the program interrupt technique. Unfortunately, a consequence of this approach is that the peripheral control unit and interface (sometimes called device **handlers**) increase in complexity and hence cost. The overall system costs can often be reduced, however, by sharing the device handler between a number of peripheral equipments.

Let us now consider a buffered IO system in more detail, and in particular the input of punched paper tape. As we have already seen, this operation may be performed one character at a time, but it would be more economical of CPU time if a complete word could be assembled before being loaded into the CPU store. Suppose, for example, we wish to input characters from a 5-bit punched-paper tape reader. Using the buffer store approach it is possible to assemble the characters into an 18-bit word (performing a parity check on each character if required) and then to input the complete word to the CPU. In this way not only is time saved by eliminating the delay due to the operating time of the device, but the need for program assembly in the computer is also obviated.

Figure 7.7 shows a block diagram for a typical IO buffered system to perform this function. As we have seen earlier, buffered systems require initialization from the main program, this puts the equipment in the correct mode and also transfers the relevant address information. In this case, since the only function of the device is to input a predetermined number of characters, a single INA type instruction will suffice. Thus the function of this order is simply to identify the equipment and cause

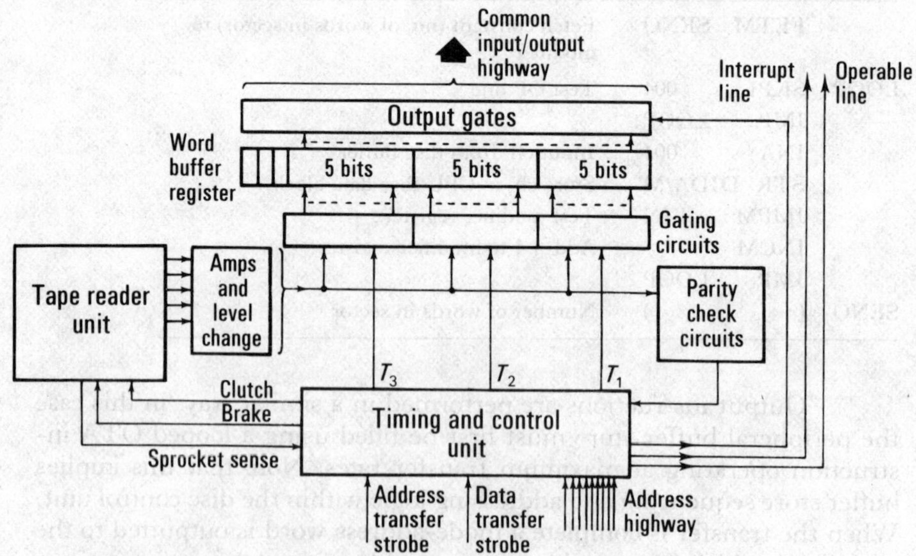

Figure 7.7 Buffered paper tape input.

it to start reading characters into the word buffer. When the buffer register is full the tape reader control unit generates an interrupt signal which results in the assembled word being inputted into the CPU in the usual way. The tape reader control unit, as well as providing the usual drive and control circuits for the reader mechanism, also contains a counter which governs the selection and routing of the characters into the word buffer register using the timing pulses T_1, T_2, and T_3.

7.4.3 Direct memory access systems

In the IO systems so far discussed all data transfers into or out of the store, including program interrupt transfers, have passed via the accumulator register using the OTA or INA instructions. Thus, if the current contents of the accumulator are required to be preserved they must be stored away, otherwise they will be overwritten in the transfer operations. Moreover, to input or output a block of data requires a considerable amount of program (see, for example, Table 7.5) and for some devices this procedure is inordinately time consuming.

To overcome this software speed limitation, **data-break** or **hesitation** facilities are included in the CPU whereby an external device may directly insert or extract data from the computer memory, bypassing all program control. This technique of **direct memory access** (DMA) is particularly well suited for devices, such as disc stores, magnetic tape stores, etc., that transfer large blocks of data in a single continuous operation. When using this mode of operation the external device effectively controls the data transfers; thus the current computer program has no cognizance of the transfers and consequently must check for the presence of the data before using it. This may easily be done by inserting an identifying tag in the last word of the data to be transferred and then checking for the presence of this tag to ensure that the data transfers have been completed.

To use the data-break facility the peripheral must first be initialized in the same way as for a buffered system. This time, however, a data break or hesitation line is used to request attention from the CPU rather than an interrupt signal. The data break requests will occur at random times relative to the CPU operations and in general, with short orders (those requiring 2 R/W cycle times or less), the peripheral can be made to wait until the end of the instruction currently being executed. If, however, the maximum waiting period of the peripheral device is shorter than the execution time of an instruction, break points must be provided as an integral part of the micro-program.

There are two basic methods of incorporating the data break mode of operation in the CPU. These differ according to whether the destination addresses of the data in the CPU store and the number of word transfers to be performed (the **word count**) are specified by the device itself, or are held in some predetermined locations in the CPU store. In the latter case the peripherals are collectively allowed to request a data break, and a particular peripheral (determined on a priority basis) must be identified, thus defining the address of its control word in the CPU. The control word, one per device, is a standard computer word (held in a software protected location) containing, for example, 12 bits of address and 6 bits defining the number of words in the data block. The contents of the control word are set, under program control, before the initialization instruction.

When the data break occurs the device control word is automatically obtained from store and the address portion used to route the contents of the IO register (which holds the data word) to its appropriate location in the store. At the same time the control word is updated by decreasing the word count by +1 and increasing the address portion by 1 and then written back to the store. After these operations have been performed, control returns to the current program and normal operation is resumed. All the data break operations are performed under micro-program control and must, of course, include the temporary storage of any control registers or information pertinent to the current program. For example, in examining and updating the device control word, it is necessary to fetch the word to a register and address the store; if the registers used in this process contain data relevant to the current program, these data must be stored away and reinstated after the data break. In some machines special registers are provided for the data break routines; for example, a duplicate instruction register and accumulator might be provided. The problem would be aggravated if a data break were allowed to occur in the middle of an instruction, when it would be necessary to preserve the contents of all the control registers. A data break facility of this kind is, in effect, a micro-programmed interrupt, but because it is executed at the basic machine level it is very much faster and does not involve any software routines.

An alternative approach is to allow the device, or rather the device interface, to specify the store address of each data transfer and to count the number of words in a data block. As described above, the peripheral device must first be initialized, a data break being requested in the usual way when the peripheral is ready to transfer data. On receipt of a data break the current instruction is temporarily halted (usually at the end of a store R/W cycle) and the peripheral device given direct access to the store control unit. In the input mode the data word is preceded by an address word which specifies the location where the data is to be placed in the CPU store; in the output mode only the address word is transmitted to the CPU. Under micro-program control the address word is routed to the store control unit and the data word to the memory register, finally a R/W cycle is initiated. In this way it is possible to read (or write) data from the CPU store with the minimum disruption of the current program by simply monopolizing the next R/W cycle (the technique is called 'cycle-stealing' for this reason). The interface unit is more complicated, however, since it is now necessary to include registers (and updating logic) for the data address and word count functions.

7.4.4 Communications interfaces

One of the most common requirements in a computer system is to effect a communications interface between terminal equipment (such as a teletype) and the CPU. Typically, this takes the form of an asynchronous serial data transmission channel (normally one per terminal, though they can be multiplexed) for low data rate operation up to 500 chs/second. Because of its extensive use, particularly in mini- and microcomputer applications, semiconductor manufacturers have developed the **UART** (universal asynchronous receiver transmitter) as a single chip LSI device.

The basic function of the serial asynchronous line interface is to perform the parallel to serial and serial to parallel conversions required to convert between characters handled as words in the computer and the data format used for serial data transmission.

The format of a serial character, shown in Fig. 7.8(a), consists of a start-bit followed by 5–8 data-bits, then a parity check-bit and finally a stop-bit. The start-bit is always zero and the stop-bit always one, thus the beginning of a character can be recognized by a negative going transition. When data is transmitted at less than the maximum rate an arbitrary number of idling-bits (all ones) will be present between characters.

In order to facilitate the synchronization of the receiver to the serial data a special clock must be provided with a period Tc given by the expression:

$$Td = \frac{1}{\text{Baud rate}} = K.Tc$$

where Td is the bit-time of the serial data and K is a constant, typically 16.

The **Baud rate** specifies the maximum modulation rate of a code in bits per second, and is useful as a measure of the bandwidth required by the transmission channel; note that it is not the same as the data rate. For instance, in our example each character format requires 11 bits and if the data is transmitted at 10 characters per second the Baud rate is given by:

11 bit-times per character $\times$ 10 characters per second $=$ 110 Baud

However the data rate is given by:

8 data bits per character $\times$ 10 characters per second $=$ 80 bits/second

$$\text{Thus, } Td = \frac{1}{\text{Baud rate}} = \frac{1}{110} = 9{\cdot}1 \text{ ms}$$

(Note that in general the Baud rate $= 1/Td'$ where Td' is the duration of the shortest bit in a character; in our example all bits are assumed to be of the same length).

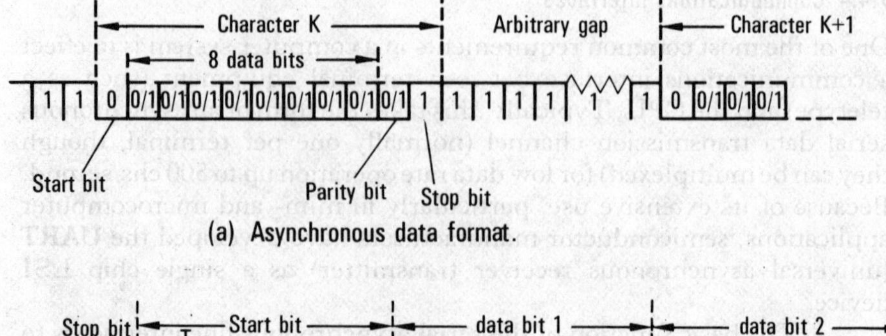

(a) Asynchronous data format.

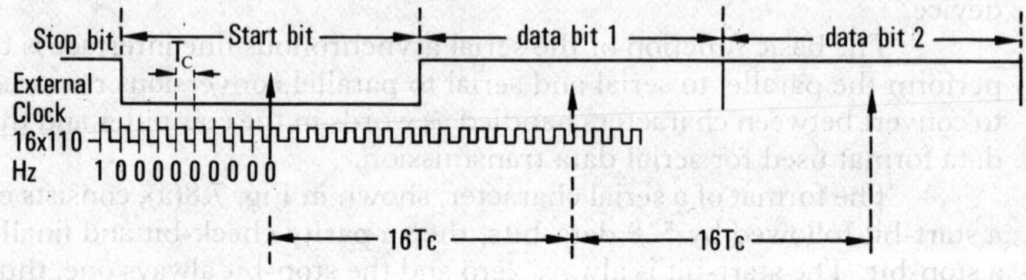

(b) Receiver timing.

Figure 7.8 Serial data transmission.

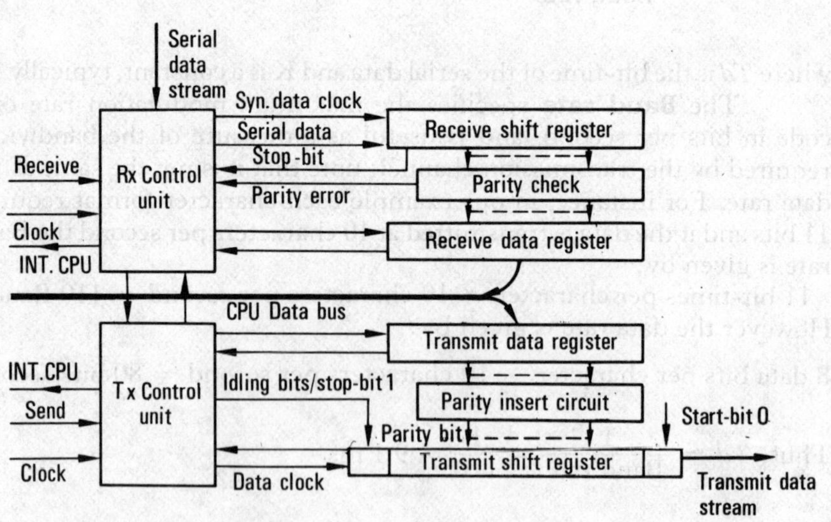

Figure 7.9 Asynchronous communications interface.

The clock is used to synchronize the receiver to enable the decoding of the serial input; this is done in the following manner. After detecting a stop-bit, or any number of idling-bits, the data input is strobed with each rising edge of the clock until nine consecutive zeros have been found to occur. As the clock is running at 16 times the data bit-time this procedure will determine the centre of the start-bit, see Fig. 7.8(b). Once the centre of the start-bit has been found subsequent sampling takes place once every 16 clock periods (that is 16 Tc or $Td/16$) until a stop-bit is encountered. In this way the data bits can be sampled with reasonable accuracy at the centre of each bit-time. (Note that the DEC-DM11 asynchronous serial multiplexer operates using 7 samples per bit-time and detects four successive zeros to detect a valid start-bit and locate its centre.)

The clock source can be provided either as an integral part of the equipment or as a separate unit. In the case of UART special LSI chips are used to provide the clock, these either count down the microcomputer clock (to give a fixed Baud rate) or are special programmable bit-rate generators with their own crystal oscillator.

So far we have only considered the receiver, but the communications interface must also transmit serial data. Moreover the device must be able to receive data from one source and transmit data to another source at the same time. Thus separate clocks, with perhaps different Baud rates, will be required for the two sections of the device; the serial data formats must however be identical. Fig. 7.9 shows a block program for a typical communications interface.

The received serial data stream is shifted into the Receive Shift Register using the synchronized clock (as described above). When the entire character has been received the data-bits are transferred to the Receive Data Register and an interrupt request sent to the CPU. Once this transfer has taken place the Receiver Control Unit begins looking for the next start-bit and the process repeats. Note that the CPU transfers the contents of the Receive Data Register during the character conversion time, some 100 ms in this case; this procedure is known as **double-buffering**.

Character transmission proceeds in a similar manner. The character (data bits only) is transferred from the Transmit Data Register to the Transmit Shift Register generating the parity bit in the process. When the entire character including stop and start-bits, has been assembled in the Transmit Shift Register serial data transmission commences; the CPU is interrupted at this point thereby allowing, as before, a full character time to service the interrupt.

The communications interface is normally under program control using interrupt procedures to effect data transfers, or, alternatively, it can be connected directly through a DMA channel. In most cases a control word is also required to set up the device, for example, to specify the character format in terms of number of data-bits, odd/even parity, etc., to select the Baud rate and setting flags to enable/inhibit the transmitter/receiver interrupt operations. In the same way a transmitter/receiver status word is required to indicate to the CPU control information such as interrupt requests, parity errors, state of data registers etc.

[Multiplexed communications interfaces are also available which can handle a number of input/output channels (typically 16). In this case care must be taken to ensure that all the lines can be serviced in 1/k of a bit-time to avoid corruption of the serial data stream. The number of lines a multiplexer can handle is obviously a function of the Baud rate.]

7.4.5 Peripheral control unit

All input/output equipment must be connected to the CPU via a unit called the **peripheral control unit** (PCU) or more commonly the **interface unit**. This unit must provide for the decoding of the address instructions and commands from the CPU (as distributed on the common address highway) and the generation of the appropriate responses to the transfer control signals (for example, the operable line). In some cases the interface unit may also be called upon to control a buffer store and perform assembly and coding operations. In addition the PCU must develop the necessary control and timing voltages required to operate the peripheral devices; for example, the clutch/brake relays must be energized in the paper tape reader.

The functions of the peripheral control unit would include, for example:

(a) The provision of circuitry (**line drivers**) to generate the drive voltages required to distribute the logic data signals along the common highways. Incoming data lines must be correctly terminated to ensure protection against interference from noise (see Section 8.3). **Logic level changers** are also required if dissimilar equipment employing various voltage levels are used.

(b) The conversion and assembly of character formats (for example, as read from magnetic tape or punched cards) into computer words, and vice versa.

(c) The control of several peripheral units, such as magnetic tape, when a single tape handler may control up to three or four tape units. At the same time the interface would also have to control the routing of data transfers between the CPU and selected units.

(d) The performance of parity and sum checking operations (including parity insertion) on data transfers between the CPU and the peripherals, and the initiation of an error interruption (usually on a special line) when errors are detected.

In the case of mini- and microcomputer circuits special provision must be made to allow for the **three-state or tristate** logic used on the bidirectional buses. A bidirectional data-bus implies that any one line of the bus can be driven by any number of devices, but only by one device at a time. This is realized by ensuring that the source that controls a given line can force that line into one of three states – logic 1, logic 0 and an off or high impedance state. In the high impedance state the line is available for other devices to use without adversely affecting the other sources that can drive the line. In practice this is done by employing special tristate buffers or invertors for the input port as shown in Fig. 7.10.

The bus outputs D_0–D_7 equal the inputs I_0–I_7 (either logic 1 or 0) when $\overline{EO}$ is low, thereby allowing the inputs to be put on the data bus. When $\overline{EO}$ goes high the outputs D_0–D_7 go to a high impedance state which does not affect the data-bus. Since it is only necessary to use tristate logic when driving the bus, output ports are much simpler and normally consist of a set of D-type bistables which are used to store the state of the bus lines on receipt of a clock pulse.

The circuitry for both input and output ports are available as integrated circuits and form an essential part of any microcomputer system.

A generalized block diagram of an interface unit is shown in Fig. 7.11; the arrangement is typical for all types of IO operations, with the exception of the DMA mode, when the address and word-count is also controlled by the interface. The common address highway is decoded in the address and command decoder (governed by the transfer address strobe) which routes the command bits to the control unit when the appropriate device address is received. The control unit then generates the necessary control signals back to the CPU (for example, the operable line) and prepares the PCU to accept (or deliver) data from the address highway; if required (for example, in an input instruction) the IO equipment is activated at the same time. Data transfers to and from the buffer register are controlled by the control unit, in conjunction with the external transfer data strobe. Thus, the responsibility of the control unit is to generate all the necessary control and timing pulses, as required by the CPU or peripheral equipment. The buffer register may hold a single character or word, or provide storage for a complete block of data; in some cases data assembly ('packing' characters into a computer word) may also be performed by the unit. The device interface circuitry includes all the line drives and output lines to the peripheral equipment, including the termination units on incoming lines. The line drivers are used to match the high impedance logic outputs to low impedance cables and output lines; in some cases voltage level changing capabilities may also be required.

It will be obvious from the above description that the PCU can, in some cases, be of considerable complexity, since it must provide all the special control signals and timing for its own internal functions (as well as external control) during data manipulation and transfer operations. Since this type of control is best provided by a micro-program approach, the PCU (and in particular the control unit) will have a similar structure to the computer itself and, indeed, in many large systems, small 'satellite'

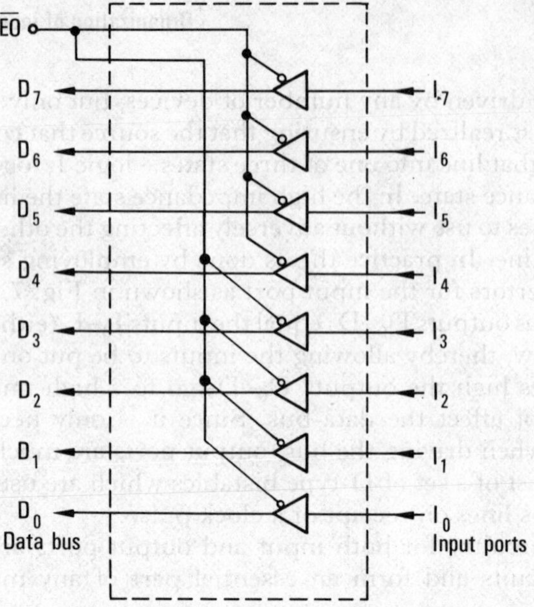

Note: The small circles indicate inversion.

Figure 7.10 Tristate 8-bit input port.

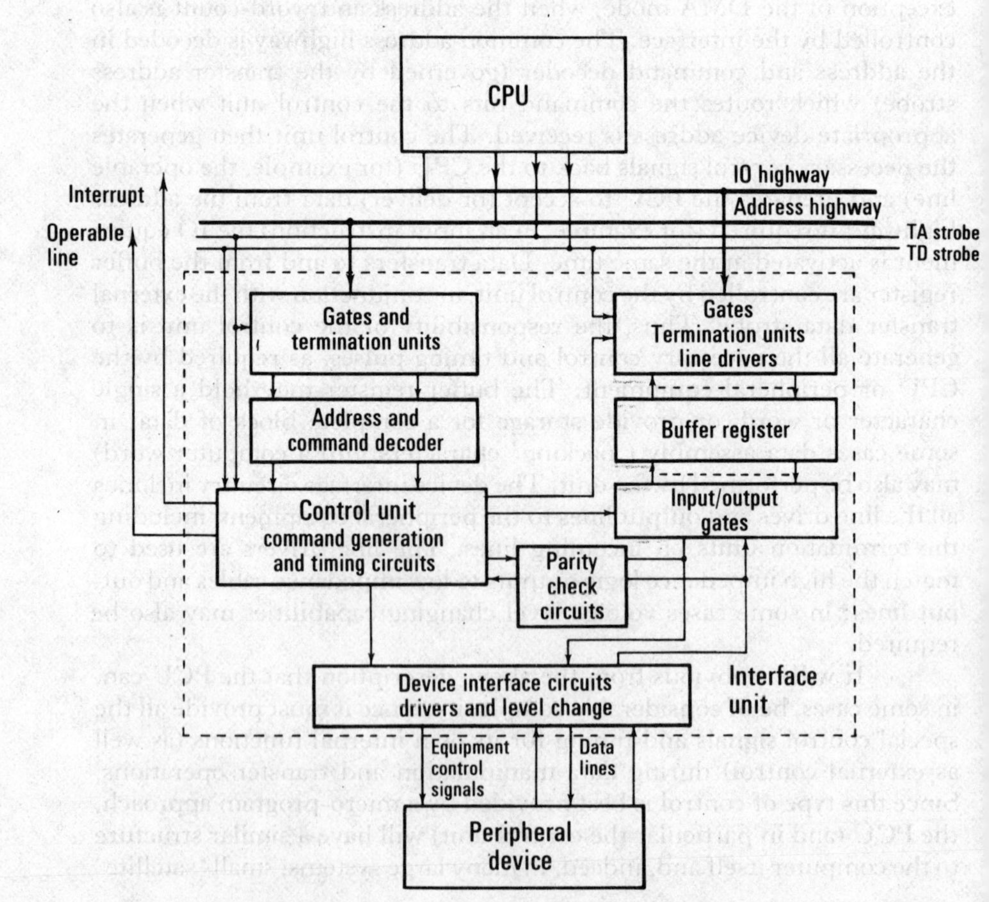

Figure 7.11 Generalized peripheral control unit.

computers are used for this purpose. In these cases, however, the satellite machine would fulfil the function of a shared general purpose interface unit, that is, controlling and multiplexing data transfers between the CPU and *all* its associated peripheral devices. However, with the availability of cheap processing power in the form of microprocessors there is less need to share computing facilities and many current systems now use microprocessors to perform interface control for individual peripherals.

In most cases the manufacturer of peripheral equipment supplies only the basic mechanism (a tape reader or magnetic tape unit) with the minimum of electronic control circuitry. This is understandable since, as there is such a diversity of computers available on the open market, to produce an interface unit for every computer model would be an economic impossibility. It will be apparent then that the problem of interfacing IO equipment to a particular computer is the responsibility of the computer systems, or main frame, manufacturer. Thus, the design of the interface units is one of the chief functions of the computer systems engineer. Moreover, to accomplish this design process effectively it is essential for the engineer to have a complete understanding of the IO requirements from the viewpoint of both the hardware and software specifications.

The design of interfaces has become of even greater importance with the arrival of the microcomputer. Most microcomputer applications involve the interfacing of peripheral equipment, and indeed the design of a microcomputer system itself is essentially one of interconnecting special purpose LSI chips to give the required performance. The realization of a microcomputer system is thus considerably simplified by the availability of a suitable 'chip family' which is compatable with the microprocessor (this is an important factor in selecting a microprocessor for a given application). Typical of such chips are the Input/Output Ports already described, these are normally packaged (as for example in the Motorola PIA, Peripheral Interface Adaptor) as programmable dual I/O Ports complete with associated handshake controls. Other devices which are available include decoders, load drivers, keyboard encoders, UARTs, termination units for signal lines, interrupt controllers, DMA controllers etc.

More recently, 'programmable' input/output controllers for peripheral devices have come on the market (for instance, controllers for floppy discs and CRT displays) which are intended to release the microcomputer of major peripheral control functions.

Some attempts have been made to standardize the logic levels, control signals, character formats, etc, of IO equipment in an effort to ease the problems of device interfacing (and also the portability of databases). Unfortunately, these proposals have met with little or no success owing, in the main, to the individualistic nature of the IO devices and the manner of their development. However this situation could change due to the influence of microcomputer technology where it would seem essential to standardize the peripheral chips, particularly the bus interface between different systems.

Some success has been achieved in designing standard interfaces[5,6,7] which allow a wide variety of different devices to be connected to any computer. In particular, the Hewlett-Packard interface[8] which defines interface bus standards for instrumentation systems has much to commend it.

7.4.6 Operating systems[9,10,11]

In large computing systems, particularly those working in real-time, special provision must be made to handle the many different input/output activities, including the processing of program interrupts. For example, inputs from teletype terminals, punched card readers, and so on, must be dealt with, simultaneously perhaps with producing output on a line-printer or refreshing a CRT graphics display unit. All of this activity must be continuously scheduled and controlled as it occurs by the computer on a priority basis with the lower priority devices being queued. Coordination and control of these input/output activities is generally assigned to a software routine, called the **supervisory program** (or the executive, monitor, or operating system – OS). The main function of this software is the scheduling of the programming work load (including, of course, the IO activities) in such a way as to optimize the overall machine throughput. Supervisory programs must be written by very able and experienced programmers, with every effort being made to produce an efficient and error free program. The responsibility for writing these programs normally resides with the computer manufacturer who would design, deliver, and maintain (update) the supervisory software as part of the overall computer system. Thus, the supervisory program is normally regarded as part of the basic machine system or architecture and as such is essential to the operation of a computer system.

Some of the major functions which must be performed by the operating system are tabulated below:

(a) *Input/output organization* – the control and scheduling of input/output equipment, including error detection and rectification.

(b) *Storage allocation* – the assignment of main memory and disc storage, as and when required, during program loading and operation. (This is in fact the operation of the 'virtual memory' system, described in Section 6.10.2.)

(c) *Communication with the operator* – facilities must be provided to allow the operator to give instructions to the computer concerning program loading and initialization, allocation of priorities and peripheral devices, 'trouble shooting' for faults, and so on. At the same time error messages, requirements for changing magnetic tape reels, etc. must be notified to the operator.

(d) *Interrupt processing* – the cause of the interrupt must be investigated, and control transferred to the appropriate routine if warranted by the priority of the device; storage and retrieval operations on the contents of registers used in the current program must also be performed.

(e) *Queue control* – queues must be formed of all requests for input/output operations, which are allocated in an optimal sequence.

(f) *Display handling* – the organization and updating of information for repetitive scanning displays, such as CRT graphics terminals.

In complex multi-programmed systems all these functions, and many others besides, would be required. Smaller systems, however, would be provided with a simpler executive program concerned primarily with program loading and the scheduling of peripheral devices.

In particular the following features will be required:

(a) *Input/Output control and device assignment* – The OS would comprise device handling software (IO routines) for the peripheral equipment and these routines must be allocated using system macros prior to running a user program. In general, peripheral equipment is referred to by **device numbers** in both the high-level language user program and the systems software; this enables modifications in device allocation in the user software to be made without major program changes – for example, magnetic tape could be substituted for disc simply by re-assigning the device numbers in the OS.

(b) *System Generation* – Operating systems software is normally written as a general systems package and it must be tailored to fit a specific machine configuration. This requires a special system package called a **system generator** which accepts the basic OS program, allows the user to specify core size, interrupt options, peripheral devices etc., and then generates a system tape which is used to install the OS system on disc.

(c) *File Manipulation* – one of the most important user functions of the OS is the organization and manipulation of files stored on disc or magnetic tape. Typical commands would include FIND, SAVE, LOAD, DELETE, RENAME, OPEN, CLOSE, etc. The OS will also assign a directory table for each disc or tape listing all files stored and their addresses – this information can be displayed by giving the command DIRECTORY followed by the device number.

The design and implementation of operating systems is a major intellectual exercise and much still remains to be achieved in this area. This is particularly so when concurrent operation in multi-processor systems is considered. Many operating systems are inadequate and contain logical errors which only become apparent when the software is thoroughly exercised, sometimes many months after final implementation! In effect the requirement is to design a large man/machine system, which is fundamentally an *engineering* task. Unfortunately, the software engineering approach is not as yet universally accepted but until it is the user will continue to be plagued with ineffectual, badly documented, and ill-produced software systems. One of the future developments we can expect to see in computer architecture is the replacement of many of the software supervisory functions by hardware logic circuits.

7.5 Graphic systems

In many applications, in particular computer aided design (CAD), an important output requirement for a computer system is to render its results in a graphical form, for example, as line drawings, diagrams, and graphs.

The **line printer**, as well as producing a standard printed page, may also be programmed to output simple line diagrams; this technique has been employed, for example, by IBM to prepare logic diagrams for maintenance engineers. This is possible because of the high speed (300–2000 lines per minute) and large character set (up to 120 character positions per line) of the equipment. Another important advantage is that a complete row of characters may be printed at a time, with the format being controlled by plugboards within the printer itself or by program from the CPU. In common with other peripheral devices the line printer may be operated in a buffered interrupt mode, using a buffer store capable of holding a complete line of characters.

An alternative form of graphic output is the **incremental plotter**, which can be either of the **drum** or **flatbed** type. Drum plotters use continuous edge-perforated stationery up to 30 inches in width, which is driven longitudinally by sprockets engaging in the perforations, from one roller to another. At the same time a pen is moved in discrete steps across the paper in the horizontal direction. Thus all moves, either by the pen or the paper, are in incremental steps controlled by the computer. Increments are usually in the order of 0·005 in and occur at a rate of approximately 300 per second with a positional accuracy of about $\pm 0{\cdot}005$ in. Flatbed plotters differ from drum plotters in that the pen is moved in both X and Y directions (again in incremental steps) over a stationary sheet of flat paper. They have the advantage of greater precision ($\pm 0{\cdot}002$ in) and also allow the use of non-standard stationery. For some applications, for example CAD, provision is also made to input positional data (in a digital form) to the computer. Moreover, it is also possible to adopt this form of output device to cut masks for integrated circuits directly from the computer derived output.

Another form of graphics output which has rapidly gained popularity, particularly in interactive systems (that is, where an intimate collaboration between man and machine is required) is the **cathode ray tube (CRT) display**.[12,13,14] There are two basic ways of using the CRT display in a computer system, they are:

(a) *Alpha-numeric tabular displays* – these devices effectively perform the function of an electronic typewriter, allowing data, manually set up from an associated keyboard, to be visually inspected on the tube face before insertion into the computer. In addition, data already stored in the computer may be called up (from the keyboard) for display by the CRT terminal.

(b) *Graphic and interactive displays* – in this case the CRT has the ability to display diagrams and line drawings as well as alpha-numeric data. The graphic terminal is usually provided with pointing and indicating devices such as a light-pen or tracker-ball, to allow direct interaction between the operator and the computer.

The keyboards associated with graphic display units usually have a full alpha-numeric character set and some form of editing and function keys. Data to be inserted into the computer store are typed out on the keyboard and simultaneously displayed on the CRT where they may be visually inspected; mistakes may be rectified by using special editing keys which have spacing and erasure functions. When the data message is correctly assembled a control key is depressed which inputs the message into the computer store.

When working in the interactive graphics mode it is necessary to be able to point to any position on the tube face, that is, a part of a diagram or an alpha-numeric control list (sometimes called a 'recipe'). One way of doing this is to use a tracker-ball mechanism, which is a large ball free to rotate about its axes within a fixed mounting (alternatively a joy-stick control may be used). The position of the ball in its mount controls incremental digital counters in the X and Y axis of the CRT screen, the contents of the counters being decoded to display an electronic marker (such as an arrow head) on the tube face. By manually rolling the ball in its mounting the marker may be moved across the tube face to any desired position, the XY coordinates of this point may then be stored by pressing a special function key. Alternatively, the pointer may be used to define the start and finish points of a line whereupon a vector function key is depressed causing a line to be drawn between the indicated points. In this way a simple line diagram may be constructed on the tube face.

The same facilities can also be provided by a photo-sensitive light-pen, which gives a more natural writing action. The pen can be used to mark a position on the tube face by placing the pen over the area of interest and then operating the appropriate function key. When the computer 'refreshes' the area at which the light-pen is pointing, the pen senses the flash and the resulting signal is used to inform the computer that the operator was pointing at a particular part of the pattern drawn on the tube face. To draw with a light-pen the usual technique is to 'collect', or latch on to, a preset mark from the corner of the display, and 'drag' it across the tube-face. Operation of the appropriate function keys enables the start and finish points of vectors to be defined, or the placing of alpha-numeric characters, etc., in specified positions.

The characters and vectors are normally generated on the CRT by means of a **stroke-writing** technique,[15] using either a dot matrix or a dot vector system. The dot matrix is the simplest form of display and normally uses a matrix of 7 × 5 positions; a combination of points within the matrix format is 'painted' in sequence on the CRT to produce a visible character or symbol. The dot vector method is an extension of this technique, in which a combination of points is defined for each character and the electron beam made to traverse a path from point to point to form the character.

The display system itself normally consists of a digital store driving a conventional CRT unit. This technique gives rise to many design problems such as refresh rates (a flicker free display is essential), character generation, and beam deflection times. If the dot vector method is used, for example, the deflecting waveforms must start and finish properly if character distortion is to be avoided; the problem is aggravated as the character generation time is reduced. One major innovation in this area has been the introduction of main and secondary deflection systems for the CRT. The main system is comparatively slow, but works over the entire area of the tube face, while the secondary is very much faster but only operates over a limited area.

The **direct view storage tube** has also been used for this purpose. These have the advantage that display information is written as a charge pattern on a storage mesh immediately behind the viewing surface, thus it is not essential to continually refresh the display. When a visible display is required a separate electron gun floods the storage mesh with low velocity electrons, which penetrate wherever a charged area has been written, and are consequently accelerated to the phosphor on the viewing screen. Though the display is steady and bright, resolution is only marginally satisfactory and there are problems in the selective erasure of parts of a diagram. Another disadvantage is that since continuous refresh is not used a light-pen will not work with this tube.

A typical graphics display would have picture repetition rates varying from 30 to 70 cycles per second, depending on the characteristics of the tube phosphor and the method of data storage. The normal alphanumeric type of terminal will display between 100–2000 characters in preset matrix positions, while an interactive graphics terminal will have the additional facility of drawing 150–10,000 inches of connected line segments.

It will be apparent that the display terminal must be backed by some comprehensive software system to enable the drawing and pointing functions to take place, as well as the generation, updating, and refreshing of the actual picture itself. To produce a picture on the CRT a sequence of command words must be generated and transmitted to the display unit. These commands are used to position the beam, to move it linearly across the tubeface to draw a line (vector generation), or to produce a character. Normally a binary-coded command word, containing a function code and XY coordinates (usually 10-bits each) would be used for this purpose. A picture is built up from a given sequence of these words contained in a **display file**, which may be part of the computer store or held in a separate display buffer. To produce a steady picture on the CRT the contents of the display file must be periodically cycled through (a word at a time) and passed to the display unit; this procedure is known as the **refresh cycle**. The display file is a very small part of a complete graphics software package. If any serious computational work is to be performed on the system or devices represented on the screen, another more comprehensive form of data storage, usually called a **data-structure**, is an essential requirement. The data-structure is used to set up a complete model of the system being displayed and will contain many different types of display file. Moreover, it is possible to generate new display files (for example, different visual interpretations of the same object) from the data-structure. With this approach it is possible to expand the picture, which means in practice that selected areas of the picture can be viewed in greater detail. This is achieved by storing the XY coordinates in the data-structure to a greater degree of accuracy than that required by the resolution of the display unit. Other facilities such as picture manipulation and deletion, production of symbols by using 'sub-pictures' etc., are also possible using the data structure concept. In some of the more sophisticated graphics software, special graphics languages are available which enable users to construct diagrams for insertion in the computer store by writing a verbal description of the required picture.

A block diagram of a typical graphics display system is shown in Fig. 7.12. The buffer store, containing the current display file, may be loaded from the computer, using the DMA mode, a complete display file at a time, or alternatively a specified data block may be updated. Some systems attempt to control the display directly from the main CPU store, but the duty cycle required to present a flicker-free display soon renders this technique uneconomic (in terms of computer time). Operating instructions from the keyboard go directly to the computer, thus allowing the basic data structure to be modified. In operation, data words are read

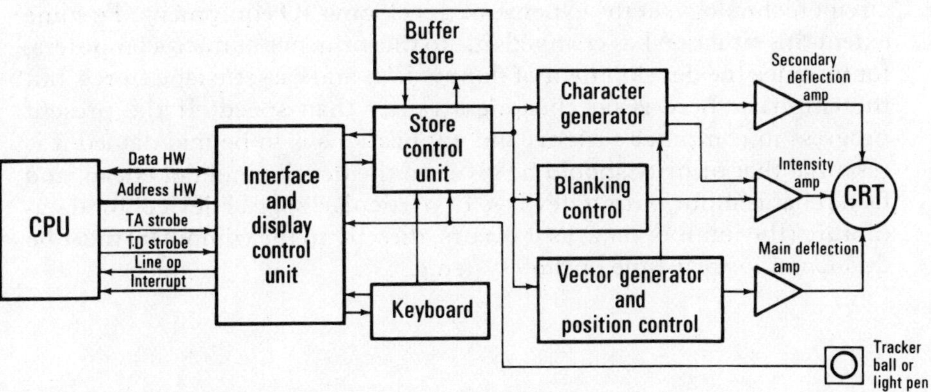

Figure 7.12 Graphics display systems.

down from the buffer store and decoded, the decoded outputs together with the *XY* coordinates being used to control the character and vector generators. The blanking control unit is used to 'black-out' the unwanted paths made by the beam in its traverse from one *XY* coordinate to another.

It will be obvious by now that a CRT graphic system is a very complex and sophisticated peripheral device and as such is a costly investment. Moreover, since many of the operations it performs are analogous to those required of a computer (and at a comparative cost) it has become common practice to replace the control and generation units with a small computer directly driving the CRT display system. In this case the entire system (including the control computer) is treated as a peripheral to the main computer system.

As one would expect microprocessors are being increasingly used to perform many of these functions[16] particularly character, symbol and vector generation. The microprocessor allows a higher level graphics terminal language to be used in the host machine which is interpreted in the microprocessor into a sequence of low level commands for the graphics display unit. For example, to draw a circle all that is required is to send the command DRAW CIRCLE together with the radius and centre coordinates of the required circle – the microprocessor will then generate the series of coordinates specifying the line segments needed to draw the circle. The use of a microprocessor does not dispense entirely with the host computer, this is still required to perform scheduling and **windowing** operations (windowing gives the effect of a zoom lens to allow the detailed examination of a certain part of the display) as well as the processing of interrupts from keyboards, cursors etc.

The major problems associated with input/output systems are directly related to the slow operating speeds of the peripheral equipment. In fact it has been said that 'peripheral devices are the Cinderellas of the computer industry' and to a large extent this is true. Too much attention perhaps has been paid to the design of the CPU, and with progressing circuit technology, at the expense of developing IO equipment. To some extent this situation has changed due to the influence of microcomputers, for instance the development of floppy discs and cassette tape stores, but the emphasis here is on cheapness rather than speed. If the present progress in computer systems and applications is to be maintained it is essential that priority should be given to the development of cheap, and fast, reliable input/output devices. In particular, special devices for data-capture (the input of data, as it occurs, directly to the computer) must be developed for real-time on-line systems.

References and bibliography

1 Hoeschele, D. F. *Analog-to-Digital/Digital-to-Analog Conversion Techniques.* John Wiley, New York, 1968.

2 Lacy, R. D. A high speed tape reader. *J. Brit. I.R.E.*, 1960, **20**, 661–668.

3 Turner, L. R., and J. H. Rawlings. Realisation of randomly timed computer input and output by means of an interrupt feature. *I.R.E. Trans. Electron. Comput.*, 1958, **EC7**, 141–149.

4 Bock, R. V. An interrupt control for the B5000 data processing system. *Proc. AFIPS F.J.C.C.*, 1963, **24**, 229–241.

5 Barber, D. L., and E. P. Woodroffe. *The NPL Standard Interface.* NPL Report AUTO 14, 1965.

6 British Standards Institution, BS 4421 : 1969A, *A Digital Input/Output Interface for Data Collection Systems.* British Standards Institution, London, 1969.

7 *Euratom CAMAC : A Modular Instrumentation System for Data Handling, Description and Specification.* Euratom Publ., EUR-4100e, 1972.

8 *IEEE Standard Digital Interface for Programmable Instrumentation.* IEEE Std., 488, 1975.

9 Katzan, H. Operating system architecture. *Proc. AFIPS S.J.C.C.*, 1970, **36**, 109–118.

10 Donovan, J. *Systems Programming.* McGraw Hill, New York, 1972.

11 Madnick, S. E. and Donovan, J. *Operating Systems.* McGraw Hill, New York, 1974.

12 Lewin, M. H. An introduction to computer graphics terminals. *Proc. IEEE*, 1967, **55**, 1544–1552.

13 Walker, B. S., Gurd, J. R. and Drawneek, E. A. *Interactive Computer Graphics.* Crane Russak, New York, 1976.

14 Newman, W. M. and Sproull, R. F. *Principles of Interactive Computer Graphics.* McGraw Hill, New York, 1973.

15 Dammann, J. E., E. J. Skiko, and E. V. Weber. A data display sub-system. *IBM J. Res. Dev.*, 1963, **7**, 325–333.

16 Raymond, J. and Banerji, D. K. Using a microprocessor in an intelligent graphics terminal, *IEEE Computer*, **9**, 18–25, April 1976.

Further reading

Cluley, J. C. *Computer Interfacing and On-Line Operation.* Crane Russak, New York, 1975.

Kilburn, T., D. J. Howarth, R. B. Payne, and F. H. Sumner. The Manchester University Atlas operating system Pt 1-internal organization. *Computer Journal*, 1961, **4**, 222–225.

Clayton, B. B., E. K. Dorff, and R. E. Fagen. An operating system and programming system for the 6600. *Proc. AFIPS, FJCC*, 1964, **26**, pt. II, 41–57.

Oestreicher, M. D., M. J. Bailey, and J. I. Strauss. George 3 – a general purpose time sharing and operating system. *Comm. ACM*, 1967, **10**, 685–693.

Edwards, D. B., D. Aspinall, and T. Kilburn. Design principles of the magnetic tape system for the atlas computer. *Radio and Electronic Engineer*, 1964, **27**, 65–73.

Nicholson, A. W. Peripheral transfer system for a fast computer. *Proc. IEEE*, 1964, **111**, 15–26.

Pooch, U. W. Computer Graphics, interactive techniques, and image processing 1970–1975: A Bibliography, *IEEE Computer*, **9**, 46–64, August 1976.

Tutorial problems

7.1* Using the computer structure and micro-order set described in Chapter 4, consider the problems involved in implementing the data break mode of operation. Write an appropriate micro-program to perform this function, defining any additional micro-orders that may be required. Describe the resulting changes, if any, on the hardware structure of the machine.

7.2 Outline, in flow-chart form, suitable software routines for identifying and queuing multiple interrupts (assume a program specified priority system).

7.3* Describe in detail the overall systems design for an interface unit required to input data, using the interrupt mode, from a teletypewriter into the main store of the CPU. Consider in particular the problems of error erasure and the provision and coding of control symbols. List any assumptions you need to make in accomplishing your design.

7.4 Perform a detailed logic design for the buffered paper-tape input unit shown in Fig. 7.7. In particular consider the design of the control unit and attempt to partition the logic into the following sections:

(a) basic interface addressing and control circuitry;

(b) special logic required for the buffering and word assembly processes.

Estimate the number and the speed of the logic packages required to implement your design.

7.5* In large teleprocessing systems, data are transmitted to a central computer installation from remote stations over a serial digital data link, using standard telephone lines operating at 50 bits/second. Discuss the problems associated with the design of such a system, assuming that the input/output organization of the computer is similar to that described in the chapter. In particular, consider the encoding and decoding of the data (including error detection and correction) and whether or not this function should be performed in the peripheral unit or the CPU (either as a software routine or micro-programmed firmware).

8

Engineering and systems aspects

8.1 Introduction

In this chapter we shall discuss some of the problems associated with the design and engineering of digital computer systems. Though it is impossible to provide specific solutions (or design rules) for many of these problems, we shall attempt to present some general guidelines and techniques which may prove helpful to the reader when faced with a specific application. In particular, after considering the basic systems design philosophy we shall deal specifically with such important topics as reliability theory, wiring and noise problems, and systems testing.

The overall and detailed design of digital computers, like any large systems project, is an iterative process. That is, starting with a tentative design specification, a solution is postulated which is then critically evaluated and the information so gained used to modify or improve the original concept. The engineering of a large system involves the following:

Defining the problem – evolving, in conjunction with the customer, a mutually acceptable specification.
Synthesis and analysis of a postulated solution, including economic and technical feasibility studies.
Development of system components.
Engineering and production of the finalized system.
System testing and evaluation.

It is essential to remember throughout these design stages that any system is essentially a man/machine system,[1,2] and that due attention must be paid to the role of the operator, production and maintenance engineers, and so on. The writing of operational and maintenance manuals (for both hardware and software), for example, should begin as soon as is feasibly possible and not postponed (or forgotten!) until after the system is completed.

Bearing these concepts in mind, the design of a digital computer system may be conveniently divided into four major engineering activities, these are:

System design. The initial specification for a computer system must be formulated with regard to speed, cost, size, reliability, maintainability, and operating environment. A design philosophy must also be established concerning operational requirements such as arithmetic and logic functions, input/output facilities, serial/parallel working, error-checking, program interrupt and any other special features dictated by customer or market demands. During this initial design phase the LSI logic family, storage media, and input/output devices would also be chosen.

Logic design. There is considerable overlap and interplay between logical and systems design, and quite often both activities can proceed in parallel. In this stage of the design the overall computer architecture would be finally decided, including the computer wordlength, the organization of the machine-code instruction set, systems software etc. With these decisions made, it is possible to proceed to the detailed design of the logic circuits and sub-systems. The end product of this phase would be an overall logic design for the machine, expressed as a set of equations or, more likely, as a detailed set of logic diagrams, flow diagrams and microprograms. Similarly, software development should also commence at this stage (particularly diagnostic and test programs) using a simulated version of the machine-code order set.

Circuit design. This stage considers the design of special purpose circuit hardware such as logic level convertors, power amplifiers and line drivers, sense and drive amplifiers for storage media and input/output devices. Circuit and logical design may proceed concurrently once the basic logic family, including interface equipment, have been fully specified.

Production and Commissioning. This is the final phase of the design process and involves the production, testing, and installation of the equipment. It is normal practice with large computer systems to forgo the manufacture of a complete prototype equipment because of the considerable expense involved. Small sub-systems may be built and tested, especially when the design may be considered critical, but generally it is the finalized design which is produced and commissioned. In most cases, however, the first production model would be retained by the manufacturer for software development.

The most important system features that must be considered in the initial design stages (particularly during the feasibility study) are the operational requirements, speed, cost, and reliability of the computer. Since these parameters are not independent, it is often necessary to 'trade-off' one feature against another – for example, an increase in speed may be paid for by a decrease in reliability.

In meeting the customer's requirement for a computer system (or, as is more usually the case, in-house sales policy based on market research), the one that exactly fits his specification is not necessarily the best in practice. If the machine is custom built for a specific application it is, to all intents and purposes, a special purpose machine with a very limited market. Moreover, all the development costs must be borne by the first few models, instead of being distributed over a large number of machines; this adversely effects the selling price and/or profit margin. Consequently, an attempt must be made to produce a general purpose machine with enough capabilities to cover a wide applications market. This approach has led to the **modular** concept of computer systems architecture (assiduously applied by IBM in their 360 series) in which a customer specification may be met by assembling different standard modules, such as slow and fast core stores, floating-point and fast arithmetic units, various input/output peripherals, etc., into a particular system configuration. The same arguments of course also apply to software packages, and in many real-time applications the cost of developing system software can far exceed the cost of the hardware. Once again the principle of general purpose modules applies, but this time the writing of particular applications programs is left to the customer, the main responsibility of the manufacturer being the provision of basic utility programs such as language compilers and operating systems.

The speed of a computer is largely determined by the speed of its components and fast devices (logic and storage elements) are very costly. It is also possible to achieve high speed by a suitable choice of computer structure, in the simplest case, by using parallel mode in place of serial mode operation, but this too can be costly. In general, increased costs are unavoidable when high speeds are required, so it is essential to question the necessity for high speed and to investigate alternative means of achieving it if the overall systems cost is to be kept at a low level. One method of accomplishing an economical yet high speed design, is to integrate low and high speed components or techniques – for example, the use of a fast slave memory in conjunction with a slow speed core store (as described in Section 6.10.1) or serial/parallel arithmetic circuits.

Due to rapid developments in semiconductor technology a major change has occurred in digital systems design. Instead of implementing designs at the basic gate level it is now common practice to use complex MSI and LSI components working at the sub-system level. Consequently much of the existing logic design theory has been rendered redundant but unfortunately a new theory has not, as yet, emerged.

Moreover, the development of the LSI microprocessor, and its associated modules, has enabled computer systems to be fabricated, almost entirely, from these sub-system components. This has changed the emphasis in computer engineering from logic design to computer architecture and software systems. Indeed the major factor in computer design and application (though not often realized by the naive user of microprocessors!) is the development of appropriate and robust software. It is also interesting to note that the process of digital systems design has now moved down to the device and component level with the development of microcomputers on a chip.

Unfortunately the structure of the available microprocessor systems is still very primitive and, except for innovations like facilities for stack processing, very similar to first generation computers. Consequently they are no easier for the user to program and exhibit many of the basic problems encountered in the early days of computing. It is to be hoped that future research in computer architecture, rather than device technology, will rectify this anomaly.

8.2 Implementation of the logic design

There are many types of logic circuit available as standard integrated circuit packages and no computer manufacturer today would dream of designing and producing his own logic modules (however, IBM do just this!). Initially, integrated circuits were very expensive but owing to large scale production techniques they are now very much cheaper and more reliable than their discrete component logic counterparts. Thus, at the present time integrated logic circuits are used exclusively for realizing digital systems.

Various forms of micro-circuit[3, 4] are in existence, such as:

Semiconductor integrated circuits in which all the passive and active elements, including conductors, required to perform a specified logic function are formed in, or on, a semiconductor substrate by diffusion and/or epitaxial growth processes; they are also called **monolithic circuits**.

Multiple-chip circuits are a variant of the above devices in which, instead of forming the circuits on one large substrate, several semiconductor substrates, or **chips**, are used; the individual chips are subsequently interconnected within a single component package. This technique enables the isolation of critical components and thus prevents the occurrence of parasitic oscillations.

Thin-film circuits have all the passive components and conductors formed on an inert substrate, such as glass, by evaporation, spluttering, or plating techniques. The active devices, transistors and diodes, are added afterwards, as discrete miniature components to this circuit.

Hybrid circuits use semiconductor integrated circuits for the active elements, with thin-film passive devices added as overlays.

Since integrated circuits are by far the most developed and widely used of the micro-circuit family, we shall confine our comments to this type of circuit.

Integrated circuits are normally encapsulated in three basic package types, the most common being the **TO-5** style. This is basically a transistor mounting can ('header') of approximately 0·325 inches diameter with the connections (8–12) brought out from the base. Another method of packaging is the so called **flat-pack**; this is a rectangular capsule and comes in two standard sizes, 0·25 × 0·25 × 0·05 in and 0·25 × 0·15 × 0·05 in with the connections some 10, 12 or 14 pins brought out straight from the edges. A cheaper method of encapsulation, which is also much easier to assemble onto printed circuit boards is the **dual-in-line** (DIL) package. This is rectangular in shape, much larger than the flat-pack, but with the connections (up to 24) brought out from each side and bent at right angles with 0·1 in spacing for easy mounting. DIL packages are also used for LSI and MSI circuits.

The principle advantages of integrated logic circuits over discrete component logic are:

Reliability, which depends mainly on the bonding and encapsulation of the device rather than the components or circuit configuration.

Considerable reduction in weight and size.

Reduced power consumption.

Speed of operation: 5–150 MHz compared to 10 kHz–5 MHz for discrete component modules.

The disadvantages of integrated circuits arise mainly from their low logic levels, in the order of 0·8–5 volts, and their consequent susceptibility to d.c. noise. This may arise in various ways: electromagnetic radiation from other equipment, such as adjacent contactor circuits and power supply transients; internal noise injection from the system, occurring as a result of bistable switching transients and large earth currents. Pick-up problems can generally be overcome by screening, the use of common earth planes, and careful component and wiring layouts (see later).

In most cases the logic function provided is the NAND/NOR operation but other variations are available, including in particular the AND/NAND and OR/NOR circuits. Many different electronic circuit configurations have been developed to reproduce the required logical functions[5] – for example, **diode transistor logic** (DTL), which uses a well proven NAND/NOR circuit capable of operating between 2 and 20 MHz with a good noise immunity; logic levels vary between 0·5 and 5 volts. **Resistor transistor logic** (RTL) is another well-tried and fast circuit (compared with DTL) but has a relatively poor fan-out factor and noise immunity. **Current mode logic** (CML) and **emitter-coupled logic** (ECL) use transistors only in a logic circuit in which a constant current is switched from one transistor to another. This type of circuit operates at very high speeds in the order of 150 MHz with switching times in the range of 1–5 ns. Since each logic unit draws a constant current there are no sudden demands on the power supply (when the logic gate is switched on) thereby removing one source of internal d.c. noise. **Transistor-transistor logic** (TTL) is very similar to DTL (the diode gates being replaced by transistors) but with the advantages of being much faster (4–65 MHz an average switching time of 10 ns) and having a better noise immunity. Figure 8.1 shows typical circuits for these basic logic elements. TTL logic is by far the most common form of integrated circuit logic family and because of this a number of modifications have been made to the basic circuit to enhance its speed and reduce power dissipation (not unfortunately in the one device since the two properties tend to be mutually exclusive). The TTL gate action can be speeded up, almost by a factor of 5, by interposing a Schottky diode

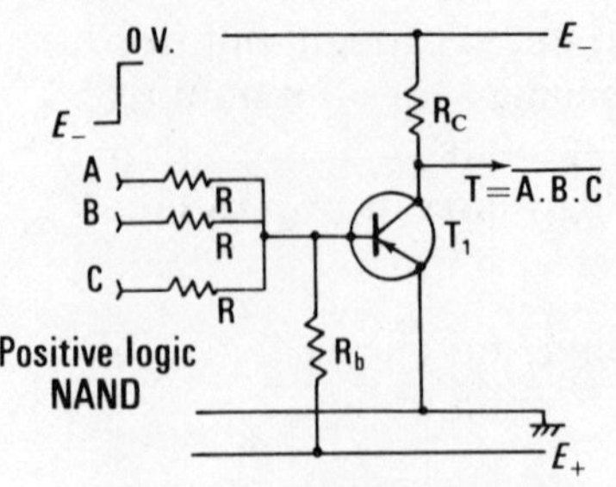

(a) RTL–resistor transistor logic.

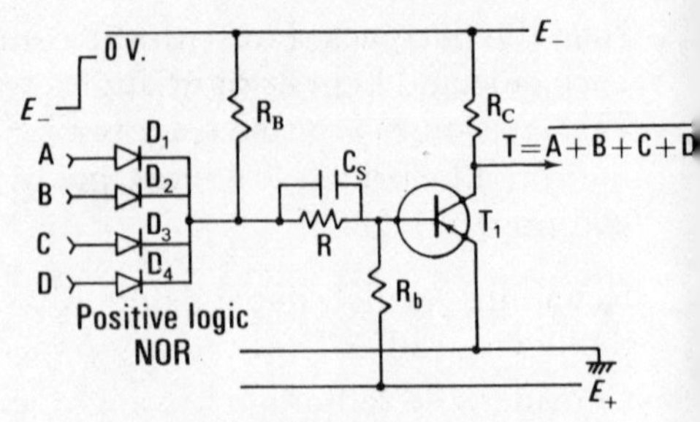

(b) DTL–diode transistor logic.

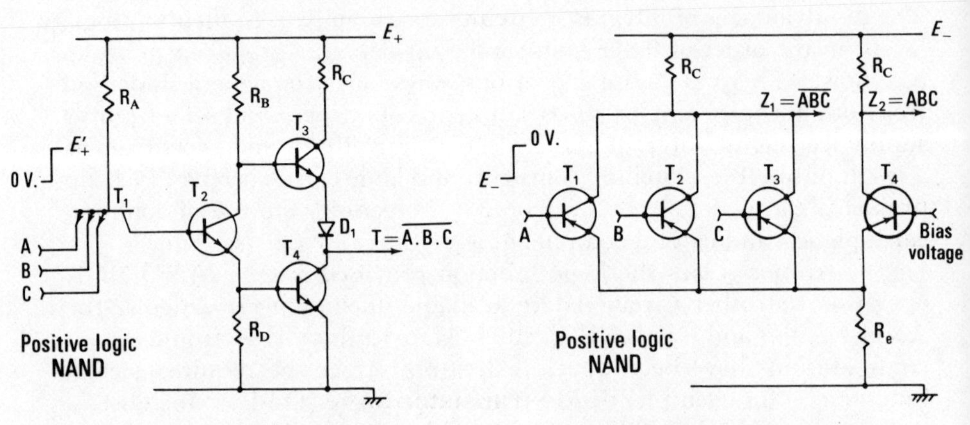

(c) TTL–transistor transistor logic.

(d) CML–current mode logic.

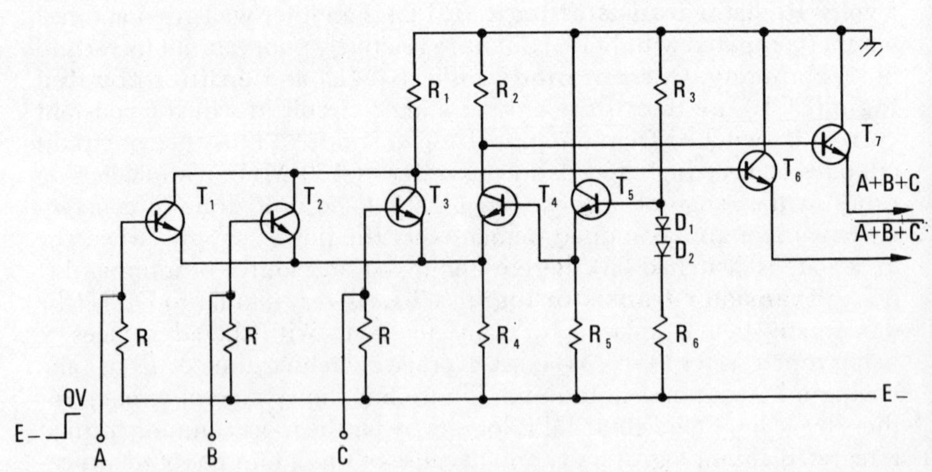

(e) ECL - Emitter coupled logic.

Figure 8.1 Logic gate circuits.

between the base and collector of the switching transistor to prevent saturation. Though **Schottky TTL** is comparable in speed with ECL the power dissipation is greater than standard TTL. **Low power Schottky TTL** can be produced however by the simple expedient of increasing the resistive path within the TTL gate circuit thereby reducing the power consumption by some 80 % of that required by a standard TTL gate, the speed unfortunately is also reduced and becomes comparable with TTL. The same technique (without the Schottky diode) can be used to produce a low power (but considerably slower) version of TTL.

Metal oxide semiconductor (MOS) **logic** circuits are very different from the circuits described above in that they are **field effect transistors** (FET) which are voltage rather than current sensitive devices.[6,7] Furthermore they have only been commercially produced in micro-electronic form and there is no equivalent discrete circuit. MOS transistor circuits are used extensively in digital design, particularly for LSI and MSI modules (such as shift registers and stores) but it is the **complementary** MOS (CMOS) circuit which is used primarily for logic gates.

There are two types of MOSFET, p-channel and n-channel. Each device consists of a drain, a source, a gate and a substrate area (normally earthed to the source) with external terminals. The drain to source resistance of a p-channel device changes from 500 M Ω to 750 Ω when a negative voltage is applied to the gate terminal and similarly for the n-channel except that a positive voltage is required. Thus the MOS transistor effectively acts as a switch or better still a relay contact (connected between drain and source) which can be opened or closed depending on whether or not a voltage is applied to the gate. Fig. 8.2(a and b) show typical MOS NAND and NOR circuits, note that the transistors are all assumed to be of the same channel type and that T_1 is functioning as a resistor. Note also that a current is required to maintain the logic 0 output.

The main advantage of the CMOS logic gate is that it has an extremely low power dissipation (no current flows in either the logic 0 or 1 static states) and a moderately high speed of operation (in the order of 25 MHz with a switching time of 25 ns).

CMOS logic gates use both p- and n-channel MOS transistors in a complementary circuit. For example, in the NOR gate shown in Fig. 8.2(c) if either input is high at least one of the n-channel transistors is shorted to ground and at least one of the p-channel transistors is open-circuit, so that the output is low (note that current only flows when the device is switching). A high output occurs only if both inputs are low enhancing the two p-channel transistors (enhancement occurs when the gate is made negative with respect to the substrate). In that case both n-channel transistors are open-circuit, isolating the output from ground. (Note that in Figs. 8.2(c and d) a different symbol is used for the CMOS transistors to show the difference between n- and p-channel devices.)

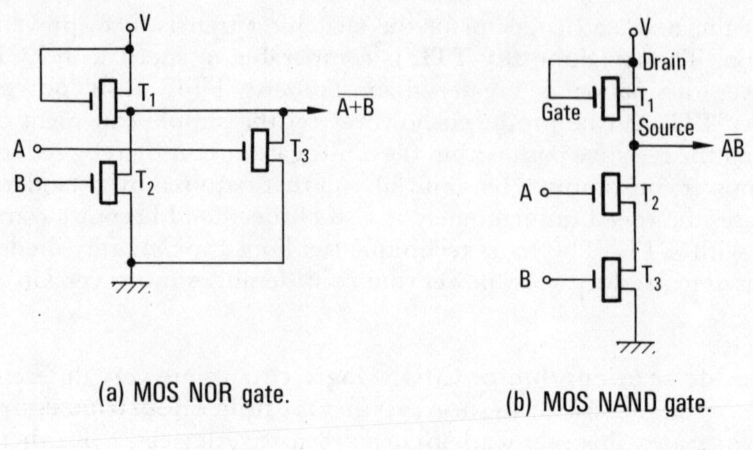

(a) MOS NOR gate.

(b) MOS NAND gate.

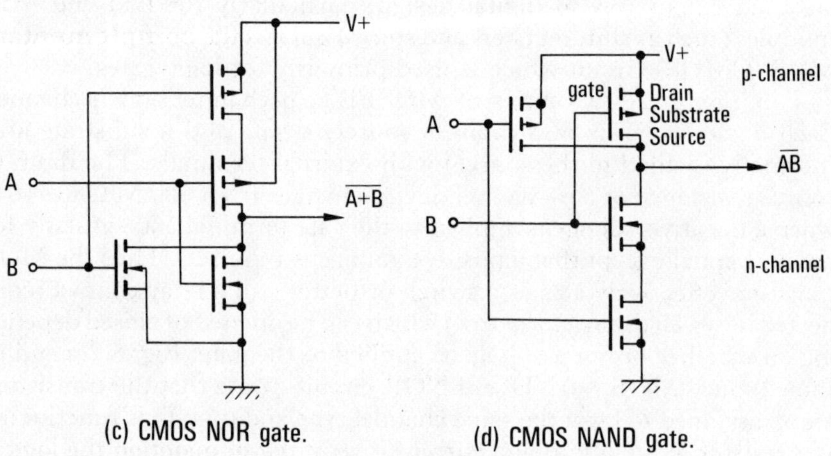

(c) CMOS NOR gate.

(d) CMOS NAND gate.

Figure 8.2 MOS logic circuits.

Logic gates are generally packaged according to their **fan-in factor** (that is, the number of inputs to a gate) – for example, four 2-input gates, three 3-input gates, two 4-input gates, etc.; this is mainly because of the pin limitation of the DIL package (nominally 14 pins). The number of inputs that can be fed from one output (called the **fan-out factor**) varies between 5 and 10, with power gates supplying up to 50 inputs.

As well as basic logic modules, such as NOR/NAND elements, most component manufacturers also market logic subsystems in an integrated circuit form. Typical medium scale integrated (MSI) subsystems would include:

Bistable elements – JK, SR, and D-type clocked bistable circuits; also master–slave (dual-latch) bistables, found particularly in TTL systems. Some systems also include d.c. SR bistables (toggles or latches) but these are more generally constructed from quad 2-input NAND/NOR packages.

Arithmetic logic half-adder circuits (exclusive OR plus carry logic) for use in basic binary arithmetic, single and multiple gated full-adder stages and look-a-head carry generators are provided by most manufacturers.

Registers – used mainly for storing and manipulating digital data; single or multiple-stage shift registers, with parallel/serial input/output, are normally available.

One-shots and timers – monostable circuits for timing control and oscillators.

Counters – binary, BCD, and decade up-down counter stages, in both synchronous and asynchronous versions.

Voltage level changers and opto-isolators – devices for interfacing one logic system to another by changing logic levels.

Analogue–digital converter elements – used for the conversion of analogue signals to logic voltage levels and vice versa.

Line drivers – power amplifiers to drive peripheral devices such as relay contactors and paper tape readers. It is also possible to obtain line termination units which match directly to twisted pair signal lines (see later).

Transceivers – bidirectional bus drivers for input-output.

UARTS – universal asynchronous receiver transmitter for interfacing I/O equipment.

Expandable gates – some logic elements have expander input terminals; in the case of a DTL NAND circuit this could be a direct connection to the base of the transistor. These terminals can be used to increase the fan-in factor of the element by connecting external clusters of diodes or resistors (called **expanders**) to the expander input. The fan-out factor can also be increased by the use of separate buffer amplifiers.

Decoder and comparator circuits – these are used to multiplex a number of input channels to a single output and for the comparison of binary numbers or parity checking etc.

ROM and PLA's – used for decoding and table-look-up.

When comparing logic systems the following parameters should be considered:

Type of logic – DTL, RTL, TTL, ECL, CMOS etc. ECL should only be used when high speed is essential; CMOS best system when power dissipation is important.

Fan-out/Fan-in factors, and if expander terminals are provided (note these increase propagation delay).

Speed

(i) Binary speed; that is, the maximum switching rate of simple set–reset bistables (also called the **toggle-rate**) this can vary from 500 kHz to 150 MHz for integrated circuits.

(ii) Propagation delay, this is the time delay through the element between input and output waveforms, and can vary with loading (fan-out) temperature and collector supply voltage. Values range from 1–100 ns with a typical average being in the order of 10 ns. (It is important to note that the propagation delay from logic 1 to logic 0 is generally less than that from logic 0 to logic 1.)

Supply voltages

(i) Logic levels, whether positive or negative logic convention and the maximum terminal rating, that is, the limiting values that can occur before damage to the component ensues.

(ii) Supply line voltages and tolerances (normally 10–20% except for ECL)

(iii) Power taken from supply line, this can vary with fan-out.

(iv) Maximum power that can be delivered to an external source.

Package – size of module and whether TO5, DIL, or flat-pack; pin spacing, number of connections, etc.; ease of assembly into hardware system; replacement problems.

DC noise sensitivity – this is defined as the difference between the logic output voltage, and the input voltage that will just maintain that level. Noise immunity is a function of temperature, power supply voltage and fan-out.

Failure rate – this is normally very difficult to assess since data are not always available, particularly on new devices.

Logic functions – whether NOR/NAND or otherwise, types of bistable available – JK, SR, master-slave, etc. What logic sub-systems are included in the system – counters, shift-registers, etc. Some logic gates (but not emitter follower or emitter coupled devices) will permit the direct connection of logic outputs to give a wired OR or AND function. Thus, if this facility exists, units can be saved simply by connecting together those outputs which are required to be OR'd together. The compatibility of different logic families must also be considered.

Capacitance drive capability – ability to drive capacitive loads is important when long cables are to be driven; depends on output impedance.

However, a direct comparison of the electrical properties does not necessarily result in the selection of the ideal system. For example, on these grounds alone, one might be tempted to choose a new device which is made by one, relatively unknown, manufacturer only, but possessing the ideal electrical characteristics. This could be very risky, and certainly unsound systems practice, because very little would be known about the reliability of the device. Furthermore, should the manufacturer discontinue production, the entire project might need to be re-engineered if no direct equivalent could be found. Thus, we must also take into consideration the following points when selecting a logic system; these remarks, of course, could apply equally well to any systems component:

Devices must have been in production long enough to have been adequately 'debugged' of design and production faults.

The delivery of devices, particularly in bulk, must be guaranteed.

Design advice and assistance from manufacturers should be readily available; also the device should be adequately documentated.

Consider if the device will become cheaper in the near future owing to increased demands and mass production techniques.

Alternative manufacturers should be available; for example, DTL and TTL logic is manufactured by most companies and it is possible in many cases to get direct equivalents.

Consider the interface requirements: one should select a logic system so that the minimum of interface circuitry is required.

After the basic switching device has been decided upon the next step is to see how this affects the final form of the system logic equations. The most obvious is that of fan-in, which determines the number of inputs, and hence the type and degree of factorization required. (The overall speed of the system is decreased by factorization since the signals must propagate through a greater number of logic levels; cascaded circuits of this type can also cause hazards.) Other factors to be considered are the existence or otherwise of a wired OR connection, and the fan-out factor which, if too small, may mean the introduction of expander units or power gates.

Using ultra-fast integrated logic circuits with propagation delays of 1 ns or less, the number (and length) of interconnections in a circuit can prove to be more important than the number of actual elements. This is because the logic speed is limited by the delays caused by the length of the interconnecting paths (any interconnection greater than 2–3 inches must be a matched line). In addition, because of lead inductance and stray capacitance, the more paths there are the greater is the risk of cross-talk. Moreover, the cost of system wiring (the interconnection of logic boards) is becoming more significant than the cost of logic gates. For instance, it is now common practice to distribute only one phase of logic signal round the system (say Q bistable outputs only) inverting as and when required at board level. Consequently, in the very near future, it will be necessary to optimize a logic circuit to conserve the number and length of connections rather than logic gates.

These concepts led quite naturally to the idea of large-scale integrated circuits (LSI) and the possibility of producing complex logic circuits, a microcomputer system, say, on a single monolithic substrate.

In this way the limitations imposed on high speed operation by interconnection and gate delays can be overcome to some extent by decreasing the physical size of the system and hence the length and number of interconnection paths. This means that the conventional logical design philosophy of thinking in terms of discrete logic units, like NOR/NAND modules, has now become obsolete and system logic and circuit design is merging into one discipline.

Current digital system design relies heavily on the use of the MSI and LSI components described above, for example full-adders, shaft registers, ROMs, PLAs etc.[8,9,10] These subsystems use either **bipolar technology** based on junction transistor theory or **MOS technology** employing field effect transistors. The use of bipolar technology for complex integrated circuits allows high speed operation with little or not interface problems but unfortunately the packing density (number of gates per chip – in the order of 30–40 gates/mm^2) is small and moreover the devices tend to be costly. A new type of bipolar technology known as **integrated injection logic** (I^2L), which uses a pnp drive transistor to a multiple collector npn transistor,[11] has been developed for use in LSI systems and shows great promise of achieving reasonable speeds (equivalent to CMOS devices) coupled with a high packing density (in the order of 100–200 gates/mm^2).

For low speed systems (up to 25 MHz) devices employing MOS technology are simpler to produce (and consequently cheaper) and have a greater packing density (80–120 gates/mm^2). The functional complexity that can be obtained on a single chip of silicon (100–5000 gates on a chip some 100–250 mils square) is about 5–10 times greater than that obtainable with standard bipolar devices (excluding I^2L). When MOS subsystems are used in conjunction with bipolar devices (such as using TTL logic to drive a ROM store) there is an interface problem which can be costly to resolve and usually results in a degradation in performance. The basic problem is that the logic levels used within the MOS chip are unsuitable for use with external circuits, and consequently require logic level conversion or special MOS output devices capable of driving external capacitive loads. MOS devices are at their highest level of efficiency when used in a regular structure and for this reason the shift register configuration and ROM matrixes are ideally suited for this technology.

It seems likely however that MOS (and in particular n-channel devices) and I^2L technologies will continue to be used for large scale integrated devices, such as microcomputer chips, well into the immediate future.

A major design constraint with LSI circuits is that they are **pin-limited**; that is, the number of connections that can be made to external circuitry is limited by the physical size of the chip and its package. Normally, DIL packs are used with a capability of 14, 16, or 24 pins, the maximum number of connections (assuming special purpose packs) being in the order of 64 pins. Thus, it is important to get as much logic as possible onto a single chip, using serial or coding techniques where possible, to increase the gate-to-pin ratio for the device.

Large scale integrated circuits are not as yet being utilized to their full capabilities, in fact we have a situation where component manufacturers are taking the initiative in deciding what types of LSI circuit to produce. This is a retrograde step since it should be the function of the computer system designers to generate new LSI circuit designs. Unfortunately, such decisions can only be made on the basis of a viable systems design philosophy, which at the moment seems non-existent. In theory it should be possible to reproduce any logic design (within a certain complexity) in the form of an LSI circuit, but in practice, this is economically impossible. To implement a computer system design in terms of LSI circuits the logic designs must first be partitioned into suitable chip-sized sub-systems, and then masks made showing the component positions and the deposition and etching patterns for each chip. Though much of this work can be performed by a computer,[12] including the actual cutting of the mask, the preparation and production of a mask is still a lengthy and very costly process (a mask for a prototype circuit could easily cost in the region of £15,000). So unless the development costs can be recovered by producing the chip in very large quantities, the process becomes prohibitively expensive.

One technique for custom designed LSI that has met with considerable success is the use of read only memories and programmable logic arrays. In both cases these devices consist of regular arrays of MOS or bipolar transistor circuits which are fabricated without interconnections using a standard diffusion mask. The final interconnection mask however is produced by the manufacturer to the customer's own specification and can be of any desired pattern. Note that though the process is considerably cheaper than producing a special chip it is still comparatively expensive for a one-off circuit (hence the reason for the development of EPROM's).

Though in the main these devices perform combinational logic functions they can in conjunction with external bistables and appropriate feedback connections be used to implement sequential circuits.[13] Moreover, some PLA devices are appearing with integral bistable elements on the chip, which allow *internal* feedback connections to be specified. Note that this configuration realizes the basic finite-state machine model, which, theoretically, can be used to implement any sequential machine. It is also interesting to observe that given sufficient ROM/RAM storage any digital system can be realized, thus logic gates as such become redundant!

Other systems, generally known as **uncommitted logic arrays**, are also available based on the same technique of a master chip with customer specified interconnections. These devices range from arrays of standard NOR/NAND gates to complex chips containing logic gates, bistables, oscillators, monostables and interface circuits.

It will be obvious that some logic configurations, for example the implementation of an asynchronous counter or shift-register circuit using NOR/NAND gates, would be uneconomical in terms of the number of gates used and it is better practice to use mixed systems incorporating sequential circuits per se. Generalized logic arrays can also be uneconomical in that they normally contain a fixed number of gates (or words in the case of a ROM) and it is not always possible to use all the available bit capacity. However, in many cases even a 50% utilization would still be economic (from an overall systems point of view) because of the low cost of producing these devices and the savings in design and development time.

One problem with all these systems is that the size of the module is limited (for example a typical PLA would have 14 inputs, 8 outputs and 48 product terms) and in general circuits must be realized within the size constraint imposed by the module. Consequently the system must first be partitioned into suitable sub-system elements which can then be implemented in LSI.

Unfortunately the partitioning of logic circuits presents a major difficulty since no formal theory exists, as yet, which can help with design at this level. Though computer programs to perform this function are available they are based on simple trial and error procedures and do not necessarily yield an optimum solution. In fact, it may not be possible to obtain the full technical and economic benefits from LSI technology until a more formal design approach is established for computer systems.[14] It seems essential to be able to specify and partition a system without regard to the constraints imposed by current thinking or technology if a major breakthrough in computer systems structure is to be achieved.[15]

The production and assembly of a computer system breaks down into three major steps:

the manufacture of the printed circuit (PC) boards (or cards) holding the basic logic packages and LSI modules;

the mounting and placing of these cards in panels;

the interconnection wiring between cards on the back panels.

Most of these tasks are now performed using computer-aided design (CAD) techniques, particularly the layout of printed circuit boards and cabinet back wiring.[16, 17] Most of these CAD programs use a topological description of the finalized logic circuit as the primary input to the program; that is, the circuit is defined explicitly by specifying the logic packages and their interconnections. A placement algorithm is then used to position the packages on the PC board in such a way as to minimize the length of the interconnection paths between individual modules. The placement procedure is basically an optimization problem with the connection length as criterion, assuming that the packages can be freely manipulated within the area of the board. The next step is to determine an optimum routing of connections within the previously defined fixed environment. This operation must be performed with due regard to the reduction of cross-overs (that is, tracks which must cross over other tracks to reach their final destination) and stray coupling between parallel signal paths.

The same routing techniques may also be used to produce the card layout and wiring pattern for the panels in the main frame cabinets (called the **backplane** wiring). A useful byproduct of these CAD methods is the automatic production of wiring schedules (specifying the interconnections between boards), maintenance and logic diagrams, etc.

8.2.1 Microcomputer systems

The implementation of a digital system design using microcomputers poses special problems, though the general principles outlined above still apply. Microcomputers are used in two main ways, as a direct replacement for random logic performing a dedicated control function, or as a general purpose computer system undertaking tasks which were previously performed by mini-computers. In both application areas the required logic processes are implemented using programmed routines, thus the main design problem is software development.

It is also important to realize that though processing power is very cheap, and storage though more expensive at present is rapidly falling in price, the main cost of a microcomputer system is the software development and the provision of input/output hardware and peripheral devices. Thus the cost of the microcomputer chip (in the order of £100) is almost negligible compared to the cost of the necessary peripheral equipment and software development (estimated at some £1–£5 per line of fully commissioned and documented real-time application program). Nevertheless, from the hardware point of view, it is possible to produce a microcomputer system with a capability equal to that of a mini-computer at considerably less cost.

In many cases however, particularly when random logic is being replaced, the amount of software development and peripheral equipment required is minimal. Moreover, the programs themselves are normally predetermined and can be held in a small ROM memory. Such applications include intelligent instruments, television games, alarm and monitoring systems, process controllers, washing machine control, calculators etc. In these cases where the microcomputer can be integrated with the product itself and mass production methods employed there are tremendous advantages to be gained. However caution should be exercised when considering designing for one-off applications since it may be better economics to use conventional digital techniques.

The main advantages of replacing random logic with microcomputers are as follows:

(a) standard hardware modules can be programmed for a variety of tasks
(b) greater processing and memory capability
(c) small size, low power consumption and increased reliability
(d) commissioning, modifications and maintenance can be easier

Note that the basic microprocessor is a one chip CPU comprising instruction decode and control, ALU, general purpose registers and program counter on a single chip; in order to produce a microcomputer it is necessary to add separate chips containing ROM/RAM and I/O circuits. Microcomputers are now available on a single chip containing a CPU and limited amounts of memory and I/O, which are ideal for small dedicated control applications.

When selecting or deciding to use a microcomputer system for a particular application the following parameters should be considered:

(i) *System Architecture* – this is determined primarily by speed requirements and the necessary accuracy of the processing. Factors such as clock rate, word-length (both data and instruction) the suitability of the instruction set, register organization including the availability of stack processing, interrupt and I/O facilities must be evaluated. It is also essential to consider the memory organization and facilities such as word-lengths, total extended storage capacity, bus requirements, availability of compatible PROM's, EPROM's and DMA's should be investigated.

(ii) *Physical Structure* – the physical characteristics of the system components will directly influence the engineering costs and final reliability. Consideration should be given to the hardware level at which the microcomputer system is to be constructed – microprocessor chip, microcomputer already assembled on a PC board or a microcomputer chip. The number of LSI packages required determines the number of PC boards to be produced, whilst the pin count on the packages limits data word-lengths and memory addressing capability. Moreover, in general reliability in LSI systems is determined primarily by the number of external interconnections.

(iii) *Software Aids*[18] – the availability of software aids to programming are essential requirements for any microcomputer system (as of course they must be for *any* computer systems application work). Though it is possible to code short programs (say less than 100 instructions) directly in machine-code language it is a tedious and error-prone task. Most applications are programmed in an assembly language and require an assembler; this is normally run on another machine (either in-house or using time-sharing bureau facilities) and hence is called a **cross-assembler**. Once the program is assembled into the object code of the microcomputer it is normally run on a **simulator** which simulates the microcomputer instruction set and allows the program to be debugged. It is also essential that editing facilities are available (either as part of the simulator software or as a separate **editor** package) to allow corrections and modifications to be made to the program. Once the program is tested and working correctly it can be loaded into a PROM or RAM store for final testing in the actual system. A major advantage of the off-line computer approach is that cross-assemblers for a variety of microcomputers can be provided on the same machine.[19]

An alternative method is to use a microcomputer system, with associated RAM and disc storage etc., and utilizing the same processor type as the target machine. This system is used to maintain a **resident assembler** and associated program editor and trace facilities sufficient to assemble and debug programs. Systems of this type are often provided by the manufacturer (for example, Intel's Microcomputer Development System for the 8080 microprocessor) and cost in the same order as a small mini-computer system (some £5000–£10000). This method has the disadvantage that the user can become 'locked-in' to a particular microcomputer system and furthermore simulation facilities of the sophistication provided by the larger off-line machines are not available.

For large systems applications the use of a high-level language should be regarded as essential, consequently the availability of cross-compilers for languages such as FORTRAN, BASIC, PASCAL Intel's PL/M[20] and CORAL 66 etc. should be investigated.

In some complex real-time applications the design problems can be simplified by allowing a number of different control programs to share a single computer. In this case the availability of a multi-task executive which allocates run-time to different tasks on a priority basis and performs the central control functions can be invaluable. Systems software of this type is gradually becoming available from the manufacturers, for example Intel's MX80 real-time executive.

If these systems are written in a high-level language (say BASIC since there are a number of microcomputers with resident compilers for this language) the software will be portable and transferable between different machines (providing of course there is adequate RAM).

(iv) *Testing Procedures* – this is an essential phase in microcomputer system design and the availability of diagnostic routines and special test equipment should be considered. Since most of the control programs will eventually be implemented in ROM it is essential to ensure that the program is correct and, moreover, can control the required peripheral configuration according to the system specification. Final testing of the microcomputer system is usually performed using an **emulator** which comprises a target microcomputer with RAM, I/O ports and a terminal input. To use this equipment the control programs are set up in the RAM and the actual devices to be controlled connected to the I/O ports; a full systems test is then performed. Changes in the control program can easily be made by entering modifications from the terminal. Equipment of this type is also provided by the manufacturer, for example the Motorola EXORciser for the 6800 microprocessor.

In addition to the software and systems testing problem there is also the question of debugging hardware faults. Many of the hardware faults in a digital system involve the relationship between a wide variety of signals during successive clock periods. The use of an oscilloscope (even the multi-beam types) is ineffective in this situation. Though this of course has always been a problem in digital fault-finding, the increased complexity of microcomputer systems and the trend towards more and more custom design has resulted in special equipment, called **logic analysers**, to be developed.[21]

The logic analyser can display digital information in the form of high and low levels (waveforms) or as binary data in the form of logic ones and zeros. Multi-channel probes (from 4 up to 32 channels) enables the analyser to look simultaneously at a large number of input points (down to the CPU chip). Triggering can be obtained from any combination of address information, data bus, register outputs etc., including malfunctioning points. The information is presented over a selected time period, or 'window', of successive memory cycles (typically a 'slice' of 20–30 executed instruction cycles can be presented simultaneously).

(v) *Cost* – this is difficult to assess but must be based on *total system costs* including software development and peripheral equipment. It is also important to realize that for the new user there is a capital expenditure in development equipment (local mini-computer or time-sharing facility, Microcomputer Development System etc.) and a considerable educational problem.

Though microcomputers have been hailed as the answer to everyone's problems in electronic systems it is worth bearing in mind that they are only a cheap source of computer *hardware*. As such they must be considered together with established practices before deciding on a final design.

8.3 Noise problems[22,23,24]

One of the major problems associated with the engineering of any large digital system is the elimination of unwanted interference (**noise**) on the logic signal lines. In this context noise may be defined as any unwanted signals which give rise to erroneous operation of the system. This is a particularly troublesome problem when fast logic is used or when logic signals are transmitted over long lines as, for example, when driving peripheral devices.

Noise sources can be divided into three major categories:

noise generated by high frequency effects within the circuitry and backplane wiring – for example, cross talk between adjacent signal lines,

extraneous pulses that may be generated internally by circuit interaction – for example, cross talk arising from perturbations on supply lines or imperfect earths.

noise which may enter the system by inductive or capacitive coupling from outside sources – that is, radiated interference.

Many of the pulse waveforms that are transmitted through the backplane wiring of computer systems contain high frequency components (generated by the fast 'edges' of the pulses) which are very much faster than the basic clock rate. For example, consider a pulse having a rise-time of

0·2 μs; this could be considered, in the simplest case, as one quarter of a sinewave with a duration of 0·8 μs (see Fig. 8.3(a)). In theory, of course, the pulse waveform consists of all its Fourier components extending from the high frequency transient down to the first harmonic, a sinewave of period two times the pulse width. Consequently, if any of the conductors in the backplane wiring have a self-resonant frequency that falls within this band of pulse frequencies, they will be susceptible to pick-up from any adjacent wires that carry the pulse. This problem is accentuated if wires are run close together in parallel, since the stray capacitance that exists between parallel conductors (in the order of 1·5 pF per foot) will couple the noise capacitively into the system (see Fig. 8.3(b)). To reduce this form of noise pick-up it is necessary to use heavily insulated wire with good dielectric properties. The wiring layout is also important and in general it is always advisable to use direct point-to-point wiring rather than a neatly cabled system.

Capacitively coupled noise generally takes the form of voltage spikes, caused by the differentiation of the driving pulse, ranging from 0·1 to 1·5 V. The magnitude of the spikes depends on the amplitude and rise-time of the pulse and the impedance of the driven line to ground; lowering the impedance at the terminating end of the line will often cure this form of pick-up.

Though it is possible to use point-to-point connections in the backplane wiring, it is often necessary to run long leads or cableforms between equipment, in such cases the following guidelines have been found useful in avoiding pick-up:

If possible do not run trigger, clock, or count input lines with high current lines. For example, in the core store the sense outputs must not be routed with the *XY* driver lines and master reset lines must be run separately.

For short cable runs (under 24 inches in length) extra vinyl sleeving may be slipped over trigger lines, etc, before cabling.

When using TTL logic, cable runs of 30 inches or more may safely be used providing each pair of signal lines is interspersed with separate earth lines.

Shielded coaxial lines must always be used for low level signals (for example, outputs from magnetic R/W heads) even over short distances. For long distances it is also necessary to use matched drivers and to terminate the line correctly with the characteristic impedance (Z_0) of the coaxial cable.[25]

When running long leads in the backplane wiring (up to 3 feet or more) 'twisted-pair' transmission lines must be used. The twisted-pair is made by twisting together a ground lead, which starts at the source of the signal, with the signal lead for the full length of the line and then earthing the additional lead at the termination end; this is shown in Fig. 8.4. The pair of leads form a transmission line that may be terminated with a 500–1000 Ω resistor. However, over long runs (say greater than 3 feet) it is necessary

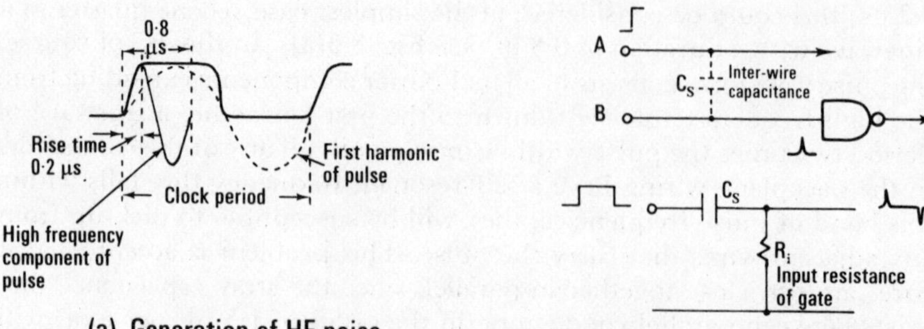

(a) Generation of HF noise.

(b) Spikes in parallel conductors.

Figure 8.3 Noise 'spikes' in digital systems.

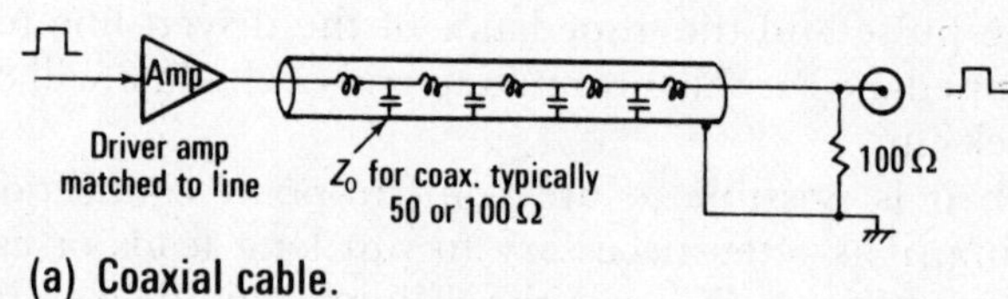

(a) Coaxial cable.

(b) Twisted-pair transmission line.

Figure 8.4 Distribution of signals over long runs.

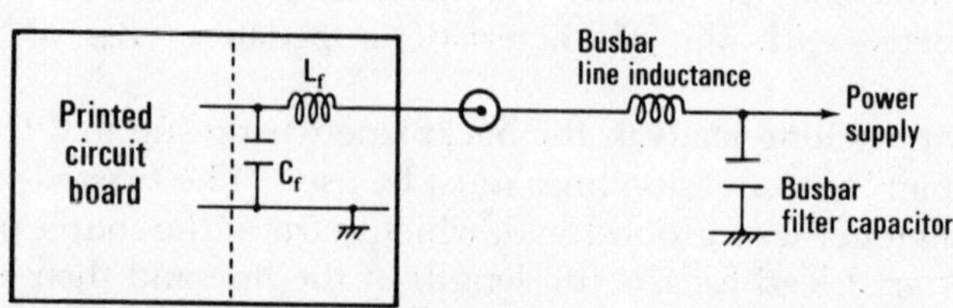

Figure 8.5 Local filtering on PC boards.

to terminate the line with its correct characteristic impedance (in the order of 600–820 Ω).

Shields on cables for low frequency signals (<1 MHz) should be grounded at one end. However for higher frequencies or where cable lengths exceed one twentieth of a wavelength, it is often necessary to ground a shield at both ends to ensure it remains at earth potential.

Another source of noise pick-up results from the direct electromagnetic coupling between wires that carry high current transients and other wires which present low impedance paths. In effect we have an air-cored transformer action in which any wire carrying a fast current pulse can act as the primary winding, inducing noise into the secondaries formed by the other wires. To overcome this it is necessary to use heavy gauge bus bars for power lines and large local filter capacitors at the end of each power bus and on each deck or row of modules. Long wires also have the property of 'ringing' at their resonant frequency (determined by the self inductance of the wire and its capacitance to ground) when 'shocked' by a fast pulse. This effect may be minimized when, for example, it is required to distribute a signal round the system by fanning out the wires from a central point rather than step from one point to another.

We will consider next the elimination of noise generated in the logic boards themselves. The best way of reducing noise generation in the power leads and earth returns is to use separate power filter networks on each of the individual logic cards, to isolate internal switching transients from the main power supply, and by providing very large gauge earth returns in the cabinets (typically $\frac{1}{4}$ in square). Referring to Fig. 8.5, the filter capacitor C_f acts as the main source of current for current surges and thus bypasses most of the surge currents from the main power supply. The series inductor L_f is designed to provide a large amount of inductance compared to the series inductance of the power buses; typical values for C_f would be at least 22 μF, and for L_f not less than 1500 μH.

In high speed digital circuitry, ground returns must be considered as just another wire with its associated inductive voltage drops. It is an erroneous, and dangerous, concept to consider all grounds as a zero reference; they are, in fact, only zero reference at the filter point of the main power supply. Whenever possible the power and ground bus bars should be run from each section of the logic system directly to the power source. In this way it is possible to avoid the build up of power and earth currents through the bus bars; this is shown in Fig. 8.6.

In general most systems require at least three separate ground returns for example, the grounds for low-level electronic signals should be kept separate from the noisy ground used for relays and motors; a separate ground must also be used for chassis, racks and cabinets. For instance in a digital cassette unit the read and write circuits, the digital interface logic, capstan and reel motor drives and power supplies would all have separate ground returns earthed only at the primary power ground.

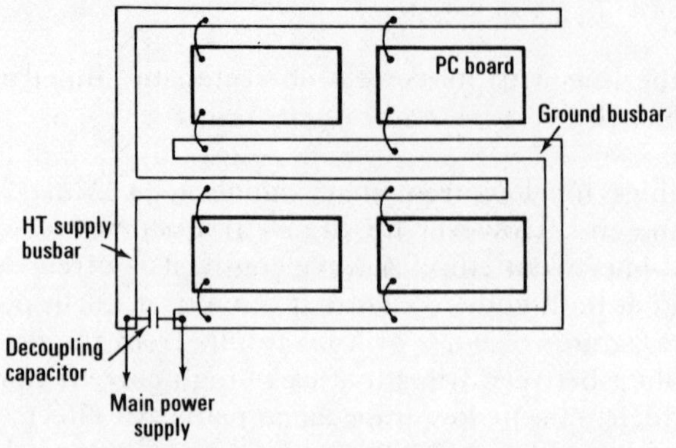

Figure 8.6 Power supply wiring in back-plane.

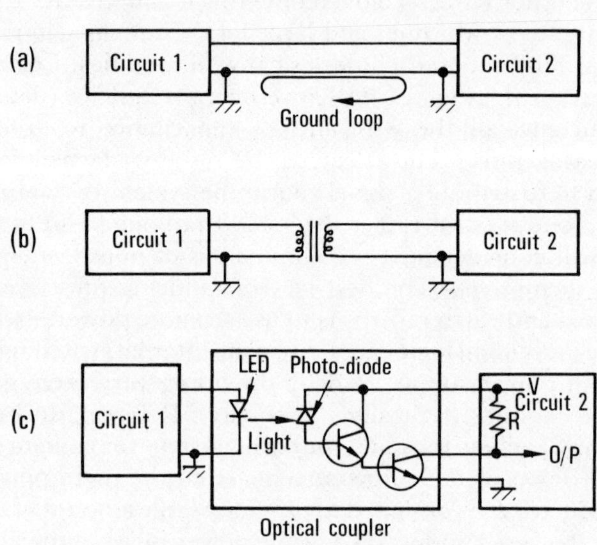

Figure 8.7 Ground loops.

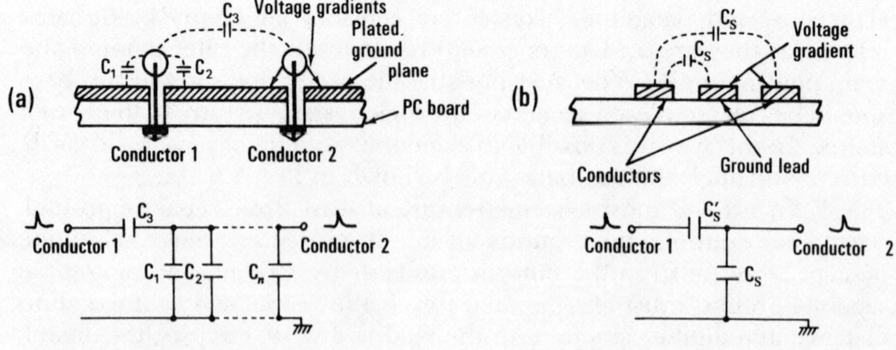

Figure 8.8 Shielding using extra conductors and ground plane.

Another serious problem in grounding equipment is the avoidance of the formation of **ground** or **earth loops** which occur when two circuits are connected together and both ends of the circuit have a common ground, this is shown in Fig. 8.7(a). A circuit grounded at both ends is highly susceptible to electro-magnetic radiation because of the large noise currents that can be induced in the ground loop. The simplest way of breaking the ground loop is to electrically isolate the equipment, this can be done by interposing an isolation transformer between the two circuits (see Fig. 8.7(b)). A much better way however for logic circuits is to use an optical coupler or **opto-isolator** as shown in Fig. 8.7(c).

The opto-isolator, which is available as an IC package, is used extensively in digital systems to couple computer inputs and outputs to other devices. The device is similar to a solid-state relay but designed using a light emitting diode (LED) and a phototransistor; the circuit is optimized for current gain and/or speed of response and is intended to drive logic rather than heavy loads.

When relay and other electromechanical devices are used in the system (for example, in on-line peripheral equipment), it is necessary to provide each solenoid with a separate damping network (series diode and resistor) which is connected directly across the coil to clamp any flyback voltages and currents that may be generated. Moreover, a separate filter point and earth return must be used for the power supply of the solenoid driver amplifiers (alternatively, a completely separate power supply could be used). It is also good practice to arrange that any data transfer operations, such as opening output gates to highways, take place after all solenoid transients have occurred, thus avoiding any likely noise troubles.

With the very high packing densities now being achieved on digital circuitry boards care must be taken to ensure that pick-up between components or tracks on the PC cards (due to inter-component capacitance) does not take place. The most effective method of screening components is to incorporate a complete ground shield (or **earth-plane**) on the component side of the module (see Fig. 8.8(a)) allowing only circular ports in the board through which the component leads can pass. The ground-plane also serves as an inter-module shield (when the cards are stacked vertically, side by side in the cabinet) cutting down coupling from one card to another. An alternative method, if space permits, is to put a grounded wire between adjacent conductors or components (see Fig. 8.8(b)).

Finally, we must consider the control of external noise sources – that is, pick-up due to electromagnetic radiation, etc. The chief sources of external interference that may be encountered in a computer system are a.c. and d.c. motors, contact breakers and buzzers, and solenoid operated devices. When digital computer systems are connected to external or remote equipment (as in the case of peripheral devices) the routing of power and logic signals (including earth returns) must be carefully organized if interference is to be avoided. For example, the power and ground wiring systems should never use the chassis or metal cabinet housing as a ground return (nor for that matter as part of the *logic* ground system). The outer housing, which should be used as a noise shield, becomes ineffective if it is also used for carrying ground currents. When several large pieces of equipment need to be connected together it is essential that the potential of the cabinet housings be stabilized in relation to each other. Thus, when interconnecting digital equipment via multicore cables, the common earth bus or cable shield must be grounded at the output connector on the cabinet housings as shown in Fig. 8.9. All other ground leads should, if possible, converge to the same common point, with the a.c. mains being earthed via a 50 mH choke.

Pick-up due to high frequency radiation is very prominent when logic signals are transmitted over long distances. It is possible to reduce this interference by the use of filters on the output lines (typically an LC π-network circuit) or by inserting a series inductor (or a ferrite bead) in the lines; alternatively, shielded coaxial cables may be used. In extremely noisy surroundings it may become necessary to use a balanced pair of signal lines with a common coaxial screen held at earth potential.

The a.c. mains power line must always be considered as the prime generator of both high and low frequency noise. Simple LC filter networks may be used at the power supply input to protect against this form of interference. Care must also be taken when connecting the a.c. mains supply to different items of equipment in the system. For example, it is very bad practice to take the mains supply to the CPU, and then loop to other peripheral devices in the system. All equipment must be connected directly to the a.c. supply and earthed only at that point, otherwise reflections down the earth returns (caused by current surges) can give rise to interference. Another common form of mains interference is line dropout. This is caused by an instantaneous current surge on the mains (such as a lift or motor starting up) causing a drastic fall in the level of the line voltage. A separate rechargeable battery supply, 'floated' across the main power supply, is a good way of overcoming temporary line dropouts, as

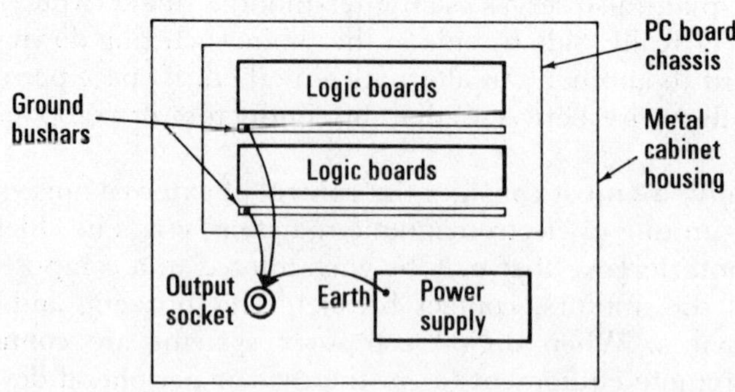

Figure 8.9 Earthing of equipment in cabinets.

well as providing what is effectively a very large capacitive filter circuit. In many cases, when the a.c. mains is considered unreliable, a separate motor-generator set is used to provide a completely isolated mains supply, which is protected against line dropout by the mechanical inertia of the generator system. An additional advantage of this technique is that if the a.c. supply is accidentally switched off the inertia of the system will maintain the correct output voltage for a sufficient time (in the order of milliseconds) to enable protective software action to be initiated. That is, the mains failure will cause an interrupt to the current program, whereupon the resultant interrupt routine stores away any important results in the core store.

8.4 System testing procedures

The periodic testing of a computer system and its components, both on- and off-line, is a vital and necessary operation if reliable performance is to be obtained. Moreover, it is essential to be able to test the operation of the computer in order to perform routine maintenance and for the detection and rectification of machine faults. One of the advantages of a general purpose computer system is that the computer itself may be used for system testing and check-out, the checking routines often being included as an integral part of the systems operational program. It will be obvious, however, that some parts of the machine (for example, the basic read/obey cycle logic) must be operative in order to perform any program tests.

Computer system testing can be broadly classified into two main categories: **programmed checks**, including **diagnostic routines**, and **machine-checks**. Programmed checks[26] can be incorporated by the programmer into his normal software routines and may assume the following forms:

Duplication of operations – A set of machine instructions may be repeated, preferably in a different way, to verify that the machine is operating correctly. For example, the product $a \times b$ is checked by calculating $b \times a$.

Feasibility or order of magnitude checks – In this case, the form and order of magnitude of a solution (including intermediate results if known) can be used to check the machine. For example, it may be impossible for a particular process variable to be negative (for instance, specific gravity) or to exceed some maximum amount.

Mathematical checks – When computing the sine of an angle, say, the cosine is also computed and used to verify that the sum of the squares is unity to within the precision of the computation.

Sum checks – In this technique the programmer sums a block of data and/or instructions as though it were a sequence of pure binary digits, the final binary sum being written with the data block as a check word. Later, the correct (error-free) arrival of the data block at a given destination (say in transferring from the disc file to the core store) can be easily verified by summing the block again and comparing the new sum with the stored checkword.

Diagnostic test programs[27] – The test programs described so far will ascertain whether or not the machine (or program) is malfunctioning but will give no indication of the actual fault condition. Diagnostic test programs are specially devised by a programmer in conjunction with the logic designer to exploit the full range of the computer's capabilities. As well as establishing the existence of a fault, they also describe it in detail and in some cases indicate the possible causes of error.

Unlike the programs described earlier, diagnostic test programs are not usually incorporated into actual software routines, except sometimes as a special case in real-time systems. They are normally used by the commissioning and maintenance engineer as a means of validating and rectifying the operation of the computer. In general, the diagnostic tests are progressive and would commence with a rudimentary test program to check the basic read/obey function and simple orders such as fetch and store. This would be followed by a comprehensive test of the main storage system (using generated pseudo-random numbers[28] or a 'worst pattern' sequence) and then a systematic check of each machine-code instruction. After this, specific tests on the arithmetic and control unit and input/output facilities would be performed. Unfortunately, these tests can never be completely exhaustive, since this would take far too long.

Should any of these tests indicate a fault, a computer print-out is produced which describes the error in detail and in some cases its possible location. In most cases, however, to find the actual faulty component or board it is necessary to set up a simple repetitive program (inserted manually from the console keyboard) which shows up the fault and allows a more detailed examination (using an oscilloscope or logic analyser) of the hardware circuits to take place.

It is possible to develop more sophisticated versions of the diagnostic programs that will locate the exact position of a fault as well as detecting it. In this case, trouble-shooting reduces simply to replacing the appropriate malfunctioning unit as indicated by the diagnostic program. Unfortunately, the development of this type of test program is a complex and costly process and, unless considered at the initial system design stages, can be very inefficient in operation. For example, the precision with which a fault can be located depends on the module size and is thus determined by the partitioning of the logic system during the initial engineering stages.

Recently off-line computers have been used to assist in the diagnosis of faults in computer installations, this allows the use of more powerful computers and data-base facilities which would not normally be available on-line. The DEC CLINIC system used to maintain the DEC SYSTEM-2020 computer enables operations staff or maintenance engineers to dial into a remote computer via a telephone line linked terminal and obtain automatic testing and diagnostic information.

Normally diagnostic and test programs would be run with the computer or system functioning under **marginal conditions**. This is a technique whereby the power supply voltages (both a.c. and d.c.) and basic clock timing are varied about their normal operating conditions while the tests are taking place. The use of voltage margins was very effective with first generation valved machines (because of the inherent loss of emission in the valves) but is less so with solid state machines which normally only use phase margins on the clock to detect timing faults. The chief advantage of marginal testing is that it shows up faults which are just about to occur or occurring intermittently.

Machine checks use special hardware and/or codes incorporated into the computer or system, and are normally designed to detect a certain class of error only. The chief forms of this type of test are as follows:

Data transfer checks detect errors which arise during the transfer of information from one place to another in a machine or system – for example, the parity checks written on magnetic tape. This type of check could vary from a simple error-detecting code, such as a parity check, to an error-detecting/correcting code of the Hamming variety.[29,30]

Arithmetic checks include various schemes for verifying the correct performance of the arithmetic logic. One possibility would be the complete duplication of the arithmetic section with cross verification between the two outputs. Another approach, already discussed in Section 5.18, is the inclusion of parity bits in the operands, or the use of modulo arithmetic units.

Overflow and out-of-range checks are used to detect and indicate whenever a register or calculation overflows, as this can often be the result of a machine fault.

In most of the above checks, the detection of an error would cause the computer to link automatically to an error routine devised by the programmer which would specify what remedial action, if any, to take. Alternatively, when a fault is detected its presence could be indicated by an audible alarm or flashing lights, whereupon suitable corrective action would be initiated by the operator.

The continuing cost and performance improvements of semiconductor storage are now making them viable for use as main frame memory. Unfortunately their reliability as yet does not match the requirements for large memory systems and error-detection and correction techniques are essential.[31] The Hamming code is used primarily for this purpose since it can provide low cost fault detection and correction circuitry; note however that only single errors can be corrected using this method though double errors can be detected. To take full advantage of this technique the structure of the storage system should be organized

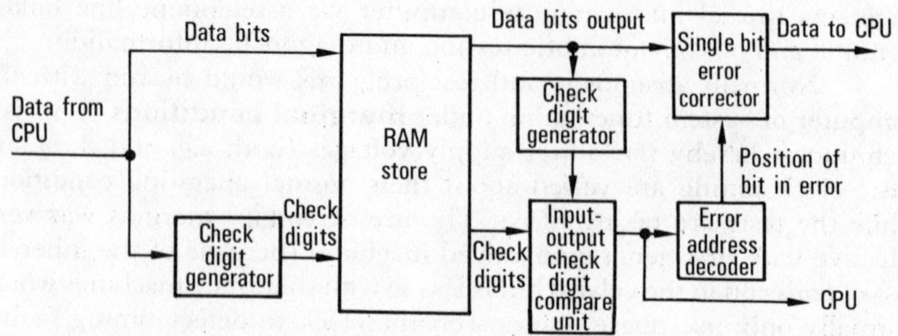

Figure 8.10 Automatic testing of semi-conductor memory.

such that each RAM is assigned to only one bit of a word in memory thus ensuring that a single RAM failure does not cause multiple faults.

A schematic for a RAM store using automatic error detection and correction is shown in Fig. 8.10. Note that the Hamming code check digits are generated separately and added on to the end of the data words, rather than distributed throughout the word as described earlier. This is possible as a consequence of the cyclic code properties of the Hamming code and the check digits can be realized using linear switching circuits.[32] On reading a word from store the check digits are re-generated and compared with the stored version, the output of the comparator is used, if necessary, to correct the data bits. In operation the procedure works automatically but the output error bits are made available to the CPU in order to keep a record of devices which malfunction for subsequent maintenance action.

Note that this method, though considerably increasing the reliability of the storage system, has the disadvantage of requiring extra bits in the data words and specialized control circuits. Moreover, the read and write times for the store are increased as a result of the extra pre- and post-processing involved.

One of the practical difficulties in computer testing is to decide whether an error is caused by a machine fault or a logical fault in the program. It is essential to exonerate the software routines (using a **trace** program) before investigating the hardware circuits. A trace routine is an interpretive type of program designed to assist the programmer in locating software errors. This is achieved by causing additional instructions to those in the main program to be executed at each program step. The effect of these trace instructions is to print out each main program instruction as it is obeyed, together with the contents of specified registers. This record may then be used to determine, for example, whether jump instructions are executed correctly or to provide a check on the contents of the arithmetic and control registers, thereby enabling the operation of the program to be traced out in detail.

Software testing poses a number of difficult problems. Unlike hardware faults, which are due to a breakdown in components due to wear or fatique, software faults are due to basic logic errors, or oversights, in the original design. Thus the repair of a software fault generally results in a re-write of the program rather than the replacement of a component or sub-system. In essence there is no such thing as a software fault per se, only hardware faults since if the program is correct it will always run correctly, unless a hardware fault develops. Consequently the emphasis should be on producing a correctly engineered software system and a new discipline called **software engineering**[33] has recently emerged to cope with this problem. Software engineering is defined as 'the practical application of scientific knowledge in the design and construction of computer programs and the associated documentation required to develop, operate and maintain them'. Nevertheless unless program correctness can be guaranteed (perhaps never!) it will still be necessary to include programmed checks etc. as an integral part of the system design.

All the tests and checks described above could be used to check out a digital computer system, both on and off-line, that is, under actual working conditions or during maintenance. In real-time control systems, test programs would be interleaved with the systems program in a cyclic sequence using a 'watch-dog' facility. This takes the form of an internal timing signal which must be cancelled periodically, usually only when an all correct check-out has been obtained, if an error alarm is not to be sounded.

Computer tests which are designed as an integral part of the operating procedures result in redundancy in the system, either in the form of the extra hardware needed for the machine tests or as increased operation times when software checks are used. We shall discuss in a later section how the principles of redundancy may be used to enhance the reliability of a system. In general, because of the complexity and cost (as well as increased running times) of producing software checks, hardware error-detection logic is preferable particularly on economic grounds.

As a conclusion it is worthwhile discussing **acceptance tests**.[34] These are tests specifically designed to ensure that the digital system is fully operational and that all the system facilities (as prescribed by the specification) are complete and working before acceptance of the system by the customer. The tests must also ensure (as far as possible) that the equipment is capable of reliably fulfilling the tasks assigned to it and should, ideally, be satisfactorily completed at both the manufacturer's and customer's site.

The acceptance procedure is generally based on a test cycle, comprising both engineering test programs and operational programs, which may well be modified, or actual versions of the application programs. A typical acceptance test would consist of:

demonstration programs to ensure that the CPU, peripherals, systems software, and so on, function according to specification;

cyclic runs, comprising the repetition of a set of programs, each of approximately 15 minutes' duration, covering both the engineering and user aspects of the system.

The duration of a complete test cycle would normally be four to eight hours, and six such cycles would be run over three working days. The success or failure of each test is judged on the basis of a comparison with previously agreed correct master data output. It is essential that operational programs as well as engineering programs are used for the acceptance trials as the latter alone would not constitute a very stringent test; it has often been said 'it is a poor machine that cannot learn its own test programs'!

8.5 Logic circuit testing and simulation[35,36]

Once a machine fault has been traced to a particular logic sub-system, the unit may either be repaired off-line or, ideally, simply thrown away – the 'disposable' LSI or MSI circuit. However, we have not as yet truly reached the expendable element stage and most logic boards are repaired and put back into operation. Thus, the logic designer has the further responsibility of devising suitable test-schedules to detect and diagnose faults in the logic sub-systems. Moreover, there is a very pressing need to test LSI chips and the manufactured boards before final assembly in the machines; this requirement will still persist (with even greater emphasis) when LSI circuits are fully utilized.

Table 8.1
Fault matrix method of fault detection

(a) F-matrix

AB	test	f_0	f_1 C1/0	f_2 C1/1	f_3 C2/0	f_4 C2/1	f_5 C3/0	f_6 C3/1	f_7 C4/0	f_8 C4/1	f_9 C5/0	f_{10} C5/1	f_{11} C6/0	f_{12} C6/1	f_{13} C7/0
00	t_0	0	0	1	0	1	0	0	0	0	1	0	1	0	0
01	t_1	1	1	0	0	1	1	1	0	1	1	1	1	0	0
10	t_2	1	0	1	1	0	0	1	1	1	1	0	1	1	0
11	t_3	0	1	0	1	0	0	1	0	1	1	0	1	0	0

(b) G_D-Matrix

	f_0f_1	f_0f_2	f_0f_3	f_0f_4	f_0f_5	f_0f_6	f_0f_7	f_0f_8	f_0f_9	f_0f_{10}	f_0f_{11}	f_0f_{12}	f_0f_{13}	f_0f_1
t_0		1		1					1		1			1
t_1		1	1				1					1	1	
t_2	1			1	1					1			1	
t_3	1		1			1		1	1		1			1

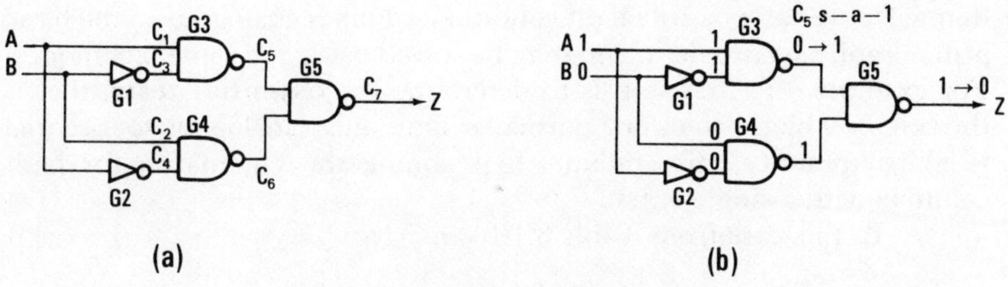

Figure 8.11 Exclusive OR circuit.

With complex circuits containing many variables, the exhaustive test approach – that of applying all possible input conditions and observing the outputs – becomes prohibitive in terms of time and cost. Consequently, it is necessary to determine in some way a minimal set of inputs which, when applied to the logic circuit, will produce a defined output indicating whether or not the circuit is operating correctly and according to specification.

A number of formal techniques, based on switching theory, have been evolved[37,38] for determining the necessary and sufficient tests for combinational, and to some extent, sequential circuits. These tests are generally limited to the detection of single logical errors of two types: those in which the signal lines are either stuck at a logical 1 (s-a-1) or at a logical 0 (s-a-0). There are four basic methods of fault detection in current use – the fault matrix, path sensitizing, boolean difference, and partitioning techniques. To convey some of the ideas involved in these techniques we shall consider the first two methods in some detail, starting with the fault matrix.[39]

The **fault matrix** relates the set of all possible input tests for a given circuit to their associated faults. The entries within the F-matrix are the output values resulting from applying a given test input under specified fault conditions. For example, consider the simple exclusive OR circuit shown in Fig. 8.11(a) and its associated fault matrix detailed in Table 8.1. There are two inputs and hence four different tests, termed t_0–t_3. Since the circuit contains seven connections (C_1–C_7) there are 14 possible fault conditions, f_1–f_{14}, referred to as C1/0, C1/1, etc., where C1/0 denotes connection C1 s-a-0, etc. The entries within the table may be calculated by hand computation (that is, the circuit is considered with test input 00, and the output function for each fault condition, C1/0, C1/1 etc. is derived; the process is then repeated for all test inputs); alternatively a computer simulation (see later) may be used.

The F-matrix is usually transformed into a G_D-matrix for ease of manipulation; this is done by performing an exclusive OR operation between the correct (no fault) output column f_0 and all the other fault columns. We are now in a position to select a minimal set of input tests that will cover all possible fault conditions. This is analogous to the basic prime implicant problem, and may be solved using the same techniques. For example, the first step is to determine the **essential tests**, that is those tests which detect one particular fault only (analogous to essential PI's). In practice this amounts to scanning the G_D matrix for fault columns with a single entry.

In this case from Table 8.1(b) we have:

t_1 – only test for f_0f_7 and f_0f_{12}; that is, C4/0 and C6/1
t_2 – only test for f_0f_5 and f_0f_{10}; that is, C3/0 and C5/1
t_3 – only test for f_0f_6 and f_0f_8; that is, C3/1 and C4/1

The essential tests are therefore t_1, t_2, and t_3, and in fact these three tests also cover all other faults. If this was not so it would be necessary to add other tests until a complete fault cover was obtained. The full test set, expressed in terms of input/output vectors is thus:

01/1, 10/1, 11/0

If these tests are applied sequentially to the input terminals of the circuit, any deviation from the defined output sequence would indicate the existence of an error. It is possible to extend this technique to allow the actual location of a fault to be diagnosed, but this involves a considerable amount of computation which, for large variable circuits, rapidly becomes excessive.

The basic **path sensitization** technique relies on three processes:

1. the postulation of a specific fault within the circuit structure, for example, C5 s-a-1;

2. the propagation of the logical effect of this fault, from its original location to the output terminals, along a **sensitive path**; this is called the **forward trace**;

3. **a backward trace** phase, in which the necessary gate conditions required to propagate the fault along the sensitive path are established.

In this technique the inputs to each circuit element on the sensitive path are grouped into a **control** input, which is part of the sensitized path and must be allowed to vary in order to detect a fault, and the **static** inputs, which are held at a constant value to maintain the sensitized path.

An example will best serve to clarify these points. Consider the fault C5 stuck at 1 in the circuit shown in Fig. 8.11(a). The first step is to determine those gates through which the fault will be propagated before it reaches the output Z; in our example this is trivial since only gate G5 needs to be considered. However, in order to detect the presence of C5 s-a-1 on the output of G5, the other input to the gate, C6, must be held at 1. Under these conditions, with C5 specified at 0 and C6 held constant at 1, the output C7 would be 1, so if C5 was s-a-1 (that is, the fault condition) the output of G5 would go to 0, indicating a fault.

We must now establish the necessary primary input conditions to ensure that C6 is held at 1 and C5 at 0, for the correctly functioning circuit, that is, the backward trace. The static inputs for the output C6 to be held at 1 is given by $\overline{C2} + \overline{C4}$ (that is, $\overline{C2} + C1$); for C5 to be 0 it is necessary for the inputs to be $C1 + C3$ (that is, $C1 + \overline{C2}$). Thus, the input combination 10 (test t_2) is the only test that will detect C5/1, as we ascertained earlier using the fault-matrix approach. The sensitive path is thus through gates G3 and G5, as shown in Fig. 8.11(b), with gate G4 being used to maintain the path.

Once an input test has been established all other faults detected by that test are derived. The process is then repeated, using the same procedure with an as yet undetected fault condition until all faults are covered. In practice, the forward and backward trace phases would be combined and a computer simulation used to establish the input test conditions.

The major disadvantage of this technique is that fan-out paths in the circuit can reconverge again, leading to difficulties in establishing the control inputs. This problem has been overcome by Roth[40] using the D-algorithm technique which was adopted as a standard procedure by IBM for circuit testing. The problem of testing sequential circuits (i.e. circuits with feedback paths) is still largely unsolved, though some progress has been made using state-table analysis[41] and in the application of a modified D-algorithm to sequential networks.[42] In the latter case the feedback loops in the circuit are first identified and then 'broken' thus allowing the sequential circuit to be represented as a cascaded connection of identical combinational circuits each one representing the sequential circuit at a given instant of time. A more empirical approach consists of applying design constraints such that the sequential (bistable elements) and combinational logic are physically separated on the PC board and hence can be tested independently.

In practice special **automatic test equipment** (ATE) is employed to perform the testing functions. These usually consist of a small mini-computer with back-up storage, coupled with comprehensive input/output and probing facilities. Testing is accomplished by applying a stored test sequence (normally generated off-line using manual or algorithmic techniques) to the circuit under test and evaluating the results against the expected response; the average rate of applying test vectors can vary from 10–200 kHz.

An alternate form of ATE, known as the comparison type, applies the same test sequence to a 'known good circuit' and the actual circuit under test, the results being compared for any discrepancies which indicate the presence of a fault. This method of testing has led to the use of random test sequences which when generated using special purpose pattern generators can achieve a very high rate of testing (some 1–40 MHz).

ATE's can be used to test both combinational and sequential circuits and in particular the random test method has met with considerable success. Care must be taken however to ensure that sequential networks are correctly initialized and to detect and prevent hazards and races which can invalidate the testing procedures.

One of the most difficult problems associated with logical design (software as well as hardware) is how to check that the final circuit satisfies the original system specification. As with error detection, a test for all possible input conditions would be prohibitive, yet an arbitrary random technique is far from satisfactory. Moreover, to attempt to do this on a hardware prototype would be virtually impossible for a large system (on economic grounds alone). Formal logic design methods demand that the designer considers all possible circuit conditions, thereby eliminating a major source of errors, but the problem of how to detect conceptual errors still remains. With intuitively designed systems, particularly when they are to be implemented as LSI circuits, the problem is particularly acute, since it is essential to 'debug' the design before manufacture.

The primary method used for the evaluation of systems and logic designs prior to actual manufacture is that of **simulation**. Simulation is a process whereby it is possible to model either functionally or mathematically, the behaviour of a real system; experiments can then be conducted and related back to the actual system. Note however that simulation does not necessarily give an exact representation of the system as it behaves in the real world owing to the constraints and necessary approximations used in constructing the model.

Digital systems may be simulated at basically three levels:

(a) *Systems level* – this consists of modelling the system in terms of subsystem components such as arithmetic and logic units, memory modules, buses, peripherals etc.

(b) *Register-transfer level* – in this case data flow at the register level is modelled thereby enabling micro-programs and instruction sets to be evaluated. This form of simulation language was dealt with in section 4.7 and of course also includes the Iverson language and microprocessor simulators.

(c) *Gate or logic level* – here the actual logic gates, or modules, and their interconnections are functionally modelled in the computer. Other forms of simulation are sometimes used but these are predominantly at the physical device and circuit design level and will not be dealt with in this book.

System simulation[43] is normally performed using general purpose simulation languages such as GPSS,[44] SIMSCRIPT[45] and GASP.[46] In general they are used to model the overall performance of computer systems in the following five areas:[47]

(i) the feasibility of performing a given workload on a certain class of computer

(ii) comparison of computer systems (and in the evaluation of software systems) against a specific benchmark or typical test problem
(iii) determination of processing capability for various system configurations and operating conditions
(iv) investigation of the effect of changes in the system structure etc. during the design process
(v) improvement of system performance by identifying and modifying the critical hardware/software components of a system – a process known as 'tuning'.

Note that the system is normally simulated at the behavioural user level rather than at the lower structural or functional levels. Typical of these investigations would be whether increasing the size of store buffers results in increasing message throughput, or the determination of the optimum size of a cache store. Note that in these cases only that part of the system under investigation needs to be accurately simulated, the remainder of the system, such as processor operation, disc storage etc., can be less accurately defined.

Logic simulators[48,49] have been developed to assist the designer in the verification of logic designs before the hardware stage is reached. The simulator software packages are normally designed so that the engineers can work directly with the computer (in a conversational mode) via a teletypewriter or CRT terminal. The logic diagram for the completed design is described to the computer in a topological form; that is, each element of the circuit is coded and its input and output conditions specified. Table 8.2 shows a typical input format for the exclusive OR circuit shown in Fig. 8.11. Once the circuit is described the designer

Table 8.2
Simulation of logic circuits

(a) Input list

	Type	*Input 1*	*Input 2*	*Output*
1	INV	C1		C4
2	INV	C2		C3
3	NAND	C1	C3	C5
4	NAND	C2	C4	C6
5	NAND	C5	C6.	C7

(b) Timing diagrams

Initial conditions C1 = 0 C2 = 0
Input C1 = 1 C2 = 0

Clock	Outputs C3	C4	C5	C6	C7
0	1	1	1	1	0
1	1	0	0	1	0
2	1	0	0	1	1
3	1	0	0	1	1

Initial conditions C1 = 1 C2 = 0
Input C1 = 0 C2 = 1

Clock	Outputs C3	C4	C5	C6	C7
0	1	0	0	1	1
1	0	1	1	1	1
2	0	1	1	0	0
3	0	1	1	0	1
4	0	1	1	0	1

may also specify the required input waveforms, initial starting states, monitor points, etc., that best suit his system.

During a simulation run, the computer calculates the output status of the logic elements at each basic machine time interval (usually taken to be in nanoseconds, or alternatively, measured in terms of gates per clock-time) and continues to advance in simulated time until the logic comes permanently to rest (a stable state) or reaches some previously specified state. Various nodes in the circuit may be selected for monitoring during a simulation run, and the status of these will be recorded on the teletypewriter at the end of each time interval. Gate propagation delays may usually be assigned and in some cases varied, either individually or collectively, during the simulation runs.

The majority of simulators are designed to operate in terms of basic elements (in the order of 3000–5000 units). However, it is possible to simulate higher level modules (for example, a JK bistable) in terms of these basic units and then refer to it in later simulations as a separate compound element (that is, a *block* structured simulator).

A simple example of the type of output format obtained with a logic simulator is shown in Table 8.2(b). In this case the initial circuit conditions have been specified as C1 = 0 and C2 = 0 and the input allowed to change from C1C2 = 00 → 10 → 01. The timing has been set to be one gate per clock – that is, with the propagation delay of the elements being equal to the clock duration. Note that in the case of the input change 10 → 01, a circuit hazard is indicated since the output C7 should remain constant at 1 during the input change. This is caused by the difference in path lengths for input signal A, being applied directly to gate G3 but via gate G2 to G4 which is on the same level.

It is possible to obtain more sophisticated output formats, such as waveform diagrams, using an incremental graph plotter. Other alternative outputs would include, for instance, loading (fan-out) statistics, and a table showing the maximum number of gate levels a signal has to pass through for each particular clock time. The major benefits of this technique are:

Time and money are saved by eliminating the logical and loading faults before hardware manufacture.

Corrections and modifications may be made during the early design stages.
The computer data-bank and listings serve as a documentation of the actual design.
When the number of inputs is small, exhaustive testing of the circuit can be performed.
The simulator can be used to devise test schedules (as described above) and to generate waveform and timing diagrams for maintenance purposes.
Timing faults and circuit hazards can easily be observed and rectified.
Simulation has become an essential software aid for the design and testing of computer systems, both at the logic and systems level. However

there are still many problems to be solved, particularly with complex LSI circuits with very large logic gate densities. In this case to attempt a simulation entirely at the logic gate level is becoming practically impossible and a 'top-down' approach to LSI design must be adopted. This necessitates using a hierarchal structured simulator which allows a behavioural description and evaluation at the sub-system component level, whilst at the same time enabling a lower level gate analysis to be performed on specified modules.[50]

8.6 Reliability and the use of redundancy[51]

It is common knowledge that computers are being increasingly applied in all fields of endeavour and in many cases the emphasis is shifting from the purely numeric and data-processing functions to those of supervising the control of vital processes. In these applications (such as real-time and military systems) where we have become dependent on computing facilities, it is essential that the machine should be reliable in operation and under no circumstances should it be allowed to breakdown completely. Thus, it has become necessary for the computer systems engineer to be able to specify and control the reliability of the machine by predicting, for example, the reliability of a given system (thus ensuring it meets the customer's specification) and, if required, to enhance this figure by the use of redundant elements.

The concept of **reliability** is difficult to define but we shall take it to mean a measure of the capability of an equipment to operate without failure when put into service. It is therefore the **probability of survival** of the equipment over some specified time interval. Before we can estimate the reliability of a given system we must first consider the **failure rate** for individual components. Suppose that among a large (ideally infinite) population of components, N of them are operative after time t; N will clearly decrease with t so that dN/dt is negative (or zero) for all t (see Fig. 8.12(a)). The ratio of the rate of decrease of N to N itself at any time t is defined as the **failure rate** at that time t and is denoted by:

$$\lambda = -\frac{dN}{dt}\bigg/N$$

Failure rate is usually expressed in the form 'x failures per cent per 1000 hours' or '$x \times 10^{-5}$ failures per component hour', thus 0·01% per 1000 hours is equivalent to 10^{-7} failures per component hour. Rearranging the expression for λ, we have:

$$\frac{dN}{N} = -\lambda\, dt$$

and if λ is assumed constant, integrating the expression yields:

$$\log N = -\lambda t + \text{constant}$$

or $$N = Ke^{-\lambda t}$$

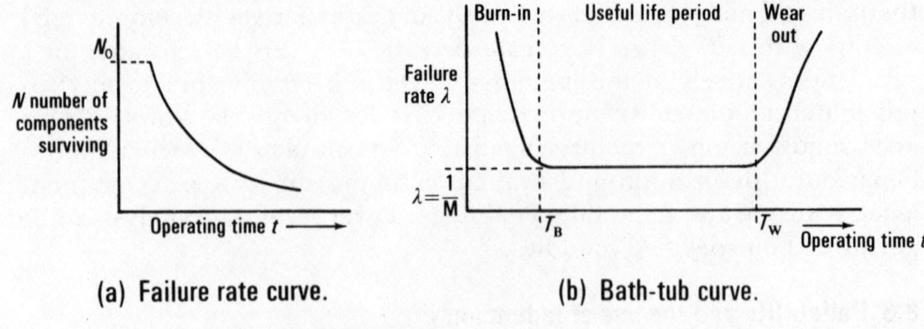

(a) Failure rate curve. (b) Bath-tub curve.

Figure 8.12 Reliability curves.

If $K = N_0$, the initial value of N, then:

$$N = N_0 \, e^{-\lambda t}$$

Thus the probability of no failure occurring during the time t is given by:

$$\frac{N}{N_0} = e^{-\lambda t}$$

If we take this function as the probability of survival, that is, the reliability, we may write:

$$R_{(t)} = e^{-\lambda t}$$

or using $m = 1/\lambda$, where m is usually known as the mean time between failures (MTBF),† we have the alternative form of the expression:

$$R_{(t)} = e^{-t/m}$$

This relationship is known as the **exponential failure law**; Fig. 8.12(b) shows the curve of failure rate against operating time (known as the 'bath-tub' curve) for a typical system. Note that during the useful life period the failure rate is constant and corresponds to random and independent failures; it is only over this portion of the curve that useful reliability predictions can be made.

To predict the reliability of a digital system, or for that matter any system, the following assumptions must be made:

All the component parts have constant failure rates; that is, the equipment is operating in its useful life period.

Failure rate data are available for the components, and pertain to the actual environment to be used.

All the component parts must be functioning for the equipment to operate.

The failure of any one part is independent of all the others.

† Some authors define $1/\lambda$ as the mean time to failure (MTTF) and consider the mean time between failures (MTBF) as including the mean time to repair a fault (MTRF) thus MTBF = MTTF + MTRF.

The prediction calculation is performed in the following way:

List the quantities of the different components used in the system, and multiply each component quantity by its failure rate. The sum of the individual products gives the overall systems failure rate. Thus:

$$R_{system} = R_1 \times R_2 \times R_3 \times \cdots \times R_n$$
$$= e^{-\lambda_1 t} \times e^{-\lambda_2 t} \times \cdots \times e^{-\lambda_n t}$$

And $\lambda_{system} = \lambda_1 + \lambda_2 + \lambda_3 + \cdots + \lambda_n$

To obtain a more realistic result, weighting factors (WF) for the environment, component ratings, and temperature range should also be included in the calculation (see Table 8.3). Using these factors the expression becomes:

$$\lambda_{system} = \text{Failure rate (\% per 1000 hours)} \times \text{WF (environment)} \times \text{WF (rating)} \times \text{WF (temp.)} \times \text{no. of devices}$$

Thus, $\text{MTBF} = \dfrac{10^5}{\lambda_{system}}$

Consider, for example, a computer that contains 2000 integrated circuits in DIL packs with a failure rate of 0·01% per 1000 hours, 5000 wrapped

Table 8.3
Typical weighting factors used in reliability predictions

Environment	*Multiply by WF*	*Rating*	*Multiply by WF*	*Temperature °C*	*Multiply by WF*
Laboratory or office (air conditioned)	0·5	Resistors $\frac{1}{10}$ max. watts	1·0	0–10	1·5
Laboratory or office (normal)	1·0	Resistors $\frac{1}{2}$ max. watts	1·5	10–20	1·0
Portable field equipment	1·5	Resistors max. watts	2·0	20–70	1·5
Mobile (vehicle mounted)	2·0	Capacitors $\frac{1}{10}$ max. volts	1·0	70–100	2·0
Airborne (civil airlines)	1·5	Capacitors $\frac{1}{2}$ max. volts	3·0		
Airborne (military)	4·0	Capacitors max. volts	6·0		
Seaborne (small craft)	2·0	Transistor/diodes $\frac{1}{10}$ max. rated power	1·0		
Seaborne (large craft)	1·5	Transistor/diodes $\frac{1}{2}$ max. rated power	1·5		
Missile (not included – special case)	—	Transistor/diodes max. rated power	2·0		

connections with a failure rate of 0·0001%, 200 resistors with a failure rate of 0·005%, and 100 capacitors with a failure rate of 0·01%. Assume that the system is to work in normal laboratory conditions, that all the components are to work at half their nominal values, and that the temperature range is 10°C → 20°C. Then we have:

$$\lambda_{\text{IC's}} = 0{\cdot}01 \times 2000 \times 1{\cdot}0 \times 1{\cdot}5 \times 1{\cdot}0 = 30$$
$$\lambda_{\text{connections}} = 0{\cdot}0001 \times 5000 \times 1{\cdot}0 \times 1{\cdot}0 \times 1{\cdot}0 = 0{\cdot}5$$
$$\lambda_{\text{resistors}} = 0{\cdot}005 \times 200 \times 1{\cdot}0 \times 1{\cdot}5 \times 1{\cdot}0 = 1{\cdot}5$$
$$\lambda_{\text{capacitors}} = 0{\cdot}01 \times 100 \times 1{\cdot}0 \times 310 \times 1{\cdot}0 = 3{\cdot}0$$

Thus $\lambda_{\text{system}} = 35\%$ per 1000 h

$$\text{MTBF} = \frac{10^5}{35} = 2860 \text{ h}$$

The reliability for a 100 hour period is:

$$R_{100} = e^{-100/2860} = e^{-0{\cdot}035} = 0{\cdot}965$$

8.6.1 Designing for reliability

The reliability should be considered at all stages of design and manufacture, from the initial systems specification right through to production and commissioning. At the initial systems design stage the trade-offs between reliability and other system parameters, such as size and weight, must be considered. For instance, the life of components is directly related to their physical size, the amount of thermal insulation used, and the temperature reached in practice. Consequently, special artificial environments to prolong component life, such as cooling and heating chambers and vibration mountings, must be considered when high reliability is essential. For large computer installations it is common practice to house the system in an air-conditioned and dust free environment. The computer room would normally have a special false floor containing inter-cabinet cable ducting, and a false ceiling for air-conditioning trunking and lighting.

Cost is always an important consideration where reliability is concerned, the most reliable system may not be the best in practice if the cost is too great. A good indication of the amount of money that should be invested in reliability is to consider the cost of a system failure. In all cases the cost of reliability should be contrasted with the full operational costs, including system failure, cost of subsequent maintenance, and the provision of spare parts.

An important contribution to system reliability is the correct choice of reliable components or sub-systems, since they determine the reliability of the ultimate system. When selecting components or circuits for a system the following factors must be borne in mind:

The availability of life test data (in the actual system environment) – these are needed for reliability prediction.

Basic component design and manufacture – these must be investigated to ensure that the manufacturing processes are well controlled.

The number of alternative sources of component suppliers – the specification must be sufficiently detailed to allow, if necessary, the replacement of one component by another.

Standardization of circuits where possible – to achieve a small number of basic components or sub-systems, thus reducing the number of spare parts and testing procedures required by the system.

Components should be derated or used conservatively with large safety factors (the effect of derating on component life must be taken into account).

The use of 'worst-case' and statistical circuit design methods[52] must be considered.

Marginal testing facilities must be evaluated, bearing in mind the cost of the extra equipment required for checking purposes.

8.6.2 Maintainability of the system

In large complex systems the time taken to repair a fault is just as important as the time that elapses between faults (the MTBF). The **maintainability** of a system is defined as a probability that, when maintenance action is initiated under stated conditions, a failed system will be restored to an operable state within a specific time. This probability is usually expressed as the mean time to repair a fault (MTRF). The availability of a system or the **uptime ratio** is given by:

$$\text{Availability} = \frac{\text{MTBF}}{\text{MTBF} + \text{MTRF}}$$

The availability may also be regarded as the probability that the system will be free from faults at any particular time during its scheduled working period. This is an important factor in assessing the amount of work the computer system can handle; that is, the actual time available for work is the total scheduled or 'switched on' time multiplied by the availability. Note that halving the MTRF will increase the availability by the same amount as that obtained by doubling the MTBF!

It is essential that the maintenance aspects associated with a computer are borne in mind during the initial system design phases. For example, the cost of maintaining the system after it is installed may greatly exceed the initial capital expenditure, thus maintenance costs are an important economic factor in considering the feasibility of a system. Moreover, it may be necessary to provide expensive standby equipment to offset the effects of prolonged downtime. In effect, the problem is one of man–machine communication and involves human aspects such as the availability and training of skilled personnel, data presentation, and fault-finding aids. All these problems of course are aggravated by the ever increasing complexity of computer systems.

Maintainability – that is, the rapid diagnosis of failure and the isolation and repair of a particular unit—may be designed into the computer system by incorporating the following facilities (where possible):

Automatic or built-in test procedures – in considering these facilities all facets of the problem must be covered, from diagnosis of system or module faults (as described in earlier sections) to the provision of the appropriate test equipment.

Use of marginal checking methods to investigate intermittent faults, etc.

Simplicity of design – this is particularly valid in logic circuitry where there are many alternative solutions to a design problem.

Adaptation of circuit and logic diagrams to the needs of the maintenance personnel – for example, system diagrams, as well as showing the logic flow, should also identify the position of boards and connections.

Provision of numerous and readily available test points – these are normally collected together, with other test aids, in the engineer's console, which allows remote monitoring and control of micro-operations, etc.

Easy accessibility to all parts of the system – this involves the cabinet design and should include, for example, a modular plug-in construction and racks which swing or pull out for easy access.

Quick and positive identification of all parts of the system such as plug-in units, terminals, and cableforms.

Provision of adequate guards for any system hazards, such as exposed high voltages, high temperatures, moving parts, etc.

8.6.3 Redundancy techniques

One method of enhancing the reliability of a computer system is to use the principle of **redundancy** by duplicating various parts or functions of the system. Note that the additional equipment is redundant only when considered in the sense of providing the *basic* system requirements, it is, of course, essential if increased reliability is required. For example, in a real-time computer system the overall system reliability may require a MTBF of 3000 hours, the problem is how to achieve this figure using computers having an average MTBF of 2000 hours. It is important to stress, however, that improvements in reliability cannot be obtained using poor quality components; the individual component or sub-system reliability must be high to begin with.

Redundancy may be incorporated into a system in three basic ways:

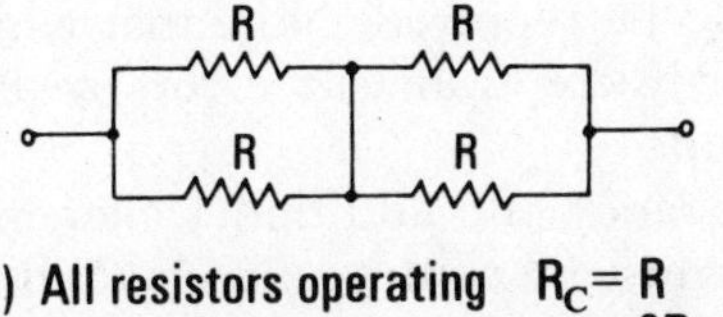

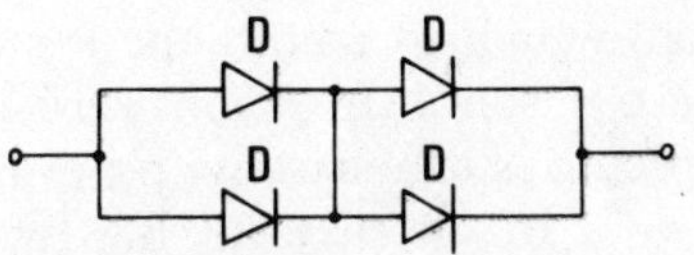

(a) All resistors operating $R_C = R$
Any resistor O/C $R_C = \frac{3R}{2}$
Any resistor S/C $R_C = \frac{R}{2}$

(b) Circuit maintained for any diode open or short-circuit.

Figure 8.13 Hammock networks.

Information redundancy – the use of error-detecting and correcting codes, software checks, etc.

Component redundancy – the replication of resistors, capacitors, etc. in 'hammock' networks,[53] as shown in Fig. 8.13.

System redundancy – the use of switched standby and parallel operating sub-systems.

Information redundancy has partly been discussed in earlier sections however it is worthwhile at this point to consider software redundancy in more detail. Software redundancy includes all additional programs, instructions and micro-instructions which would not be required in a fault-free system. There are three major forms of software redundancy:

(a) multiple storage of critical programs and data
(b) test and diagnostic programs or micro-programs
(c) special programs which respond to hardware error checks and initiate error routines and program restarts.

One of the advantages of software redundancy is that it is possible to add on fault-tolerance features after the hardware structure has been designed. However there are problems in ensuring that the software fault routines function correctly after the occurrence of a fault.

In a sense the use of extra program to obtain fault-tolerance is hardware redundancy in that extra storage is generally incurred. An alternative way of employing software redundancy, which results in increased running times, is to repeat the execution of a program should an error be detected. This technique, known generally as **rollback and recovery**,[54] is normally associated with some form of error correction, usually retransmission of data in the case of storage and I/O errors. Rollback can be of single instructions, program segments or entire programs, in the latter case protected storage for the rollback addresses and a running record of current system parameters (to establish check-points from which to restart the system) would be required. Note that while these systems function very satisfactorily for transient errors severe difficulties can arise with permanent faults.

In all cases of hardware redundancy the additional elements (whether components or sub-systems) must be continually checked to ensure they are still operative (and hence maintaining the enhanced level of reliability); unfortunately, this is only really practicable on a sub-system basis. Furthermore, in component replication there are problems with circuit tolerancing; for example, as shown in Fig. 8.13, a circuit must be designed to function with a resistor in the range $R/2 \rightarrow 3R/2$.

If the system has redundancy in the form of standby equipment this can be switched over manually (or automatically in a more sophisticated design) to ensure minimum loss of system control. If there is no standby equipment, two courses remain open: either to operate on a reduced requirement basis, or to set the system to some safe condition before resorting to manual control.

One method of operation would be to use, say, three computers to control a given process.[55] Two of these (working together) would be allocated to fulfill the system operational requirements, whilst the third would be used as the standby and system-testing computer. Should any one of the operational computers go faulty, it can be detected and replaced by the third computer, maintaining full system facilities but with reduced testing. If one of the remaining two computers becomes inoperational, the system can be kept going by ensuring the continuance of essential operations. In this way we have obtained two levels of standby under fault conditions.

Note that the preceding remarks infer a distributed method of control whereby it becomes more advantageous to use two or more smaller computers in a system than one large one. Having established by automatic testing that a computer is faulty, it is replaced and taken out of the system, the fault being located and repaired off-line in the usual way by using diagnostic programs in conjunction with marginal testing. Note the need to find and repair the fault as quickly as possible – a high maintainability is required – since the computer system would be operating without testing or full system capabilities during this repair period.

Parallel redundancy is simple to implement (a trivial example is the use of double rear tyres on heavy duty lorries) and the gain in reliability easily calculated. Fig. 8.14(a) shows a system consisting of four sub-systems, each having a reliability of $R = 0{\cdot}96$, connected in series. The overall system reliability may be calculated using simple probability theory and is given by:

$$\begin{aligned} R_s &= R_1 \times R_2 \times R_3 \times R_4 = R^4 \\ &= (0{\cdot}96)^4 = 0{\cdot}85 \end{aligned}$$

In the case of two identical devices in parallel the probability that at least one or both devices will continue to operate is given by:

$$R_s = 2R - R^2$$

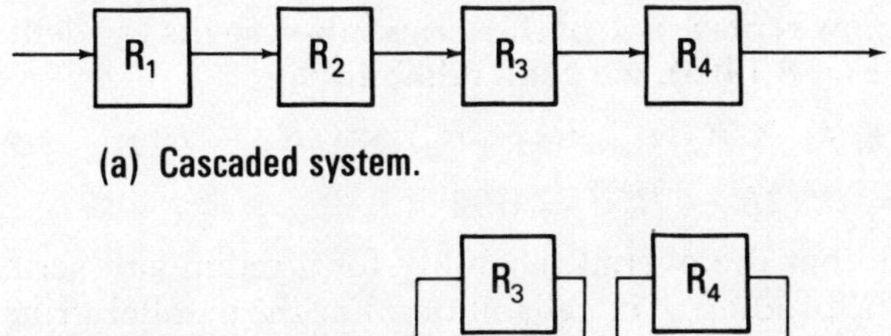

(a) Cascaded system.

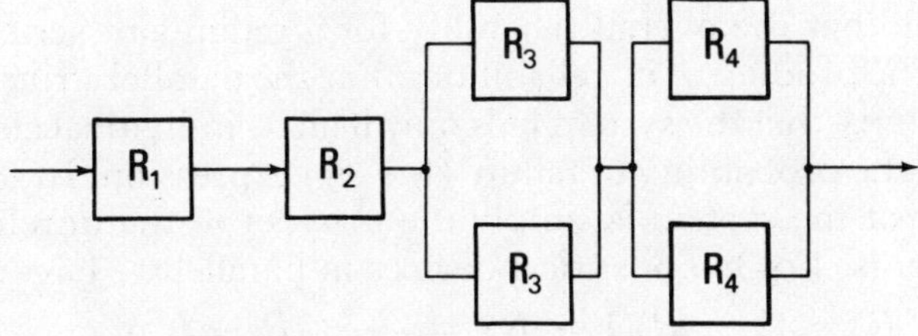

(b) Parallel redundancy.

Figure 8.14 System reliability.

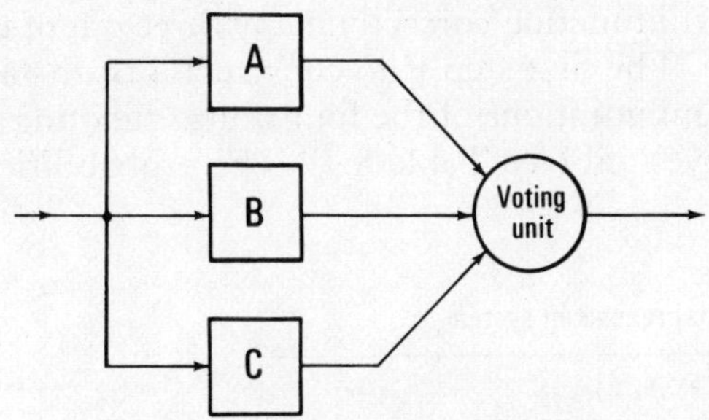

Figure 8.15 Majority voting system.

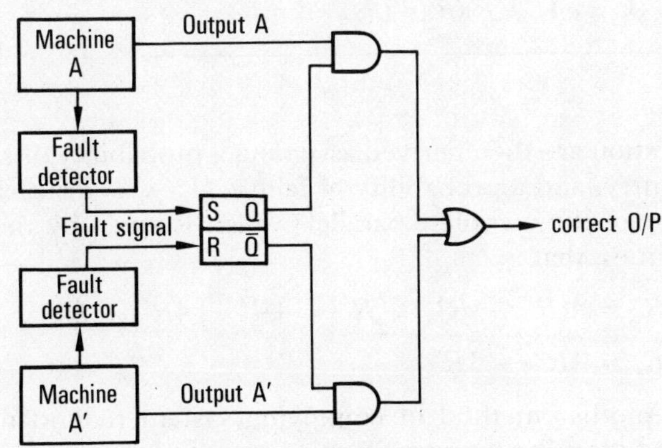

Figure 8.16 DMR fault tolerant circuit.

Thus, if we now replace two of the series sub-systems by identical parallel units, as in Fig. 8.14(b), we get a reliability of:

$$R_s = R_1 \times R_2 \times (2R_3 - R_3^2) \times (2R_4 - R_4^2)$$

So $\quad R_s = R^2(2R - R^2)^2 = 0{\cdot}92$

Note that the overall reliability for a redundant series structure is given by the product of its reliabilities. For the parallel structure, which has the property that the system fails only if all its individual elements fail, we may use the probability of failure $(1 - R)$ expression. In this case the *unreliability* of the system is simply the product of the unreliabilities of its components. For two identical devices in parallel we have:

$$1 - R_s = (1 - R)(1 - R) = 1 - (2R - R^2)$$

Thus $\quad R_s = 2R - R^2$

An alternative technique that may be used to calculate the reliability of redundant parallel structures is to express the individual probabilities in the form of a logic truth-table. For example, consider a parallel redundant system comprising four identical units in parallel; the system is such that it will function correctly if any three out of the total of four units is operative. The first step is to construct a truth-table listing all possible working configurations of the four units, denoting by a 1 the presence of an operative unit (see Table 8.4). The probabilities for each

Table 8.4

Truth-table evaluation of parallel redundant system

Possible combinations	*Probabilities*
1 1 1 0	$(R)(R)(R)(1 - R) = R^3 - R^4$
1 1 0 1	$(R)(R)(1 - R)(R) = R^3 - R^4$
1 0 1 1	$(R)(1 - R)(R)(R) = R^3 - R^4$
0 1 1 1	$(1 - R)(R)(R)(R) = R^3 - R^4$
1 1 1 1	$(R)(R)(R)(R) = R^4$

configuration are then derived assigning a probability of survival (R) for each 1 entry, and a probability of failure $(1 - R)$ for each 0 entry. The reliability of the complete parallel system is given by the sum of these probabilities, that is:

$$R_s = 4(R^3 - R^4) + R^4 = 4R^3 - 4R^4 + R^4$$

Thus $\quad R_s = 4R^3 - 3R^4$

Another method of employing system redundancy is to use a **switched standby system**;[56] this approach has already been discussed generally in an earlier paragraph. If we assume that there are n components standing by, and a total of $n + 1$ components in the system, n failures may occur before the system fails. Let us assume that we have a perfect switch and that all the components have the same failure rate λ, then the probability of one or no failures occurring is:

$$R_b = e^{-\lambda t} + e^{-\lambda t}\,\lambda t = e^{-\lambda t}(1 + \lambda t)$$

(this method assumes a Poisson distribution, that is $e^{-\lambda t}$ is the probability that no failures happen in time t, and $\lambda t\, e^{-\lambda t}$ the probability that a failure happens once). Now the MTBF for the switched standby system is given by:

$$\mathrm{MTBF}_b = \int_0^\infty R_b\,dt = \int_0^\infty e^{-\lambda t}\,dt + \int_0^\infty e^{-\lambda t}\,\lambda t\,dt$$

Therefore

$$\mathrm{MTBF}_b = \frac{1}{\lambda} + \frac{\lambda}{\lambda^2} = \frac{2}{\lambda}$$

Let us compare these results with those for the parallel system:

$$R_p = e^{-\lambda t}(2 - e^{-\lambda t})$$

$$\mathrm{MTBF}_p = \int_0^\infty R_p\,dt = 2\int_0^\infty e^{-\lambda t}\,dt - \int_0^\infty e^{-2\lambda t}\,dt$$

$$\mathrm{MTBF}_p = 3/2\lambda$$

Thus, a switched system is only slightly more reliable than a system using parallel redundancy but it has a longer MTBF. The advantages are lost, however, if the reliability of the switching device (R_{ss}) is less than 100%, when the expression for reliability becomes:

$$R_b = e^{-\lambda t} + R_{ss}e^{-\lambda t}\lambda t$$

An alternative scheme is the **voting or majority system** shown in Fig. 8.15. Here the sub-systems are triplicated (the next possible structure must contain five sub-systems) and the correct output chosen by the voting or majority logic, that is, 2 out of 3. This scheme called **triple mode redundancy** (TMR) has been used in practice for some military systems but is very expensive to implement. Moreover, the overall reliability depends on the reliability of the majority element, which is given by:

$$R = 3R^2 - 2R^3$$

for sub-systems with equal reliability.

As we have seen fault tolerance in digital systems can be obtained by incorporating protective redundancy in the form of additional hardware or software. This can be achieved either by the replication of existing non-redundant elements (as in TMR systems) or by including the redundancy as an integral part of the basic element. The underlying principle of **fault-tolerant computer design**[57] is that the machine must be able to tolerate a fault in the sense that the behaviour under fault conditions relates very closely to its normal behaviour.

Considerable research is in progress on this topic, so far, from the hardware point of view, the most promising work has been in the use of error-correcting codes for the state assignment of sequential machines. In particular the work of Sawin and Maki who have produced a general design procedure for single fault tolerant machines[58] is worth mentioning. The circuit structure is shown in Fig. 8.16; note that it functions as a double mode redundancy (DMR) system and requires a minimum of hard core (that is not replicated) logic.

References and bibliography

1 Simon, H. A. *The Sciences of the Artificial*. MIT Press, Massachusetts, 1969.

2 Singleton, W. T. *Man-Machine Systems*. Penguin Books, 1974.

3 Ghandi, S. K. *The Theory and Practice of Microelectronics*. Wiley, New York, 1968.

4 Hamilton, D. and Howard, W. *Basic Integrated Circuit Engineering*. McGraw Hill, New York, 1975.

5 Garrett, L. S. Integrated circuit digital logic families, *IEEE Spectrum*, **7**, 46–58 Oct.; 63–72 Nov.; 30–32 Dec., 1970.

6 Gosling, W., Townsend, W. G. and Watson, J. *Field Effect Electronics*. Wiley, New York, 1971.

7 Penney, W. and Lau, L. (editors) *MOS Integrated Circuits: Theory, fabrication, design and systems applications of MOS LSI*. Van Nostrand Reinhold, 1972.

8 Greenfield, J. D. *Practical Digital Design using ICs*. Wiley, New York, 1977.

9 Barna, A. and Porat, D. *Integrated Circuits in Digital Electronics*. Wiley, New York, 1973.

10 Blakeslee, T. R. *Digital Design with Standard LSI and MSI*. Wiley, New York, 1975.

11 Horton, R. L., Englade, J. and McGee, G. I^2L takes Bipolar Integration a significant step forward, *Electronics*, Feb. 6th, 1975, 83–90.

12 Mays, C. H. A Brief Survey of computer-aided integrated circuit layout, *IEEE Trans Circuit Theory*, **CT–18**, 1971, 10–13.

13 Sholl, H. A. and Yang, S. C. Design of Asynchronous Sequential Networks using read-only memory; *IEEE Trans. Computers*, **C24**, 1975, 195–206.

14 Glazer, E. L. Computer aided design for computing systems. *IEE Conference on Computer Science and Technology*, IEE pub. no. 55, 1969, 13–19.

15 Lewin, D. *Computer Aided Design of Digital Systems*. Crane Russak, New York, 1977, Chapter 2.

16 Case, P. W., *et al.* Solid logic design automation. *IBM J. Res. Dev.*, 1964, **8**, 127–140.

17 Breuer, M. A. (ed) *Design Automation of Digital Systems*, Vol. 1, Theory and Techniques. Prentice Hall, Englewood Cliffs, N.J., 1972.

18 Martinez, R. A Look at Trends in Microprocessor/Microcomputer Software Systems, *Computer Design*, **14**, June 1975, 51–57.

19 Conley, S. W. Portable Microcomputer Cross-Assemblers in BASIC, *IEEE Computer*, **8**, Oct. 1975, 32–42.

20 *A Guide to PL/M Programming*, Intel Corp., Santa Clara, California, 1974.

21 Down, R. L. Understanding Logic Analysers, *Computer Design*, **16**, June 1977, 188–191.

22 Ricketts, L. W., Bridges, J. E. and Miletta, J. *EMP Radiation and Protective Techniques*. Wiley, New York, 1976.

23 Ott, H. W. *Noise Reduction Techniques in Electronic Systems*. Wiley, New York, 1976.

24 Jones, J. P. Causes and cures of noise in digital systems. *Computer Design*, 1964, Sept. 24–29; Oct. 20–27; Nov. 24–31.

25 Matick, R. E. *Transmission Lines for Digital and Communication Networks*. McGraw-Hill, New York, 1969.

26 Gruenberger, F. Program testing and validating. *Datamation*, 1968, July, 39–47.

27 Bashow, T., J. Friets, and A. Karson. Programming systems for machine malfunctions. *IEEE Trans. Electron. Comput.*, 1963, **EC12**, 10–17.

28 Chambers, R. P. Random number generation on digital computers. *IEEE Spectrum*, 1967, Feb., **4**, 48–56.

29 Peterson, W. W. and Weldon, E. J. *Error Correcting Codes*. MIT Press, Cambridge, Mass., 1972.

30 Lin, S. *Introduction to Error Correcting Codes*. Prentice Hall, Englewood Cliffs, N.J., 1970.

31 Levine, L. and Meyers, W. Semiconductor Memory Reliability with Error Detecting and Correcting Codes, *IEEE Computer*, **9**, Oct. 1976, 43–50.

32 Lewin, D. Theory of Linear Switching Circuits, *Control*, **13**, March 1969, 196–203.

33 Boehm, B. W. Software Engineering, *IEEE Trans. Computer*, **C25**, 1976, 1226–1241.

34 Brock, P., and S. Rook. The problem of preparing acceptance tests. *J.A.C.M.*, 1954, **1**, 82–87.

35 Bennetts, R. G., and D. W. Lewin. Fault diagnosis of digital systems – a review. *The Computer Journal*, 1971, **14**, 199–206.

36 Lake, D. W. Logic Simulation in Digital Systems, *Computer Design*, **9** 1970, 77–83.

37 Breuer, M. A. and Friedman, A. D. *Diagnosis and Reliable Design of Digital Systems*. Computer Science Press Inc., Calif., 1976.

38 Chang, H. Y., E. G. Manning, and G. Metze. *Fault Diagnosis of Digital Systems*. John Wiley, New York, 1970.

39 Kautz, W. H. Fault testing and diagnosis in combinational digital circuits. *IEEE Trans. Comput*. 1968, **C17**, 352–366.

40 Roth, J. P. Diagnosis of automata failure: a calculus and a method. *IBM J. Res. Dev.*, 1966, **10**, 278–291.

41 Hennie, F. C. Fault-detecting experiments for sequential circuits. *Proceedings 5th Annual Switching Theory and Logical Design Symposium*, 1964, S-164, 95–110.

42 Kriz, T. A. A Path Sensitizing Algorithm for diagnosis of binary sequential logic, *Proc. IEEE Computer Group Conf.*, **70-C-23-C**, 1970, 250–259.

43 MacDougall, H. H. Computer System Simulation: an introduction, *Computing Surveys*, **2**, 1970, 191–210.

44 Efron, R. and Gordon, G. General purpose digital simulation and examples of its applications, *IBM Systems J*, **3**, 1964, 22–34.

45 Wyman, F. D. *Simulation Modelling: a guide to using SIMSCRIPT*. Wiley, New York, 1970.

46 Pritsker, A. A. B. *The GASP IV Simulation Language*. Wiley, New York, 1974.

47 Bell, T. E. Objectives and problems in simulating computers, *AFIPS FJCC*, **41**, 1972, 287–297.

48 Hays, G. G. Computer-aided design: simulation of digital design logic. *IEEE Trans. Comput.*, 1969, **C18**, 1–10.

49 Reynolds, J. S. A conversational logic simulator for use with a time-sharing computer. *IEE Conference on CAD*, IEE pub. no. 51, 1969, 608–615.

50 Flake, P. L., Musgrave, G. and White, I. J. A Digital System Simulator – HILO, *Digital Processes*, **1**, 1975, 39–53.

51 Amstadter, B. *Reliability Mathematics*. McGraw Hill, New York, 1971.

52 Atkins, J. B. Worst case circuit design. *IEEE Spectrum*, 1965, March, **2**, 152–161.

53 Moore, E. F., and C. E. Shannon. Reliable circuits using less reliable elements. *J. Franklin Inst.*, 1956, **262**, 191–208, 281–297.

54 Chandy, K. M. A Survey of Analytic models of rollback and recovery strategies, *IEEE Computer*, **8**, May 1975, 40–47.

55 Buchman, A. S. The digital computer in real-time control systems: pt. 4 – reliability; pt. 5 – redundancy techniques. *Computer Design*, 1964, **3**, Sept., 12–18; Nov., 12–15.

56 Flehinger, B. J. Reliability improvement through redundancy at various system levels. *IBM J. Res. Dev.*, 1958, **2**, 148–158.

57 Avizienis, A. Fault-Tolerant Systems, *IEEE Trans. Computers*, **C25**, 1976, 1304–1312.

58 Maki, G. K. and Sawin III, D. H. Fault-Tolerant Asynchronous Sequential Machines, *IEEE Trans. Computers*, **C23**, 1974, 651–657.

Further reading

Breuer, M. A. Techniques for the simulation of computer logic. *Comm. A.C.M.*, 1964, **7**, 443–446.

Wilcox, R., and Mann, W. *Redundancy Techniques for Computer Systems.* Spartan Books, Washington, 1962.

Dent, J. Diagnostic engineering. *IEEE Spectrum*, 1967, July, **4**, 99–104.

Podraza, G., Clegg, R. and Slager, J. Efficient MSI partitioning for a digital computer. *IEEE Trans. Comput.*, 1970, **C19**, 1020–1028.

Defalco, J. Reflections and cross-talk in logic circuit interconnections. *IEEE Spectrum*, 1970, **7**, 44–50.

Ross, D. T., Goodenough, J. B. and Irvine, C. A. Software Engineering: Process, principles and goals, *IEEE Computer*, **8**, May 1975, 17–27.

Adkins, G. and Pooch, U. W. Computer Simulation – A Tutorial, *IEEE Computer*, **10**, April 1977, 12–17.

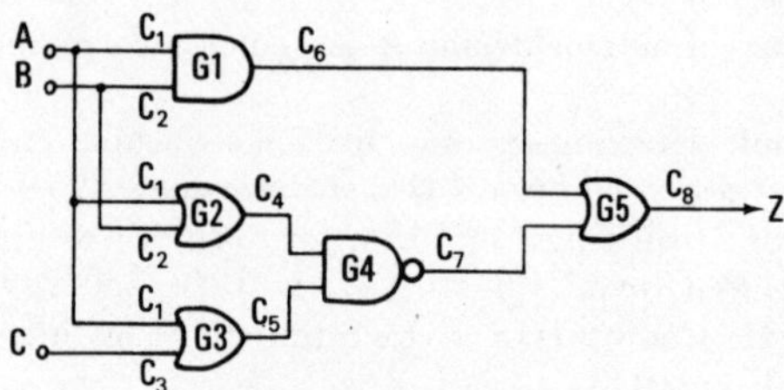

Figure 8.17

Tutorial problems

8.1* Derive the fault matrix for the circuit shown in Fig. 8.17 and hence deduce the minimum set of test inputs that will detect the occurrence of a logical fault.

8.2 A small computer has the following number of components with failure rates as stated:

Component	*Number*	*Failure rate % per 1000 h*
A	2,500	0·01
B	8,000	0·0005
C	5,000	0·002
D	60,000	0·00005
E	10,000	0·002

Predict the reliability of the system over a 100-hour period. What is the MTBF? What effect would environmental conditions have on the predicted reliability?

If the equipment was used with two identical units in parallel, ascertain how this would affect the reliability. Substantiate all formulae used and state any assumptions made in the calculations.

8.3* A computing system has the following sub-system reliabilities:

Arithmetic unit	$r_1 = 0{\cdot}95$
Core store	$r_2 = 0{\cdot}92$
Input/output channel	$r_3 = 0{\cdot}85$
Disc file	$r_4 = 0{\cdot}80$
Teletype terminal	$r_5 = 0{\cdot}72$

For this system to function, the arithmetic unit, at least one core store, the input/output channel, the teletype channel, and at least two disc files must be operational; the system structure is shown in Fig. 8.18.

Derive an expression for the overall reliability of the system and then calculate the MTBF for an operating time of 1000 hours.

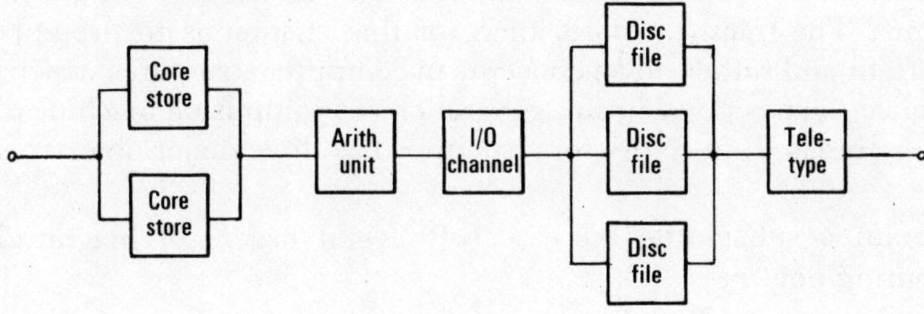

Figure 8.18

9

Highly parallel processing systems

9.1 Introduction

The preceding chapters have been primarily concerned with the theory and design of conventional digital computer systems. The architecture and techniques that have been described are, of course, fundamental to any commercially available machine and as such form an essential prerequisite to this chapter, which is concerned with the study of unorthodox systems. The main purpose, then, of this chapter is to present those important and rather novel concepts in computer structures which could influence the design of future generations of computing machines.[1]

Research in computer structures has three major objectives:

to obtain a substantial increase (of several orders of magnitude) in computing power;

to take full advantage of the batch fabrication of large scale integrated circuits using MOS or bipolar devices; and

to develop machines which emulate the high level algorithmic procedures required in numerical and non-numerical computing techniques.

Increases in computing power (usually expressed as the number of bits processed per second) can be obtained either by using more components or by a better utilization of these components. Moreover, because of the progressing developments in device technology and the availability of cheap and reliable components the possibility of designing highly complex computer structures becomes a feasible proposition. For maximum efficacy a regular, iterative array of basic elements would be the ideal structure; this has naturally led to the idea of designing **homogenous cellular machines** with distributed logic and storage functions.

When comparing the computing power of various systems a better measure than operations per second is **throughput**. Since the execution of a specific task may require 10 instructions in one machine as compared to say 25 in another, a comparison of the time required to perform a specified task (throughput) is a much better method of evaluation.

Another way of enhancing the power of computation systems is to organize the solution to a problem such that separate calculations (each pertaining to the same problem) may be performed at the same time.

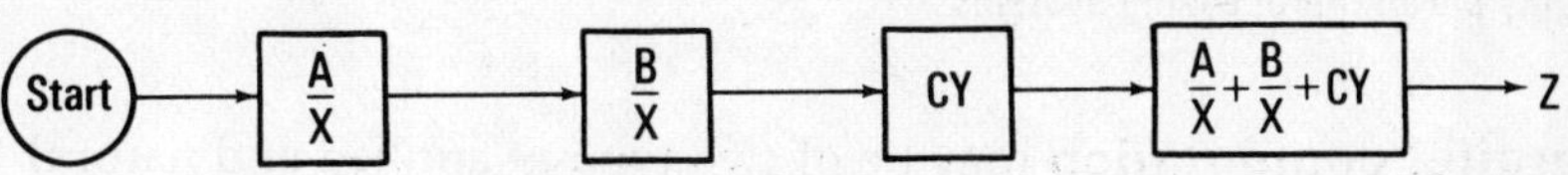

(a) Sequential computation.

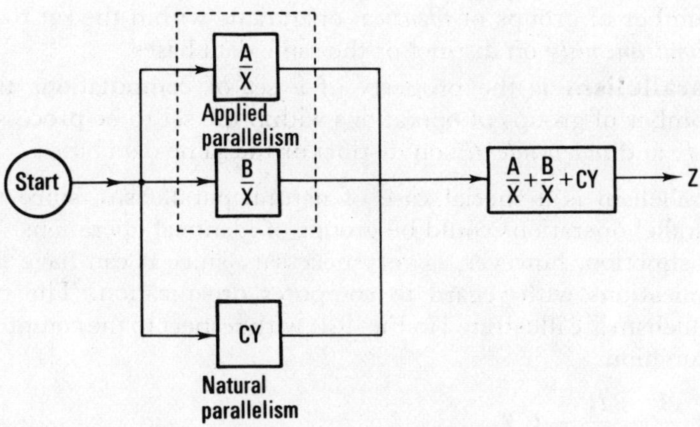

(b) Parallel computation.

Figure 9.1 Computation methods.

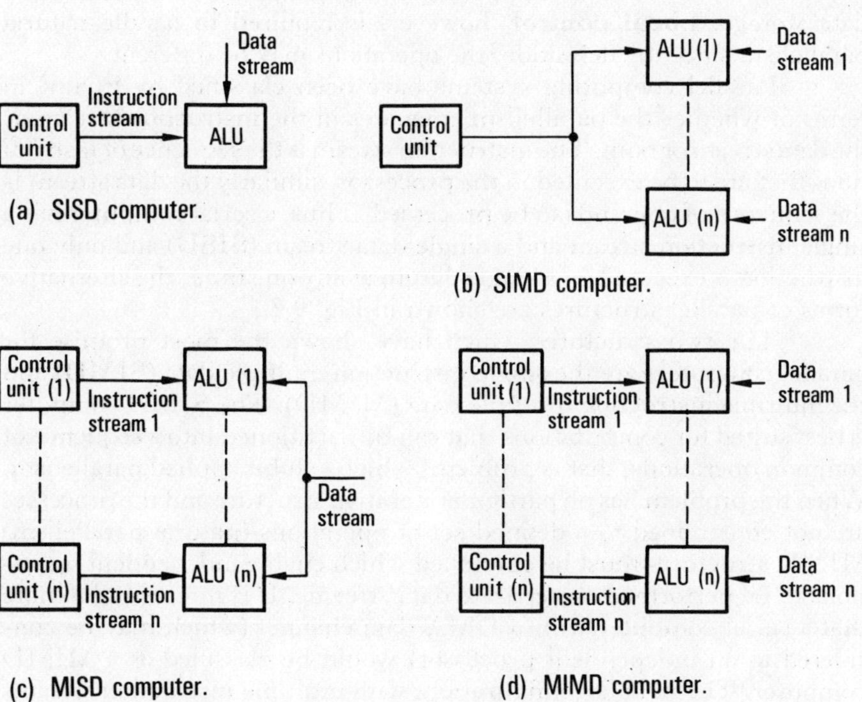

(a) SISD computer.

(b) SIMD computer.

(c) MISD computer.

(d) MIMD computer.

Figure 9.2 Parallel processing computer systems.

Parallel computation may be of two types – applied and natural – and these are defined as follows:[2]

Applied parallelism is the property of a set of computations which enables a number of groups of *identical* operations within the set to be processed *simultaneously* on distinct or the same data bases.
Natural parallelism is the property of a set of computations that enables a number of groups of operations within the set to be processed *simultaneously* and *independently* on distinct or the same data bases.

Applied parallelism is a special case of natural parallelism, since the naturally parallel operations could be groups of identical operations. To draw this distinction, however, is very necessary since it can have important implications with regard to computer organization. The two types of parallelism are illustrated in Fig. 9.1, with respect to the computation of the function:

$$Z = \frac{A}{X} + \frac{B}{X} + CY$$

It is very obvious that by taking advantage of both natural and applied parallelism computation times can be drastically reduced.

Applied parallelism may be handled in a computer organization by using **global control**, in which a number of processing units may be controlled by a central control unit providing common instruction and data storage. **Local control**, however, is required to handle natural parallelism since, by definition, the operations may be different.

Parallel computing systems have been classified by Flynn[3] in terms of whether the parallelism emanates in the instruction stream or the data stream or both. The instruction stream is the sequence of instructions that are to be executed in the processor, similarly the data stream is the sequence of operands to be processed. Thus, a serial computer has a single instruction stream and a single data stream (SISD) and only one instruction is executed on a single datum at any one time; the alternative forms of parallel structures are shown in Fig. 9.2.

The two structures which have shown the most promise for parallel computers are the single instruction multiple data (SIMD) and the multiple instruction multiple data (MIMD). The SIMD computer is best suited for computations that can be partitioned into a sequence of common operations, that is problems which exhibit applied parallelism. When the problem has no particular iterative structure and the processes are not constrained to a defined set of operations (natural parallelism) MIMD structures must be employed which enable independent operations to be performed on separate data streams. It is interesting to note that a serial computer with a DMA data channel (which may be considered as an independent processor) would be classified as a MIMD computer. The MISD configuration, with multiple instruction streams operating on a single data stream, is generally considered unrealistic for parallel computing systems.

Development of SIMD computers (also called **array** computers) has progressed more rapidly than MIMD computers, the major problem with the later being the difficulty of partitioning computations into parallel processes which can be executed concurrently. However work is in progress on multi-processor computer systems[4] and this will increase rapidly in the future with the availability of cheap processing power.

One of the major problems associated with parallel processing is the assignment and sequencing of the parallel operations – that is, the programming of the system. In addition there are problems of resource and task allocation (both in static and dynamic modes) and synchronization. To obtain full efficiency it is essential to break down the computation into appropriate parallel sections but unfortunately there is, as yet, no general solution to this problem. Moreover, there is only a limited number of problems where the added complexity and cost of parallel computation can be justified (or for that matter, only a limited number of problems which lend themselves to the parallel approach). This situation could radically change however with the availability of cheap LSI microcomputer circuits making distributed concurrent processing a practical proposition. Typical examples of such problems are large matrix computations, such as those involved in modelling economic and ecological systems, weather prediction calculations, and step-by-step integration of complex differential equations.

In the following sections of this chapter we shall consider in detail various machines which have been proposed to implement some of the general principles described above.

9.2 Parallel processing systems[5]

One of the first machines to be proposed for parallel processing was the **Solomon computer**,[6] a simplified version of which was actually constructed. The Solomon machine is based primarily on the principles of applied parallel operations, thus allowing the speed to increase linearly with the number of processing units. Consequently, the objective of the Solomon structure is to be able to control the processing of a number of different data streams with a single instruction stream (SIMD computer). The machine has four principal features:

A large rectangular array of processing elements (PE's) is controlled by a single control unit, so that a single instruction stream sequences the processing of many data streams.

Store addresses, and data common to all of the data processors, are distributed on a common highway from the central control unit.

Limited local control at the PE level is obtained by permitting each element to enable or inhibit the execution of instructions in the common stream according to locally controlled tests.

Each processing element in the array is connected to its nearest neighbour to provide data exchanges.

A simplified block diagram of the Solomon organization is shown in Fig. 9.3. The **program store** contains the program instructions which are decoded by the **network control unit** (acting in much the same way as a conventional computer control unit) to generate the micro-orders needed to control the operations of the PE array. Thus, the PE network executes one instruction at a time, with each PE operating on different data, but all performing the same operation. The PE array consists nominally of 1024 units arranged in a 32 × 32 matrix, with each processing element containing its own arithmetic and logic unit, plus a 4096 24-bit word operand store. Instructions are only executed by those PE's

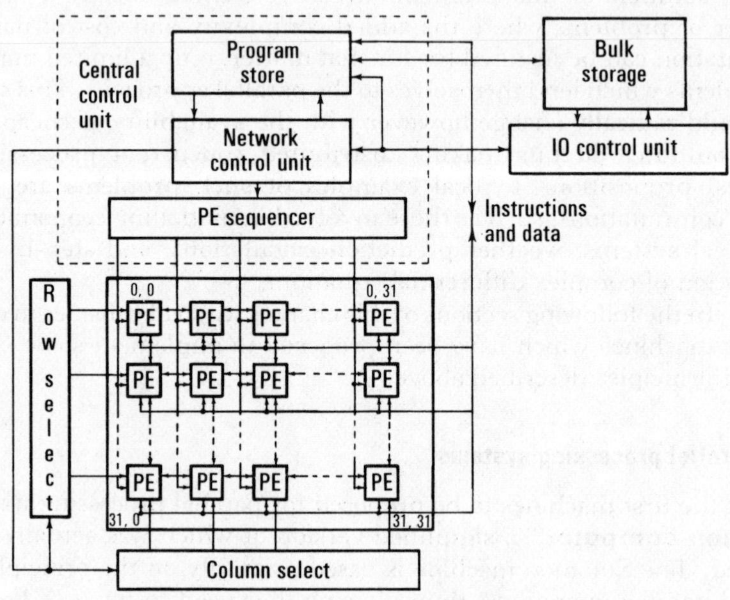

Figure 9.3 Organization of Solomon computer system.

which are selected by both **mode** and **geometric** control. The mode control is concerned with the internal conditions of the PE and is a function of the stored data. Each PE contains a 2-bit mode register, identifying one out of four possible states, which may be set or cleared under various internal conditions. The processing element executes an instruction only if its mode corresponds to that specified by the instruction, thus enabling 'conditional jumps' to be programmed. The geometric control is used to select a particular configuration of PE's, via row and column registers, which may be set under program control.

There are five basic array interconnection patterns that can be selected by the programmer, these are:

A vertical cylinder, in which communication paths are established between the leftmost and the rightmost columns of the array.
A horizontal cylinder, in which paths are established between the top and bottom rows of the array.
The first two options may be combined together to form a torus.
All the PE's may be combined together to form a single straight line of 1024 units.

A circular arrangement can also be obtained by connecting together the ends of the straight-line configuration.
Note that though information in the program store is available to all PE's, it is only the selected elements that actually make use of it; information can also be exchanged directly between neighbouring PE's, as shown in Fig. 9.3.

The processing element is the basic module of the Solomon system, and is a complete computer in itself, possessing a full repertoire of arithmetic and logical operations (employing a 24-bit wordlength) plus the usual store unit. The instruction set, which is basically single-address, includes the usual range of data transfer (including transfers to the main program store) fixed-point arithmetic, boolean and conditional instructions, etc. The latter instructions are effected via the mode control, which can be set unconditionally, or by the result of comparing data stored in the PE. Thus, the data conditional jumps of conventional computers are accomplished by PE tests, the results of which enable or inhibit local execution of subsequent commands in the instruction stream.

9.2.1 The Illiac IV system

Design studies on the Solomon system led to the development of the **Illiac IV** computer system[7, 8, 9] (again with a SIMD structure) which was constructed by the University of Illinois for the Burroughs Corporation in the U.S.A. The structure of the Illiac IV system (shown in Fig. 9.4) comprised 256 64-bit word processing elements, arranged in four reconfigurable SOLOMON type arrays each consisting of 64 PE's and a control unit (called a **quadrant**). The four arrays could be connected together under program control to allow multi- or single-processing operations. The system program resided in a Burroughs B6500 general purpose computer, which supervised program loading, array configuration changes, and IO operations (both internal and external). Back-up storage for the arrays was provided by a large, directly coupled, parallel access disc system.

The internal structure of an array, shown in Fig. 9.5, consists of 64 PE's arranged in a string, the end connections of which are folded back to form a circular configuration. It is also possible to consider the array

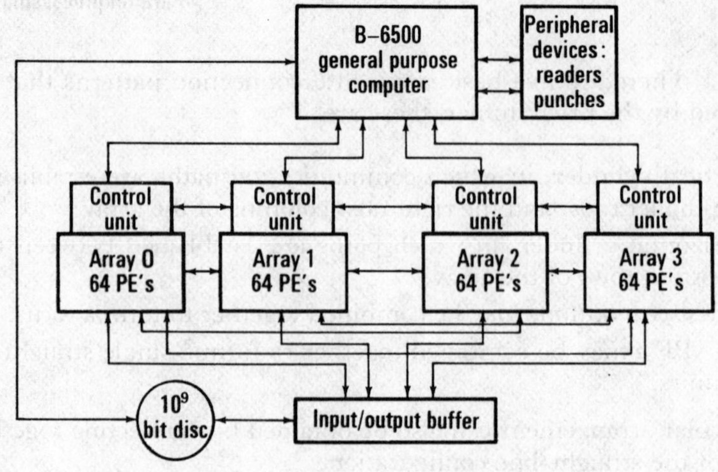

Figure 9.4 Illiac IV system organization.

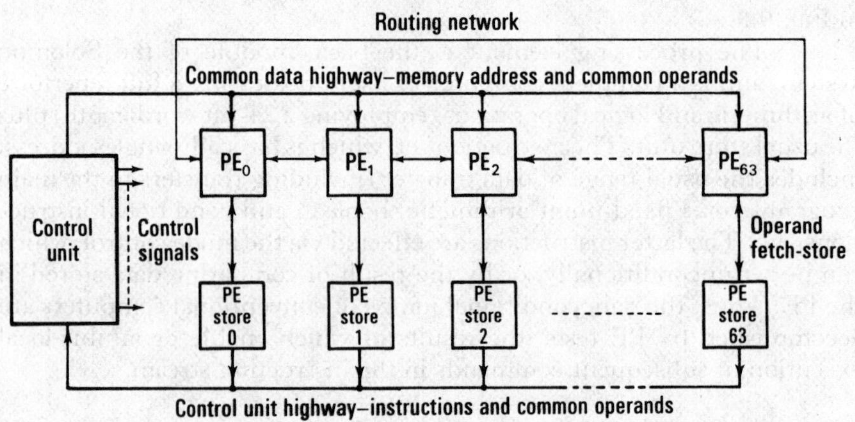

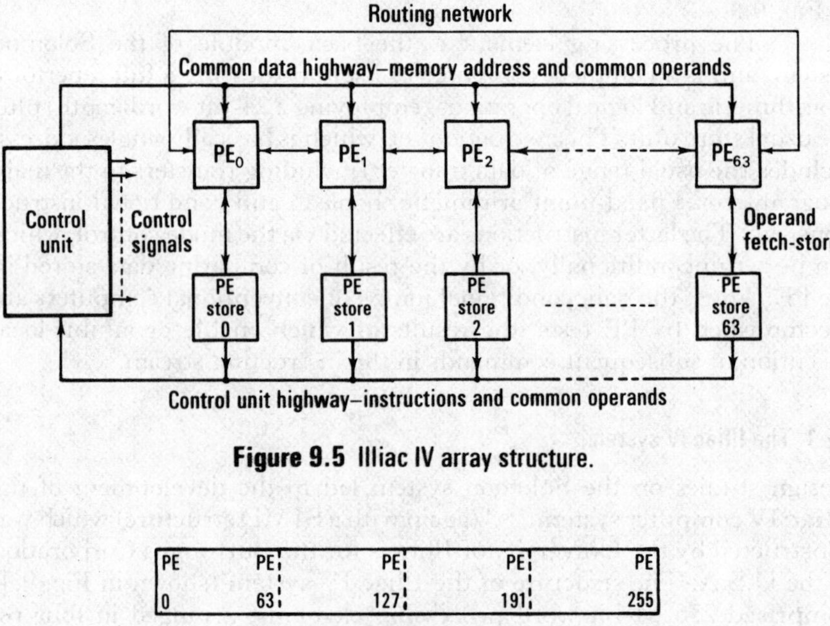

Figure 9.5 Illiac IV array structure.

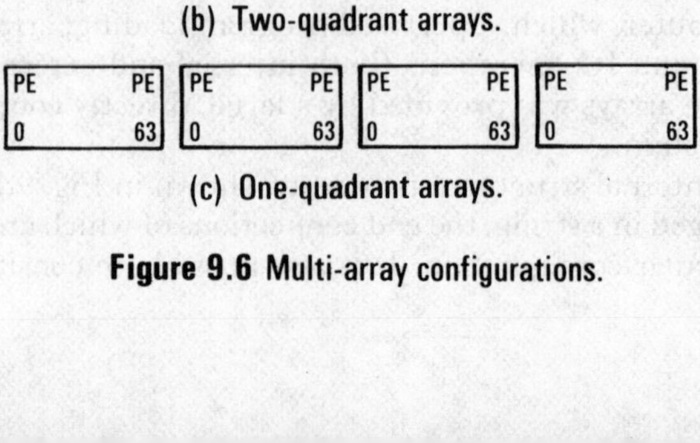

PE 0 | PE 63 | PE 127 | PE 191 | PE 255

(a) Four-quadrant arrays.

PE 0 | PE 63 | PE 127   PE 0 | PE 63 | PE 127

(b) Two-quadrant arrays.

PE 0 | PE 63   PE 0 | PE 63   PE 0 | PE 63   PE 0 | PE 63

(c) One-quadrant arrays.

Figure 9.6 Multi-array configurations.

structure as a square matrix or, alternatively, as a cylinder when the top and bottom rows of PE's are connected together. Data communication between PE's takes place via a common 64-bit data highway, with additional direct routing between the nearest neighbours of the string (it is also possible to communicate between any PE's eight units apart). Furthermore, the end connections of a circular string array may be broken and connected to the ends of other arrays to perform multi-array processing. In this way the array size may be matched to a particular problem; the four arrays may be united in three different configurations, as shown in Fig. 9.6. In multi-array processing all the control units receive the same instruction stream, but the execution of the instructions proceeds independently.

The **array control unit** decodes the instructions and generates control signals for all the PE's in an array, thus ensuring that all the PE's execute the same instruction in unison. As with the SOLOMON machine it is sometimes necessary to exclude some data streams or to process them differently, and this is accomplished in Illiac IV in much the same way using a mode control register. In particular an **enable** bit is used to control instruction execution at the processor level. For example, suppose at the start of a problem all the PE enable bits are set to 1, and that the system program causes the control unit to 'broadcast' to all 64 PE's the instruction: 'search through the store for the quantity X'. Each PE carries out the search, and any element finding the value X sets its enable bit to 0. The control unit may now issue a sequence of instructions to be performed only by those PE's whose enable bit is still 1. Similarly, the contents of two registers within a PE can be compared and the enable bit set according to the outcome of the comparison. Common constants or other operands used by all the PE's can be fetched and stored locally by the central control unit and then broadcast to the processors in conjunction with the relevant instruction.

Thus the array control unit has five basic functions:

to control and decode the instruction stream;

to generate the micro-orders needed to execute the instructions in the PE's;

to generate and broadcast common memory addresses;

to process and broadcast common data words; and

to receive and process control signals such as those received from the IO equipment, or the B6500 computer.

The principle components of the control unit are two fast access, 64-word, buffer stores, one of which is associatively addressed and used to hold the current program, while the other acts as a local data buffer. The control unit only has basic arithmetic facilities – addition, subtraction and boolean operations – since the more complex functions are performed by the PE's. Four 64-bit accummulators (CAR) are used to hold address indexing information and for data manipulation operations which are performed on a selected accummulator, and the result returned to the same accummulator.

The specification and control of the array configurations is governed by three 4-bit registers, which may be set either by the B6500 computer or by a control unit instruction. The three configuration control registers (CFC) are used for the following purposes:

CFCO is used to specify the array configuration by placing a 1 in the appropriate stage of the register.
CFC1 specifies the instruction addressing to be used within the array.
CFC2 specifies the control unit data address form.

The control unit can fetch either individual words or blocks of eight words from the array memory to the local data buffer. In addition, it can fetch a single bit from each of the PE mode registers to form a 64-bit word which is read into an accummulator in the control unit. The program counter and configuration registers of the control unit are also directly addressable.

The **processing element** executes the data computations and performs any local address indexing required to fetch the operands from store. It contains the following units:

Four 64-bit registers to hold operands and results.
A floating-point adder/multiplier circuit and a logic unit, capable of performing arithmetic, boolean, and shifting operations.
An 18-bit index register and adder for store address modification.
An eight-bit mode control register.

Table 9.1
P.E. operation times

Operation	*Time* (ns)
Add, subtract	350
Multiply	450
Divide	2750
Boolean	80
Shift	80
Fetch	350

PE's can be operated in either 64-bit or 32-bit mode, in the latter case each 64-bit is considered as two 32-bit words. In 64-bit mode, floating-point numbers are represented using a 48-bit mantissa with 16-bits for exponent and sign; 32-bit mode floating-point numbers are restricted to a 24-bit mantissa. Typical operation times for 64-bit word processing are shown in Table 9.1.

The index register and address adder of the PE's allow independent operand addressing. For example, the final operand address A_i for processing element i is determined by:

$$A_i = A + (B) + (C_i)$$

where A is a base address specified in the instruction, (B) is the contents of a central index register in the array control unit and (C_i) is the contents of the local index register in processing element i.

Both data and instructions are held in the combined stores of the array, but the control unit has access to the entire memory, while the PE's can only reference their own 2048 word store. A variable-structure 20-bit address word is used in the control unit (see Fig. 9.7). The least significant 6 bits identify the PE column within a given array, the next 2 bits the array number, and the remaining most significant bits the row value. Thus addresses used by the PE consist of three components: a fixed address contained in the instruction; a control unit index number value, added from one of the control unit accummulators; and a local PE index value added in the PE before transmission to its own store.

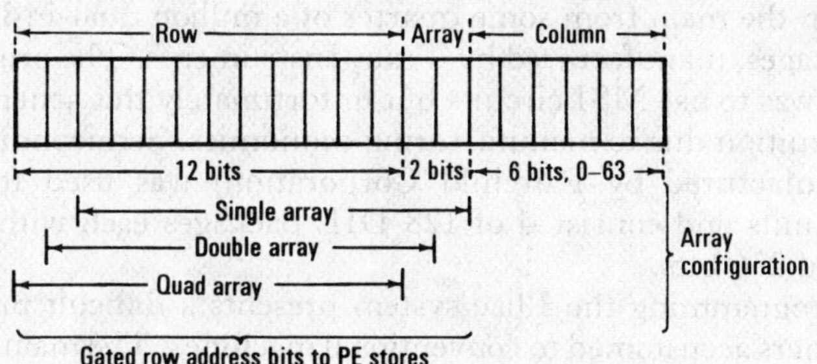

Figure 9.7 Store address structure.

All instructions are 32-bits in length and belong to one of two categories – control unit instructions which generate local operations, and PE instructions which are decoded in the control unit and then transmitted as micro-orders to all the PE's. Instructions flow from the array store, upon demand, in blocks of 8 words (16 instructions) into the instruction buffer of the control unit.

The B6500 computer, which has overall control of the system, is assigned the following tasks:

executive control over the execution of array programs;
control of the multiple-array configuration operations;
supervision of the internal IO operations;
processing of the disc file system;
independent data processing, including compilation of Illiac IV programs.

In order to control the array operations, there is a single interrupt line and a bidirectional 16-bit data highway between the B6500 and each of the control units as well as direct connections to the array memories.

The reliability and maintenance of such a complex system is in itself a major problem in engineering design. The reliability of the greater part of the system, that is, the 256 processing elements plus storage, is expected to be of the order of 10^5 hours per element. This order of reliability can only be achieved by the use of large-scale integrated circuits wherever possible.

The organization of the Illiac IV system as a collection of identical plugable units does, however, simplify the maintenance problems. Thus, a faulty sub-system can be diagnosed using the B6500 computer (assumed fault free after successfully completing self-testing routines) and the unit replaced. The B6500 tests the control units, which in turn check out all the processing elements; the memory units can also be checked independently via the disc file channels. Once the faulty unit has been replaced, the exact cause of the failure may be established off-line using special diagnostic computers (called **exercisers**).

Because of economic considerations only one quadrant of the four quadrant Illiac system was actually built. The system was constructed in the main from some quarter of a million dual-in-line 16 pin logic packages, manufactured by Texas Instruments. The original plan, however, was to use MSI circuits but unfortunately this scheme did not come to fruition due to manufacturing economics. Semiconductor storage (manufactured by Fairchild Corporation) was used for the PE memory units and consisted of 128 DIL packages each with a storage capacity of 256 bits.

Programming the Illiac system presents a difficult problem for programmers accustomed to conventional machines. The main difference is that, in addition to devising the computing algorithms, it is also necessary to produce a memory allocation scheme for storing data which allows the algorithms to be implemented in a parallel mode. Thus, the programmer must consider the memory allocation while he is preparing the program, rather than regard it as a subsidiary problem which can be defined later. Moreover, at present it is not possible for the programmer to write in some high level language and remain blissfully ignorant of the hardware operations. It is essential to have a knowledge of the computer at the machine language level in order to exploit fully the parallelism of the system. Work was commenced on the development of a high level array processor language (called TRANQUIL[10]) which was similar in some respects to ALGOL, but was in fact a superset containing ALGOL. It is very unlikely, though, that TRANQUIL will ever be used as the primary language to write Illiac IV software.

9.3 Pipe-line computers

An alternative way of increasing computing speed, again employing what is basically a parallel concept, is to divide the conventional arithmetic and logic unit into a number of functionally independent sub-units which can be operated autonomously by the control unit. In this way separate instructions are allowed to be executed concurrently in the individual functional units rather than sequentially. This idea of concurrent computing has been used in present generation machines such as the CDC 6600. However, increasing the number of operands processed by the arithmetic unit necessitates increasing the speed at which operands are fetched (and stored) from the memory. To overcome this operand 'bottleneck' the total storage must be divided into several sets of memory banks and operated simultaneously. For example, a slow-speed (1 μs cycle-time) store can be made to provide a high-speed operand stream by using **multi-phase** or **time-interlacing** techniques. That is, if K independent slow speed stores are read successively, but out of phase by T/K microseconds, where T is the cycle-time, the K stores will furnish K operands in 1 cycle-time.

These ideas have lead to the general concept of a **pipe-line computer** such as the CDC STAR machine. As we have seen, a conventional computer is limited by the time required to fetch each operand from the store, execute the operation, and return the result back to store. The pipe-line processor, shown in Fig. 9.8, overcomes this limitation by starting the retrieval of the second set of operands (relevant operands are held in adjacent memory locations) *before* the first result has been returned to store.

For the processor to be economic the memory pipe-line must be full for a substantial part of the overall computation time. This means that the arithmetic unit must deliver the result back to the store at the same time and rate that it is receiving new operands. Thus, the arithmetic unit itself must be constructed on a pipe-line principle in that it must be capable of receiving and starting work on a second set of operands before finishing the calculations for the first set. For example, floating-

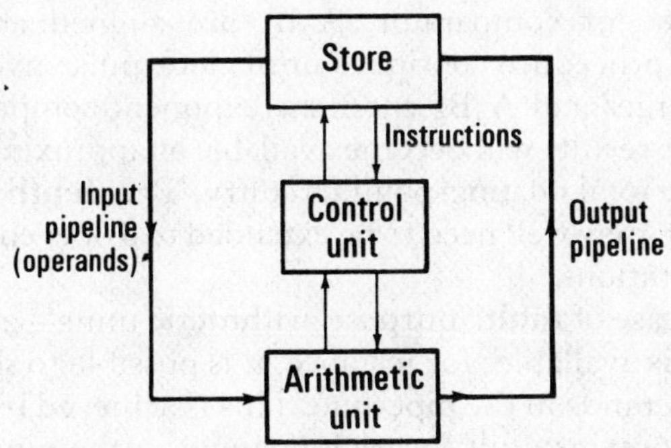

Figure 9.8 Pipeline processor organization.

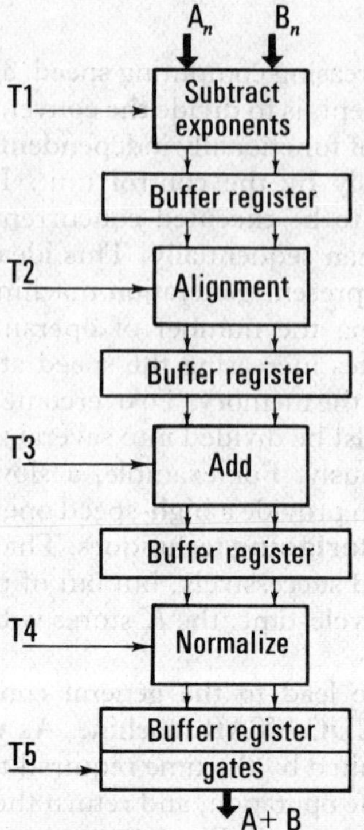

Figure 9.9 Pipelined FP addition.

point addition may be partitioned as shown in Fig. 9.9. The four basic operations associated with this instruction (that is, comparison of exponents, alignment, add, and normalize) are each separated by inter-stage buffer registers. Let us assume we are adding together two streams of floating-point numbers A_n and B_n. At the first clock time A_1 and B_1 enter the exponent comparator, are subtracted and the result latched in the inter-stage buffer register. During the second clock period the operands A_1 and B_1 proceed to the alignment stage, simultaneously with A_2B_2 entering the exponent comparator. On the next clock period A_3B_3 enter the exponent comparator, A_2B_2 are aligned and A_1B_1 added together. This procedure continues until clock pulse five, when the first result, S_1, emerges and A_5B_5 enter the exponent comparator. Once the pipe-line is full, results will become available at approximately four times the speed of the total floating-point circuitry. The **depth** of the pipe-line, in this case four, may well need to be extended to 8 or even 16, for multiply and divide operations.

In the case of multi-purpose arithmetic units[11] a limited form of logical control is available; for instance, it is possible to skip an operation on specified operands in the pipe-line. This is achieved by using a control vector which contains a bit for each operand in the pipe-line (stored in logical correspondence with each operand). The state of the control vector may be tested and, depending on whether the relevant operand bit is 1 or 0, the operation is either performed or skipped. If no operation is performed, the memory location for the result is left unchanged.

However, the complexity of the logical control processes needed to ensure a smooth flow of data in the pipe-line (referred to as a Boolean orgy!) would appear to limit pipe-line machines to one type of arithmetic operation per stream, and only one stream at a time.

Pipeline processors are rather difficult to classify using the Flynn taxonomy. If we consider that a single instruction operates on n different items of data (as for example in a n-stage floating-point addition) then the processor could be called SIMD. On the other hand the processor could be labelled MISD if it was considered that the instructions in consecutive stages were different but processing the same data stream.

9.3.1 Overlap design

The pipe-lining principle may also be employed in the control unit of the CPU where the preparation and execution of instructions can be **overlapped** by the use of suitable timing and logic circuits. Though the overall time for an individual instruction is not changed by this technique, the rate of execution of instructions (throughput) can be greatly improved. For instance, the Read/Obey cycle can be broken down into basic steps, such as instruction fetch, operation decoding, address modification, operand fetch and execution, that are relatively independent of each other. In a similar manner to pipe-lined arithmetic, separate units can be employed to perform each operation. These units pass a partially processed instruction onto the next stage and then commence the processing of the succeeding instruction. As an example of this process consider Fig. 9.10. The operation R corresponds to the Read processes of instruc-

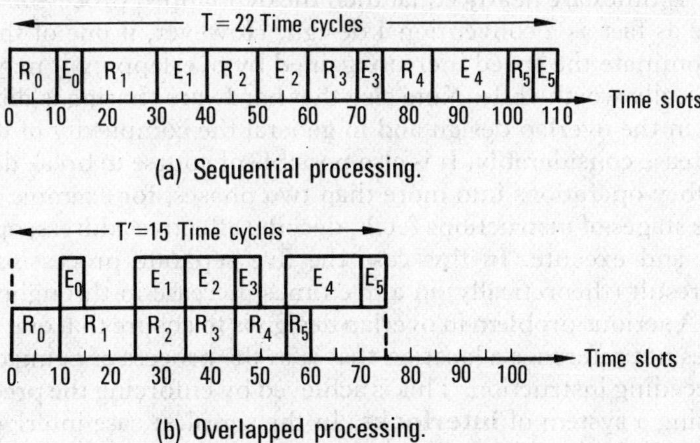

(a) Sequential processing.

(b) Overlapped processing.

Figure 9.10 Instruction pipe-lining.

tion fetch, incrementing the control register, decoding and effective address generation, whilst E is the Obey sequence.

When executing a program of instructions there are definite rules of procedure which must be followed. For instance, in a conventional sequential processor, if R_n represents the instruction preparation and E_n the associated execution operations then the following rules apply:

(a) The preparation of R_n must precede the execution of E_n
(b) R_n precedes R_{n+1} and
(c) E_n precedes R_{n+1}

These rules lead to a simple design and allows hardware sharing to take place between the R and E operations. If T (R_n) and T (E_n) are the handling times for instruction preparation and execution respectively then the overall processing time for k instructions is given by the expression:

$$T = \Sigma_{n=1}^{k}[T(R_n) + T(E_n)]$$

Now if the overlapping of the R and E phases is required the procedure rules must be changed to allow for the concurrent processing of E_n and R_{n+1}. This may be done by modifying condition (c) above to read:

(c) E_n precedes R_{n+2}

In this case the overall processing time becomes:

$$T' = \Sigma_{n=0}^{k} \max\,[T(E_n), T(R_{n+1})]$$

with the convention that $T(E_0) = T(R_{k+1}) = 0$

If the R/E times are nearly equal then the overlapped processor tends to be twice as fast as a conventional design. However, if one of these two times dominate the speed increase gained by overlapping it may not be economically worthwhile. Note also that hardware sharing is difficult to achieve in the overlap design and in general the complexity of the logic will increase considerably. It is also possible of course to break down the Read/Obey operations into more than two phases, for example into the separate stages of instructions fetch, decode, effective address, operation decode, and execute. In this case the five separate processing stages should result (theoretically) in a five times increase in throughput.

A serious problem in overlap design is to ensure that one instruction does not reference a location that is in the process of being changed by a preceding instruction. This is achieved by enforcing the precedence rules using a system of **interlocks**. In the simplest case interlocks take the form of signals which indicate the completion of a task in a particular processing unit (as in asynchronous circuits) and the validation of registers at the unit interfaces. In more complex pipe-lined systems special synchronizing clock pulses must be used which require all stages to produce results at fixed time intervals. Another problem exists when branch instructions are encountered since it is impossible to know ahead of execution which path the program is going to take – this is particularly difficult when a stream of instructions are being pipe-lined at the same time. Though **lookahead** circuits are employed[12,13] which attempt to prepare the pipe-line for any eventuality when a test or jump instruction is encountered, there are still situations which can only be rectified by clearing the pipe-line and starting again (with a consequent loss of time).

9.3.2 Control of pipe-line operations

The operation of a pipe-line processor is such that each individual segment of the pipe is only capable of performing a specific function, and a task is processed by sequencing the operands through the pipe-line in a synchronous manner. Each segment will have a fixed processing speed (the **delay** of a pipe-line is the time between the initiation and termination of a task) and in general there will also be internal feedback loops, that is the output of a segment is fed back as an input to another segment in the pipeline. Temporary buffer storage is sometimes used between segments, however this can lead to an inefficient design due to increased costs and delay times. A block diagram of a typical pipe-line processor is shown in Fig. 9.11(a).

The basic control problem encountered in the design of a pipe-line is the determination of the times when new inputs can be introduced into a segment without causing a **collision** (that is when two or more tasks attempt to use the same segment simultaneously). Thus the design objective is to schedule queued tasks awaiting initiation in the pipe-line in such a way as to achieve high throughput without collisions taking place. Davidson *et al* have described a method,[14, 15] based on a **reservation table**, which allows operands to be efficiently sequenced through a pipeline without risk of collision. The reservation table, shown in Fig. 9.11(b) is used to represent the flow of operands through the pipeline segments. The rows of the table correspond to processing segments and the columns to units of time following the initiation of a task; each segment S_i requires one unit of time for processing. In the reservation table an X is placed in cell (i, j) whenever the task requires segment i at time j; any pattern of Xs is legal, multiple Xs in a column indicates parallel operations and in a row, the existence of feedback paths or repeated operations. Note that the pipe-line illustrated in Fig. 9.11(a and b) requires the operands to be processed in the sequence S1, S1, S2, S3, S4, S5, (S5S2), S1. The table may be analysed to determine at which future times a new computation may be initiated without causing a collision. Collision will occur when two tasks are initiated with a time interval equal to the column distance between two Xs in the same row of the reservation table; the number of time units between successive initiation is called the **latency**. The set of distances between each pair of Xs in the same row of the table, over all rows, is called the **forbidden initiation interval set** or more simply, the **forbidden latency set**. Thus in Fig. 9.11(b) we have for row 1 forbidden latencies of 1, 6 and 7, and for row 2, a forbidden latency of 4, thus the forbidden list is 1,4,6,7. From this forbidden list it is possible to construct a **collision vector**, which is a binary vector of *n*-bits, $C_nC_{n-1} \ldots C_2C_1$, where *n* is the largest set element and $C_i = 1$ if i is in the set and $C_i = 0$ otherwise. In our example the collision vector is given by 1101001. Note that for the case of a linear pipe-line, that is one without feedback or parallel operations, the collision vector is empty ($n = 0$) and there can be no collision.

The collision vector may be used as the basis of a very simple control algorithm by checking, before a new computation is initiated, that there are zero's in every location of the vector corresponding to the number of time units that have elapsed since the previous computation was initiated. Davidson has shown how this algorithm may be implemented by using an *n*-bit logical shift register which shifts right once each time unit (introducing zero's on the left). A new computation may be initiated if and only if the rightmost digit of the shift register is zero;

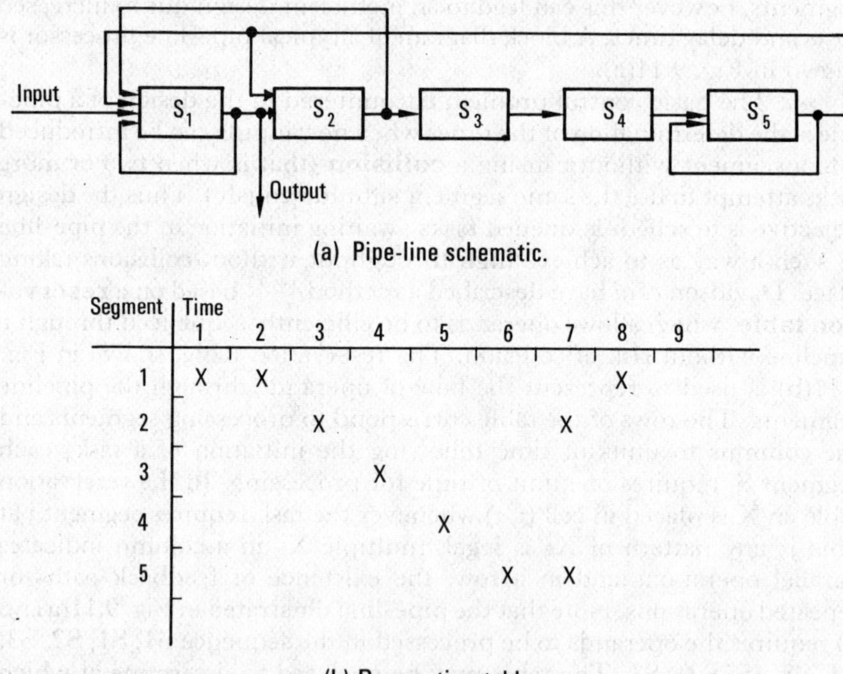

(a) Pipe-line schematic.

Segment	Time 1	2	3	4	5	6	7	8	9
1	X	X						X	
2			X				X		
3				X					
4					X				
5						X	X		

(b) Reservation table.

Figure 9.11 Control of pipe-line sequencing.

Table 9.2
Shift register sequence

Collision Vector 1 1 0 1 0 0 1

Time	*Operation*	D_6	D_5	D_4	D_3	D_2	D_1	D_0 (Output)	*Action*	*State*
0		0	0	0	0	0	0	0	Initiate S1	
1	Shift/OR	1	1	0	1	0	0	1		1101001
2	Shift	0	1	1	0	1	0	0	Initiate S1	
3	Shift/OR	1	1	1	1	0	1	1		1111011
4	Shift	0	1	1	1	1	0	1		
5	Shift	0	0	1	1	1	1	0	Initiate S2	
6	Shift/OR	1	1	0	1	1	1	1		1101111
7	Shift	0	1	1	0	1	1	1		
8	Shift	0	0	1	1	0	1	1		
9	Shift	0	0	0	1	1	0	1		
10	Shift	0	0	0	0	1	1	0	Initiate S3	
11	Shift/OR	1	1	0	1	0	1	1		1101011
12	Shift	0	1	1	0	1	0	1		
13	Shift	0	0	1	1	0	1	0	Initiate S4	
14	Shift/OR	1	1	0	1	1	0	1		1101101
15	Shift	0	1	1	0	1	1	0	Initiate S5	
16	Shift/OR	1	1	1	1	0	1	1		1111011
17	Shift	0	1	1	1	1	0	1		
18	Shift	0	0	1	1	1	1	0	Initiate S5, S2	1101111
19	Shift/OR	1	1	0	1	1	1	1		

etc.

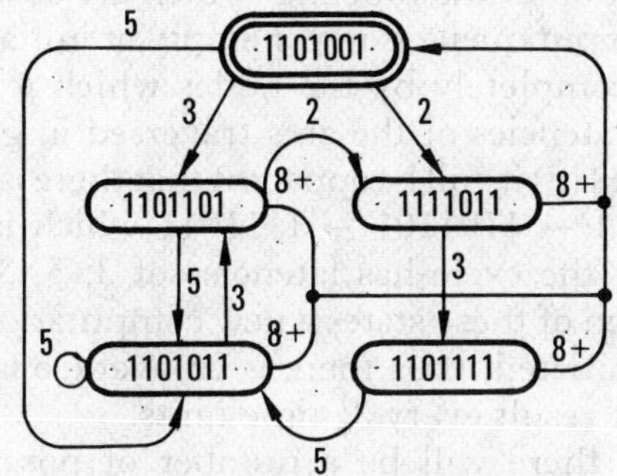

(d) Modified state diagram.

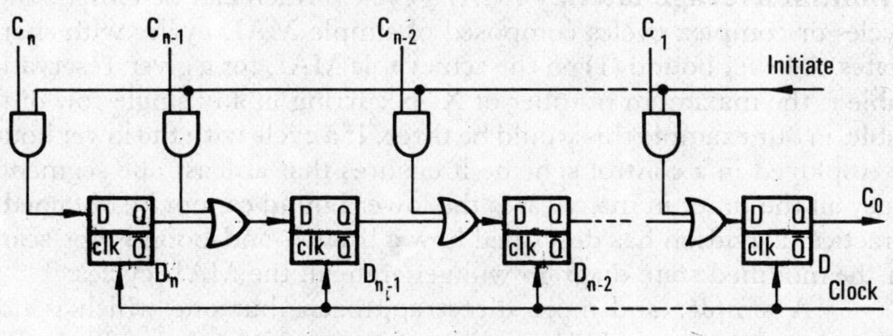

(c) Shift register controller.

Figure 9.11 contd. Control of pipe-line sequencing.

immediately after the right shift following the initiation the collision vector is OR-ed into the shift register and the procedure continued. A schematic of the shift register circuit is shown in Fig. 9.11(c) and the actual shift register sequence for the collision vector 1101001 in Table 9.2.

Since the shift register controller is a sequential machine it may be conveniently represented by a state-diagram; the **modified state-diagram** for the controller is shown in Fig. 9.11(d). The states in the diagram represent *all* the shift-register states that can immediately follow an initiation, each one corresponds to the initiation of a computation and is labelled with the latency of the initiation (the number of time units since the previous initiation).

The initial node is coded with the collision vector itself (the state of the shift register after the initiation of the first computation) and each node has an output arc for every zero in the state code. The arcs are labelled with the position subscript of its corresponding zero in the state code. Thus, an arc with label i leaving state S leads to state S′, the code for which is obtained by shifting the code for S i-places right and then OR-ing the result with the collision vector. In addition, each state has an outbound arc, leading to the initial node, labelled $(n + 1)^+$, which indicates that if more than n units of time elapse between the initiations then the shift register returns to its initial state.

Cycles in the state-diagram correspond to possible cycles of collision free initiation in the pipeline (which are sustainable sequences provided that the input queue is never empty at initiation time). A cycle may be specified completely by the nodes which it passes through in sequence and the latencies of the arcs traversed in going from node to node. From Fig. 9.11(d) it will be apparent that there is a cycle 1111011 → 1101111 → 1101011 → 1101101 → 1111011 which is entered through the state 1101001; the cycle has latencies of 3, 5, 3 and 2 time units respectively. At each of these states a new computation with a new set of operands can be initiated, thus there is an average latency of $3\frac{1}{4}$ which corresponds to one result every $3\frac{1}{4}$ time units.

In general there will be a number of possible cycles and to achieve efficient sequencing control of a pipe-line requires the use of **minimal average latency** (MAL) cycles which can be either simple cycles or complex cycles composed of simple MAL cycles with shared states. A lower bound (1) on the achievable MAL for a given reservation table is the maximum number of Xs occurring in any single row of the table, in our example this would be three. If a cycle with the lower bound is employed in a control scheme it ensures that at least one segment is busy all the time; in many cases the lower bound cannot be obtained in practice. Davidson has described how a branch-and-bound type search of the modified state diagram will generate all the MAL cycles.

A simpler and more direct approach, but one which is non-optimum under heavy load conditions, is to use **greedy control** which corresponds to initiating waiting tasks at the first opportunity. Greedy control is based on **greedy cycles** which are derived by using the out-

Segment	Time 1	2	3	4	5	6
1	A				A	
2		A				A
3			A			
4				A		

(a) Reservation table function A.

Segment	Time 1	2	3	4	5	6
1	B	B			B	
2			B			B
3				B		
4				B		

(b) Reservation table function B.

Segment	Time 1	2	3	4	5	6
1	AB	B			AB	
2		A	B			AB
3			A	B		
4				AB	B	

(c) Composite table function A,B.

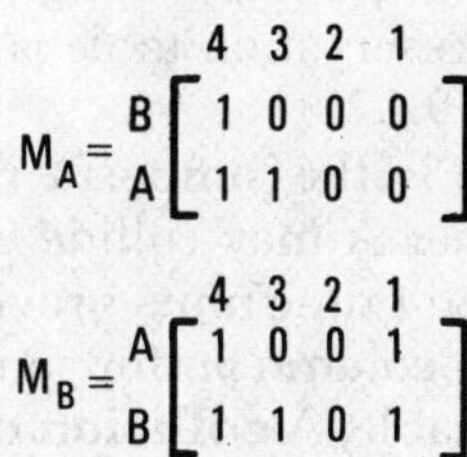

(d) Collision matrices.

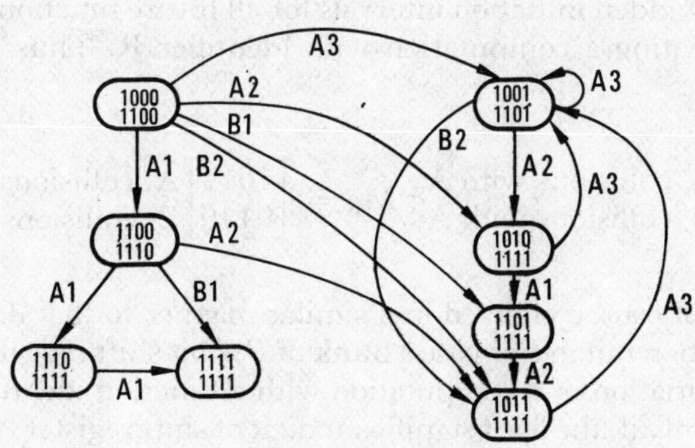

(e) Unified state-diagram.

Figure 9.12 Multi-function pipe-lined units.

bound arc with the lowest latency from each of the states to form the cycle; note that the cycle used for the example is in fact a greedy cycle. It can be shown that an upper bound on the average latency of any greedy cycle is given by the number of ones in the collision vector plus one, for instance in the case of our example the upper bound would be four. It can also be shown that for any modified state table the lower bound $\leqslant$ MAL $\leqslant$ all greedy cycle average latencies $\leqslant$ upper bound.

The greedy cycle is the simplest to implement using the shift register controller, as demonstrated in the example. An optimum controller would require an end-around shift register with zero's spaced to correspond to the desired MAL cycles latency sequence.

So far only pipelines with single function units have been considered, there are instances however when multi-functional (programable) units are required. In this case the general method is still applicable but it is necessary to use P reservation tables where P is the number of separate function units. The technique used is to overlay the separate reservation tables to form a composite table the cells being marked by an identifier corresponding to the particular function units; a typical composite reservation table is shown for a two function pipe-line (A and B) in Fig. 9.12.

In the composite reservation table a computation with a function identifier A may collide with a computation with function identifier B initiated t time units previously if, and only if, some row of the table has an A in column m (for some m) and a B in column $m + t$ from the composite table. A **collision matrix** M_R of dimension $P \times n$ can be formed where the rows correspond to functions (A,B) and the columns are numbered 1 to n from the right. Thus P collision matrices are formed and n is the largest overall collision causing value of t. Note that M_R indicates the forbidden initiation intervals for all future functions which result from initiating a computation with identifier R. Thus, for our example we have:

$$M_A = \begin{bmatrix} 0110 \\ 1010 \end{bmatrix} \begin{matrix} \text{A, collisions with A} \\ \text{B, collisions with A} \end{matrix} ; \quad M_B = \begin{bmatrix} 1011 \\ 0110 \end{bmatrix} \begin{matrix} \text{A, collisions with B} \\ \text{B, collisions with B} \end{matrix}$$

Sequence control can be effected in a similar manner to that described for a single function but in this case a **bank** of P n-bit shift registers must be used. The initiation of a computation with a function identifier A is allowed if, and only if, the least significant digit of shift register A is zero. M_A is OR-ed into the shift register bank immediately after the shift following an initiation of function A. A **unified state-diagram**, shown in Fig. 9.12(e) can be constructed which shows the states of the shift registers in the bank immediately following initiation; arcs are labelled with the latency and function identifier of the initiated computation. Note that in deriving the successive states the rows of the state matrix must both be shifted right (according to the identifier) and then OR-ed with the corresponding row of the appropriate collision matrix. For example, in state $\begin{bmatrix}1100\\1110\end{bmatrix}$

for identifier B1 we shift each row one place right to give $\begin{bmatrix}0110\\0111\end{bmatrix}$

and then OR it with the M_B collision matrix to give $\begin{bmatrix}1111\\1111\end{bmatrix}$.

As before the unified state diagram may be examined for MAL cycles, but this time the required function mix must also be taken into account; the determination of the minimum latency cycles can easily become a very difficult problem. The problem of obtaining an efficient MAL cycle for a given pipe-line can generally be overcome by inserting **non-compute (pure) delays** between segments. In this way the reservation table can be modified to allow the lower bound to be achieved; alternatively temporary storage buffers may be employed.

It is important to realize that the control methods described above are not just restricted to the sequencing of pipe-line processors. The techniques can also be employed generally to avoid any resource or facility conflict due to the existence of multiple paths or operations; in most cases it is convenient to think of tasks not as computations but as general processes.

The main difference between the two highly parallel systems described above, Illiac IV and pipe-line machines,[16,17] comes about because of the different methods used for storage allocation. The array computer requires that the operands for each stage of the computation be distributed throughout the storage unit of the PE's in the two-dimensional arrangement, whereas the pipe-line machine requires the operands to be packed together in contiguous locations of an apparently one-dimensional store. Another difference is, of course, the ability of the processing elements in an array computer to communicate directly with each other.

9.4 Multiple-processor systems[18,19,20]

A multiple-processor system can be broadly defined as a system composed of two (or more) processor units (not necessarily identical) under integrated control and normally sharing common memory. Thus, strictly speaking, all the systems discussed above fall under this heading, however it is convenient to describe the specific class of MIMD architecture separately. Multiple-processor systems may be further classified into two types:

(a) *Distributed systems*: where each of the processors has its own specific executive (or operating system) held in local memory, and performs a **dedicated** system function. This necessitates the overall system requirements being partitioned into distinct tasks at the design stage – called **static allocation**.

(b) *Multi-processor systems*: in this configuration the processors are controlled by a single integrated operating system (usually stored in shared memory) which is capable of **dynamic allocation** of system tasks.

The static allocation of tasks in distributed systems allow special purpose programs to be devised for individual processors and thus considerably simplifies software development. However there are problems with optimizing the overall system load, since requirements may change radically during real-time operation. Moreover, the expansion of system facilities may necessitate extensive structural modifications (similarly the failure of a system module could seriously degrade (or abort) the performance).

The control of multi-processor systems is usually accomplished using an autonomous operating system (written in re-entrant code) which can be executed by any of the processors (there will of course be some critical processes which can only be executed by one processor at a time). Thus in this **floating control** mode, each processor has access to the operating system and can schedule itself. This scheme has the merit of giving greater reliability and user protection, and can be organized to give a **graceful degradation** of system performance under fault conditions. An alternative method of control is to dedicate one or more processors (masters) to execute the operating system and dynamically allocate tasks to all the remaining processors (slaves). Thus it is the sole responsibility of the master processor(s) to schedule, terminate and initiate processes. Though this is a simpler organization to develop and one which can easily accommodate different processor types, it has the disadvantage that a failure in the master processor can bring the whole system to a halt.

9.4.1. Bus structures

The method of differentiating between the different MIMD architectures is by considering the **coupling** or switching of the processor units and memories, and the homogeneity of the constituent processing units. In **tightly-coupled** multiple processing systems the number of processing units is fixed at the design stage and they operate under the direction of a well defined control scheme, generally a hardware unit. Moreover the processing units are heterogenous in the sense that they consist of specialized modules such as address, multipliers etc., supervised by a control unit. Typical examples of such systems would be the CDC 6600, CDC 7600 and the IBM 360/91. Pipe-lined processors such as the CDC STAR-100[21] and Texas Instruments ASC[22] can also be considered as falling in this category. Tightly-coupled systems cannot in general be dynamically reconfigured and the structure is primarily used when maximum-computing power is required. To obtain structural flexibility in a multiple processing system (in order to handle a wide range of applications) it is necessary to employ a much looser and more modular type of connection.

The interconnection mode or bus between elements of a computing system is a critical design parameter which if improperly determined can result in a serious degradation of overall system performance. The main factors to consider in selecting a bus[23] are the type and number of buses, control and communication techniques, data transfer methods and bus width. There are four basic types of bus organization which are generally employed in **loosely-coupled systems** for the processor-memory – I/O switching functions. These are the **dedicated bus, crossbar switch**, **time-shared buses** and **multiported systems**.

(a) *Dedicated bus* – This method of interconnection can be either a permanent link between the devices or a bus used solely for one function, such as an address highway. Figure 9.13 shows a typical system employing

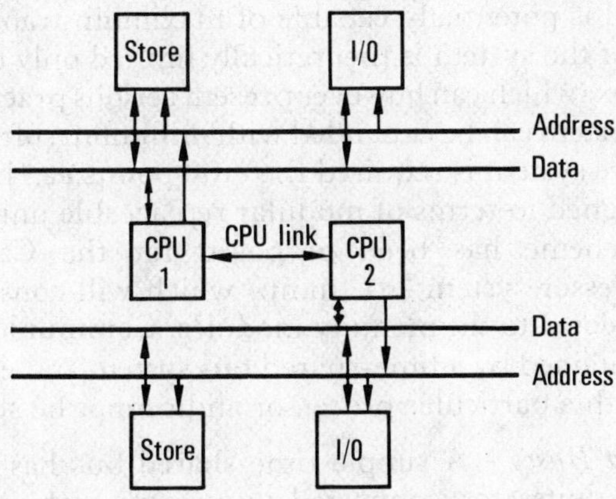

Figure 9.13 Dedicated bus.

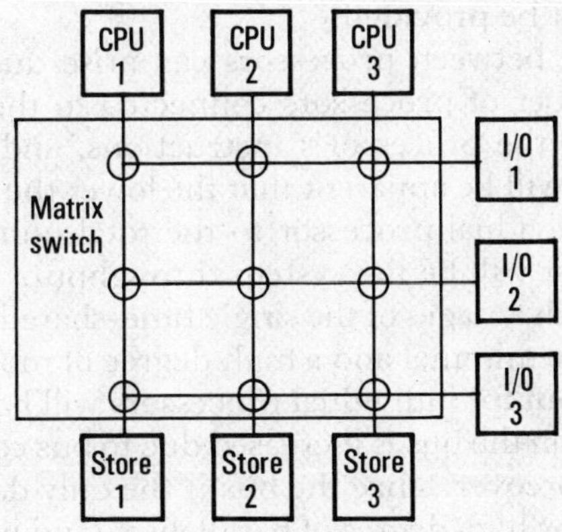

Figure 9.14 Crossbar switch.

dedicated buses; note that the CPU's are connected directly and can access both memory and I/O units via separate address and data buses. The advantages of high throughput and simple bus control are somewhat outweighed by cost and the lack of flexibility, for example in expanding the system. Moreover, if high reliability is required it would be necessary to replicate the buses.

(b) *Crossbar Switch* – The crossbar matrix shown in Fig. 9.14 allows a direct transmission path between processors and memories, with the possibility of simultaneous connections, providing the paths are mutually exclusive. If there are m processors and n memories the crossbar requires $m \times n$ switches, thus if $m \approx n$ the number of cross points increases as n^2. Since each crosspoint must have control logic capable of switching parallel transmissions and resolving conflicting requests for a given module, the switching hardware is the dominant cost factor in this system. This type of interconnection scheme can achieve a very high throughput and is potentially capable of maximum transfer rates. The expandability of the system is theoretically limited only by the physical size of the matrix (which can however present serious practical problems) moreover the system can be expanded with minimum software modifications. If a secure system is required the crosspoints can be replicated or the matrix designed in terms of modular replaceable units.

This scheme has been proposed for the Carnegie-Mellon multimini-processor system[24] (C.mmp) which will consist of 16 DEC PDP-11's connected to the memory modules. Communication between processors is obtained by a time-shared bus system and each I/O device is associated with a particular processor and cannot be shared.

(c) *Time-shared Buses* – A simple time-shared bus has all processors, memories, I/O units etc., connected to a single highway as shown in Fig. 9.15(a). The bus is considered to be a shared system resource (and hence concurrent transactions are not possible) and consequently some method of resolving conflicts between processors requesting simultaneous use of the bus must be provided.

Contention between processors can arise due to the following factors,[25] the number of processors connected to the bus, the relative execution time for the processor's instructions, and the memory and I/O cycle times. It will be apparent that the lower the ratio of bus cycles required by an individual processor to the total number of bus cycles available, the higher will be the system throughput.

The main advantages of the single time-shared bus are that interconnection costs are minimal and a high degree of modularity is obtainable. The throughput for individual processors will however drop as the number of devices on the bus is increased due to bus conflicts and allocation overheads. Moreover, since the bus is the only data highway in the system, to obtain any high degree of parallelism (and incidentally greater system reliability, since a bus fault will be catastrophic) multiple time-shared buses (either uni- or multi-directional) must be used.

A multiple time-shared bus, (a typical example is given in Fig. 9.15(b) requires the active devices (usually a processor or an intelligent peripheral such as a DMA module) to be capable of selecting a bus and passive devices (such as memories) to be able to resolve simultaneous requests. A separate bus controller (or arbiter) is also required to allocate the buses and resolve priorities (priorities can of course be assigned to specific units) and if necessary to perform an error recovery procedure.

The time-shared bus in one form or another seems to be the most favoured mode of connecting together a large number of processors and small memory modules, as for example, in a distributed microprocessor system.

(d) *Multiported Systems* – In this system shown in Fig. 9.16 multiple dedicated buses are employed and the device switching is incorporated in the peripheral modules. Each processor has access through its own bus to all peripheral devices and conflicts are in general resolved through hardwired fixed priorities (this can pose problems when expanding the system unless the arbiter circuits in the peripheral devices have been designed with spare ports). It will also be apparent that the cabling and connector costs can soon become prohibitive for large systems.

With the availability of very cheap processing power future general purpose computing systems will almost certainly consist of a loosely-coupled connection of a large number of processing modules and

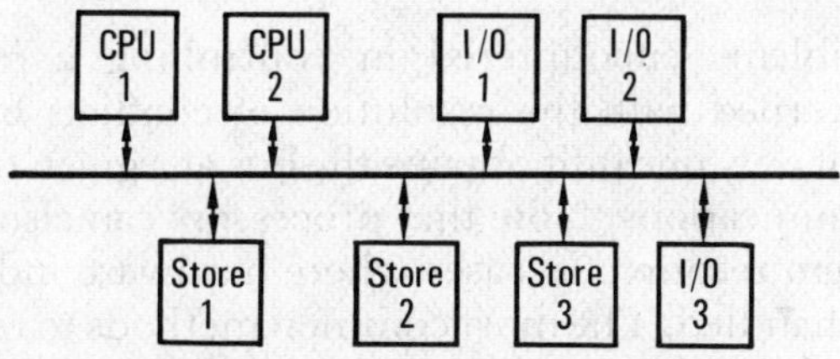

(a) Single time-shared bus.

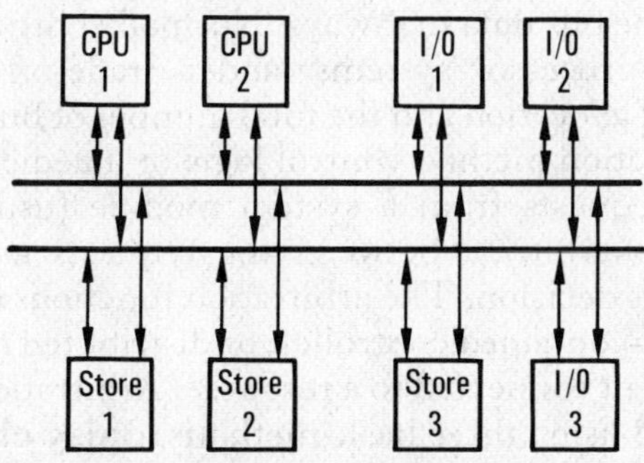

(b) Multiple time-shared bus.

Figure 9.15 Time-shared buses.

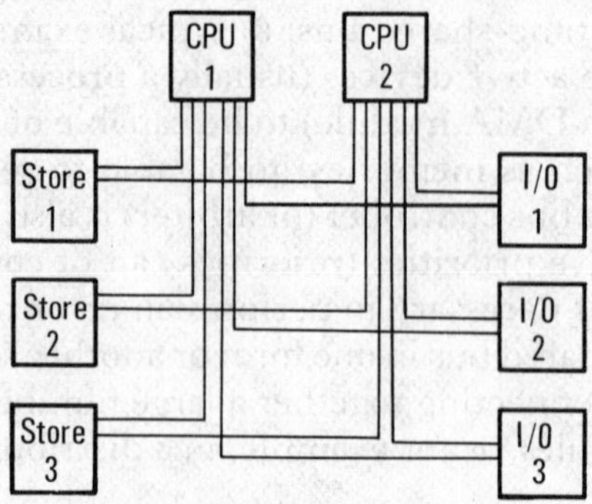

Figure 9.16 Multiported system.

memories. Processors should be able to communicate with each other and access memories (and I/O units) with as much concurrency as possible. Moreover, it should also be possible to dynamically reconfigure the processor organization according to specific applications. It is worth noting in this context that if storage becomes as cheap as processing power (and there is every indication that this will soon be the case) the necessity to share memory resources will vanish.

9.4.2 System control

The major problems encountered in controlling a multi-processor system are concerned with the resolution of conflicts between shared resources (so that only one unit obtains the bus at a given time) and inter-processor communications. Note that processors can also be considered as a shared system resource in cases where hardware independent processes are being handled. The most common methods to resolve resource allocations are **arbitration** and **semaphore** or **status flags**. Resolution techniques can either be implemented in terms of physical control lines (as described below) or the equivalent functions can be performed using coded transfers on the bus data highways. Normally both techniques are employed in multi-processor systems, and a trade-off can be made between the speed of allocation and the total number of bus control lines.

In the arbitration method control logic or a dedicated processor is used to accept requests from a system module (usually an active device) and then to arbitrate between the requests and inform the selected module of its decision. The arbitration functions may be carried out centrally in a self-contained controller, or distributed throughout the system in each element connected to a resource. Arbitration schemes are usually implemented using three basic methods, **daisy chaining**, **polling**, and **asynchronous requests**.

The daisy chain method of control is illustrated in Fig. 9.17 for both the centralized and decentralized modes. In the centralized system, Fig. 9.17(a) each device can generate a request via the common bus

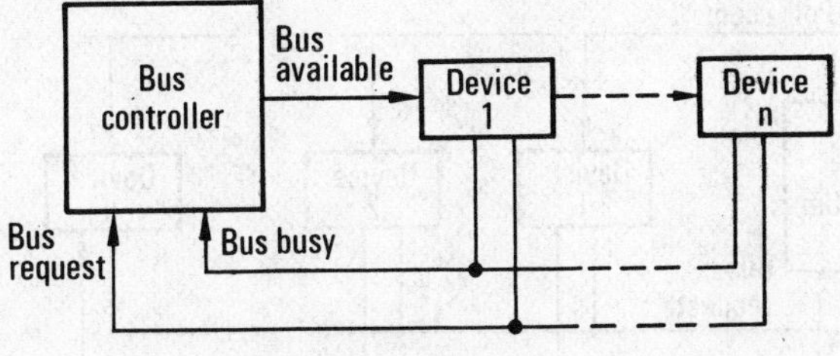

(a) Centralized mode.

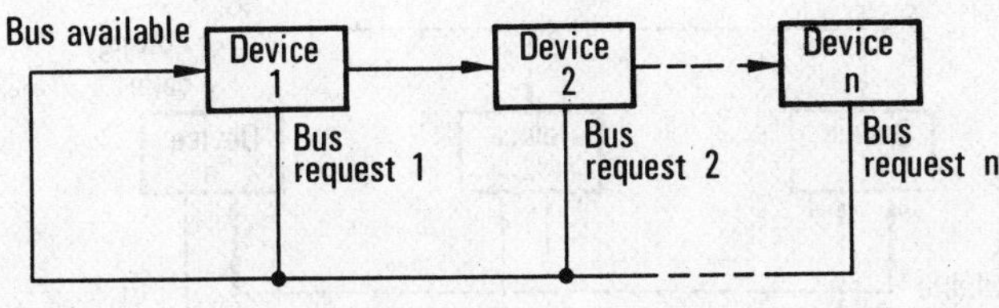

(b) Decentralized mode.

Figure 9.17 Daisy chain method of control.

request line. When the Bus Controller receives a request it acknowledges on the bus available line which is linked through to each device (daisy chained). If a device requesting the bus receives a bus available signal it responds by placing a signal on the bus busy line, cancels its own bus request signal and commences data transmission. However, if the unit did not request the bus, it simply passes the bus available signal down the chain to the next unit in line. The bus busy signal keeps the bus available line up until transmission has ceased when if the bus request line is again up the procedure is repeated.

Note the fixed priority structure which results from daisy chaining; the devices which are physically closer to the controller will always gain control of the bus before those further down the chain. This can of course result in lower order devices being locked out if there is a high demand from devices which are close to the controller. If the bus busy line is omitted and the common bus request connected to the bus available line of the first unit in the chain we have the decentralized mode of operation as shown in Fig. 9.17(b). In this case a device would request the bus by raising its bus request line if the incoming bus available is low; devices which do not require the bus propagate the bus available signal down the chain. The device requiring the bus on receiving the bus available signal (which is now high) inhibits its propagation down the chain and keeps its bus request signal high for the duration of the bus transactions. When the device has finished with the bus it lowers its bus request

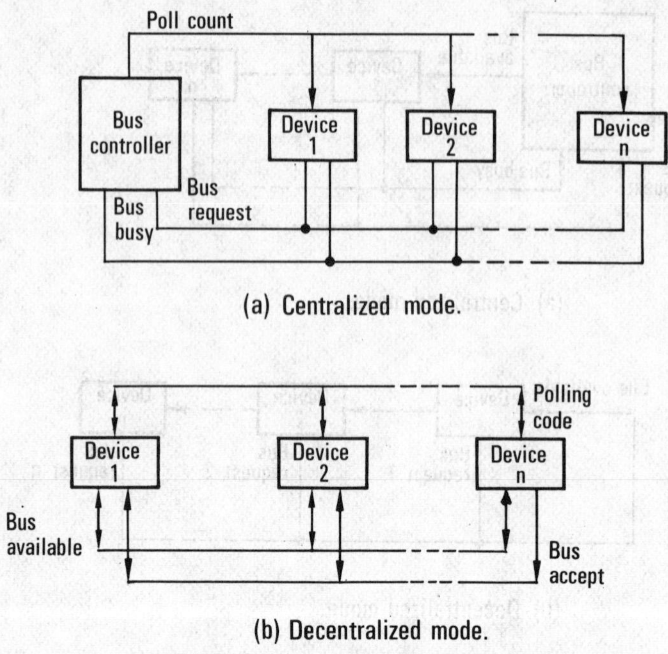

(a) Centralized mode.

(b) Decentralized mode.

Figure 9.18 Polling method of control.

signal which also lowers the bus available line (providing of course no other, successive, devices have initiated bus requests). In this case the bus available signal (which will still be high) will be propagated down the chain to the first device in sequence requesting the bus; note that allocation is on a *round-robin* basis with each unit having equal opportunity since it is considered in rotation. Since a structure of this type can give rise to undesirable race conditions it is necessary to either synchronize the system or use one-shot latches in each device.

The major advantage of the daisy chain method is that very few control lines are required and moreover they are independent of the number of devices, hence the system can be easily expanded. Unfortunately, due to the inherent cascaded nature of the daisy chain scheme it is very susceptible to failures. A fault in the bus available logic of a device (or a power supply failure) could prevent succeeding units from ever gaining control of the bus. Similarly, bus assignment can be slow in operation due to the need for the bus available signal to ripple through each device. The centralized polling method of control, shown in Fig. 9.18(a) is similar to the daisy chain technique in that each device can request the bus by placing a signal on the bus request line. In this case however the Bus Controller responds by polling each device in turn to establish which unit is making the request; the polling is done by counting each unit using the polling lines. When the count corresponds to the predetermined number of a requesting device the unit raises the bus busy line and commences data transmission, at the same time the Controller stops the polling procedure. When the data transmission has been completed, the device removes its busy signal and the Bus Controller reverts back to the counting mode if a new bus request is waiting. The counting may be performed in two ways.

(a) restarting the count from zero each time a bus request occurs; this procedure gives the same priority as the daisy chain method.

(b) continuing the count cyclically (without resetting) after each request has been satisfied which is effectively a round-robin approach.

Note that priorities need not be fixed because the counting sequence can easily be modified. Polling does not have the reliability and structural disadvantages inherent with the daisy chain method but the number of devices connected to the bus is limited by the number of polling lines. This can be overcome however by placing a counter in each device and effectively simplifying the Bus Controller to a clock generator.

Polling may also be performed on a centralized basis as shown in Fig. 9.18(b). In this case every device must have the same allocation hardware (basically a counter circuit) as a centralized Bus Controller and in addition the system must be initialized such that one device is given control of the bus. When a unit is ready to relinquish control of the bus it puts the address of a device (from the counter) onto the polling lines and raises the bus available signal. If the address matches that of another device which is requesting the bus, that device responds with a bus accept signal whereupon the polling device relinquishes the polling operation and lowers the bus available signal. The device which has accepted the bus now lowers its bus accept signal and begins data transmission. If the polling device does not receive a bus accept signal it changes the address and tries again. The counter may be arranged to give either a round-robin or an ordered priority allocation procedure, depending on whether the counter starts by incrementing its own address code or from zero. Note that with this system the failure of a single device does not necessarily affect the operation of the bus.

Both the polling and daisy chain can dispense with the bus available signals by propagating addresses between the devices on a common line. This tends to add flexibility at the expense of extra logic and lines.

The allocation method known as asynchronous or independent requests is shown in Fig. 9.19(a) for the centralized mode of operation. Note that each device has an independent pair of bus request and bus granted lines which communicate directly with the Bus Controller. Devices which require access to the bus send a bus request signal direct to the Controller. The Controller, working on pre-specified priorities, a round-robin or both, selects the next device to be serviced and sends a bus granted signal. The selected device, on receiving this signal, cancels its bus request and raises its bus assigned signal which indicates to the system that the bus is now busy. After the data transmission is completed the device lowers its bus assigned signal, the Bus Controller then removes the bus granted signal and prepares to select the next device.

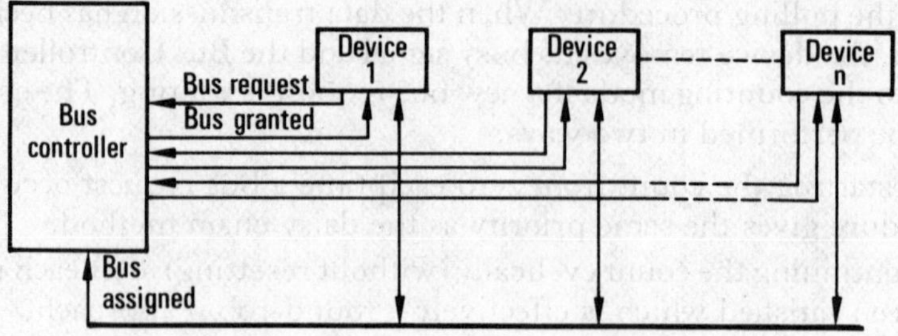

(a) Centralized mode.

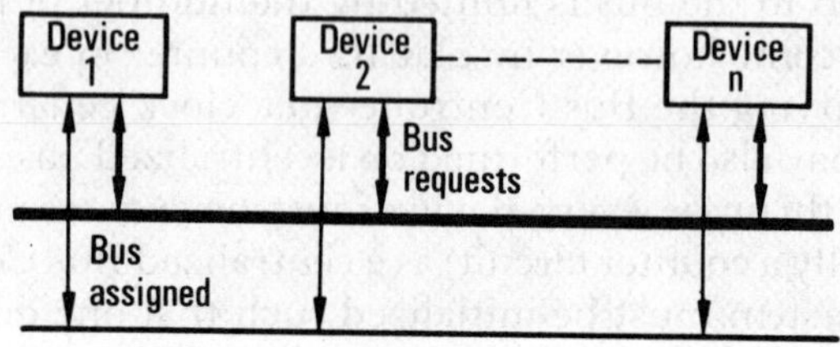

(b) Decentralized mode.

Figure 9.19 Asynchronous requests.

Since all bus requests are presented simultaneously allocation times are appreciably shorter than with other methods. Moreover, there is complete flexibility on how and which device is selected, though this will of course be reflected in the complexity of the Controller hardware. The major disadvantage is the number of lines and connections required to implement the control functions.

The asynchronous control method can also be used in a decentralized mode as shown in Fig. 9.19(b). In this case units are normally assigned a fixed priority so that when devices request the bus by raising their bus request lines the priority is automatically indicated. When the current user releases the bus by lowering the bus assigned signal, all the requesting devices examine all the active bus request lines. The device which has the highest priority (compared to all other current requests) obtains control of the bus by raising its bus assigned signal; this causes all the remaining requesting devices to lower their bus requests. If it is required to operate a round-robin scheme the units must store the priority of the successful device and eliminate it from the next round of priority checking until all other units have been examined.

Though the priority logic is in general simpler than for the centralized version of the scheme the number of lines and connections is higher. Moreover, the system is susceptible to noise and clock failure and the problems of timing and synchronization of the control signals limit the process to small compact systems.

The selection of a particular arbitration scheme depends on several criteria, the importance of which is dictated by the overall system design. The main parameters to be considered are simplicity of design, device allocation procedures, expandability, susceptability to failure, number of interconnections and lines, and control line restrictions. Arbitration speed is usually taken to be inversely proportional to the number of control lines; a general rule is that the arbiter speed should be chosen such that the time required to access a device is a small fraction of the operation time of the device.

9.5 Computer systems with content-addressable storage[26]

The parallel computer systems described so far have been, in the main, orientated towards large numerical problems where the speed of computation is at a premium. There are, however, many problems which involve the processing of non-numerical data; these include, for example, information retrieval, pattern recognition, language translation and, in general, the entire field of artificial intelligence machines. Consequently, a class of highly parallel, iteratively structured machines have emerged, the basis of which is an associative or content-addressable store. These stores are designed to retrieve data by specifying a set of logical conditions, or **attributes**, which when fulfilled identifies the required information.

The usual method of retrieving data from store by absolute addressing and the consequent loading of the required data into an accumulator in the CPU is largely the consequence of existing computer organization. There is no practical merit in this scheme and in fact it can lead to unnecessary complexity since, though the address of any item of data must be known (since it is required in the machine-code), it bears no useful relationship to the actual data contained in the location. Moreover, the allocation of storage space is a local problem which ideally should not enter into the formulation of a computing algorithm. (However, as we have seen, this is not usually the case, even with the Illiac IV machine.)

Thus, the conventional design philosophy of having an absolute addressed store with a separate arithmetic and logic control unit has very little to commend it. Indeed, since the physical separation of the logical units and the finite length signal paths impose a severe constraint on the eventual speed of a system, any reduction in the length and number of interconnection paths must prove advantageous.

Consequently, it is necessary to take a fresh look at the requirements of information processing, particularly at the method of accessing the stored information. For example, suppose the following data were held in a computer store:

SMITH W. R.   AGE 63   HEIGHT 5 FT 11 IN
WEIGHT 150 LB   SEX MALE

The conventional techniques of extracting data would be to search (using some form of table-look-up routine) through the contents of the store, accessing each file by an absolute address and individually comparing the contents of each location; this is a time consuming operation which in some cases would be prohibitively so. If, however, we can ask the stored data direct questions such as 'What is the age of W. R. Smith?' or 'What are the names of men over 63 years of age?' without the requirement of scanning consecutively through the stored data, we not only decrease the search time but also eliminate the need for elaborate search routines based on absolute store addresses.

Note that using this technique we are identifying the required information by specifying one or more of its *own* attributes. For instance, in the case of establishing the age of W. R. Smith, we must first specify SMITH, then the initials W. R. followed by AGE, to extract the required information: 63. Thus, by a progressive sequence of questions (analogous to a logical **tree** structure) we can narrow down the exact location of the required data in the memory without the necessity of knowing where the information is physically stored. At the beginning of the retrieval process the answers will be quite large (there are a considerable number of Smiths!) but will rapidly diminish. Thus, we can think of the process as the elimination of irrelevant information rather than searching for useful information.

It is also important to note the hierarchal nature of this scheme – for example, the argument of the **name/argument** pair SMITH/(data) divides into similar pairs, such as AGE/63. Similarly, the SMITH/(data) pair may itself be part of some larger pair such as would be required in a street or town classification.

An Associative Processor may be defined as a processor which possesses the following characteristics:

(a) stored data items can be accessed and retrieved using content addressing.
(b) data transformation operations (including arithmetic functions) can be performed over many sets of arguments with a single instruction.

Thus, from the architectural viewpoint associative processors can be classified as SIMD machines. Because of the inherent parallel processing properties associative processors have a much faster data processing rate than conventional sequential computers and hence are much more efficient in handling information processing problems.[27]

The fundamental difference between an associative processor and the conventional Von Neumann type of machine is that an associative memory is used in place of the contiguous location addressed memory; this can be seen in Fig. 9.20. However, this basic difference brings about major structural changes to the constituent blocks of the processor, and the architecture is normally based on the type of associative memory used and its organization. For instance in Fig. 9.20 the program and data are

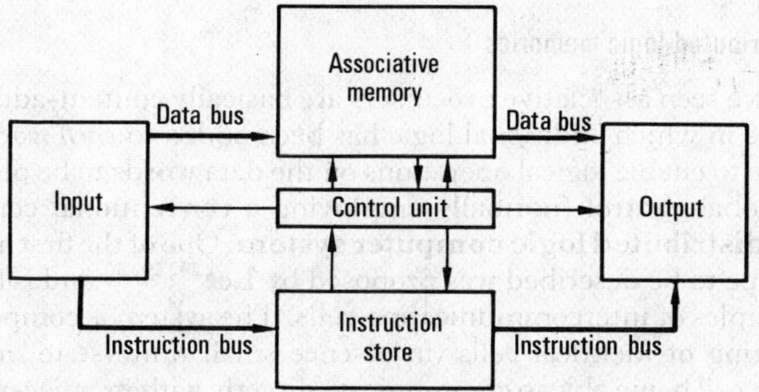

Figure 9.20 Block diagram of associative processor.

stored in physically distinct units with the data being processed *in situ* within the associative memory itself thereby dispensing with independent processing units (a separate unit may sometimes be required for complex arithmetic functions). Note that in general the word-sequential bit-parallel (arithmetical) processor of the conventional machine can be replaced by a word-parallel bit-parallel (logical) processor that is the associative memory.

Associative processors can be classified into four categories according to the comparison (matching) mode of the associative memory. These are:

(a) *Fully parallel*, which can be further divided into

(i) *Word-organized*, in which the comparison logic is associated with each bit-cell of every word and the logical decision is available at the output of each word; this is the form of content addressable store described in Chapter 6.

(ii) *Distributed logic*, where the comparison logic is associated with each character-cell (a fixed length *n*-bit byte or group of cells.

(b) *Bit-serial* systems operate on only one bit-column (also called a bit-slice) of all the words at a time; an associative processor based on this principal is also called a **bit-serial word-parallel** system.

(c) *Word-serial* associative processors effectively implement in hardware a software search routine as a single macro-instruction. In this way the instruction execution times can be considerably reduced compared to a conventional sequential machine.

(d) *Block orientated* schemes employ some form of rotating storage with matching logic associated with each track, for example the CAFS (Content Addressable File System) system developed by ICL. Of all these categories the fully parallel and bit-serial associative processors are the most important developments and show considerable promise for the future. Because of their relatively high implementation cost associative processors are normally used in conjunction with conventional sequential computer systems to enhance their performance when performing information processing tasks.

9.5.1 Distributed logic memories

As we have seen associative processors are basically content-addressable memories in which additional logic has been added to *each word or byte* of storage to enable logical operations on the data words to be performed under global control (normally employing a conventional computer), that is, a **distributed logic computer system**. One of the first machines of this type to be described was proposed by Lee[28, 29, 30] and is based on the principles of intercommunicating cells. The system is composed of a linear string of identical cells (in essence small finite-state machines) with each cell being able to communicate directly with its adjacent neighbours. The cells consist of two main sections, a storage register which holds a data symbol and a logic circuit for control and matching operations. Information to be stored and processed is held in this iterative cellular array in the form of a string of symbols. Each string consists of a name/argument pair with all these quantities being represented by an arbitrary number of symbols – that is, the device has a completely variable wordlength.

The memory may be interrogated in two basic modes:

Direct retrieval – a name string is presented to the memory, whereupon all the argument strings associated with the specified name appear as outputs.

Cross retrieval – in this case the argument is used to extract the relevant name strings.

The overall structure of the distributed logic memory (DLM) is shown in Fig. 9.21.

Each basic cell in the DLM contains an n-bit symbol register, which may be of any size but is usually restricted to six bits, a match bistable M, a control bistable C, and appropriate matching and input/output logic. The operations of the cells are governed by seven control inputs which are defined by the set of micro-operations (or commands) each cell can execute, these are shown in Table 9.3. When the SET input is energized either the C or M conditions, or the symbol present on the input

Table 9.3
Micro-operations for cellular store

MATCH S	Set the match bistable in each cell (i.e., activate each cell) in which the pattern S and the condition on C (if specified) are stored.
SET S	Store the pattern S or the conditions on C or M in every cell.
STORE S	Store the pattern S or the conditions on C in each active cell.
READ	Read out the pattern bits of any active cell
MARK	Activate all cells to the right of each active cell up to the first cell whose C bistable is set to 1
LEFT	Activate the left neighbour of each cell for which C = 1
RIGHT	Activate the right neighbour of each cell for which C = 1

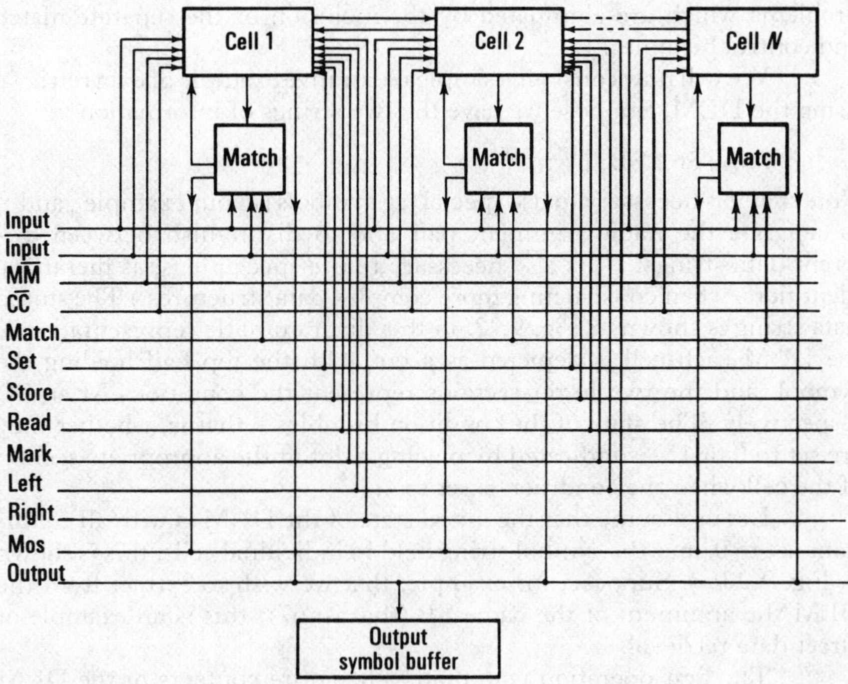

Figure 9.21 Block diagram of the string memory

lines, are stored in the selected cell. Note that it is necessary to have a separate line for each state of the input variables (C and $\bar{C}$, etc.) which must be set to 1 for the required condition. The STORE operation is very similar except that in this case the symbol or the C condition is only stored in active cells, that is, those cells which have their match bistables set to 1. Symbols may be outputted, but only from an active cell, by energizing the READ control line.

The most important function performed by the DLM is the simultaneous matching of the contents of each of the cells with some specified input and, if required, the C condition. If the comparison is successful the match bistable of the appropriate cell is set to 1 and the match output signal (MOS) generated; the MOS line is used to determine whether or not a multiple match has occurred. The function of the MARK operation is to activate all those cells to the right of cells which are already active, up to the first cell whose C bistable is set to 1. The activity of a cell may also be propagated left or right, depending on whether or not the C variable is set, by giving the commands LEFT or RIGHT.

The match function is sometimes combined with an automatic propagate activity left or right operation.[28] This command causes the activity bistable of the neighbouring left- or right-hand cell to be set to that of the matched cell, the activity of which is then set to 0. The combined match and propagate function can, however, give rise to timing problems which are eliminated by the inclusion of the separate match and control bistables.

We will now consider a simple illustrative example of data retrieval using the DLM. Suppose we have the two strings of information:

: SA; 56 : SB; 67 :

Note that it is necessary to use special tag symbols (in our example ; and :) to separate the name/argument pair and to distinguish between different data strings. (It is also necessary to use special tags, as hierarchal identifiers, when constructing more complex data structures.) The stored data string is shown in Fig. 9.22, in this diagrammatic representation of the DLM each cell is depicted as a box with the top half holding the symbol, and the two lower sections represent the conditions M and C respectively. The state of the condition bistables – that is, whether they are set to 0 or 1 – is indicated by placing a dot in the appropriate section of the cell when the condition is set to 1.

Let us assume that the initial state of the DLM is with all conditions set to 0, and the symbol string held in individual cells, this is shown in Fig. 9.23(a). Suppose, for example, that we wish to extract from the DLM the argument of the name SB (that is, 67); this is an example of direct data retrieval.

The first operation is to match the entire contents of the DLM with the start symbol for a data string (:), using the micro-operation:

MATCH :

This results in the M bistables of those cells containing ':' being set to 1, that is, they become active (see Fig. 9.22(b)). The next step is to check the following symbol to see if it corresponds to the first symbol of the required name (S). To perform this operation it is essential that only specific symbols be matched otherwise an incorrect sequence of interrogation will result. For example, if all the cells were interrogated any occurrence of the symbol S, which could appear either in a name or argument, would generate a response. Thus, the next cell in the data structure must be appropriately marked by setting the condition bistables. This incidentally is the reason for having a combined match and propagate operation as described earlier. In this case we must identify the next cells to be matched by setting their C conditions; this is done by using the following sequence of commands:

SET C = 0	All the C bistables are set to 0
STORE C = 1	The C bistables of all active cells are set to 1
SET M = 0	All the M bistables are set to 0
RIGHT	The right-hand cells of all cells with C = 1 are made active
SET C = 0	All the C bistables are set to 0
STORE C = 1	The C bistables of all active cells are set to 1
SET M = 0	All the M bistables are set to 0

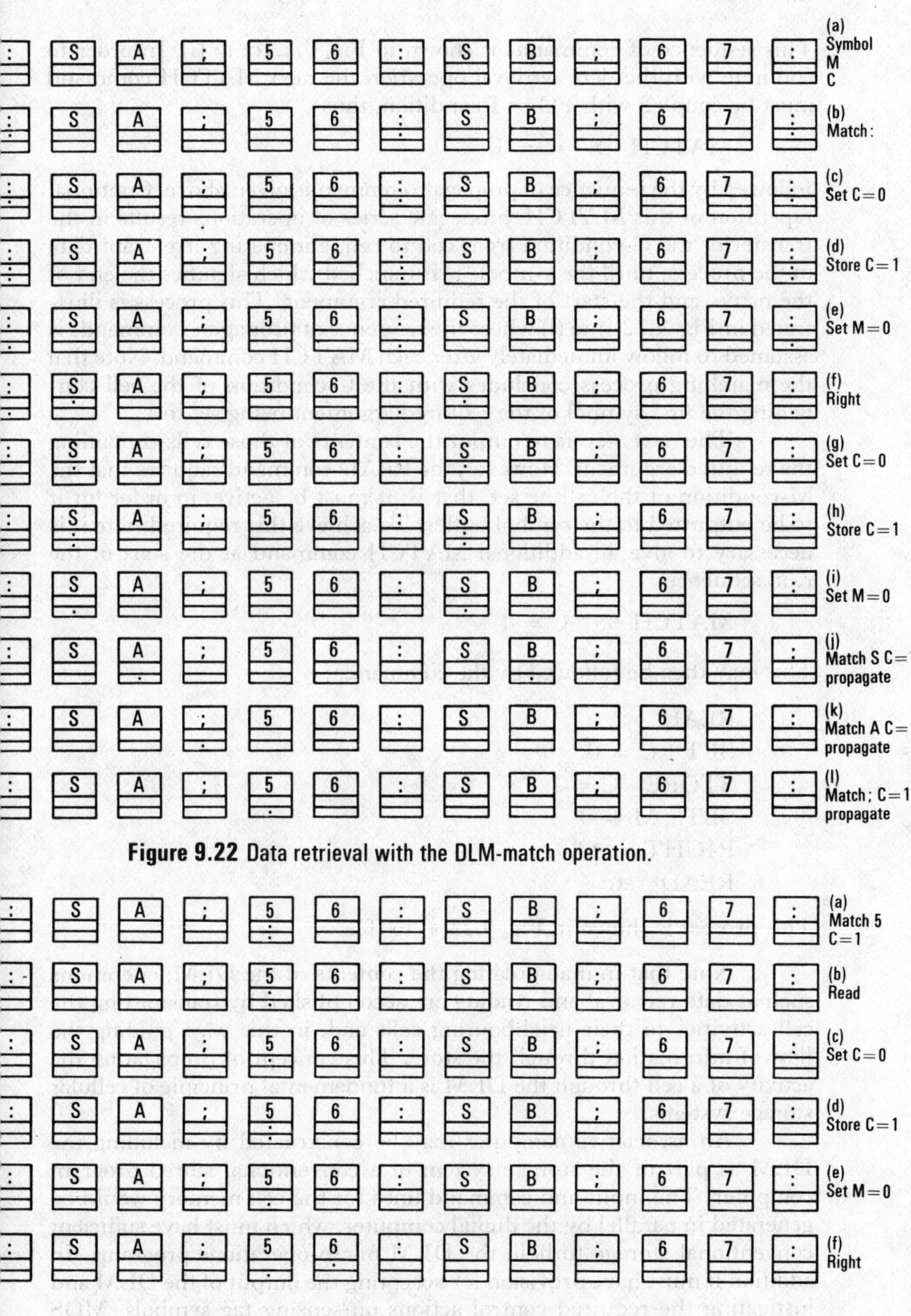

Figure 9.22 Data retrieval with the DLM-match operation.

Figure 9.23 Data retrieval with the DLM-read operation.

This sequence of commands is shown in Fig. 9.22(c) to (i). In order to continue with the data retrieval operation the next MATCH command must be coupled with a C = 1 condition, thus:

MATCH S C = 1

followed by the sequence of propagate commands given above. Continual repetition of this MATCH–propagate series of operations results in the transfer of the C-condition from cell to cell, eliminating irrelevant data in the process, until the symbol ';' is matched which signifies the end of the name, and the start of the required argument. This process is illustrated in Fig. 9.22(j) to (l) where the sequence of propagate commands is assumed to follow immediately after each MATCH command. Note that the matching process concludes with the C-condition of the cell containing the first symbol of the required argument being set to 1.

The next step is to output the contents of those cells containing the required argument. However, the READ command requires that the M-condition of the cell be set, that is, it must be active, in order for it to be outputted to the symbol buffer. To achieve this required state it is necessary to give an additional MATCH command at the start of the read sequence:

MATCH 5 C = 1

This may then be followed by the commands:

READ
SET C = 0
STORE C = 1
SET M = 0
RIGHT
READ etc.

This process is shown in Fig. 9.23(a) to (f).

Note that in manipulating the contents of the DLM, operations such as data retrieval and reading are accomplished by transmitting the cell activities to their neighbouring cells and, in this way, guiding the flow of information through the store. This concept of propagating the activity of a cell through the DLM is a fundamental principle of cellular storage systems.

An associative computer may be constructed by including the DLM as part of the storage system of a conventional stored program computer. The input and command lines for the cell memory would be generated in parallel by the digital computer, which must have sufficient conventional storage to hold the DLM micro-operations programs. In addition it must have provision for accepting the output of the DLM and instigating the required control actions on sensing tag symbols, MOS signals, and so on.

The main advantages of the DLM system are:

There is no restriction on the length of data strings that can be stored, that is, the DLM has a completely variable wordlength.
The time required to retrieve data from the DLM is independent of the amount of information held in the store.
The increase in speed is obtained by the inherent parallel structure rather than any advances in component technology.
The DLM has a uniform modular structure and may be extended by simply adding extra storage cells.

Unfortunately, if the DLM is to be of any practical use it must contain many thousands, perhaps millions, of cells, so that a low cost cell is essential. However, since the cells are identical and the interconnections regular and repeated, it seems highly probable that future developments in LSI circuits (particularly MOS technology, since high speed is not of paramount importance) will make this type of machine a viable economic proposition.

9.5.2 Fully parallel associative processors

The ideas described above have been extended by Crane and Githens[31] to produce a highly parallel content addressable computer with the ability to perform simultaneous arithmetic operations on many sets of data. In this machine the normal DLM structure, with one symbol stored per cell, has been modified to allow operation on a bit per cell basis in order to increase the efficiency of the arithmetic processes. This type of organization is shown in Fig. 9.24, where each bit of a data word is stored in separate cells; for example, the *n*-bit words A, B, C, etc. are stored in *n* consecutive cells. The data cells are also used for control purposes; for instance, the data arrays X_1, X_2, and X_3 are reserved for this function, and would be used for marking individual bits of the words.

By storing the data words in this fashion it is possible to use one match bistable for each bit of a word, thus allowing a simultaneous search to be performed on all bits of a word. Moreover, if two words are stored side by side, all the corresponding bits can be simultaneously searched. This type of data organization is particularly effective where there is a natural grouping of the data to be processed, for example, in the processing of radar data tracks,[32,33] where the majority of the processing occurs between data words contained in the same group.

The basic commands for the machine (a typical set is shown in Table 9.4) are conceptually very similar to those of Lee's DLM described above, and include the fundamental comparison, writing, and reading operations. Note, however, that in this machine a combined match and propagate activity left or right operation is included, that is, the MATCH LEFT (RIGHT) command. The CLEAR MATCH LEFT (RIGHT)

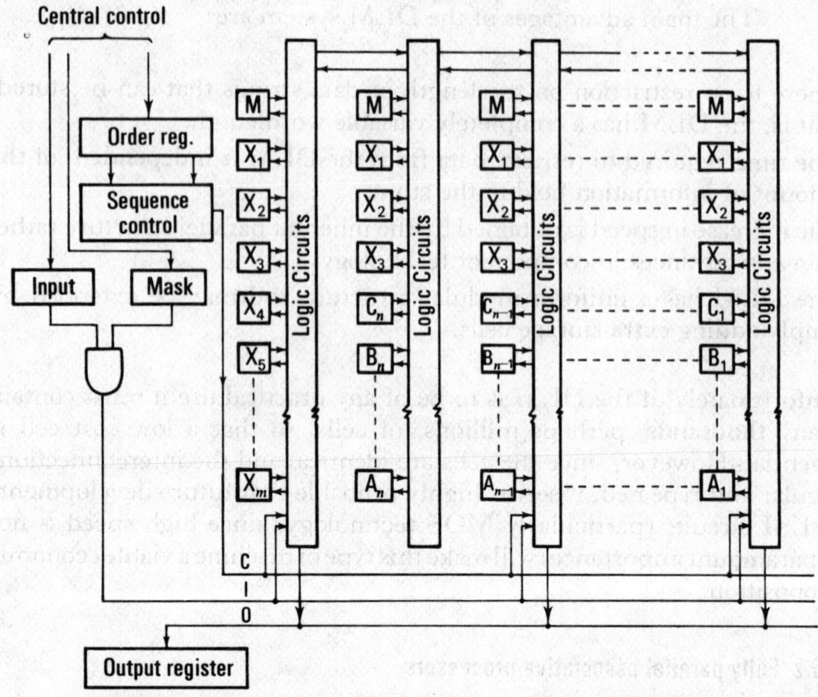

Figure 9.24 Block diagram of DLM using bit per cell storage.

command is similar in operation, except that in this case all the M-bistables are set to 0 except those for which a match occurs; if no match occurs all the cells will be deactivated. The PROPAGATE LEFT (RIGHT) command causes an already active cell to propagate its activity through 'strings' of adjoining cells, all of whose contents match the input pattern.

Any combination of the data cells X_1–X_n and the M-bistables can be individually specified in the input pattern to the machine for the write and comparison functions. The X_1, X_2, and X_3 data words are used in a similar capacity to the C-bistable in the DLM, but of course the extra bistables add considerable flexibility to the system.

Let us now consider how arithmetic operations, in particular binary addition, may be performed simultaneously on many different data sets. Suppose, for example, we wish to add together the contents of the data words stored in the X_4 and X_5 data arrays, and that the words contain binary numbers represented in the usual way with the least significant digit at the extreme right of the word. The bits or words which enter into the calculation are marked by setting $X_1 = 1$; this would normally occur as the result of a previous associative search through the stored data. A typical addition algorithm in terms of a sequence of basic commands is shown in Table 9.5. Each command in the program has an input pattern (or 'address') associated with it which specifies the position and value of the bits which enter into the operation; data word X_2 is

Table 9.4
Command structure for Crane and Githens machine

MATCH	Activate (set M bistable to 1) all cells whose contents match the input pattern.
CLEAR MATCH	Activate all cells whose contents match the input pattern, and deactivate (set M bistable to zero) all other cells.
MATCH LEFT/RIGHT	Activate the next cell to the left (or right) of each cell whose contents match the input pattern.
CLEAR MATCH LEFT/RIGHT	Activate the next cell to the left (or right) of each cell whose contents match the input pattern, and deactivate all other cells.
PROPAGATE LEFT/RIGHT	Activate all cells between an already active cell and the first cell to its left (or right) that *does not* match the input pattern.
STORE CONDITIONALLY	Write the input pattern in all active cells.
STORE UNCONDITIONALLY	Write the input pattern in all cells.
READ	Read out the contents of any active cell.

Table 9.5
Binary addition algorithm for DLM

	Command	*Input address*
1	CLEAR MATCH LEFT	$X_1 = 1, X_4 = 0, X_5 = 0$
2	STORE CONDITIONALLY	$X_2 = 1$
3	CLEAR MATCH LEFT	$X_1 = 1, X_4 = 1, X_5 = 1$
4	PROPAGATE LEFT	$X_1 = 1, X_2 = 0$
5	STORE UNCONDITIONALLY	$X_2 = 0$
6	STORE CONDITIONALLY	$X_2 = 1$
7	CLEAR MATCH	$X_1 = 1, X_2 = 0, X_4 = 0, X_5 = 1$
8	MATCH	$X_1 = 1, X_2 = 0, X_4 = 1, X_5 = 0$
9	MATCH	$X_1 = 1, X_2 = 1, X_4 = 0, X_5 = 0$
10	MATCH	$X_1 = 1, X_2 = 1, X_4 = 1, X_5 = 1$

Table 9.6
Shifting algorithm for DLM

	Command	*Input address*
1	CLEAR MATCH	$X_1 = 1, X_n = 1$
2	STORE CONDITIONALLY	$X_n = 0$
3	CLEAR MATCH LEFT (RIGHT)	$M = 1$
4	STORE CONDITIONALLY	$X_n = 1$

assumed to be cleared to the all-zero state before the program starts. The basic logic behind this program is initially to determine those cells (or stages) which have a carry input, marking them accordingly using the control data cells, and then using this information to generate the value of the individual sum bits for each cell.

The initial steps of the program are concerned with locating the 'no-carry generated' and the 'carry generated' terms, that is, the conditions $X_4 = 0, X_5 = 0$ and $X_4 = 1, X_5 = 1$, respectively. This operation is begun by searching the array for those cells in the marked groups ($X_1 = 1$) that hold $X_4 = 0$ and $X_5 = 0$, using the command

CLEAR MATCH LEFT $(X_1 = 1, X_4 = 0, X_5 = 0)$

This results in the M-bistables of those cells to the left of the cells whose contents match the input 'address' being set to 1; the states of the M-bistables are temporarily stored in X_2 data cells by the next command:

STORE CONDITIONALLY $(X_2 = 1)$

in order to free the match bistables for subsequent commands. The carry generate terms are located in the same manner using the command:

CLEAR MATCH LEFT $(X_1 = 1, X_4 = 1, X_5 = 1)$

which results in the start of each carry chain being marked by an active cell. At this stage, then, the M-bistables are set for those cells which always receive a carry input (generated from the previous stage) and X_2 is set for those stages which never receive a carry input. We must now determine and mark all the carry inputs by activating all those stages which can receive a propagated carry (from the cells containing $X_4 = 0, X_5 = 1$ and $X_4 = 1, X_5 = 0$) until a 'no carry generated' cell (marked $X_2 = 1$) is encountered. This is achieved by the compare and propagate command:

PROPAGATE LEFT $(X_1 = 1, X_2 = 0)$

which will start at an active cell ($M = 1$) and activate all other cells to its left until $X_2 = 1$ is recognized. The final result of this operation is that the M-bistables are set for all those cells which have a carry input. The next commands (5 and 6 in Table 9.5) store the carry inputs in the X_2 position, after first clearing it down to zero. The sum bits are determined by a final sequence of four MATCH commands that search through the marked cells for the addend-augend-carry input combinations that produce a 1 output (these correspond to the sum output terms derived from the full-adder truth-table) and activate the cells accordingly. Thus, the addition operation concludes with the M-bistables of those stages which produce a 1 output being set to 1.

It is important to note that, though the addition example given above was performed on two components of each member of the designated set of data groups, there is no reason why the set should not consist of *all* the stored data groups. Thus, the machine has the capability for parallel-by-bit and parallel-by-word operation; moreover, the duration of these operations is logically independent of the number of members in the set.

Subtraction can of course be performed in exactly the same way, except that the address inputs for the first and third commands in Table 9.5 need to be changed in order to account for the borrow combinations required to form the difference $X_4 - X_5$. The modified commands are:

1. CLEAR MATCH LEFT $(X_1 = 1, X_4 = 1, X_5 = 0)$
3. CLEAR MATCH LEFT $(X_1 = 1, X_4 = 0, X_5 = 1)$

The contents of data words may also be shifted left or right using the command sequence shown in Table 9.6. The action of the algorithm is to shift the contents of all X_n bistables marked by $X_1 = 1$, one place (cell) to the left or right. As we have seen in Chapter 5, these basic abilities are sufficient to allow any complex arithmetic operation to be performed.

One of the disadvantages of this type of machine is that, though arithmetic and logical operations are convenient to perform, the input and output of data words is relatively inefficient. This shortcoming has led to an extended form of the machine called a **two-dimensional DLM**, which consists essentially of two linear DLM arrays, one of which is used as a parallel-by-bit input/output buffer and the other as a processing array operating on a bit-per-cell basis as described above. The two-dimensional version of the machine has the form therefore of a highly parallel processor consisting of identical processing units operated under common control.

The concept of a cellular structured associative processor has also been investigated by other authors, notably Sturman[34] who proposed a 3-bit symbol machine operated in an autonomous (without a central control unit) asynchronous mode, and Wright[35] who has described a 7-bit content-addressable string memory (CASM) capable of manipulative operations.

This later work has been developed by Beaven and Lewin[36] and Lea[37,38] into a viable associative parallel processor for information processing[39]. A more ambitious machine based on the distributed logic concept is PEPE (Parallel Element Processing Ensemble) which has been developed by Bell Telephones for the US Army Advanced Ballistics Missile Defence Agency for radar data processing.[40,41] A block diagram of the system is shown in Fig. 9.25. Each processing element consists of an arithmetic unit, a correlation unit, an associative output unit and a 1024 × 32-bit random access memory, called the element memory; associative processing is performed by time-sharing the element memory between the other units. Note that the number of PEs in the system can be varied according to the requirements of the application. Note also that PEPE is used in conjunction with a conventional host computer (the CDC 7600) which employs the associative processing ensemble to perform certain tasks more efficiently.

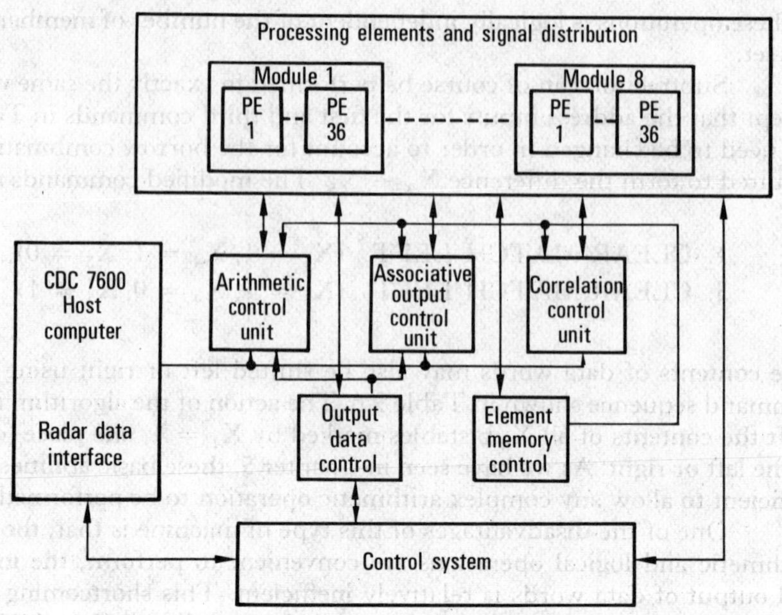

Figure 9.25 Block diagram of PEPE.

To and from control unit
Control signals
Interrogation drive amps
1 2 3 4 5 255 256
Word 1
W
R
1 word logic
Word lines
Word 2
2 word logic
Data input/ output
Word 256
plated wires
256 word logic
Word read and write amplifiers
Bit lines

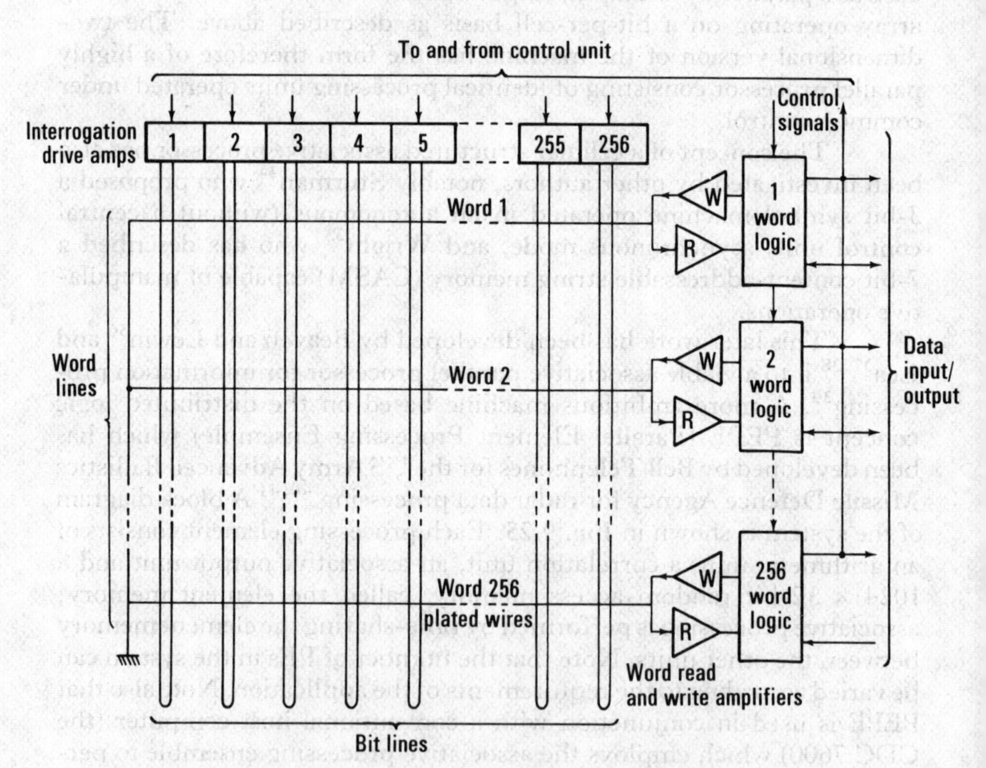

Figure 9.26 Plated-wire associative memory.

9.5.3 Bit-serial associative processors

It is also possible to construct an associative processor using the more conventional bit-serial word-organized content-addressable memory.[42] A machine of this type was in fact constructed by the Goodyear Aerospace Corporation for the Federal Aviation Administration in the U.S.A. as part of a large air traffic control system.[43, 44] The computer consists of a square array of 256 words each of 256 bits, with the basic storage element being a plated-wire word-accessed device; this is shown in Fig. 9.26. Each word has associated with it a separate arithmetic and logic unit operating in a serial-by-bit mode which is capable of executing instructions generated by a common control unit. Thus the machine is, in effect, 256 simple digital processors each containing a one-word store and an arithmetic unit, all operated simultaneously from a central control unit.

The processor has a variable wordlength (up to a maximum of 256 bits) which can be divided, under program control, into sub-words or fields. This is achieved by employing tag-bits in each 256-bit word which serve as identifiers to separate the fields. Thus arithmetic operations may be performed on fields either within the same word or in different words.

The plated-wire store has read and write cycle times of 100 ns and 300 ns respectively, giving a match time of 0·1 μs per bit, and an add time of approximately 1 μs per bit. Storage for one bit is provided at each intersection of a word and a bit line, and only one bit column may be operated upon at a time. An interrogation pulse on the selected bit line causes a signal to be emitted by each bit on that line; these signals are transmitted through the word lines to the sense amplifiers where they are compared with a specified reference bit in the associated logic. Each word has associated with it a **storage bistable** which is initially set to 1, and then alternately set to 1 or 0 depending on the outcome of the bit comparisons.

For example, suppose the first access is to bit 0 in all words, the output of the sense amplifiers would be compared with bit 0 of the desired pattern (the reference bit) in the **word logic unit**. The storage bistables of those words in which bit 0 equals the reference bit are left set to 1; if, however, a mismatch occurs the bistable is reset to zero. This process continues until all relevant bits have been compared, when only those words which are identical with the descriptor bits will have their storage bistables set to 1. At the conclusion of the comparison operation an equivalence signal is generated for those words which still have their storage bistables set to 1. All such signals are OR'd together and fed to the control unit which, upon receipt of the signal, would initiate a subroutine to identify and process the word, or if necessary isolate it from further searches (by setting a **shield bit** to 1) before proceeding to identify other responses. Data may be written into the store in the usual way by energizing the appropriate bit and word lines; similarly, the words may be 'tagged' (divided into fields) by simultaneously energizing all the wordlines and the required bit line.

The control unit of the associative processor (see Fig. 9.27) can be functionally divided into a **sequential instruction** part and an **associative control** part. The sequential instruction part of the control unit is very similar to the control unit of a conventional computer and includes an absolutely addressed random-access store which holds the instructions for the associative processor, an arithmetic and logic unit for instruction processing, and the usual IO facilities. The associative control section consists of a set of common counters and registers shared by all the modules (since only one instruction is executed at a time in all modules) and individual word logic, module driver, and response amplifiers.

Instructions are read down from the random access store and transferred to the control register, where they are held while being executed by the control unit. The accumulator register serves as a temporary store for operands participating in the associative operations, input/output buffering, etc. Bit select counters are used to determine which bit column is to be interrogated and by suitably incrementing and decrementing the counters a complete field may be operated upon. The principal components of the word logic unit are the necessary logic circuits for performing the equivalence and control operations and the storage bistable. Outputs from the sense amplifiers are used in conjunction with the control logic to determine the state of the storage bistable; moreover, the sense amplifier itself is bistable and remembers the match state from one interrogation to the next.

Each associative instruction controls the processing during a single bit time and is divided into a number of fields specifying the bit column, interrogation condition, storage bistable, and bit counter controls, etc. Since it is only possible to interrogate one bit line at a time, the search algorithms must be organized so that successive bit instructions leave the storage bistable in a state which defines the locations of those words which meet the conditions of the search. Note that the time required to identify a word is directly related to the number of bits that must be searched. The execution times for the associative processor described above are given in Table 9.7; note that the execution times are proportional to the number of bits in the field and are independent of the total number of words. For example, to search an unordered list of say 3000 items a conventional computer with a cycle time of 0·5 μs would require approximately 3000 μs, but the associative processor can search the list in less than 3 μs, representing a 1000 to 1 improvement. The same machine would retrieve and add two 24-bit words and store the sum in approximately 3 μs compared to 28 μs for the associative processor. However, if 3000 such data pairs are required to be added together the associative processor would still take 28 μs, compared to 9000 μs for the conventional machine!

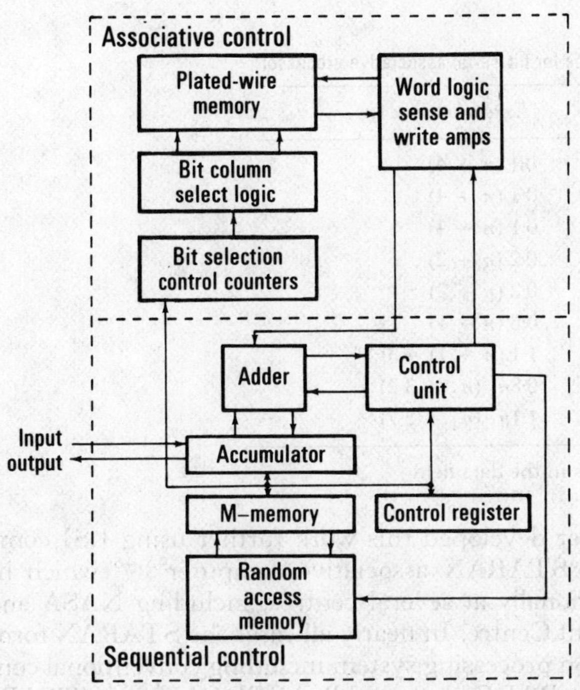

Figure 9.27 Block diagram of bit-serial associative processor.

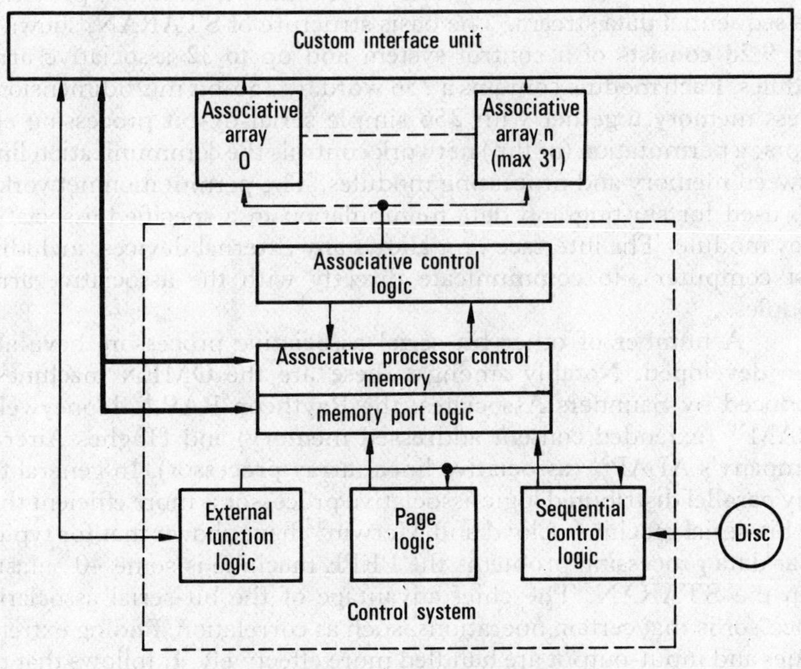

Figure 9.28 Block diagram of STARAN.

Table 9.7
Typical execution times for bit-serial associative processor

Instruction	*Execution time* (μs)
Match	$0{\cdot}1\ (n + 4)$
Less than	$0{\cdot}1\ (n + 4)$
Greater than	$0{\cdot}1\ (n + 4)$
Maximum value	$0{\cdot}2\ (n + 2)$
Minimum value	$0{\cdot}2\ (n + 2)$
Add common	$0{\cdot}8\ (n + 1)$
Add fields	$1{\cdot}1\ (n + 1) + 0{\cdot}1$
Multiply common	$0{\cdot}8n_1\ (n_2 + 3{\cdot}2)$
Multiply fields	$1{\cdot}1n_1\ (n_2 + 2{\cdot}7)$

n = number of bits in the data field

Goodyear developed this work further using LSI components to produce the STARAN associative computer[45,46] which has been installed operationally at several centres, including NASA and Rome Air Development Centre. In nearly all cases the STARAN formed part of an information processing system including conventional computers. In the case of the RADC system (called REDCAP[47]) the STARAN was interfaced with a Honeywell Information Systems 645 machine. In these hybrid configurations STARAN handles all the parallel processing tasks, and the conventional machine those tasks which can be processed as a sequential data stream. The basis structure of STARAN, shown in Fig. 9.28 consists of a control system and up to 32 associative array modules. Each module contains a 256 word by 256-bit multidimensional access memory together with 256 simple serial-by-bit processing elements; a permutation (or flip) network controls the communication lines between memory and processing modules. The permutation network is also used for shifting and data manipulation in a specified associative array module. The interface unit allows any external devices, including host computers, to communicate directly with the associative array modules.

A number of other bit-serial associative processors have also been developed. Notably amongst these are the OMEN machines[48] produced by Saunders Associates, the Raytheon RAP,[49] Honeywell's ECAM[50] (extended content addressed memory) and Hughes Aircraft Company's ALAP[51] (associative linear array processor). In general the fully parallel distributed logic associative processor is more efficient than the bit-serial machine. Lloyd and Merwin[52] have shown that for typical radar data processing problems the PEPE machine is some 40% faster than the STARAN. The chief advantage of the bit-serial associative processor is that certain operations, such as correlation, finding extreme values and input-output are handled more effectively. It follows that the choice of processor is problem dependent, and to obtain maximum efficiency the task must be suited to the hardware structure. It is also important to realize that though associative processors can give considerable increases in processing efficiency they also present new problems in task allocation and programming. It is essential that algorithms be efficiently mapped onto the hardware structures and that effective software procedures are available.

The ideas presented in this Chapter represent a significant breakaway from the fundamental Von Neumann type structures that have long been accepted as the standard computer configuration. The time has come, however, for computer designers to review the situation, particularly with regard to user requirements, if a significant breakthrough in computer architecture and ease of programming is to be achieved. In the past this has been difficult due to the high cost of implementing such systems, but with the availability of cheap LSI modules, particularly microprocessors, highly parallel structures are rapidly becoming cost-effective. The computer system of the future will almost certainly be fully distributed and will comprise both associative and conventional microprocessing elements. Moreover, it is highly probable that they will be custom built to meet specific customer specifications.

References and bibliography

1 Joseph, E. C. Computers: trends towards the future. *IFIPS Conference, Edinburgh 1968*. North Holland Publishing Co. Invited Papers, pp. 145–157.

2 Koczela, L., and G. Wang. The design of a highly parallel computer organization. *IEEE Trans. Comput.*, 1969, **C18**, 520–529.

3 Flynn M. J. Very high-speed computing systems, *Proc. IEEE*, **54**, 1966, 1901–1909.

4 Thurber, K. J. and Wald, L. D. Associative and Parallel Processors, *Computing Surveys*, **7**, 1975, 215–255.

5 Murtha, J. C. Highly parallel information processing systems. *Advances in Computers*, 1966, **7**, 1–116.

6 Slotnick, D. L., W. C. Borck, and R. C. McReynolds. The Solomon computer. *AFIPS Proc. FJCC*, 1962, **22**, 97–107.

7 Barnes, G., *et al.* The Illiac IV computer. *IEEE Trans. Comput.*, 1968, **C17**, 746–757.

8 Slotnick, D. L. The fastest computer. *Scientific American*, 1971, **224**, 76–87.

9 McIntyre, D. E. An introduction to the Illiac IV computer. *Datamation*, 1970, **16**, 60–67.

10 Abel, N. E., *et al.* TRANQUIL: a language for an array processing computer. *AFIPS Proc. SJCC*, 1969, **34**, 57–75.

11 Hallin, T. G. and Flynn M. J. Pipelining of Arithmetic functions, *IEEE Trans. Computers*, **C21**, 1972, 880–886.

12 Ibbett, R. N. The MU5 Instruction Pipeline, *Computer Journal*, **15**, 1972, 43–50.

13 Brown, J. L. *et al.* IBM System/360 Engineering, *AFIPS FJCC*, **26**, 1964, 205–32.

14 Davidson, E. S., Thampy Thomas, A., Shor, L. E. and Patel, J. H. Effective Control for Pipelined Computers. *COMPCON 75, IEEE*, New York, 1975, 181–184.

15 Ramamoorthy, C. V. and Li, H. F. Pipeline Architecture, *Computing Surveys*, **9**, 1977, 61–102.

16 Chen, T. C. Parallelism, pipelining and computer efficiency, *Computer Design*, Jan. 1971, 69–74.

17 Graham, W. R. The parallel and the pipe-line computers. *Datamation*, 1970, **16**, 68–71.

18 Searle, B. C. and Freberg, D. E. Tutorial: Microprocessor Applications in Multiple Processor Systems, *IEEE Computer*, **8**, No. 10, 1975, 22–30.

19 Baer, J. L. Multiprocessing Systems, *IEEE Trans. Computer*, **C25**, 1976, 1271–1277.

20 Enslow, P. H. Multiprocessor Organisation – A Survey, *ACM Comp. Surveys*, **9**, No. 1, 1977, 103–129.

21 Control Data Corp. *Control Data STAR- 100 Computer Hardware Reference Manual*, 1974.

22 Texas Instruments Inc. *A Description of the Advanced Scientific Computer System*. Austin, Texas, April 1973.

23 Thurber, K. J., Douglas Jeuson, E. and Jack, L. A. A systematic Approach to the design of Digital Bussing Structures, *AFIPS Proc. FJCC*, **41**, 1972, 719–740.

24 Wulf, W. A. and Bell, C. G. C-mmp A Multiminiprocessor. *AFIPS Proc. FJCC*, **41**, 1972, 765–777.

25 Reyling, G. Performance and Control of multiple microprocessor systems, *Computer Design*, March 1974, 81–87.

26 Yau, S. S. and Fung, H. S. Associative Processor Architecture – A Survey, *ACM Computing Surveys*, **9**, No. 1, 1977, 3–27.

27 Lea, R. M. Information Processing with the Associative Parallel Processor, *IEEE Computer*, **8**, November 1975, 25–32.

28 Lee, C. Y. Intercommunicating cells, basis for a distributed computer. *AFIPS Proc. FJCC*, 1962, **22**, 130–136.

29 Lee, C. Y., and M. C. Paull. A content addressable distributed logic memory with application to information retrieval. *Proc. IEEE*, 1963, **51**, 924–932.

30 Gaines, R. S., and C. Y. Lee. An improved cell memory. *IEEE Trans. Electron. Comput.*, 1965, **EC14**, 72–75.

31 Crane, B. A., and J. A. Githens. Bulk processing in distributed logic memory. *IEEE Trans. Electron. Comput.*, 1965, **EC14**, 186–196.

32 Fuller, R. H., and G. Estrin. Some applications for content addressable memories. *AFIPS Proc. FJCC*, 1963, **24**, 495–508.

33 Eddey, E. E. The use of associative processors in radar tracking and correlation. Proceedings of the National Aerospace Electronics Conference 1967.

34 Sturman, J. N. An Iteratively-structured G.P. digital computer. *IEEE Trans. Comput.*, 1968, **C17**, 2–17.

35 Wright, J., and D. W. Lewin. A draft specification of a symbol processor. IEE Conference on Computer Science and Technology. *IEE Pub. No. 55*, 1969, 282–295.

36 Beaven, P. A. and Lewin, D. W. An Associative Parallel Processing System for Non-Numerical Computation, *Computer J*, **15**, 1973, 343–349.

37 Lea, R. M. and Wright J. S. A Novel Memory Concept for Information Processing, *Data Fair Research Papers*, Vol. II, 1973, 413–417.

38 Lea, R. M. *An Associative Parallel Processor for Information Processing Systems*. Brunel University, Uxbridge (UK) Tech. Memo. C/RS/021, 1975.

39 Dyke, J. G. and Lea, R. M. An Associative Parallel Processor for Local Editing Applications, *Digital Processors*, **1**, 1975, 89–101.

40 Crane, B. A., Gilmartin, M. J. Huttenhoff, J. H., Rux, D. T. and Shively, R. R. PEPE Computer Architecture, *IEEE COMPCON*, 1972, 57–60.

41 Evensen, A. J. and Troy, J. L. Introduction to the Architecture of a 288-element PEPE, *Proc. 1973. Sagamore Computer Conf. on Parallel Processing*. Springer-Verlag, New York, 1973, 162–169.

42 Estrin, G., and R. Fuller. Algorithms for content addressable memories. *Proceedings of the Pacific Computer Conference*, March, 1963, pp. 118–128.

43 Ewing, R. E., and P. M. Davies. An associative processor. *AFIPS Proc. FJCC*, 1964, **26**, 147–158.

44 Rudolph, J., L. Fulmer, and W. Meilander. The coming of age of the associative processor. *Electronics*, 1971, **44**, 91–96.

45 Rudolph, J. A. A Production Implementation of an Associative Array Processor: STARAN, *Proc. AFIPS FJCC*, **41**, Pt. I, 1972, 229–241.

46 Batcher, K. E. STARAN Parallel Processor System Hardware, *Proc. AFIPS FJCC*, **43**, 1974, 405–410.

47 Feldman, J. D. and Fulmer, L. C. REDCAP an operational parallel processing facility, *Proc. AFIPS FJCC*, **43**, 1974, 7–15.

48 Higbie, L. C. The OMEN Computers: Associative Array Processors, *IEEE Compcon*, 1972, 287–290.

49 Couranz, G. R., Gerhardt, M. S. and Young, C. J. Programmable Radar Signal Processing using the RAP, *Proc. Sagamore Comp. Conf. on Parallel Processing*. Springer-Verlag, New York, 1974, 37–52.

50 Anderson, G. A. and Kain R. Y. A Content Addressed Memory Design for Data-Base Application. *Proc. 1976 International Conf. on Parallel Processing*, New York, 1976, 191–195.

51 Finnila, C. A. and Love, H. H. The Associative Linear Array Processor, *IEEE Trans. on Computers*, **C26**, 1977, 112–125.

52 Lloyd, G. R. and Merwin, R. E. Evaluation of performance of Parallel Processors in a Real-Time environment. *AFIPS SJCC*, **42**, 1973, 101–108.

Further reading

Koczela, L. J. The distributed processor organization. *Advance in Computers*, 1968, **9**, 286–353.

Riley, W. B. Wanted for the '70s: easier-to-program computers. *Electronics*, 1971, **45**, 62–84.

Lewin, D. Whither Data Processing? *The Radio and Electronic Eng.*, **45**, 1975, 627–630.

Worked solutions to selected problems

Chapter 1

1.5 (a) 010 000 011 000 000 111
0·511772 or 67079 × 2^{-17}
(b) 000 000 000 000 000 110
0·000046 or 6 × 2^{-17}
(c) 111 000 000 111 110 011
−0·146194 or −32269 × 2^{-17}
(d) 100 010 110 000 001 010
−0·885861 or −119798 × 2^{-17}

1.6 The program is shown in Table S1, note that it is necessary to use temporary storage for the contents of location 1732. The program also demonstrates the need for an instruction of the form 'Exchange contents of accumulator and the location specified in the address'.

1.7 A machine-code program for this problem is given in Table S2.

Table S1
Problem 1.6

Location	*Instruction*	*Mnemonic*	*Binary Word*
1500	22 1732	FET 1732	100 100 001 111 011 010
1501	23 1507	STR 1507	100 110 001 101 000 111
1502	22 1607	FET 1607	100 100 001 110 000 111
1503	23 1732	STR 1732	100 110 001 111 011 010
1504	22 1507	FET 1507	100 100 001 101 000 111
1505	23 1607	STR 1607	000 000 000 000 000 000
1506	000000	STOP	
1507	( )		

Table S2
Problem 1.7

Location	*Instruction*	*Mnemonic*	*Comment*
1500	22 1001	FET 1001	Fetch A to the accumulator
1501	20 1003	ADD 1003	Add B
1502	21 1007	SUB 1007	Sub C
1503	23 0073	STR 0073	Store A + B − C in location 73
1504	000000	STOP	Stop the computer

Table S3
Problem 2.2(a)

(i)

Location	*Instruction*		*Comment*
0000	FETM	0006	Fetch loop constant to modifier
0001	CLRA		Clear accumulator
0002	STR	1777/M	Store accumulator in modified address
0003	JMPM	0003	Jump if modifier zero to own location, i.e., stop
0004	INCM		Add +1 to modifier
0005	JMP	0002	Jump to start of loop
0006		−777	Loop constant

(ii)

Location	*Instruction*		*Comment*
0000	CLRA		Clear accumulator
0001	STR	0010/I	Store accumulator in indirect address
0002	FET	0010	Fetch operand address
0003	SUB	0011	Subtract 1777, i.e., last location cleared
0004	JMPO	0004	Jump if accumulator zero to self, i.e., stop
0005	ADD	0012	Add 2000, i.e. 1777 + 1, (incrementing by +1)
0006	STR	0010	Replace modified address
0007	JMP	0000	Jump back to start of loop
0010		1000	Operand address
0011		1777	Loop constant
0012		2000	Increment constant

Chapter 2

2.2 Machine-code programs for this problem are given in Tables S3 and S4.

2.3 A typical program for this problem is shown in Table S5. Note that in this program we are handicapped by having only one modifier register and by not being able to increment a store location directly. The 'test for zero' loop employs the modifier register to increase the speed of operation of the program and the 'unpacking' of the address is performed using indirect addressing. An alternative approach if main storage (rather than speed) is at a premium is to use the indirect address function for the test loop, since the actual address could then be held in the indirectly addressed location.

2.4 In this program, given in Table S6, we can take advantage of the fact that the same modifier constant can be used; therefore we can employ a modified instruction for both transfers.

2.7 In this program, shown in Table S7, we assume that the most significant BCD digit is stored at the least significant end of the word. This is, in fact, the usual way since decimal digits would be read MS digit first from the input devices, for instance, using the 'Input to accumulator from paper tape' instruction.

Table S4
Problem 2.2(b)

(i)

Location	*Instruction*		*Comment*
0000	FETM	0010	Fetch loop constant to modifier
0001	CLRA		Clear accumulator
0002	ADD	0007	Add +1 to accumulator
0003	STR	1000/M	Store accumulator in modified address
0004	JMPM	0004	Jump to self, i.e., stop
0005	INCM		Add +1 to modifier
0006	JMP	0002	Jump back to start of loop
0007		1	Constant +1
0010		−500	Loop constant

(ii)

Location	*Instruction*		*Comment*
0000	FET	0012	Fetch count to accumulator
0001	ADD	0013	Add +1
0002	STR	0012	Replace in store
0003	STR	0014/I	Store count in indirect address
0004	FET	0014	Fetch operand address
0005	SUB	0015	Test for end of loop
0006	JMPO	0006	Jump to self if zero
0007	ADD	0016	Increment operand address by +1
0010	STR	0014	Replace in store
0011	JMP	0000	Jump back to start of loop
0012	(	)	Accumulating number
0013		1	Constant +1
0014		500	Operand address
0015		1000	Loop constant
0016		1001	Increment constant

The basic logic of the program is to extract the 4-bit BCD digits (by collation) and then multiply by binary 10, the result is stored away and added to the next BCD digit, which is also multiplied by binary 10, and so on.

Note that multiplying two integers $\times 2^{-17}$ results in the product being placed in the least significant end of the X-register (the sign digit of the X-register is normally not used).

Table S5
Problem 2.3

Location	*Instruction*		*Comment*
0000	FETM	0020	
0001	FET	1700/M	Test for zero loop
0002	JMPO	0006	
0003	JMPM	0003	
0004	INCM		
0005	JMP	0001	
0006	EXAM		Unpacking of address of zero location and storing in table
0007	STR	0021	
0010	EXAM		
0011	FET	0021	
0012	ADD	0022	
0013	STR	0023/I	
0014	FET	0023	
0015	ADD	0024	
0016	STR	0023	
0017	JMP	0003	
0020		−700	Test loop constant
0021	(	)	Temporary store
0022		1700	Address table constants
0023		500	
0024		1	

Table S6
Problem 2.4

Location	*Instruction*		*Comment*
0000	FETM	0012	
0001	FET	1200/M	Exchange instructions
0002	STR	0013	
0003	FET	0700/M	
0004	STR	1200/M	
0005	FET	0013	
0006	STR	0700/M	
0007	JMPM	0007	Loop test
0010	INCM		
0011	JMP	0001	
0012		−200	Loop count constant
0013	(	)	Temporary store

Table S7
Problem 2.7

Location	*Instruction*		*Comment*
0000	FETM	0016	Fetch constant to modifier
0001	FET	0017	
0002	JMP	0006	
0003	FET	0017	Fetch BCD word to accumulator
0004	SRL	4	Shift right 4 places
0005	STR	0017	Store shifted word
0006	COL	0021	Extract BCD digit
0007	ADD	0020	Add contents of temporary store
0010	JMPM	0010	Stop with binary word in accumulator
0011	INCM		
0012	MULT	0022	Multiply by binary 10, result in X-register
0013	EXAC		Exchange accumulator with X-register
0014	STR	0020	
0015	JMP	0003	
0016		−3	Count constant
0017	(	)	BCD word
0020	000000		Temporary storage
0021		0017	Collation constant binary 15
0022		10	Binary 10

Table S8
Problem 2.8

Location	*Instruction*		*Comment*
0000	FET	0004	Fetch a to accumulator
0001	EXCO	0005	Form $a \oplus b$
0002	EXCO	0006	Form $a \oplus b \oplus c$
0003	STOP		
0004	(	)	a
0005	(	)	b
0006	(	)	c

Table S9
Problem 2.10

Location	*Instruction*	
0000	FETM	0012
0001	FET	0014/IM
0002	STR	0013
0003	FET	0015/IM
0004	STR	0014/IM
0005	FET	0013
0006	STR	0015/IM
0007	JMPM	0007
0010	INCM	
0011	JMP	0001
0012		−200
0013	(	)
0014		4700
0015		2200

2.8 This is a trivial example if one manipulates the original expression correctly; for example:

$$S = c(\bar{a}\bar{b} + ab) + \bar{c}(\bar{a}b + a\bar{b})$$

$$S = c\bar{z} + \bar{c}z, \quad \text{where } z = \bar{a}b + a\bar{b} = a \oplus b$$

$$S = c \oplus z = c \oplus a \oplus b$$

A suitable program is shown in Table S8.

2.10 The main difference with this program (shown in Table S9) compared to that of Problem 2.4 is that we need to use indirect addressing because the available address bits have been exceeded, that is, 2000–2200 (sector 1) and 4500–4700 (sector 2). Since we are working in sector 0 we may specify directly within the sector, but locations outside the sector must be specified indirectly. This means including two extra ‘pointer’ locations; the addresses in these locations may be modified in the usual way by setting the modifier bit in the instruction to 1. Note that when the modifier and indirect address bits are both set to 1, it is the contents of the indirect address that is modified.

In some systems, however, as described in Chapter 2, the indirect addressing and the modification functions may be used in a multiple-level system. In this case the modifier and indirect address bits of the actual address in the pointer location are set to 1. Note that if this facility is to be used, the control bits (IMS) should be placed at the most significant end of the word to allow maximum usage of the available bits.

Chapter 3

3.1 This program, as shown in Table S10, assumes that the (n, k) data is inserted from the keyboard, using the most significant 9 digits for n and the next 9 least significant digits for k. Note that it is only possible to output a block of words between the limits 0000 → 1776.

The IO instructions are specified according to Table 7.4(b), but in this example these could be arbitrarily defined. Note, however, that the paper tape output order is assumed to output only the five least significant digits of the accumulator.

When outputting the most significant code group, which due to using an 18-bit

Table S10
Problem 3.1

Location	*Instruction*		*Comment*
0000	INA	007	Read keyboard
0001	COL	0035	Extract *k*
0002	STR	0036	Temporarily store *k*
0003	ADD	0037	Add +1
0004	INV		Form $-(k + 1)$
0005	EXAM		Exchange with modifier register
0006	INA	007	Read keyboard
0007	SRL	9	Shift right 9 places
0010	COL	0035	Extract *n*
0011	STR	0040	Temporarily store *n*
0012	ADD	0036	Form $n + k$
0013	ADD	0021	Form FET $(n + k)$/M
0014	STR	0021	Store modify instruction in program
0015	FET	0035	
0016	OTA	010	Output Start code
0017	FET	0040	Fetch *n* before outputting
0020	JMP	0022	
0021	FET	$(n + k)$/M	Blank modify instruction held as FET 0000/M
0022	EAS	15	Output word from most significant end using end-around shift instruction
0023	OTA	010	
0024	EAS	13	
0025	OTA	010	
0026	EAS	13	
0027	OTA	010	
0030	EAS	13	
0031	OTA	010	
0032	JMPM	0032	Loop test. Note that the operation is performed $(k + 1)$ times to allow for initial output of *n*
0033	INCM		
0034	JMP	0021	
0035		000777	Collation constant
0036	(	)	Temporary storage
0037		1	
0040	(	)	Temporary storage

word consists of only three relevant digits, it is assumed that the two most significant digits of this 5-bit group can be arbitrary digits, since they will be shifted out of the accumulator during the assembly procedure. Note the process of constructing the modify order and then inserting it at the appropriate place in the program; this is common practice in data handling operations. The program also illustrates an inadequacy of the instruction set, that is, the lack of a double length shift using the accumulator and X-register. If this were available the process of outputting from the most significant end of a word would be considerably easier.

3.3 This input routine is not really a compiler, but just a simple translator since there is a 1:1 correspondence between the coded and the machine formats, and the absolute store addresses are specified by the block address.

The main points to consider are:

1. It is convenient to be able to type (or print out) a page listing of the program with one instruction per line. Thus the instructions must be terminated with a new line, or carriage-return/line-feed symbol.

2. Suitable instruction control and delimiter symbols must be chosen; for example:

(a) Instructions must always start on a new line.

(b) The order must be separated (**delimited**) from the store address, using for instance a prime, that is, 20′ 1460. With this technique (known as **free format**) the input routine must be organized in such a way that everything following a prime is treated as an address. The address digits can either be punched with leading zeros (e.g., 0072) or alternatively using spaces, when the number of character positions must be counted (e.g., sp.sp.72).

(c) The slash mark is used to separate the address digits from the IMS digits.

(d) The block address must start on a new line and be preceded by a control symbol for example: @ 0172, which can also indicate the start of the program.

(e) If comments are required they must always start on a new line and be contained within open and closed brackets.

(f) Control symbols will also be required if constants are to be represented in other than a pseudo-instruction format.

(g) The end of the program must also be indicated, say by an asterisk.

3. The major requirements of the input routine will be checking for control symbols, the conversion of octal to binary, assembly of the binary instruction word and its consequent loading into contiguous locations of the store.

4. It is normal practice to jump automatically to the start of the program after the assembly is complete, this may be done by arranging that the last instruction of the program is a jump instruction.

5. Spaces are usually ignored.

6. Comments, block addresses, and instructions must always start on a new line, that is, preceded by CRLF.

7. Errors in syntax (e.g. incorrect use of control symbols, etc.) must be determined and signalled to the user. This would only be included in a sophisticated version of the program.

8. An alternative to using a free format is to divide the instruction into **fields**, where groups of character positions are reserved for each section, that is, order, address, IMS bits, etc.

Figure S1 shows a flow-chart for performing this process. Note the use of repeated tests and the setting of indicator bits. This is, in fact, analogous to a finite-state machine, since past inputs must be stored and used, together with the present input, to determine the outputs.

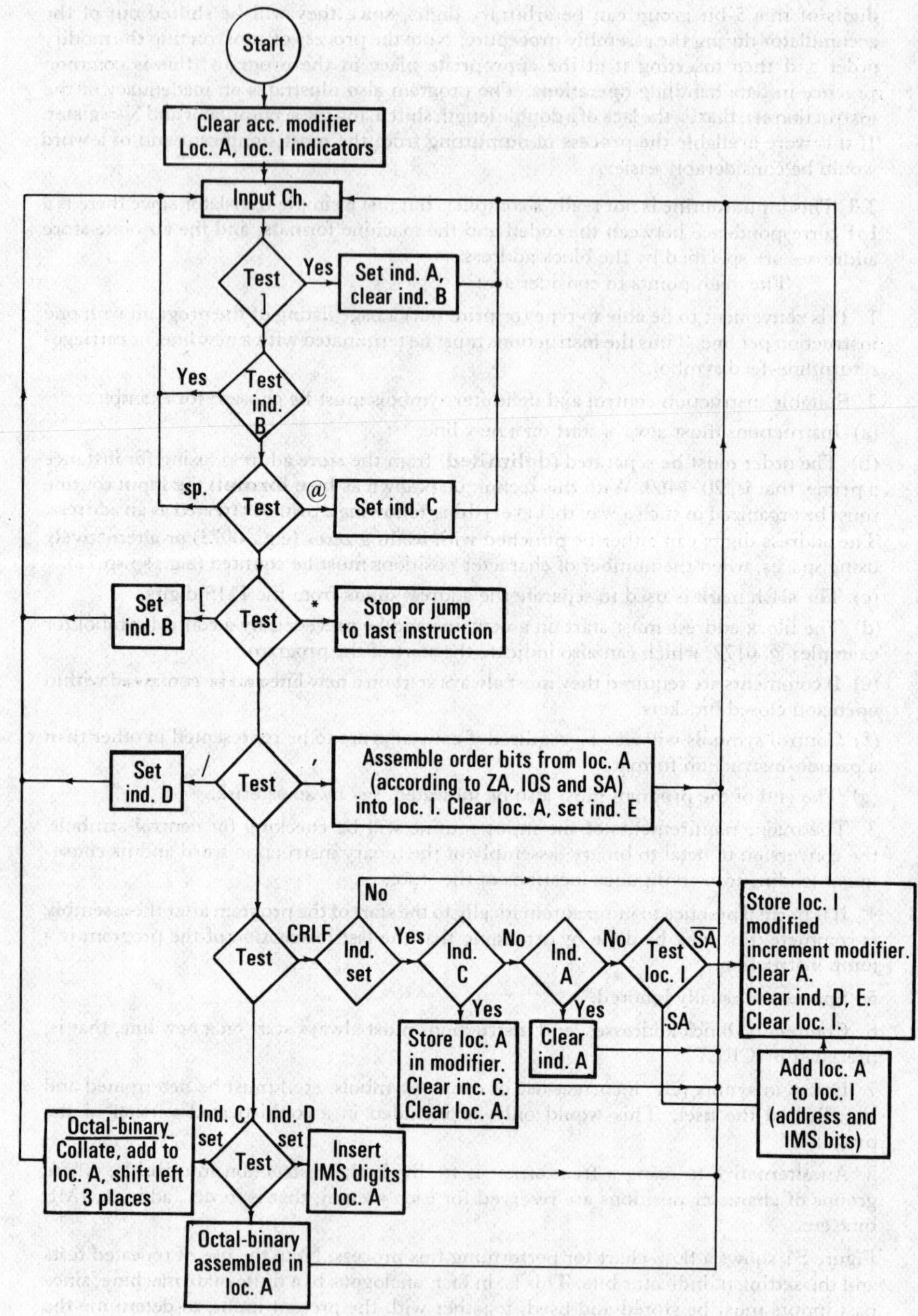

Figure S1 Problem 3.3.

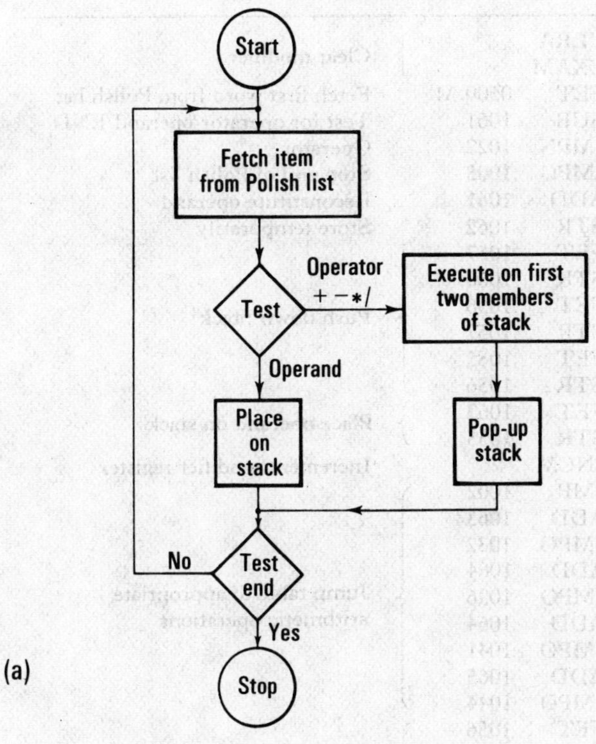

(b)

1000	A	End
1001	B	1000
1002	A*B	End
1003	C	1002
1004	D	1003
1005	C–D	1002
1006	E	1005
1007	E/C–D	1002
1010	E/(C – D) + (A*B)	End

Figure S2 Problem 3.5.

Table S11
Problem 3.5

Location	*Instruction*		*Comment*
1000	CLRA		Clear modifier
1001	EXAM		
1002	FET	0500/M	Fetch first word from Polish list
1003	SUB	1061	Test for operator/operand/END
1004	JMPN	1022	Operator
1005	JMPO	1005	Stop end of Polish list
1006	ADD	1061	Reconstitute operand
1007	STR	1062	Store temporarily
1010	FET	1057	Push down Stack
1011	STR	1060	
1012	FET	1056	
1013	STR	1057	
1014	FET	1055	
1015	STR	1056	
1016	FET	1062	Place operand on stack
1017	STR	1055	
1020	INCM		Increment modifier register
1021	JMP	1002	Jump table to appropriate arithmetic operations
1022	ADD	1063	
1023	JMPO	1032	
1024	ADD	1064	
1025	JMPO	1036	
1026	ADD	1064	
1027	JMPO	1041	
1030	ADD	1065	
1031	JMPO	1044	
1032	FET	1056	Divide
1033	DIV	1055	
1034	STR	1055	
1035	JMP	1050	
1036	FET	1056	Subtract
1037	SUB	1055	
1040	JMP	1034	
1041	FET	1056	Add
1042	ADD	1055	
1043	JMP	1034	
1044	FET	1056	Multiply
1045	MULT	1055	
1046	EXAC		
1047	JMP	1034	
1050	FET	1057	Pop-up stack
1051	STR	1056	
1052	FET	1060	
1053	STR	1057	
1054	JMP	1020	
1055	(	)	Operand stack of depth four
1056	(	)	
1057	(	)	
1060	(	)	
1061		0300	Teletype symbol
1062	(	)	Temporary store
1063		0021	Operator constants
1064		0002	
1065		0001	

3.5 Assume initially that the Polish list consists of fixed maximum length expressions, from 0500 upwards and terminated with the marker @. In this case it would be very simple to construct a push-down/pop-up store, or stack, of finite depth to perform the evaluation. Figure S2(a) shows a typical flow-chart for such a routine and a simple machine-code program is given in Table S11. In this program the stack has been programmed by physically moving the contents of the store locations; this is not the best way since, in general, it is always better to manipulate addresses rather than words. An alternative approach is shown in Table S12. In this case the modifier register acts as a **pointer** (the same results could be achieved by indirect addressing) pointing to the address of the top member of the stack. Note that only the 'top' of the stack is moved during the stacking and unstacking processes. In this case it is also necessary to make special provisions to ensure that the arithmetic operations are performed in the correct order (A ∗ B) since the normal application of this technique gives B ∗ A if B is at the top of the stack. It is still necessary with this type of stack to know its precise size beforehand; a perfectly general stack for handling Polish lists of any size may be constructed using the list structure as described in the text. In this case the operations of push-down and pop-up become simply a matter of altering the list pointers, as shown in Fig. S2(b). If at the start of the process a block of continuous words is allocated in the store for the list cells, sooner or later these will become exhausted. Moreover, the store will contain deleted cells as a result of amending the lists. In order to ensure maximum utilization an **available space list** is created (using a stack principle) which is employed to note the addresses of all free cells. When a new cell is required it is obtained from the available space list and, conversely, when no longer required it is returned to the list. Note that the top of the list is continually changing, and is given by the pointer obtained from the available space list, in this particular case taken in sequence from 1000 → 1010.

Table S12
Problem 3.5

Location	*Instruction*		*Comment*
1000	CLRA		Initial clearing of modifier register
1001	EXAM		to zero
1050	FET	(Operand A)	
1051	STR	1055/M	Store on stack and push-down
1052	INCM		
1053	FET	(Operand B)	
1054	STR	1055/M	
1055	INCM		
1070	EXAM		
1071	SUB	(+1)	Read off top of stack
1072	EXAM		
1073	FET	1055/M	
1100	EXAM		
1101	SUB	(+1)	
1102	EXAM		
1103	FET	1055/M	Add and push-up stack
1104	EXAM		(note this gives B + A
1105	SUB	(+1)	since B at top of list)
1106	EXAM		
1107	ADD	1055/M	
1110	STR	1055/M	
1111	INCM		

Table S13
Problem 3.7

Step no.	*Statement*	*Comment*
0	$\mathbf{i} \leftarrow \bar{\boldsymbol{\varepsilon}}(5)$	Input buffer register set to zero
1	$\mathbf{a} \leftarrow \bar{\boldsymbol{\varepsilon}}(18)$	Accumulator set to zero
2	$c \leftarrow -4$	Character count
3	$\mathbf{m} \leftarrow \bar{\boldsymbol{\varepsilon}}(18)$	Modifier register set to zero
4	$\mathbf{i} \leftarrow \text{(ch)}$	Read ch. to vector **i**
5	$\vee/\mathbf{i} : 0; (=) \rightarrow (4)$	Check **i** for zero content
6	$\mathbf{a} \leftarrow 5 \uparrow \mathbf{a}$	Shift **a** 5 places left
7	$\mathbf{i} \leftarrow \text{(ch)}$	
8	$\mathbf{a}_{13,14,15,16,17} \leftarrow \mathbf{i}$	Transfer **i** to least significant end of **a**
9	$c \leftarrow c + 1$	Increment ch. count
10	$c : 0; (<) \rightarrow (6)$	Test ch. count
11	$\vee/\mathbf{m} : 0; (\neq) \rightarrow (15)$	Check modifier for zero content
12	$\mathbf{m} \leftarrow \mathbf{a}$	Transfer accumulator to modifier
13	$c \leftarrow -4$	Set ch. count
14	$\rightarrow (6)$	Jump to 6
15	$\mathbf{M}^{\perp \mathbf{m}} \leftarrow \mathbf{a}$	Write into store
16	$\mathbf{m} \leftarrow 18 \top (1) + \perp \mathbf{m}$	Increment modifier by +1
17	$\rightarrow (13)$	Jump to 13

3.7 Since there are no explicit input/output instructions in the Iverson notation it is necessary to invent some means of input (in practice this would be determined by the actual machine used to implement the language). In this case we have used the notation (ch) to indicate reading a character, thus $\mathbf{i} \leftarrow$ (ch) means inputting a character to the 5-bit binary vector **i** (representing a buffer register). A possible program to perform the simple binary input routine is given in Table S13. Note that in statement 15 **M** represents a matrix (or store) of vectors where $\mathbf{M}^j$ is the *j*th word. Since **m** is a vector it is necessary to use its base 2 binary value to specify the relevant word in store. To increment the modifier register **m** (statement 16) its base 2 value must be added to +1 represented as an 18-bit binary number and the result expressed as a vector.

3.9 An Iverson program for serial to parallel data conversion, assuming 4-bit numbers, is given in Table S14. In this program **c** is assumed to be the clock vector which shifts through the sequence 1000, 0100, 0010, 0001, controlled by the count **i**, while the vector **q** is the parallel output, assumed to be a 4-bit register; the serial input is represented by the scaler *s*. Statement 4 is a compound expression which has the effect of ANDing together the operands *s* and **c**, the result of which is then OR'd with **q**. Note that the scaler *s* is assumed to have varying values of 0 and 1, representing the serial input. Moreover in step 4 above, since the operation refers to both scaler and vector quantities, the scaler is treated as a vector of dimension 4 whose components are all equal. Note also the usual Iverson right to left evaluation rule, that is, $\mathbf{c} \wedge s$, is performed first and the result OR'd with **q**. Figure S3 shows a possible implementation of the algorithm in hardware; note the analogy between the circular shift and a ring counter. There are, however, many ways of implementing the circuit since there is no exact correspondence between the Iverson algorithm and the required micro-orders.

Table S14
Problem 3.9

Step no.	*Statement*	*Comment*
0	$\mathbf{q} \leftarrow \bar{\varepsilon}\,(4)$	*q* is set to all zeros
1	$\mathbf{c} \leftarrow \boldsymbol{\alpha}^1\,(4)$	Clock vector set to 1000
2	$i \leftarrow 0$	Count set to zero
3	$i : 4;\ (=) \rightarrow n$	Jump out when $i = 4$
4	$\mathbf{q} \leftarrow \mathbf{q} \vee \mathbf{c} \wedge \mathrm{s}$	
5	$\mathbf{c} \leftarrow 1 \uparrow \mathbf{c}$	Right circular shift of clock
6	$i = i + 1;\ \rightarrow(3)$	

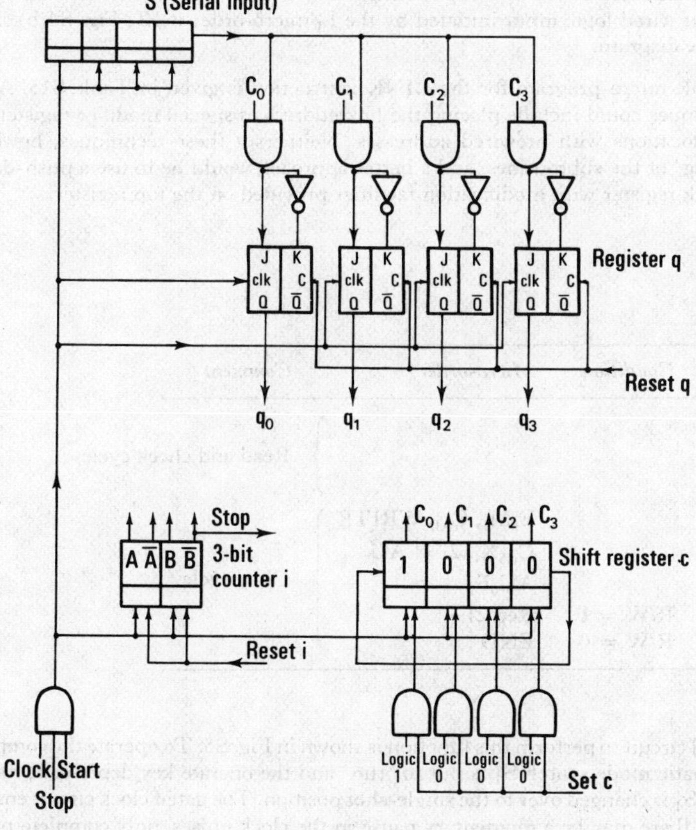

Figure S3 Problem 3.9.

Chapter 4

4.1 The first point to realize in this problem is that the micro-order Rep *n* has no physical significance! In order to set the operations counter it is necessary to generate the actual signals, thus we must specify the binary digit pattern to be set into the counter. This may be done using the K digits normally used for the K-counter, thus the micro-order would be specified as:

$$\text{Rep. } (K_1 K_2 K_4 K_8 K_{16})$$

where the K digits give the actual *n* setting in Gray-code. The Rep. micro-order must be generated separately from the K digits, as this is used as a gating signal to route the digits to the control counter. Figure S4(a) shows a typical logic arrangement to perform this function. Note that it is not advisable to use the set and clear inputs to set the pattern into the counter as this should give rise to timing problems, since the counter states are required to maintain the setting signals themselves! The timing problems could be eliminated, however, if the Rep. *n* micro-order was the *only* order in a given timing slot. The technique used is to inhibit the normal counter feedback signals (using $\overline{\text{Rep.}}$) and to replace them with the desired K values gated with Rep. In this way the next clock pulse will set the counter accordingly, when the Rep. signal will disappear, allowing normal operations to continue. Note that only the first three stages are given (the same technique is repeated) and that the END signal (required to reset the counter to zero) must be applied in the same way.

The address of the interrupt routine may be set into the control register by using a preset wired logic input initiated by the E_I micro-order itself; Fig. S4(b) shows a typical logic diagram.

4.3 A possible micro-program for the LINK instruction is given in Table S15. Alternative techniques could include placing the link address in special modifier registers, or using store locations with prewired addresses. Neither of these techniques, however, allow 'nesting' of the subroutines, and a better approach would be to use a push-down/push-up stack register with modification facilities provided on the top register.

Table S15
Problem 4.3

Timing slot	*Condition*	*Micro-order*	*Comment*
1			Read and check cycle
⋮			
17			
18		E_oM_i, C_{AD} WRITE	Obey cycle
19		$C_{oA}X_i$, 2^{-17} AU	
20		AU_oE_i	
21	R/W = 1	Rep 21	
	R/W = 0	END	

4.5 A typical circuit to perform this function is shown in Fig. S5. To operate the computer in the automatic mode switch S_1 is put to 'run' and the operate key depressed; to stop the machine S_1 is changed over to the single-shot position. The gated clock circuit ensures that, though there may be a momentary pause in the clock pulses, only complete pulses reach the counter. Note that all the switches must be buffered (by a bistable) to eliminate contact bounce.

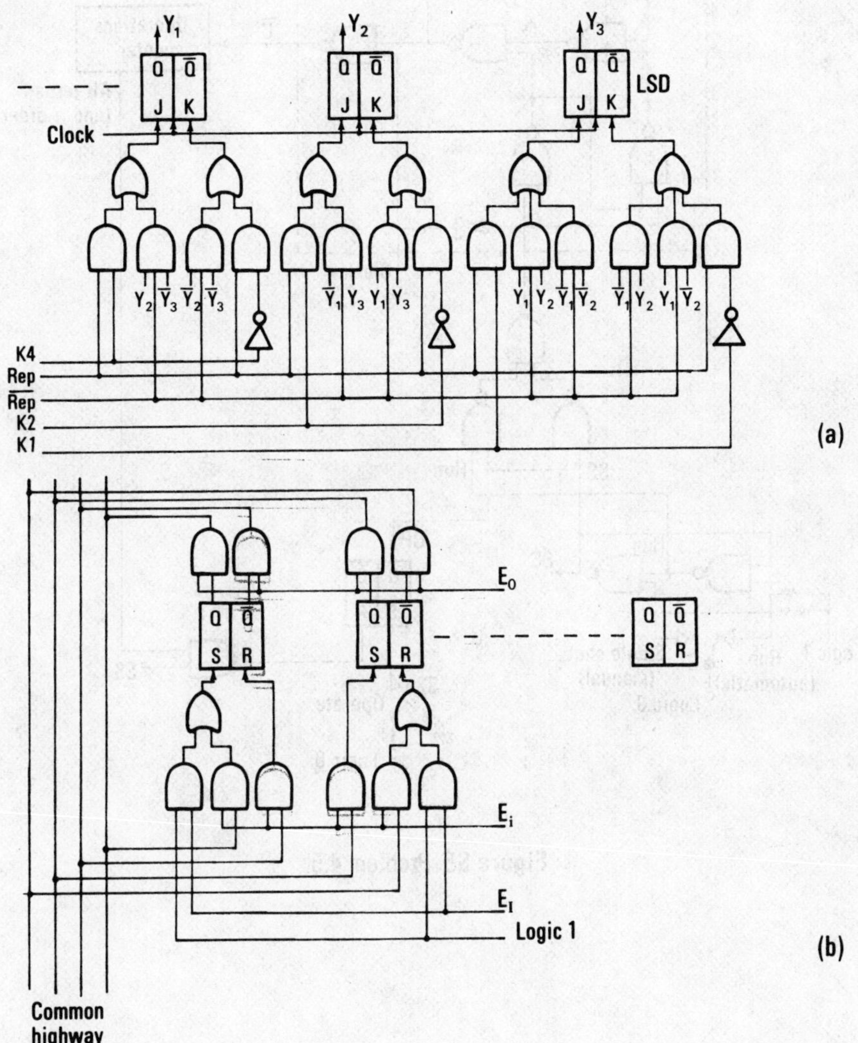

Figure S4 Problem 4.1.

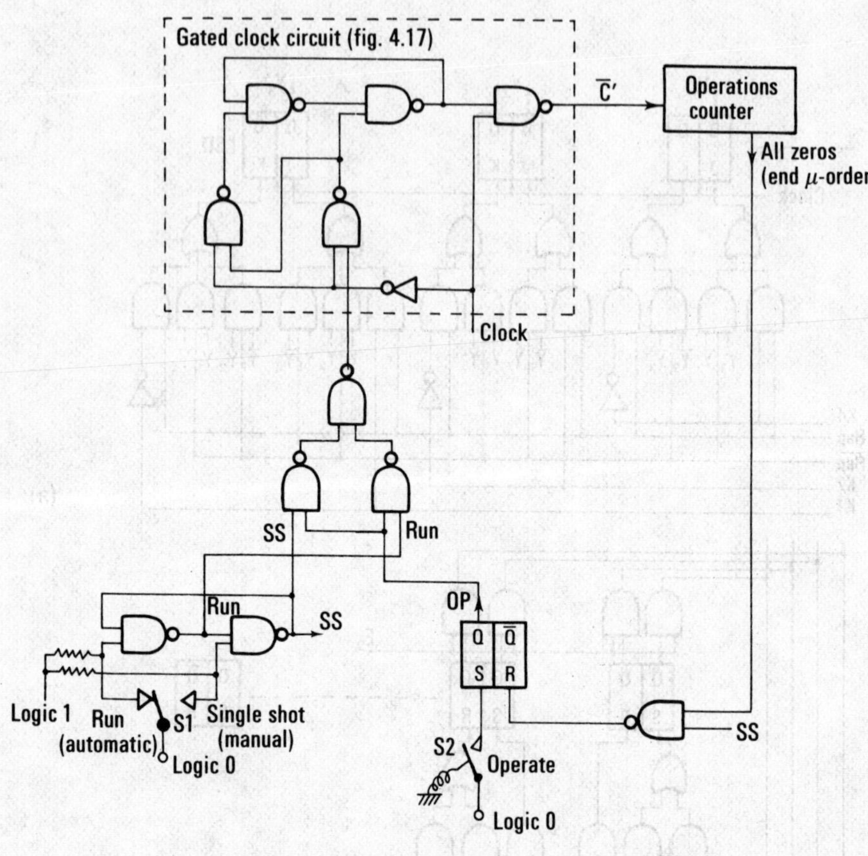

Figure S5 Problem 4.5.

When working in the manual mode it is necessary to depress the operate key (which is spring loaded) in order to initiate the execution of each instruction. The operate key sets the OP bistable, which is reset in turn by the micro-order END. Note that when the operate key is depressed in the single-shot mode the bistable has both the S and R inputs equal to 1, that is, the 'indeterminate' condition. However, the effect of this is to put both the Q and $\overline{Q}$ outputs to logic 1 (using NAND logic) which will allow the machine to operate, thereby changing the counter and causing AZ to go to 0; this assumes of course that the machine responds before the key is released.

4.7 Since multiplication as such is not available in the computer, the BCD to binary conversion (compare problem 2.7) must be performed using only shift and add operations. This is done by shifting the operand in the accumulator left by two places (× 4) then adding the unshifted value (× 5) finally shifting by one place left (× 10). In the program shown in Table S16 it is assumed that the accumulator (**a**) initially contains four 4-bit BCD digits with the least significant digit at the right-hand end. The M register (**m**) is used as temporary storage, the binary result being left in the accumulator; note that the 5-bit K-counter (**k**) is used to control the operation. A typical algorithm for BCD to binary conversion is shown in Table S16.

Binary to BCD conversion can be performed in a similar manner; in this case, however, successive division by binary 10 is used, the remainder giving the low order BCD digit. Note that it is not possible to implement the algorithm directly in hardware since the individual micro-orders are not specified; note also the similarity of the Iverson description to that employed by Register Transfer languages.

Table S16
Problem 4.7

Step no.	*Statement*	*Comment*
0	$\mathbf{m} \leftarrow \mathbf{a}$	Transfer accumulator to M-register
1	$\mathbf{a} \leftarrow \bar{\boldsymbol{\varepsilon}}(18)$	Clear accumulator
2	$\mathbf{k} \leftarrow 5 \top (-3)$	Set k-counter to -3
3	$\mathbf{a} \leftarrow 18 \top (\perp \mathbf{a}) + \perp 2 \downarrow \alpha^4/\mathbf{m}$	Add binary value of MS BCD digit to binary value of **a**
4	$\mathbf{k} \leftarrow 5 \top (1) + \perp \mathbf{k}$	Add +1 to k-counter
5	$\vee/\mathbf{k} : 0; (=) \rightarrow (5)$	Check $\mathbf{k} = 0$, stop
6	$\mathbf{a} \leftarrow 1 \underset{\circ}{\uparrow} (18 \top (\perp \mathbf{a}) + \perp 2 \underset{\circ}{\uparrow} \mathbf{a})$	Effective mult. by binary 10
7	$\mathbf{m} \leftarrow 4 \underset{\circ}{\uparrow} \mathbf{m}$	Shift BCD in M-register 4 places left
8	$\rightarrow (3)$	

Chapter 5

5.1 (a) Arithmetic using sign and magnitude numbers is very simple for multiplication and division (it is only necessary to compare signs – if different, the result is negative, if the same, the result is positive). However, addition and subtraction becomes a complicated process – the magnitude of their operands themselves must be compared as well as the signs, that is,

1. Signs the same – add magnitudes, duplicate sign in result.

2. Signs different – (i) *Augend larger* – add 1's complement of addend to augend (with end around carry) result given sign of augend.

(ii) *Addend larger* (or equal) – add 1's complement of addend to augend, complement result, result given sign of addend.

To implement the algorithm in hardware it is necessary to be able to compare the augend and addend for the >, <, or equal relationships using, for example, the circuit shown in Fig. 5.36. A conventional parallel full-adder can be used, with the overflow (that is, carry from MS stage) being fed back to the carry input of the LS stage (to give an end-around carry). The complemented result can be obtained by gating out the $\overline{Q}$ outputs of the sum register in place of the usual Q outputs. Similarly the $\overline{Q}$ register outputs can be gated to the full-adder circuits to obtain the complement of the addend. However, it may be shown that an end-around carry only occurs if the augend is larger than the addend; consequently, the final carry output may be used as a control signal thus allowing the comparator to be dispensed with. If this technique is used special precautions must be taken with the overflow resulting from the addition of two like sign numbers. Subtraction may be performed by changing the sign of the subtrahend. A suggested block diagram for a sign and magnitude arithmetic unit is shown in Fig. S6.

(b) The 1's complemented arithmetic unit is quite straightforward, the arithmetic operations being performed in much the same way as for the normal 2's complemented system, with the exception that the end-around carry scheme must be incorporated.

5.3 The design tables for this problem, treated as a switching network, are shown in Table S17; note that only mod 3 numbers (0, 1, 2) are allowed. The equations, shown below, may be implemented in the usual way:

$$S_1 = \bar{m}_1\bar{m}_2 n_1 + m_2 n_2 + m_1 \bar{n}_1 \bar{n}_2$$
$$S_2 = \bar{m}_1\bar{m}_2 n_2 + m_1 n_1 + m_2 \bar{n}_1 \bar{n}_2$$

An alternative design approach is to use binary full-adders and to correct the resulting binary sum by adding 1 into the least significant stage. This may be done by detecting the overflow (carry-out) and $S_1S_2 = 1$ conditions; a logic diagram is shown in Fig. S7.

5.4 For the parallel implementation we assume that the registers (RA, RB and RC) are connected to a common parallel highway system, complete with shifting gates, similar to that discussed in the main text; a typical logic diagram is shown in Fig. S8, and a suitable micro-program in Table S18. Note that the same symbolism as used in the text is employed for the micro-orders, that is, RA_0 puts the output of register RA to the common highway; $\overline{RA_0}$ puts the complemented outputs to the highway, etc.

A schematic logic diagram for the serial system is shown in Fig. S9. The basis of the circuit is a serial shift register RA (described in Chapter 4) combined with a serial full-adder circuit. The other registers, RB and RC, are normal end-around shifting circuits with input and output gating. The concept of a common (serial) highway may be retained, as may the fundamental ideas of microprogramming. The micro-operations required are:

1. Set (RS, TRUE, ADD, READ RB), maintain, and clock for one word-time.
2. Set (N, INV, ADD, CARRY BS, LOAD RC), maintain, and clock for one word-time.

Table S17
Problem 5.3

(a) Mod 3 addition

	0	1	2
+	00	01	10
0 00	00	01	10
1 01	01	10	00
2 10	10	00	01

(b) K-Maps

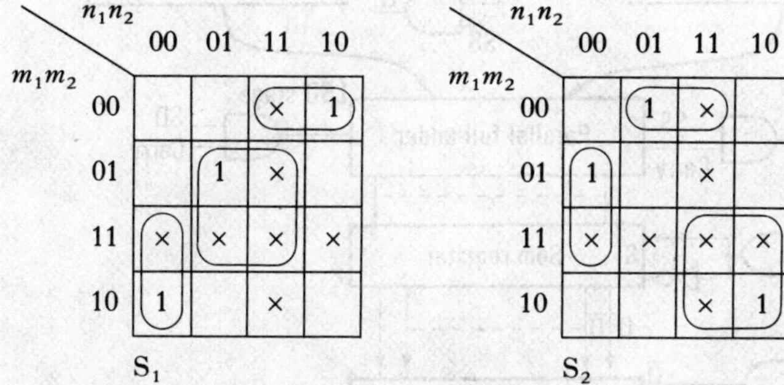

Table S18
Problem 5.4

Step	*Micro-order*	*Comment*
1	RA_oX_i, R, RB_{AU}	RA to X-register shifted one place right. Output of RB taken to input of arithmetic unit
2	AU_oRA_i	Output of AU to RA
3	RA_oX_i, $2^{-17}AU$	Complemented output RA to X-register
4	$AU_o\,RC_i$	Output of AU to RC

The operation of the circuit in the first word-time is to add serially the contents of RA, shifted right by one place, to the contents of RB, putting the result back into RA. In the next word-time the inverted output of RA is loaded into RC via the adder (with the carry bistable set to 1 – the equivalent of adding +1) to form the two's complement.

5.6(a) The first part of this problem is fairly straightforward, and a suggested flow-chart is shown in Fig. S10. The main difficulties are caused through a lack of double-length mode operation (this becomes even more apparent when the micro-program is considered later). Consequently, it is necessary to program the double-length shift operation required for multiplication. This is done by extracting the least significant digit of the most significant half of the partial product, before shifting, and inserting it in the most significant digit position of the least significant half of the partial product, *after* shifting. Note also the formation of a code word (TAG) from the inspection of the sign digits of the multiplicand and multiplier; the codeword is used in the program to determine the appropriate correction operations.

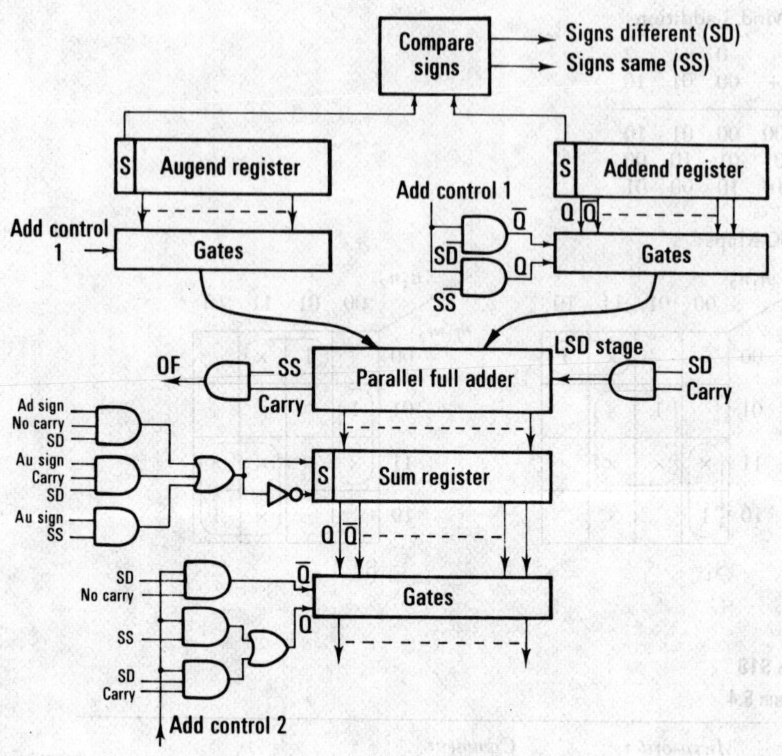

Figure S6 Problem 5.1.

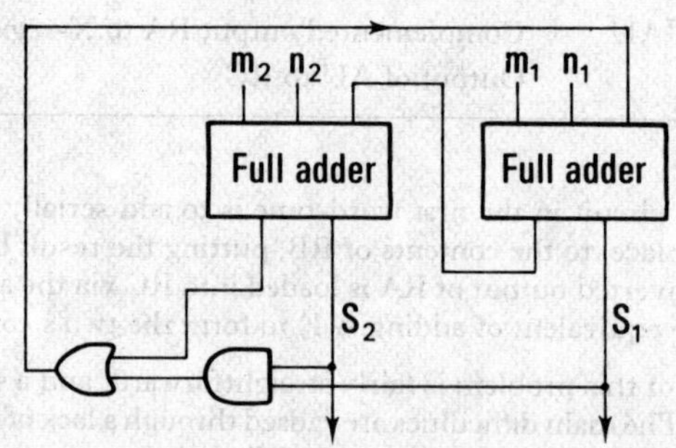

Figure S7 Problem 5.3.

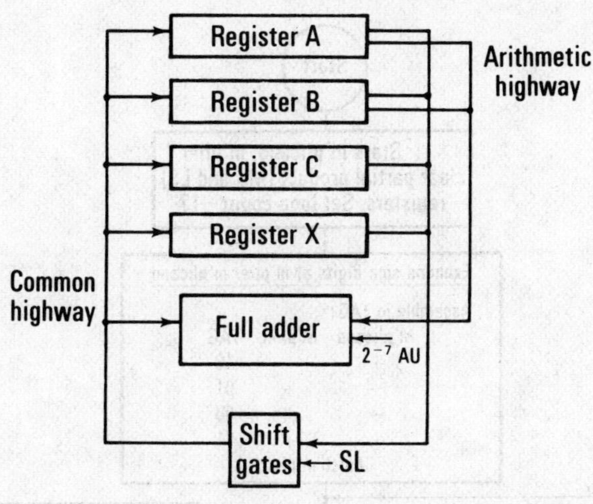

Figure S8 Problem 5.4.

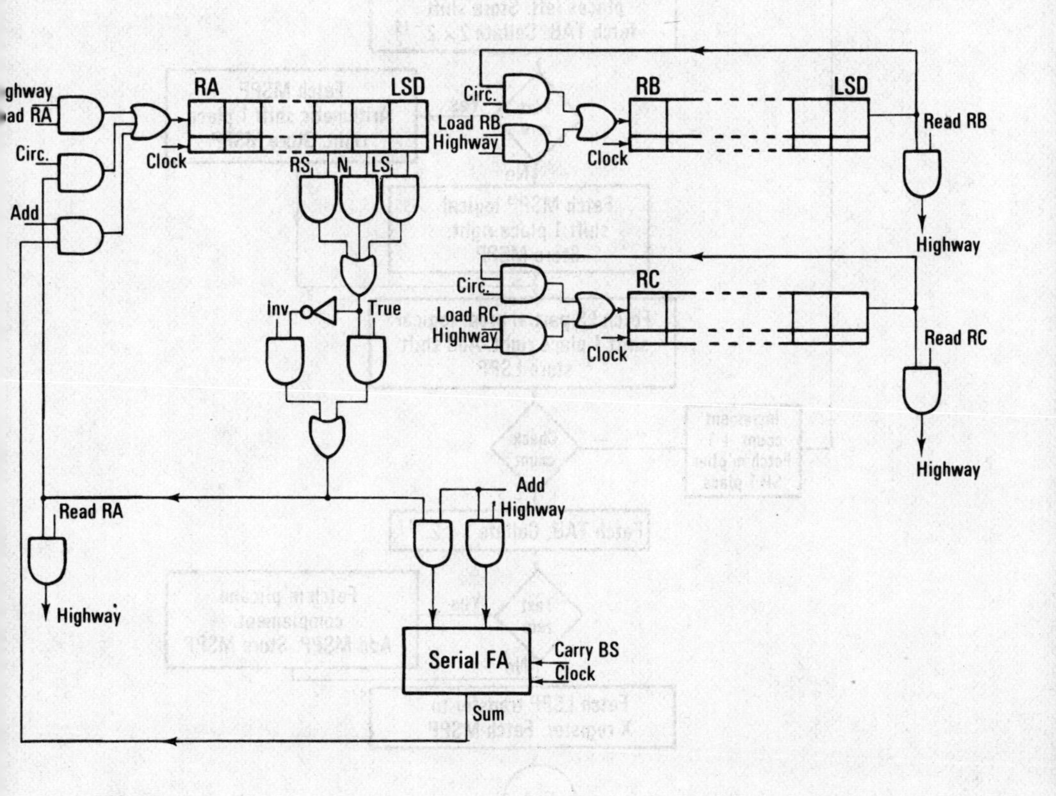

Figure S9 Problem 5.4.

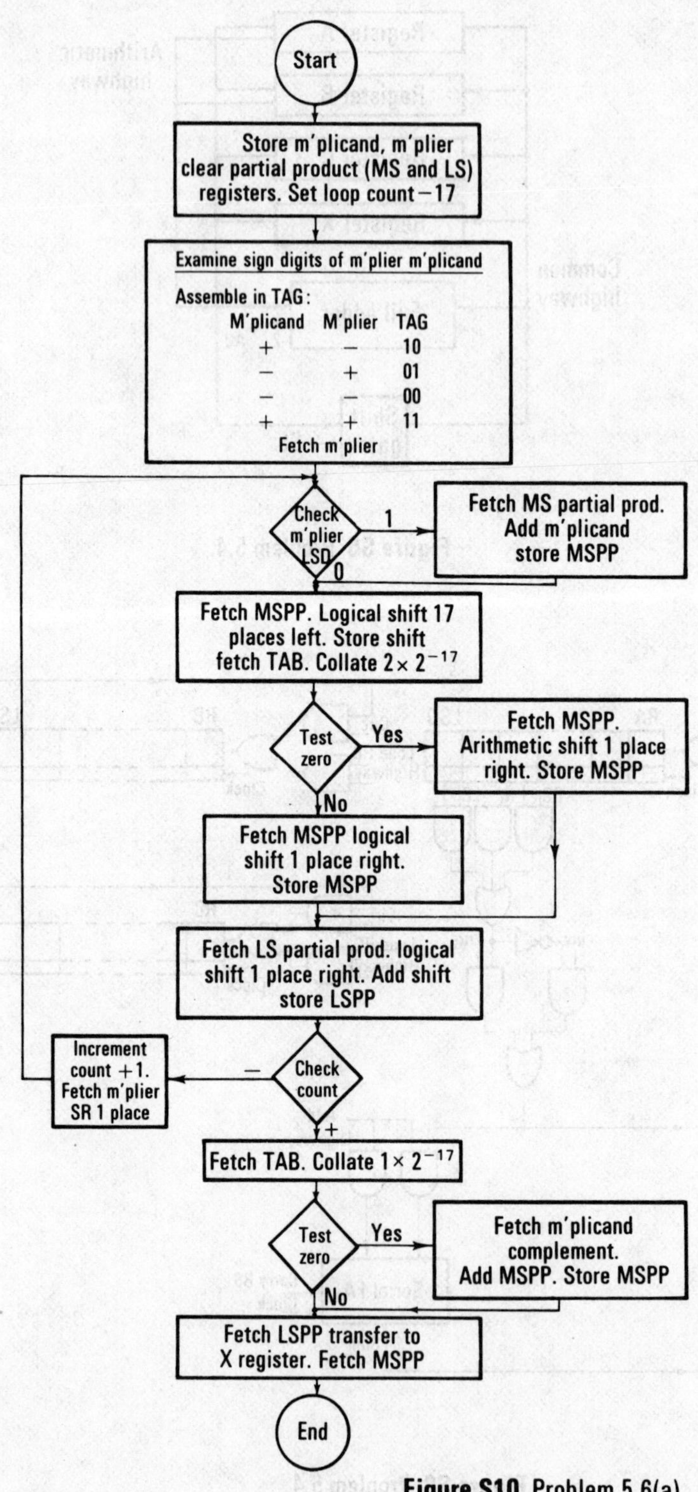

Figure S10 Problem 5.6(a).

5.6(b) To implement the algorithm in hardware presents a number of problems. Strictly speaking the machine structure is too primitive (due mainly to an insufficient number of registers) to incorporate multiplication/division algorithms effectively. In this example it is necessary to use the modifier register as part of the multiplication logic, this limits the operation of the machine since the programmer must store away the contents of the modifier register (which must be retrieved later) before executing a multiplication instruction. This function could also be included as part of the micro-program, by initially writing away the contents of the modifier (at micro-order level) into a predetermined location using a wired address technique. A suitable micro-program for the multiplication algorithm is shown in Table S19, and follows a similar logic to the flow chart given in Fig. S10.

The initial micro-orders are concerned with loading the registers; the multiplier is held in the modifier register, and the X-register contains the multiplicand. Partial products are accumulated in the accumulator (most significant part only) which is initially cleared to zero; register M is left free and is used as a temporary buffer store. The least significant part of the partial product is inserted bit by bit into the most significant end of the modifier register during the double length shift operations. Note that it is necessary to invent some new micro-orders concerned primarily with bit manipulations. These are represented by $A^n_i\ A^j_o$ which refers to input or output from a particular bit stage (n, j) of a register, for example $A^1_o\ A^0_i$ sets the bit contained in digit stage 1 into digit 0 (MSD) of register A; this operation is required for the arithmetic right shift function. In

Table S19
Problem 5.6

Step	*Condition*	*Micro-order*	*Comment*
1		C_{AD}, R/W	
2		A_oX_i	Transfer M'plicand to X-register
3		$A_c(K_{16}\overline{K}_8\overline{K}_4\overline{K}_2K_1)$	Clear accumulator. Set K-counter
4	R/W = 0	M_oMD_i, $MD^0_oG_i$	Transfer M'plier to modifier register, set flag G
	R/W = 1	Rep 4	
5	$MD_{LSD} = 0$	Rep 7	
	$MD_{LSD} = 1$	A_oM_i, M_{AU}	Add M'plicand to most significant half of partial product
6		AU_oA_i, $A^{17}_oF_i$	Set flag F
7		A_oM_i, R	Shift P.P. right
8		M_oA_i	
9	$X_{MSD} = 1$	$A^1_oA^0_i$	Arithmetic shift
10		MD_oM_i, R	Shift multiplier right
11		M_oMD_i	
12		$F_oMD^0_i$	Set MSD of M'plier to flag F
13	$K \neq 0$	K_o Rep 5	Decrement counter K
	K = 0 G = 0	Rep 17	
	K = 0 G = 1	$X_oM_i\ \overline{M}_{AU}2^{-17}AU$	
14		AU_oX_i	Flag G set add 2's complement of M'plicand to MSPP
15		A_oM_i, M_{AU}	
16		AU_oA_i	
17		MD_oX_i END	Transfer LSPP to X-register

addition it is necessary to be able to set **flags** or bistables to store certain transitory conditions, for example the sign digit of the multiplier; in the example flag G is used for this purpose. Flag F is used to store the least significant digit of the partial product, this is required during the right shifting of the multiplier. The most significant digit of the multiplicand (held in the X-register) is treated as a condition (X_{MSD}) in the usual way.

5.8 The control logic for the multiplier follows conventional lines and is based on the micro-program principle. For example, the decoded output of the multiplier compare logic would be used in conjunction with an operations counter to generate the necessary switching waveforms. The control unit must perform the following operations in each time interval:

1. Load multiplicand and multiplier registers. Clear input register. Set cycle counter to n (number of bits).
2. Examine decoder outputs from multiplier compare logic:

   $\overline{M}_2M_3$ – generate normal and gate signals
   M_2 – generate left shift and gate signals

(At this stage the contents of the input register would be added to the contents of the multiplicand register.)

3. Transfer contents of partial product register to input register.
4. Examine decoder outputs:

if $M_2M_3 \neq 1$ jump to 6
$M_2M_3 = 1$ generate normal and gate signals

5. Transfer partial product register to input register.
6. Generate: multiplicand shift 2 places left, multiplier shift 2 places right.
7. Check end of multiplication instruction

   $n = 0$ stop
   $n \neq 0$ decrement cycle counter, jump to 2

Note that two counters are required:

(a) An N-bit cycle counter (where $n = 2^N$) capable of being preset to n (generally by setting to the all 1's condition) and counting down for each input pulse. The count-down requirement is easily achieved by inverting the literals of the normal count-up bistable terms. For example, in the 3-bit binary counter of Fig. 4.7(d) the input terms would become:

$$J_A = K_A = 1; \quad J_B = K_B = \overline{A}; \quad J_C = K_C = \overline{A}\overline{B}$$

(b) A 3-bit clocked control counter capable of being reset to allow jump operations; this is similar to the Rep n counter previously described.

The logical circuits required for register shifting, gating, etc. are shown in Fig. S11.

5.10 Assume the normal parallel computer structure described in the text, register transfers being effected via a common parallel highway.

(a) The collate or AND operation may be performed using special gating logic ($n \times 3$ input gates) connecting A and B to the common highway. It is necessary to form A.B transferring the result first to the X-register, and then back to the A-register. A much better method of performing the AND function, however, is to allow only the Q outputs of register B to go to register A via the common highway (the Q outputs all being 0). This has the effect of only resetting those stages of register A which do not contain a 1 in both the A and B registers. The micro-order can be expressed as 'Inhibit Q outputs of B from

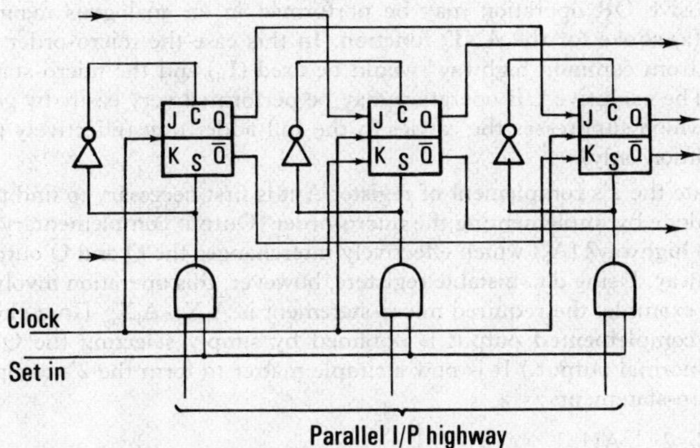

(a) Register shifting right 2 places.

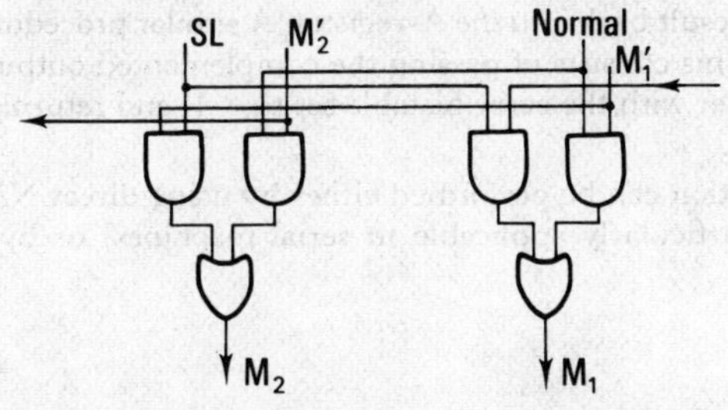

(b) Gates shifting left 1 place.

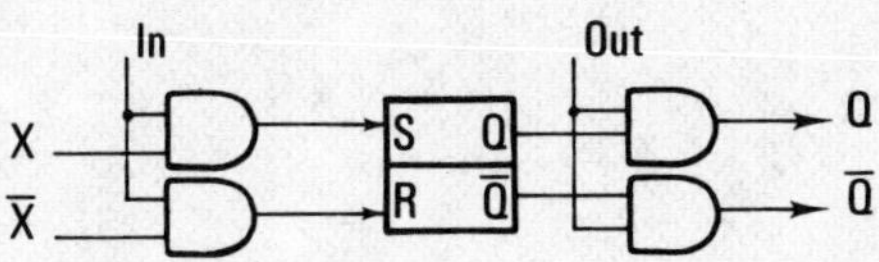

(c) Buffer register and gates.

Figure S11 Problem 5.8.

common highway'. and the micro-program statement as simply A_iI_B. In the computer discussed in the text this function is performed between the M-register and any other (micro-order I_M); $n \times 2$ input gates are required to implement the function. In a serial machine it is better, of course, to use the straightforward gating technique.

(b) The inclusive OR operation may be performed in an analogous manner to that described in (a) above for the AND function. In this case the micro-order 'Inhibit $\overline{Q}$ outputs of B from common highway' would be used ($\overline{I}_B$) and the micro-statement becomes $A_i\overline{I}_B$. The exclusive OR operation may be performed very easily by generating a micro-order which suppresses the carries in the full-adder unit (effectively performing modulo 2 addition only).

(c) To generate the 2's complement of register A it is first necessary to find the inverse. This may be done by implementing the micro-order 'Output complementary outputs of A to common highway' ($\overline{A}_o$) which effectively interchanges the Q and $\overline{Q}$ outputs on the common highway. Using d.c. bistable registers, however, this operation involves register transfers; for example, the required micro-statement is $\overline{A}_oX_i$, A_iX_o. (In a clocked serial machine the complemented output is obtained by simply selecting the $\overline{Q}$ output or inverting the normal output.) It is now a simple matter to form the 2's complement by using the micro-statements:

$\overline{A}_oX_i$, $2^{-17}AU$

AU_oA_i

which has the effect of forming the complement of register A in the X-register, adding +1, and then putting the result back into the A-register. A similar procedure may be followed for a serial machine; this consists of passing the complemented output of the A-register through the serial adder with the carry bistable set to +1, and returning the sum output back to the A-register.

(d) The NAND function can be performed either by using direct NAND gating to the common highway (particularly applicable to serial machines) or by using the micro-statements:

A_iI_B

$\overline{A}_oX_i$

X_oA_i

The micro-program first forms the AND of registers A and B in the A-register, the contents of which are then complemented using the X-register as a buffer.

Chapter 6

6.2 (a) An 8-in diameter drum has a track length of 8π in, which at 100 bits/in will allow $(8.\pi.100)/40 = 20\pi$ words (assuming a bit-serial organization). Since this is not an exact power of 2 it would be more practical to use a slightly larger bpi and record 64 words per track; in this case 64 tracks would be required to store 4096 words. The recording density would be increased to $(64.40)/8\pi \approx 102$ bits/in.

(b) The maximum access time, that is, the time for one complete revolution of the drum, is given by $60/1800 \approx 33$ ms. Therefore the average access time is approximately 16·5 ms. The word transfer rate is:

$$(64 \text{ words/rev})(30 \text{ rev/s}) = 1920 \text{ words/s}$$

(c) The clock or bit-rate is given by:

$$(64 \times 40 \text{ bits/rev})(30 \text{ rev/s}) = 76{,}800 \text{ bits/s}$$

6.3 *Timing Considerations* – The drum rotates at 3600 rev/min therefore the time for one revolution is 17 ms with an average access time of 8·5 ms. The total time required to access and read a complete track is in the order of 25·5 ms with a bit rate of 242,800 bits/s. Thus it seems convenient to use a total core store retrieval time of 25 ms. If 8-bit parallel words are read down from the store and serialized in a shift register using 10 μs pulses then, assuming a 10 μs R/W cycle, a total read time of 90 μs per word is required. This gives a total 'track' read time of $90 \times 10^{-6} \times 250 = 23$ ms which agrees very well with our suggested figure of 25 ms. Moreover, it also seems feasible to use 10 μs pulses as the basic system clock.

Addressing the Core Store – The original drum instruction will contain a read/write order and track number. The 5-bit track number can be used to address the Y coordinate drives of the 3D store, but the X drives must be generated in sequence using an 8-bit counter. The track number is effectively a block address specifying the start of the stored data, assumed to be held in contiguous locations.

Control Logic – The following units will be required:

An 8-bit address counter for the X coordinate, this counts in sequence from 0–255 thus reading down 256 words, that is, equivalent to 1 track of the drum.

The core store M-register must be a parallel access, serial shifting register.

Core-store – 10 μs R/W cycle time, $32 \times 250 \times 8$ planes, with standard control logic.

A divide by nine control counter is required to generate shifting, R/W signals, etc.

A 10 μs pulse generator is used as the basic system clock.

A general control unit which in conjunction with the $\div 9$ counter governs register transfers the sequencing of R/W operations etc.

A block diagram of the proposed system is shown in Fig. S12.

System Operation – Instructions from the external equipment are decoded, the read/write order bits going to the control unit and the track address bits to the store Y address register; the busy signal is set to 1. When the synchronizing pulse arrives (or if necessary some fixed time after the receipt of the drum instruction) the cycle start signal is generated which enables clock entry to the $\div 9$ counter, the M-register, and the external circuitry. The M-register is shifted 8 places (the $\div 9$ pulse inhibits shifting during the store R/W cycles) either reading in the first word to be written or clearing the register to zero (in later cycles outputting the serial data). The occurrence of the $\div 9$ pulse (that is, the R/W

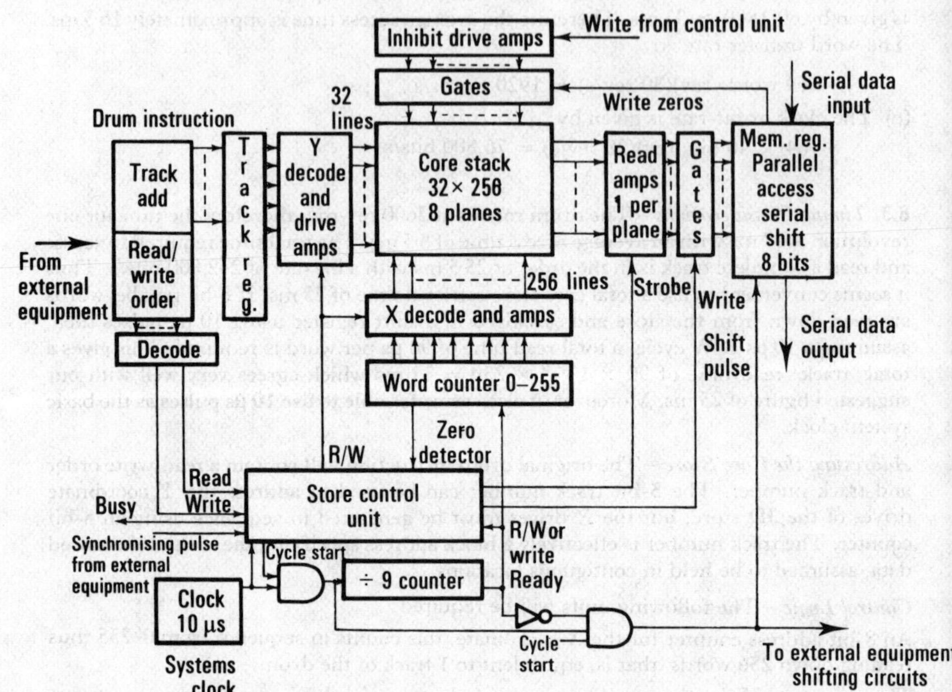

Figure S12 Problem 6.3.

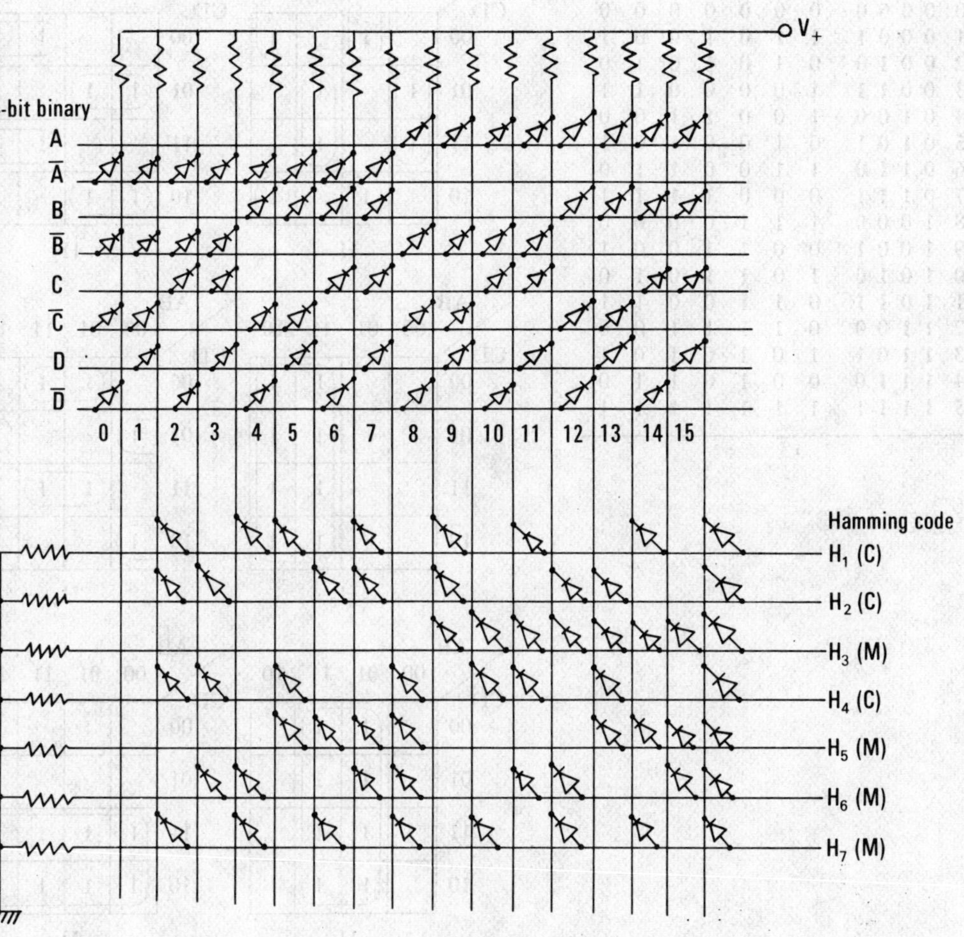

Figure S13 Problem 6.4.

Table S20
Problem 6.4

	A B C D	H_1	H_2	H_3	H_4	H_5	H_6	H_7
0	0 0 0 0	0	0	0	0	0	0	0
1	0 0 0 1	1	1	0	1	0	0	1
2	0 0 1 0	0	1	0	1	0	1	0
3	0 0 1 1	1	0	0	0	0	1	1
4	0 1 0 0	1	0	0	1	1	0	0
5	0 1 0 1	0	1	0	0	1	0	1
6	0 1 1 0	1	1	0	0	1	1	0
7	0 1 1 1	0	0	0	1	1	1	1
8	1 0 0 0	1	1	1	0	0	0	0
9	1 0 0 1	0	0	1	1	0	0	1
10	1 0 1 0	1	0	1	1	0	1	0
11	1 0 1 1	0	1	1	0	0	1	1
12	1 1 0 0	0	1	1	1	1	0	0
13	1 1 0 1	1	0	1	0	1	0	1
14	1 1 1 0	0	0	1	0	1	1	0
15	1 1 1 1	1	1	1	1	1	1	1

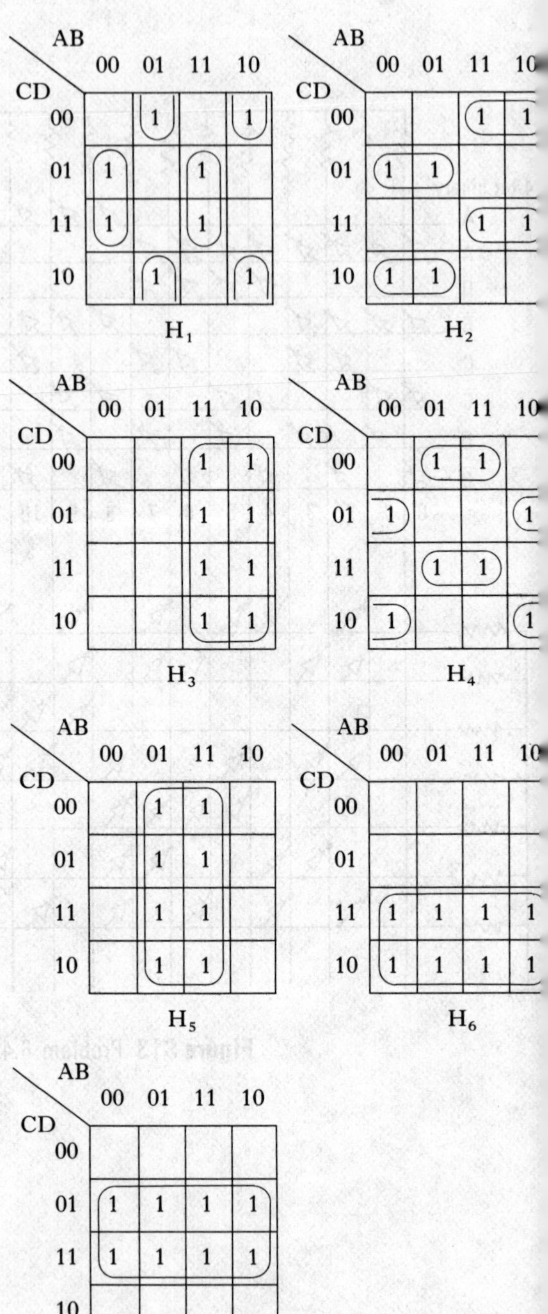

or write ready) initiates the store read or write signals and increments the X counter. The sequence is repeated until the X counter overflows to zero; the busy signal going down indicates to the external equipment that the device is available for further use.

6.4 The diode matrix for the binary to Hamming encoder is shown in Fig. S13; note that this is a many-to-many circuit representing an AND/OR configuration. An alternative approach is to use normal NAND/NOR logic gating, the K-maps for this problem are shown in Table S20; the reduced equations are given below:

$$
\begin{aligned}
H_1 &= \bar{A}B\bar{D} + A\bar{B}\bar{D} + ABD + \bar{A}\bar{B}D \\
&= \bar{D}(\bar{A}B + A\bar{B}) + D(AB + \bar{A}\bar{B}) \\
H_2 &= \bar{A}(\bar{C}D + C\bar{D}) + A(CD + \bar{C}\bar{D}) \\
H_3 &= A \\
H_4 &= \bar{B}(\bar{C}D + C\bar{D}) + B(CD + \bar{C}\bar{D}) \\
H_5 &= B \\
H_6 &= C \\
H_7 &= D
\end{aligned}
$$

Where H_1, H_2, and H_4 are the check digits and H_3, H_5, H_6, and H_7 are the message digits. The most usual form for this type of read-only memory is an LSI array – for example, a 6-bit input package capable of producing any desired binary output of 64 combinations of 24-bit words. The binary outputs are determined during the last phases of manufacture by the customer specifying the required output patterns which are used to make the final connection mask.

The technique may easily be extended to function generation, in this case the network is constructed such that when the argument is entered on the input lines the function values appear (generally in a BCD form) on the output lines. At the present time this is an impractical method of storing large tables, but with LSI becoming increasingly cheaper it could well become a practical design technique.

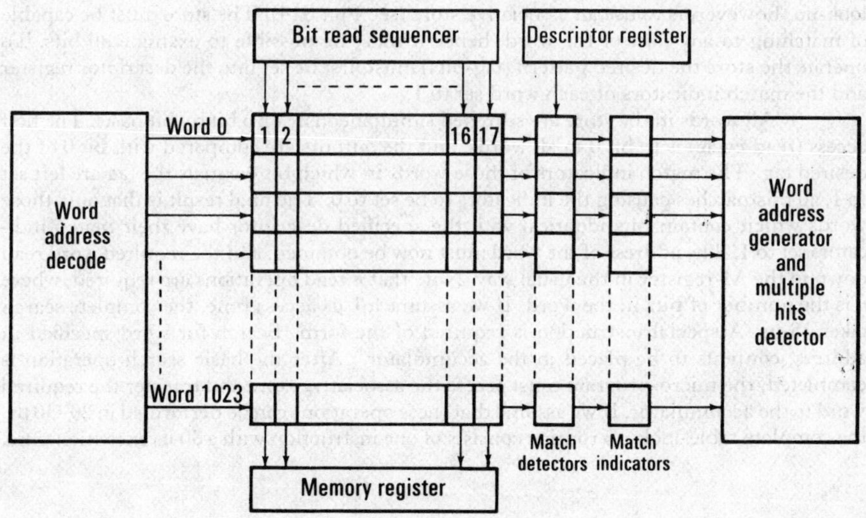

Figure S14 Problem 6.7.

6.7 In the worst case of randomly stored data the table-look-up program is required to search for any bit combination within the word, and each word must be examined in turn; a typical program is shown in Table S21. If we assume a single-address machine with an average 2 μs execution time (1 μs R/W store cycle) the maximum time required to search through the table is $2 + (6 \times 1024 \times 2) \approx 12{\cdot}3$ ms which on average gives a 6 ms search time. A special table-look-up instruction can be provided by combining the instructions:

```
COL     CONS
SUB     VARS
JMPO
INCM
JMP     BACK
```

The instruction would need to be a specified address type order, the address specifying the collation constant, with the tag word being stored immediately after it. A suitable micro-program for this instruction is shown in Table S22. If we assume a 250 ns time slot and a preliminary 5 micro-order steps for the read cycle, we have a total execution time for the search instruction of 22×250 ns that is, 5·5 μs. Using this instruction the table-look-up program now becomes:

```
  FETM   (−1024)
┌→FET    WORD/M
└─SEAR   CONS
  FET    WORD/M
  STOP
```

The number of store locations required for the program has been reduced from 11 to 7, and the maximum table-look-up time to $2 + 1024 \times 7{\cdot}5 + 4 \approx 7{\cdot}7$ ms. Thus, the use of the special instruction has effectively halved the search time; this could easily be improved upon by optimizing the micro-program. However, there is a practical limit to the improvement in speed that can be obtained using this technique. Moreover, in practice the search times could be improved by suitable structuring of the stored data and accessing the table by means of a generated address pointer. The most efficient technique for table-look-up, however, is to use an associative store (see Fig. S14). The store must be capable of matching to any part of the word, hence it must be possible to examine all bits. To operate the store the desired pattern (tag-bits) must first be set into the descriptor register and the match indicators of each word set to 1.

All words in the store are searched simultaneously on a bit by bit basis. The first access (read cycle) is to bit 0 in all words, and the outputs are compared with bit 0 of the desired tag. The match indicators of those words in which bit 0 equals the tag are left set to 1, all mismatches causing the indicators to be set to 0. The final result is that only those words which contain bits identical with the specified descriptor have their match indicators set to 1. The address of the word must now be obtained, and the required word read down to the M-register in the usual way. Note that n read operations are required, where n is the number of bits in the word. If we assume a 1 μs access time, the complete search takes 18 μs. A special instruction is required of the form 'Search for word specified in address, contents to be placed in the accumulator'. After the basic search operation is completed, the micro-program must access the associative store and transfer the required word to the accumulator. If we assume that these operations can be performed in 20–30 μs, the complete table-look-up routine consists of one instruction with a 30 μs execution time.

Table S21
Problem 6.7

Label	*Instruction*		*Comment*
	FETM	(–1024)	Fetch count constant to modifier
BACK	FET	WORD/M	Fetch word to accumulator
	COL	CONS	AND with collation constant
	SUB	VARS	Subtract required tag-bits
	JMPO	FND	Test accumulator if zero jump to FND
	INCM		Add +1 to modifier
	JMP	BACK	
FND	FET	WORD/M	
	STOP		
CONS	(	)	Collation constant
VARS	(	)	tag-bits

Table S22
Problem 6.7

Step	*Condition*	*Micro-order*	*Comment*
1		C_{AD}, R/W	Read collation constant
2		C_oX_i, $2^{-17}AU$	
3		AU_oC_i	Increment address by +1
4		I_MA_i	AND accumulator with M-register
5		C_{AD}, R/W	Read tag-bits
6	R/W = 1	Rep 6	
	R/W = 0	A_oX_i, $\overline{M}_{AU}$, $2^{-17}AU$	
7		AU_oA_i	Subtract tag-bits
8	A = 0	END	
	A < 0	MD_oX_i, $2^{-17}AU$	Increment MOD
9		$AU_oMD_iM_c$	Clear M-register
10		M_oX_i, $2^{-17}AU$	Add +2 to M-register
11		AU_oM_i	
12		M_oX_i, $2^{-17}AU$	
13		AU_oM_i	
14		E_oX_i, $\overline{M}_{AU}$, $2^{-17}AU$	Sub. +2 from control register
15		AU_oE_i, END	

Chapter 7

7.1 The data-break mode of operation may be implemented using the form of micro-program shown in Table S.23. It is assumed that multiple data-breaks are allowed (on a hardwired priority basis) with each peripheral device setting up a particular stage of a data-break register (DBR). At the conclusion of each instruction micro-program the DB register must be examined for zero contents (using the condition DBR = 0) to determine if a data-break request has been made. The technique of using a dedicated store location to hold the data-break control word (6-bits word count and 12-bits address) has been adopted in this example. The address of the control word is obtained by encoding the contents of the data-break register (using a one-to-many encoder) the outputs of the encoder circuit being gated directly to the address highway (using the micro-order DBC_{AD}). The peripheral equipment must be program initiated, and the contents of the DB control word primed (with the number of words, $-n$, and the start address of the data-block) before the data-break mode of operation takes place. Both parts of the control word must be incremented by 1 in the micro-program for each word transfer. When the word count goes to 0, a data-break flag is set (DBFS) indicating to the external device that the data transfers have finished.

One of the difficulties encountered in designing complex micro-program functions using the simple machine structure described in the text is that the number of micro-order steps required is greater than that obtainable with the 5-bit control counter. This difficulty could easily be overcome by simply using a larger counter; alternatively, two interlaced counters (out of phase by 90°) can be used. The later technique has been used very effectively in practice, and provides a much faster execution time for mutually exclusive operations.

Table S23
Problem 7.1

Step	*Condition*	*Micro-order*	*Comment*
n	DBR = 0	END	Test for data break
	DBR ≠ 0	DBC_{AD}, R/W	Output of encoder to address HW
$n+1$	RW = 1	Rep $n+1$	
	RW = 0	M_oC_i	
$n+2$		IO_oM_i	Write away data word
$n+3$		C_{AD} WRITE	
$n+4$		C_oX_i, 2^{-17}AU, 2^{-5}AU	Add +1 to word count
$n+5$		C_iAU_o	and address
$n+6$	A < 0	END	Check word count and if zero
	A = 0	DBFS END	set data break end flag

7.3 The normal teletype machine operates at some 10 characters per second so that the maximum rate of input is one 8-bit character every 100 ms; this data rate, however, would be appreciably lower in the case of operator inserted data, so it is safe to assume a one character at a time input, thereby dispensing with the need for anything larger than a one character buffer store. In actual practice it is safe to assume that the character remains on the parallel output lines of the teletype long enough for a data transfer to take place. It is also assumed that the characters must be parity checked and then packed, two to a word, in the CPU store. The input message must be preceded and terminated by suitable control signals, say @ and *; note that since data may extend over more than one line CRLF cannot be used for this purpose. The erase symbol may be used either to erase the character immediately proceeding it, or to erase the entire line; spaces are ignored.

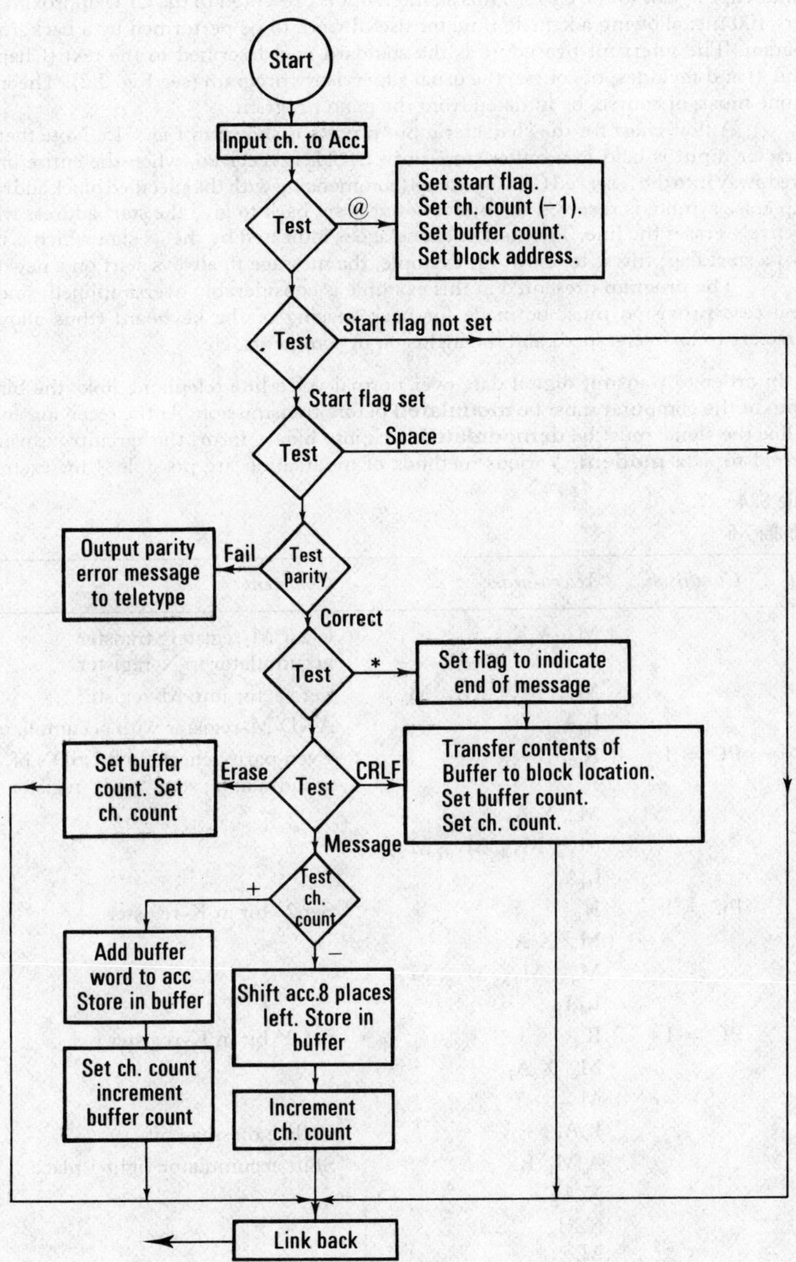

Figure S15 Problem 7.3.

Characters are inputted on an interrupt basis; that is, each time a key is depressed an interrupt is sent to the CPU. Thus an interrupt is presented to the CPU approximately every 100 ms, allowing adequate time for useful work to be performed by a background program. The interrupt procedure is the standard one described in the text (Chapters 3 and 4) and includes, of course, the usual supervisory program (see Fig. 3.2). The input routine must, of course, be initiated from the main program.

A flow chart for the character input process is shown in Fig. S15. Note that the character input is held in a buffer store until CRLF is received, when the entire line is stored away into the reserved CPU locations (commencing with the specified block address). If an erase symbol is received the buffer count is set back to give the start address which effectively erases the line. The start of a message is indicated by the @ sign which is used to set a start flag; this is to allow, for example, the message to always start on a new line.

The program presented in this example is considerably oversimplified, since in a real case provision must be made for back spacing of the keyboard (thus allowing characters to be over-typed) and the inclusion of comments, etc.

7.5 In order to transmit digital data over normal voice line telephone links the binary output of the computer must be **modulated** before transmission. At the receiving end of the line the signal must be **demodulated** back into binary form; this circuitry is usually referred to as a **modem**. Various methods of modulation are possible – for example,

Table S24

Problem 7.5

Step	*Condition*	*Micro-order*	*Comment*
1		M_c, A_oX_i	Clear M-register; transfer accumulator to X-register
2		M_{11}, M_{13}, M_{15}, M_{17}	Set vector into M-register
3		I_MA_i	AND M-register with accumulator
4	PC = 1	K_1	Even parity check on 8 LSD's of accumulator; set 2^0 in K-register
5		M_c, X_oA_i	
6		M_{12}, M_{13}, M_{16}, M_{17}	
7		I_MA_i	
8	PC = 1	K_2	Set 2^1 bit in K-register
9		M_c, X_oA_i	
10		M_{14}, M_{15}, M_{16}, M_{17}	
11		I_MA_i	
12	PC = 1	K_4	Set 2^2 bit in K-register
13		M_c, X_oA_i	
14		M_{13}	
15		I_MA_i	Collate message bit
16		A_oM_i, R	Shift accumulator right 1 place
17		X_oA_i	Assemble message word in accumulator (steps 17–23)
18		X_iM_o	
19		M_c	
20		M_{17}, M_{16}, M_{15}	
21		I_MA_i	
22		M_iX_o	
23		$\bar{I}_MA_i$	
24		K_o MD_i END	Error bits in modifier

phase amplitude and pulse coded modulation (PCM) (see reference 56, Chapter 6). The transmission speeds obtainable vary from 2400 bits per second (Bauds) on private lines to 600–1200 Bauds on public lines. (The British Post Office Datel 48k system provides 48k bits per second over a national wideband transmission network.) A typical error rate over a good quality private line is in the order of 1 in 10^6 characters in error, public lines can be appreciably worse.

Data transmission can be either in the **synchronous** or **asynchronous** mode. In the case of synchronous systems continuous transmission occurs with the transmitter and receiver being exactly synchronized to the same frequency (the oscillators are normally synchronized at the start of each message). The synchronous system is more efficient in operation than asynchronous working but is also more complicated, and hence costly, to produce. Asynchronous transmission is the more usual mode with one character being sent at a time and each character preceded by, and terminated with, a stop and start signal. The serial message is clocked by an oscillator (initiated by the start signal) which decodes the message into the required binary format.

With the 50 Baud system of the problem, asynchronous operation would be best. Moreover, the 20 ms bit-rate allows each individual message bit to be registered into a single bistable stage and inputted a bit at a time on an interrupt basis. (Using this technique it is possible to have up to 18 separate channels working into a single word buffer.) Though this technique requires the minimum of external equipment, the software load (assembly and checking of the message) is very considerable. Alternatively, the message bits may be clocked serially into a shift register and then transferred as a parallel word directly into the CPU. Thus if each character message was 12-bits long (including start–stop and parity checks) a complete word could be assembled every 240 ms. An interrupt signal to the CPU could be initiated by the stop signal with the data transfer (requiring in the order of 20 ms) occurring at the start of the next message start signal; alternatively a timing gap or dummy message could be allowed between each actual message.

One of the major decisions that has to be made by the computer systems engineer is the allocation of tasks between software and hardware. In this case we are primarily concerned with the assembly and checking functions. The final decision depends primarily on cost, speed, and availability of equipment. For example, one solution in a large teleprocessing system would be to use a small computer dedicated to the assembly and decoding (including channel multiplexing) of all incoming messages. Alternatively, special Firmware (micro-programmed) assembly and decoding routines could be included as part of the instruction repertoire of a standard machine. Table S24 shows a micro-programmed routine to decode a 7-bit Hamming code held in the least significant end of the accumulator. Note the need to incorporate special micro-orders and logic to perform a parity check on the contents of the accumulator and to set individual bits of the M register to 1. Moreover, provision must also be made to output the contents of the K register to the common highway and to set up a parity condition.

Table S25
Problem 8.1

		f_0	f_1 C1/0	f_2 C1/1	f_3 C2/0	f_4 C2/1	f_5 C3/0	f_6 C3/1	f_7 C4/0	f_8 C4/1	f_9 C5/0	f_{10} C5/1	f_{11} C6/0	f_{12} C6/1	f_{13} C7/0	f_{14} C7/1	f_{15} C8/0	f_{16} C8/1
000	t_0	1	1	0	1	1	1	1	1	1	1	1	1	1	0	1	0	1
001	t_1	1	1	0	1	0	1	1	1	0	1	1	1	1	0	1	0	1
010	t_2	1	1	1	1	1	1	0	1	1	1	0	1	1	0	1	0	1
011	t_3	0	0	1	1	0	1	0	1	0	1	0	0	1	0	1	0	1
100	t_4	0	1	0	0	1	0	0	1	0	1	0	0	1	0	1	0	1
101	t_5	0	1	0	0	1	0	0	1	0	1	0	0	1	0	1	0	1
110	t_6	1	1	1	0	1	1	1	1	1	1	1	0	1	1	1	0	1
111	t_7	1	0	1	0	1	1	1	1	1	1	1	0	1	1	1	0	1

Table S26
Problem 8.1

	(f_0f_1)	(f_0f_2)	(f_0f_3)	(f_0f_4)	(f_0f_5)	(f_0f_6)	(f_0f_7)	(f_0f_8)	(f_0f_9)	(f_0f_{10})	(f_0f_{11})	(f_0f_{12})	(f_0f_{13})	(f_0f_{14})	(f_0f_{15})	(f_0f_{16})
t_0		1											1		1	
t_1		1		1				1					1		1	
t_2						1				1			1		1	
t_3		1	1		1		1		1			1		1		1
t_4	1			1			1		1			1		1		1
t_5	1			1			1		1			1		1		1
t_6			1								1				1	
t_7	1		1								1				1	

Chapter 8

8.1 In the case of Fig. 8.17 there are three input terminals and therefore eight different tests, designated t_0–t_7. The 16 faults f_1–f_{16} are all the possible s-a-0 and s-a-1 faults occurring on the eight connections C1–C8 and are referred to in the usual way as C1/0, C1/1, etc. Thus, the fault matrix consists of 8 rows and 17 columns including the correct output; this is shown in Table S25. Examining the derived G_D matrix, shown in Table S26, we see that the essential tests are t_3 (testing f_5, that is, C3/0), t_1 (for f_8) and t_2 (for f_6 and f_{10}). These essential tests will cover all the faults except f_1 and f_{11}, which may be covered by including test t_7. A possible test set (in fact the minimal) is thus t_1, t_2, t_3, and t_7, which in terms of input/output values becomes:

001/1, 010/1, 011/0, 111/1

8.3 The reliability of the core stores is given by:

$$[1-(1-r_2)(1-r_2)] = 1-(1-r_2)^2$$

The reliability of the 2-out-of-3 disc system may be calculated in the simplified way using a truth-table format:

D_1	D_2	D_3	R
0	1	1	$(1-r_4)(r_4)(r_4) = r_4^2 - r_4^3$
1	0	1	$(r_4)(1-r_4)(r_4) = r_4^2 - r_4^3$
1	1	0	$(r_4)(r_4)(1-r_4) = r_4^2 - r_4^3$
1	1	1	$(r_4)(r_4)(r_4) = r_4^3$

The overall reliability is the sum of the individual reliabilities:

$$R_D = 3r_4^2 - 2r_4^3$$

The system reliability is given by the expression:

$$R_s = [1-(1-r_2)^2]r_1r_2[r_4^2(3-2r_4)]r_5$$
$$= [1-(1-0{\cdot}92)^2]0{\cdot}95 \times 0{\cdot}85[(0{\cdot}8)^2(3-2\times 0{\cdot}80)]0{\cdot}72$$

Thus $\quad R_s = 0{\cdot}49$

The mean time between failures may be calculated using the expression:

$$R_s = e^{-\lambda t} = e^{-\lambda 1000} = e^{-1000/M}$$

where λ = failure rate, M = mean time between failures, and t is the mission time. Expressing $e^{-\lambda 1000}$ as a series expansion we have:

$$R_s = 1 - \lambda 1000 + \frac{(\lambda 1000)^2}{2} + \cdots$$

The failure rate λ is very small so the expression approximates to:

$$R_s = 1 - \lambda 1000$$

Now $\quad \mathrm{M} = \dfrac{1000}{1-R_s} = 1/\lambda$

Thus $\quad \mathrm{M} = \dfrac{1000}{0{\cdot}51} = 1960$ hours

Therefore the mean time between failures is 1960 hours.

APPENDIX–LOGIC SYMBOLS
Based on MIL–STD–806B (U.S. Dept. of Defence)

1. AND gates

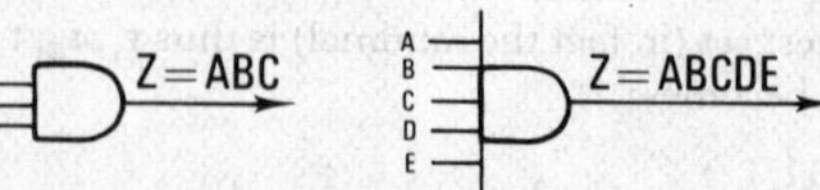

2. NAND gates

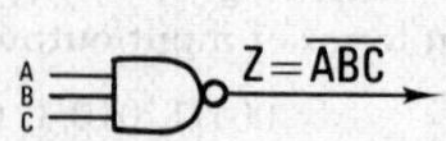

3. OR gates

A B C — $Z=A+B+C$

A B C D E F G — $Z=A+B+C+D+E+F+G$

4. Exclusive OR gates

A B — $Z=\overline{A}B+A\overline{B}$

5. NOR gates

A B C — $Z=\overline{A+B+C}$

6. Expander input gates

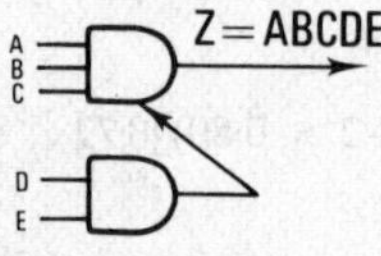

7. Invertor amplifiers

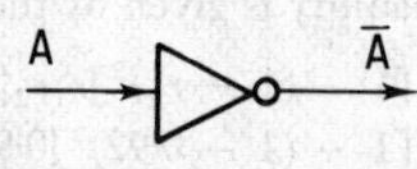

8. Bistable circuits

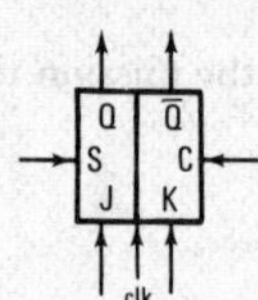

JK-bistable

Q $\overline{Q}$ S R

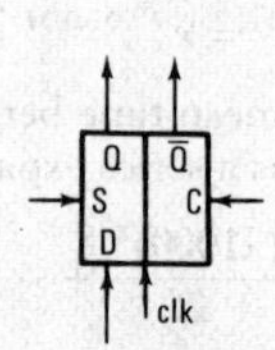

DC latch or toggle

Q $\overline{Q}$ S C D clk

D-bistable

9. Delay units

10 μs

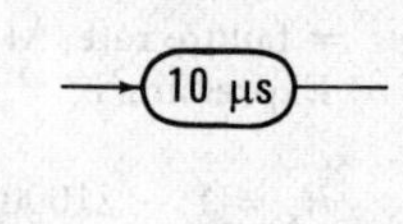

10. Register stages

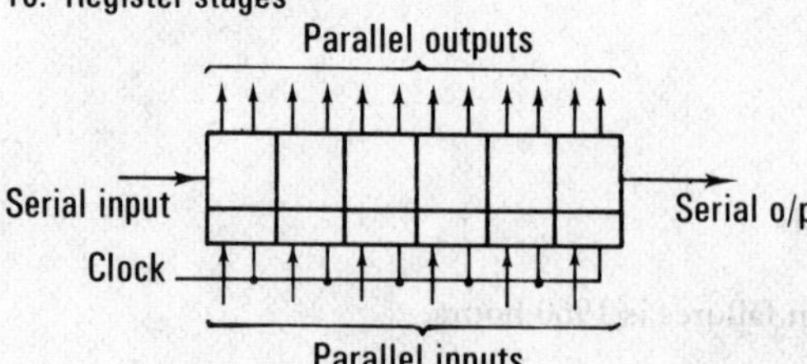

Index